Applied Chemoinformatics

Applied Chemoinformatics

Achievements and Future Opportunities

Edited by Thomas Engel and Johann Gasteiger

Editors

Dr. Thomas Engel
LMU Munich
Department of Chemistry
Butenandtstraße 5-13
81377 München
Germany

Prof. Dr. Johann Gasteiger
University of Erlangen-Nürnberg
Computer-Chemie-Centrum
Nägelsbachstr. 25
91052 Erlangen
Germany

Cover Design
Dr. Christian R. Wick
University of Erlangen-Nürnberg
Institute for Theoretical Physics I
Nägelsbachstr. 49b (EAM)
91052 Erlangen
Germany

All books published by **Wiley-VCH** are carefully produced. Nevertheless, authors, editors, and publisher do not warrant the information contained in these books, including this book, to be free of errors. Readers are advised to keep in mind that statements, data, illustrations, procedural details or other items may inadvertently be inaccurate.

Library of Congress Card No.: applied for

British Library Cataloguing-in-Publication Data
A catalogue record for this book is available from the British Library.

Bibliographic information published by the Deutsche Nationalbibliothek
The Deutsche Nationalbibliothek lists this publication in the Deutsche Nationalbibliografie; detailed bibliographic data are available on the Internet at <http://dnb.d-nb.de>.

© 2018 Wiley-VCH Verlag GmbH & Co. KGaA, Boschstr. 12, 69469 Weinheim, Germany

All rights reserved (including those of translation into other languages). No part of this book may be reproduced in any form—by photoprinting, microfilm, or any other means—nor transmitted or translated into a machine language without written permission from the publishers. Registered names, trademarks, etc. used in this book, even when not specifically marked as such, are not to be considered unprotected by law.

Print ISBN: 978-3-527-34201-3
ePDF ISBN: 978-3-527-80652-2
ePub ISBN: 978-3-527-80654-6
Mobi ISBN: 978-3-527-80655-3
oBook ISBN: 978-3-527-80653-9

Cover Design Grafik-Design Schulz, Fußgönheim, Germany
Typesetting SPi Global, Chennai, India
Printing and Binding betz-druck GmbH, Darmstadt, Germany
Printed on acid-free paper

10 9 8 7 6 5 4 3 2 1

Thomas Engel

To my family especially Benedikt.

Johann Gasteiger

To all the friends, colleagues and coworkers that ventured with me into the exciting field of chemoinformatics.
And to my wife Uli for never complaining about my long working hours.

If you want to build a ship, don't drum up people to collect wood and don't assign them tasks and work, but rather teach them to long for the endless immensity of the sea.

Antoine de Saint-Exupéry

Contents

Foreword *xvii*
List of Contributors *xxi*

1 Introduction *1*
Thomas Engel and Johann Gasteiger
1.1 The Rationale for the Books *1*
1.2 Development of the Field *2*
1.3 The Basis of Chemoinformatics and the Diversity of Applications *3*
1.3.1 Databases *3*
1.3.2 Fundamental Questions of a Chemist *4*
1.3.3 Drug Discovery *5*
1.3.4 Additional Fields of Application *6*
Reference *7*

2 QSAR/QSPR *9*
Wolfgang Sippl and Dina Robaa
2.1 Introduction *9*
2.2 Data Handling and Curation *13*
2.2.1 Structural Data *13*
2.2.2 Biological Data *14*
2.3 Molecular Descriptors *14*
2.3.1 Structural Keys (1D) *15*
2.3.2 Topological Descriptors (2D) *16*
2.3.3 Geometric Descriptors (3D) *16*
2.4 Methods for Data Analysis *17*
2.4.1 Overview *17*
2.4.2 Unsupervised Learning *17*
2.4.3 Supervised Learning *18*
2.5 Classification Methods *19*
2.5.1 Principal Component Analysis *19*
2.5.2 Linear Discriminant Analysis *19*
2.5.3 Kohonen Neural Network *19*
2.5.4 Other Classification Methods *20*
2.6 Methods for Data Modeling *20*

2.6.1	Regression-Based QSAR Approaches	20
2.6.2	3D QSAR	22
2.6.3	Nonlinear Models	25
2.7	Summary on Data Analysis Methods	30
2.8	Model Validation	30
2.8.1	Proper Use of Validation Routines	31
2.8.2	Modeling/Validation Workflow	32
2.8.3	Splitting of Datasets	32
2.8.4	Compilation of Modeling, Training, Validation, Test, and External Sets	34
2.8.5	Cross-Validation	36
2.8.6	Bootstrapping	37
2.8.7	Y-Randomization (Y-Scrambling)	38
2.8.8	Goodness of Prediction and Quality Criteria	39
2.8.9	Applicability Domain and Model Acceptability Criteria	41
2.8.10	Scope of External and Internal Validation	43
2.8.11	Validation of Classification Models	45
2.9	Regulatory Use of QSARs	46
	Selected Reading	48
	References	49

3 Prediction of Physicochemical Properties of Compounds 53
Igor V. Tetko, Aixia Yan, and Johann Gasteiger

3.1	Introduction	53
3.2	Overview of Modeling Approaches to Predict Physicochemical Properties	54
3.2.1	Prediction of Properties Based on Other Properties	55
3.2.2	Prediction of Properties Based on Theoretical Calculations	55
3.2.3	Additivity Schemes for Property Prediction	56
3.2.4	Statistical Quantitative Structure–Property Relationships (QSPRs)	59
3.3	Methods for the Prediction of Individual Properties	59
3.3.1	Mean Molecular Polarizability	59
3.3.2	Thermodynamic Properties	60
3.3.3	Octanol/Water Partition Coefficient (Log P)	63
3.3.4	Octanol/Water Distribution Coefficient (log D)	67
3.3.5	Estimation of Water Solubility (log S)	69
3.3.6	Melting Point (MP)	71
3.3.7	Acid Ionization Constants	73
3.4	Limitations of Statistical Methods	76
3.5	Outlook and Perspectives	76
	Selected Reading	78
	References	78

4	**Chemical Reactions** *83*
4.1	**Chemical Reactions – An Introduction** *84*
	Johann Gasteiger
	References *85*
4.2	**Reaction Prediction and Synthesis Design** *86*
	Jonathan M. Goodman
4.2.1	Introduction *86*
4.2.2	Reaction Prediction *87*
4.2.3	Synthesis Design *94*
4.2.4	Conclusion *102*
	References *103*
4.3	**Explorations into Biochemical Pathways** *106*
	Oliver Sacher and Johann Gasteiger
4.3.1	Introduction *106*
4.3.2	The BioPath.Database *110*
4.3.3	BioPath.Explore *111*
4.3.4	Search Results *112*
4.3.5	Exploitation of the Information in BioPath.Database *117*
4.3.6	Summary *129*
	Selected Reading *130*
	References *130*
5	**Structure–Spectrum Correlations and Computer-Assisted Structure Elucidation** *133*
	Joao Aires de Sousa
5.1	Introduction *133*
5.2	Molecular Descriptors *135*
5.2.1	Fragment-Based Descriptors *135*
5.2.2	Topological Structure Codes *135*
5.2.3	Three-Dimensional Molecular Descriptors *137*
5.3	Infrared Spectra *137*
5.3.1	Overview *137*
5.3.2	Infrared Spectra Simulation *138*
5.4	NMR Spectra *140*
5.4.1	Quantum Chemistry Prediction of NMR Properties *142*
5.4.2	NMR Spectra Prediction by Database Searching *142*
5.4.3	NMR Spectra Prediction by Increment-Based Methods *143*
5.4.4	NMR Spectra Prediction by Machine Learning Methods *144*
5.5	Mass Spectra *150*
5.5.1	Identification of Structures and Interpretation of MS *150*
5.5.2	Prediction of MS *151*
5.5.3	Metabolomics and Natural Products *151*
5.6	Computer-Aided Structure Elucidation (CASE) *153*

Selected Reading 157
Acknowledgement 157
References 158

6.1   **Drug Discovery: An Overview** 165
Lothar Terfloth, Simon Spycher, and Johann Gasteiger
6.1.1   Introduction 165
6.1.2   Definitions of Some Terms Used in Drug Design 167
6.1.3   The Drug Discovery Process 167
6.1.4   Bio- and Chemoinformatics Tools for Drug Design 168
6.1.5   Structure-based and Ligand-Based Drug Design 168
6.1.6   Target Identification and Validation 169
6.1.7   Lead Finding 171
6.1.8   Lead Optimization 182
6.1.9   Preclinical and Clinical Trials 188
6.1.10   Outlook: Future Perspectives 189
Selected Reading 191
References 191

6.2   **Bridging Information on Drugs, Targets, and Diseases** 195
Andreas Steffen and Bertram Weiss
6.2.1   Introduction 195
6.2.2   Existing Data Sources 196
6.2.3   Drug Discovery Use Cases in Computational Life Sciences 196
6.2.4   Discussion and Outlook 201
Selected Reading 202
References 202

6.3   **Chemoinformatics in Natural Product Research** 207
Teresa Kaserer, Daniela Schuster, and Judith M. Rollinger
6.3.1   Introduction 207
6.3.2   Potential and Challenges 208
6.3.3   Access to Software and Data 211
6.3.4   *In Silico* Driven Pharmacognosy-Hyphenated Strategies 219
6.3.5   Opportunities 220
6.3.6   Miscellaneous Applications 228
6.3.7   Limits 228
6.3.8   Conclusion and Outlook 229
Selected Reading 231
References 231

6.4   **Chemoinformatics of Chinese Herbal Medicines** 237
Jun Xu
6.4.1   Introduction 237
6.4.2   Type 2 Diabetes: The Western Approach 237
6.4.3   Type 2 Diabetes: The Chinese Herbal Medicines Approach 238

6.4.4	Building a Bridge 238
6.4.5	Screening Approach 240
	Selected Reading 244
	References 244

6.5 PubChem 245
Wolf-D. Ihlenfeldt

6.5.1	Introduction 245
6.5.2	Objectives 246
6.5.3	Architecture 246
6.5.4	Data Sources 247
6.5.5	Submission Processing and Structure Representation 248
6.5.6	Data Augmentation 249
6.5.7	Preparation for Database Storage 249
6.5.8	Query Data Preparation and Structure Searching 250
6.5.9	Structure Query Input 253
6.5.10	Query Processing 254
6.5.11	Getting Started with PubChem 254
6.5.12	Web Services 255
6.5.13	Conclusion 255
	References 256

6.6 Pharmacophore Perception and Applications 259
Thomas Seidel, Gerhard Wolber, and Manuela S. Murgueitio

6.6.1	Introduction 259
6.6.2	Historical Development of the Modern Pharmacophore Concept 260
6.6.3	Representation of Pharmacophores 262
6.6.4	Pharmacophore Modeling 268
6.6.5	Application of Pharmacophores in Drug Design 272
6.6.6	Software for Computer-Aided Pharmacophore Modeling and Screening 278
6.6.7	Summary 278
	Selected Reading 279
	References 280

6.7 Prediction, Analysis, and Comparison of Active Sites 283
Andrea Volkamer, Mathias M. von Behren, Stefan Bietz, and Matthias Rarey

6.7.1	Introduction 283
6.7.2	Active Site Prediction Algorithms 284
6.7.3	Target Prioritization: Druggability Prediction 292
6.7.4	Search for Sequentially Homologous Pockets 296
6.7.5	Target Comparison: Virtual Active Site Screening 298
6.7.6	Summary and Outlook 304
	Selected Reading 306
	References 306

Contents

6.8 Structure-Based Virtual Screening 313
Adrian Kolodzik, Nadine Schneider, and Matthias Rarey
6.8.1 Introduction 313
6.8.2 Docking Algorithms 315
6.8.3 Scoring 317
6.8.4 Structure-Based Virtual Screening Workflow 321
6.8.5 Protein-Based Pharmacophoric Filters 323
6.8.6 Validation 323
6.8.7 Summary and Outlook 326
Selected Reading 328
References 328

6.9 Prediction of ADME Properties 333
Aixia Yan
6.9.1 Introduction 333
6.9.2 General Consideration on SPR/QSPR Models 334
6.9.3 Estimation of Aqueous Solubility (log S) 336
6.9.4 Estimation of Blood–Brain Barrier Permeability (log BB) 342
6.9.5 Estimation of Human Intestinal Absorption (HIA) 346
6.9.6 Other ADME Properties 349
6.9.7 Summary 354
Selected Reading 355
References 355

6.10 Prediction of Xenobiotic Metabolism 359
Anthony Long and Ernest Murray
6.10.1 Introduction: The Importance of Xenobiotic Biotransformation in the Life Sciences 359
6.10.2 Biotransformation Types 362
6.10.3 Brief Review of Methods 364
6.10.4 User Needs: Scientists Use Metabolism Information in Different Ways 370
6.10.5 Case Studies 372
Selected Reading 382
References 383

6.11 Chemoinformatics at the CADD Group of the National Cancer Institute 385
Megan L. Peach and Marc C. Nicklaus
6.11.1 Introduction and History 385
6.11.2 Chemical Information Services 386
6.11.3 Tools and Software 388
6.11.4 Synthesis and Activity Predictions 391
6.11.5 Downloadable Datasets 391
References 392

6.12	**Uncommon Data Sources for QSAR Modeling** *395*	

Alexander Tropsha

- 6.12.1 Introduction *395*
- 6.12.2 Observational Metadata and QSAR Modeling *397*
- 6.12.3 Pharmacovigilance and QSAR *398*
- 6.12.4 Conclusions *401*
  Selected Reading *402*
  References *402*

6.13 Future Perspectives of Computational Drug Design *405*

Gisbert Schneider

- 6.13.1 Where Do the Medicines of the Future Come from? *405*
- 6.13.2 Integrating Design, Synthesis, and Testing *408*
- 6.13.3 Toward Precision Medicine *409*
- 6.13.4 Learning from Nature: From Complex Templates to Simple Designs *411*
- 6.13.5 Conclusions *413*
  Selected Reading *414*
  References *414*

7 Computational Approaches in Agricultural Research *417*

Klaus-Jürgen Schleifer

- 7.1 Introduction *417*
- 7.2 Research Strategies *418*
- 7.2.1 Ligand-Based Approaches *419*
- 7.2.2 Structure-Based Approaches *422*
- 7.3 Estimation of Adverse Effects *429*
- 7.3.1 *In Silico* Toxicology *429*
- 7.3.2 Programs and Databases *430*
- 7.3.3 *In Silico* Toxicology Models *432*
- 7.4 Conclusion *435*
  Selected Reading *436*
  References *436*

8 Chemoinformatics in Modern Regulatory Science *439*

Chihae Yang, James F. Rathman, Aleksey Tarkhov, Oliver Sacher, Thomas Kleinoeder, Jie Liu, Thomas Magdziarz, Aleksandra Mostraq, Joerg Marusczyk, Darshan Mehta, Christof Schwab, and Bruno Bienfait

- 8.1 Introduction *439*
- 8.1.1 Science and Technology Progress *439*
- 8.1.2 Regulatory Science in Twenty-First Century *440*
- 8.2 Data Gap Filling Methods in Risk Assessment *441*
- 8.2.1 QSAR and Structural Knowledge *442*
- 8.2.2 Threshold of Toxicological Concern (TTC) *443*
- 8.2.3 Read-Across (RA) *445*
- 8.3 Database and Knowledge Base *448*
- 8.3.1 Architecture of Structure-Searchable Toxicity Database *448*

8.3.2	Data Model for Chemistry-Centered Toxicity Database *449*
8.3.3	Inventories *452*
8.4	New Approach Descriptors *453*
8.4.1	ToxPrint Chemotypes *453*
8.4.2	Liver BioPath Chemotypes *458*
8.4.3	Dynamic Generation of Annotated Linear Paths *459*
8.4.4	Other Examples of Descriptors *461*
8.5	Chemical Space Analysis *462*
8.5.1	Principal Component Analysis *462*
8.6	Summary *464*
	Selected Reading *466*
	References *466*

9 Chemometrics in Analytical Chemistry *471*
Anita Rácz, Dávid Bajusz, and Károly Héberger

9.1	Introduction *471*
9.2	Sources of Data: Data Preprocessing *472*
9.3	Data Analysis Methods *475*
9.3.1	Qualitative Methods *475*
9.3.2	Quantitative Methods *483*
9.4	Validation *488*
9.5	Applications *492*
9.6	Outlook and Prospects *492*
	Selected Reading *496*
	References *496*

10 Chemoinformatics in Food Science *501*
Andrea Peña-Castillo, Oscar Méndez-Lucio, John R. Owen, Karina Martínez-Mayorga, and José L. Medina-Franco

10.1	Introduction *501*
10.2	Scope of Chemoinformatics in Food Chemistry *502*
10.3	Molecular Databases of Food Chemicals *503*
10.4	Chemical Space of Food Chemicals *506*
10.4.1	General Considerations *506*
10.4.2	Chemical Space Analysis of Food Chemical Databases *508*
10.5	Structure–Property Relationships *510*
10.5.1	Structure–Flavor Relationships and Flavor Cliffs *511*
10.5.2	Quantitative Structure–Odor Relationships *512*
10.6	Computational Screening and Data Mining of Food Chemicals Libraries *513*
10.6.1	Anticonvulsant Effect of Sweeteners and Pharmaceutical and Food Preservatives *514*
10.6.2	Mining Food Chemicals as Potential Epigenetic Modulators *516*
10.7	Conclusion *521*
	Selected Reading *522*
	References *523*

11	**Computational Approaches to Cosmetics Products Discovery** *527*	
	Soheila Anzali, Frank Pflücker, Lilia Heider, and Alfred Jonczyk	
11.1	Introduction: Cosmetics Demands on Computational Approaches *527*	
11.2	Case I: The Multifunctional Role of Ectoine as a Natural Cell Protectant (Product: Ectoine, "Cell Protection Factor", and Moisturizer) *528*	
11.2.1	Molecular Dynamics (MD) Simulations *530*	
11.2.2	Results and Discussion: Ectoine Retains the Power of Water *531*	
11.3	Case II: A Smart Cyclopeptide Mimics the RGD Containing Cell Adhesion Proteins at the Right Site (Product: Cyclopeptide-5: Antiaging) *533*	
11.3.1	Methods *536*	
11.3.2	Results and Discussion *536*	
11.4	Conclusions: Cases I and II *542*	
	References *545*	
12	**Applications in Materials Science** *547*	
	Tu C. Le, and David A. Winkler	
12.1	Introduction *547*	
12.2	Why Materials Are Harder to Model than Molecules *548*	
12.3	Why Are Chemoinformatics Methods Important Now? *548*	
12.4	How Do You Describe Materials Mathematically? *549*	
12.5	How Well do Chemoinformatics Methods Work on Materials? *551*	
12.6	What Are the Pitfalls when Modeling Materials? *551*	
12.7	How Do You Make Good Models and Avoid the Pitfalls? *553*	
12.8	Materials Examples *554*	
12.8.1	Inorganic Materials and Nanomaterials *554*	
12.8.2	Polymers *557*	
12.8.3	Catalysts *558*	
12.8.4	Metal–Organic Frameworks (MOFs) *560*	
12.9	Biomaterials Examples *561*	
12.9.1	Bioactive Polymers *561*	
12.9.2	Microarrays *564*	
12.10	Perspectives *566*	
	Selected Reading *567*	
	References *567*	
13	**Process Control and Soft Sensors** *571*	
	Kimito Funatsu	
13.1	Introduction *571*	
13.2	Roles of Soft Sensors *573*	
13.3	Problems with Soft Sensors *574*	
13.4	Adaptive Soft Sensors *576*	

13.5	Database Monitoring for Soft Sensors 578
13.6	Efficient Process Control Using Soft Sensors 581
13.7	Conclusions 582
	Selected Readings 583
	References 583

14   Future Directions *585*
Johann Gasteiger

14.1	Well-Established Fields of Application 585
14.2	Emerging Fields of Application 586
14.3	Renaissance of Some Fields 587
14.4	Combined Use of Chemoinformatics Methods 588
14.5	Impact on Chemical Research 589

Index *591*

Foreword

Chemistry began with magic. Who but a wizard could, with a puff of smoke, turn one thing into another? The alchemists believed that the ability to transform materials was a valuable skill, so valuable in fact that they devised complex descriptions and alchemical symbols, known only to them, to represent their secret methods. Information was encoded and hidden, suffused with allegorical and religious symbolism, slowing progress. Medicinal chemists today may be particularly interested in a legendary stone called a Bezoar, found in the bodies of animals (if you knew which animal to dissect), that had universal curative properties. I'm still looking. However, to plagiarise a recent Nobel prize-winner for literature, the times they were a changin'. Departing from the secretive "alchemist" approach Berzelius (1779–1848) suggested compounds should be named from the elements which made them up and Archibald Scott-Couper (1831–1892) devised the "connections" between "atoms," which gave rise to structural diagrams (1858). In 1887, the symbols created by Jean Henri Hassenfratz and Pierre Auguste Adet to complement the *Methode de Nomenclature Chimique* were a revolutionary approach to chemical information. A jumbled, confused and incorrect nomenclature was replaced by our modern day designations such as oxygen, hydrogen and sodium chloride. The new chemistry of Lavoisier was becoming systematised. The "Age of Enlightenment" created a new philosophy of science where information, validated by experiment, could be tested by an expanding community of "scientists" (a term coined by William Whewell in 1833), placing data at the core of chemistry.

With the accumulation of knowledge, and a language to communicate chemistry, the stage was set for the creation of a new science of information in the domain of chemistry. Up stepped Friedrich Beilstein (1838–1906), who systematically collected chemical data on substances, reactions and properties of chemical compounds in the *Handbuch der organischen Chemie* (*Handbook of Organic Chemistry*, published in 1881). The naming of compounds was a key feature which enabled the storage and retrieval of chemical information on a "grand" scale (1500 compounds). The indexing of chemical information meant chemistry could be reliably stored, common links between data established, and – most importantly – the information could be retrieved without loss. This drive for efficient indexing was the dominant feature of chemical information research for the next half century. As chemistry (and its many related disciplines) continued on an ever upward trajectory of innovation (and data collection) the paper trail required to go from perfectly reasonable questions like "how do I synthesise this compound" to "has this compound been made before" became rather complex and time consuming. I remember many happy hours spent in the library of the Wellcome Foundation trawling through the multitude of bookshelves of Chemical Abstracts to find one compound, and if lucky, a synthesis simple enough that I could perform with a yield better than my usual ten percent. Of course things got worse (or better if you were a librarian), and I recall an interesting RSC symposium in 1994 called "*The Chemical Information Explosion: Chaos, Chemists, and Computers.*" We had clearly reached a point where someone had to invent Chemoinformatics.

Although the "someone" is of course a worldwide community of scientists interested in chemical data, the term was coined by Frank Brown in 1998, and he defined it as "the mixing of those information resources to transform data into information and information into knowledge for the intended purpose of making better decisions faster in the area of drug lead identification and optimization." The combination of multi-disciplinarity, the reduction of data to knowledge and the driving force of the pharmaceutical industry have been key features of the advance of Chemoinformatics. The enabling technologies have been the availability of unprecedented amounts of chemical data (increasingly pubicly available) and the continuous development of new algorithms, designed specifically for chemistry, to achieve the goal of turning information into knowledge. Of course Moore's law (an observation by Gordon Moore at Intel), that the density of computer components (and the computation power offered) doubles every 2 years has underpinned the hardware necessary to keep pace with the data explosion. But perhaps some of the chaos remains, hence the popularity of software such as Babel, which converts many data formats to many data formats!). Some numbers here are interesting. If we recall that the first edition of Beilstein's Handbuch contained 1500 compounds, the Chemical Abstracts Service of the American Chemical Society reported in 2015 that they had registered their 100 millionth chemical substance. What is truly transformational (if you think about it) is that a new student, with a basic knowledge of chemistry, when asked to search for a single compound from the 100M registry, gets the correct result in a microsecond. Not only that, a host of measured and predicted chemical properties, synthesis strategies, available reagents, structurally similar compounds and internet links to a multitude of other diverse, information rich databases.

Clearly, Chemoinformatics has come of age. In fact the term "Chemoinformatics" has gained a certain elastic quality. The methodologies and data analysis tools developed for chemical information have evolved and extended to the data analytics of essentially any data that includes chemistry. Examples include for example, the simulations of large systems of molecules such as proteins, machine learning (and the recent resurgence in Artificial Intelligence) to create predictive models, for example, for metabolism, ADME properties and Quantitative Structure Activity Relationships of drugs (including quantum chemistry, bioinformatics, and analytical chemistry) and the detection and analysis of drug binding sites. Although much of the early work in Chemoinformatics has been applied to problems of the pharmaceutical industry, the subject has been embraced across the sciences wherever chemistry is required for example, in agricultural and food research, cosmetics, and materials science.

But in an age when computers can do "magic," ("*Any sufficiently advanced technology is indistinguishable from magic*" – Arthur C. Clark) it is tempting to return to where we were in the time of Berzelius and hide the technology behind an alchemical mask of symbols for example, a simple interface which hides highly complex search and retrieval algorithms or a machine learning application to predict metabolism. The antidote to this is of course education. A firm grounding in the principles and practice of Chemoinformatics provides students and expert practitioners alike with the knowledge of the underlying algorithms, how they are implemented, their availability and of course limitations of software for a given purpose as well as future challenges for those with a keen interest in developing the field.

The best textbooks are naturally written from the viewpoint of those who are intimately connected to their subject. The Chemoinformatics group at the Computer-Chemie-Centrum (CCC) at the University of Erlangen-Nuremberg have been pioneers in Chemoinformatics for over 30 years and are recognised as both innovators and experts at applying these methods to a large variety of chemical problems. However it is as educators that perhaps their greatest impact on the field may accrue over time. The new book "Applied Chemoinformatics – Achievements and Future Opportunities" shows the many fields chemoinformatics is now applied to and builds on the successful first edition "Chemoinformatics – A Textbook" (published in 2003) and is again edited by Johann Gasteiger and Thomas Engel. This volume is complemented by an additional textbook "Chemoinformatics – Basic Concepts and Methods" which is an introduction into this field. Johann Gasteiger has had a distinguished career in Chemistry and is well known for his seminal contributions to Chemoinformatics. He was the recipient of the 1991 Gmelin-Beilstein Medal of the German Chemical Society for Achievements in Computer Chemistry; the 2005 Mike Lynch Award of the Chemical Structure Association; the 2006 ACS Award for Computers in Chemical and Pharmaceutical Research for his outstanding achievements in research and education in the field of Chemoinformatics and the 1997 Herman Skolnik Award of the Division of Chemical Information of the American Chemical Society. Thomas Engel is a specialist in Chemoinformatics who studied chemistry and education at the University of Würzburg and spent a significant tenure at the Computer-Chemie-Centrum at the University of Erlangen-Nürnberg, followed by the Chemical Computing Group AG in Cologne and is presently at the Ludwig-Maximilians-Universität, Munich.

As editors, they have brought together a wide range of experts and topics which will inform, educate and motivate the reader to delve deeper into the subject of Chemoinformatics. The new edition provides both the foundations for Chemoinformatics and also a range of developing topics of active research, providing the reader with an introduction to the subject as well as advanced topics and future directions. This new edition is complemented by the "*Handbook of Chemoinformatics: From Data to Knowledge*" (by the same editors). It belongs on the bookshelf of students and experts alike, all who have an interest in the field of Chemoinformatics, and especially those who see "magic" in chemistry.

Robert C. Glen
Professor of Molecular Sciences Informatics
Director of the Centre for Molecular Informatics
Department of Chemistry
University of Cambridge
Cambridge
United Kingdom

List of Contributors

Joao Aires de Sousa
Universidade Nova de Lisboa
Departamento de Quimica
2829-516 Caparica
Portugal

Soheila Anzali
InnoSA GmbH
Georg-Dascher-Str. 2
64846 Groß-Zimmern
Germany

Dávid Bajusz
Hungarian Academy of Sciences
Research Centre for Natural Sciences
Plasma Chemistry Research Group
Magyar tudósok krt. 2
1117 Budapest
Hungary

Bruno Bienfait
Molecular Networks GmbH
Neumeyerstr. 28
90411 Nürnberg
Germany

Stefan Bietz
University of Hamburg
Center for Bioinformatics
Bundesstraße 43
20146 Hamburg
Germany

Thomas Engel
Ludwig-Maximilians-University
Munich
Department of Chemistry
Butenandtstr. 5-13
81377 Munich
Germany

Kimito Funatsu
The University of Tokyo
Department of Chemical System
Engineering
7-3-1 Hongo, Bunkyo-ku
113-8656, Tokyo
Japan

Johann Gasteiger
Computer-Chemie-Centrum
Universität Erlangen-Nürnberg
Nägelsbachstr. 25
91052 Erlangen
Germany

Jonathan M. Goodman
Department of Chemistry
Lensfield Road
Cambridge, CB2 1EW
UK

Károly Héberger
Hungarian Academy of Sciences
Research Centre for Natural Sciences
Plasma Chemistry Research Group
Magyar tudósok krt. 2
1117 Budapest
Hungary

List of Contributors

Lilia Heider
Merck KGaA
Frankfurter Strasse 250
64293 Darmstadt
Germany

Wolf-D. Ihlenfeldt
Xemistry GmbH
Hainholzweg 11
D-61462 Königstein
Germany

Alfred Jonczyk
Merck KGaA
Frankfurter Strasse 250
64293 Darmstadt
Germany

Teresa Kaserer
University of Innsbruck
Center for Molecular Biosciences
Innsbruck
Institute of Pharmacy/Pharmaceutical
Chemistry
Computer-Aided Molecular Design
Group
Innrain 80-82
6020 Innsbruck
Austria

Thomas Kleinoeder
Molecular Networks GmbH
Neumeyerstr. 28
90411 Nürnberg
Germany

Adrian Kolodzik
University of Hamburg
ZBG - Center for Bioinformatics
Bundesstraße 43
20146 Hamburg
Germany

Tu C. Le
RMIT University
Melbourne 3000
Australia

Anthony Long
Lhasa Limited
Granary Wharf House, 2 Canal
Wharf, Holbeck
Leeds, LS11 5PS, West Yorkshire
UK

Karina Martinez-Mayorga
Universidad Nacional Autónoma de
México
Instituto de Química
Departamento de Fisicoquímica
Avenida Universidad 3000
Mexico City 04510
Mexico

Joerg Marusczyk
Molecular Networks GmbH
Neumeyerstr. 28
90411 Nürnberg
Germany

José L. Medina-Franco
Universidad Nacional Autónoma de
México
Departamento de Farmacia
Avenida Universidad 3000
Mexico City 04510
Mexico

Darshan Mehta
The Ohio State University
Department of Chemical and
Biomolecular Engineering
Columbus, OH 43210
USA

Oscar Méndez-Lucio
Universidad Nacional Autónoma de México
Departamento de Farmacia
Avenida Universidad 3000
Mexico City 04510
Mexico

Manuela S. Murgueitio
Freie Universität Berlin
Institute of Pharmacy,
Computer-Aided Drug Design,
Pharmaceutical and Medicinal Chemistry
Königin-Luisestr. 2+4
14195 Berlin
Germany

Ernest Murray
Lhasa Limited
Granary Wharf House, 2 Canal Wharf, Holbeck
Leeds, LS11 5PS, West Yorkshire
UK

Marc C. Nicklaus
NIH
National Cancer Institute
NCI-Frederick, 376 Boyles Street
Frederick
MD 21702
USA

John R. Owen
Northern Ireland Science Park
ECIT Institute
High-Performance Computing Research Group
Queens Road
Belfast BT3 9DT
Northern Ireland

Megan L. Peach
NIH
National Cancer Institute
NCI-Frederick, 376 Boyles Street
Frederick, MD 21702
USA

Andrea Peña-Castillo
Universidad Nacional Autónoma de México
Departamento de Farmacia
Avenida Universidad 3000
Mexico City 04510
Mexico

Frank Pflücker
Merck KGaA
Frankfurter Strasse 250
64293 Darmstadt
Germany

Anita Rácz
Hungarian Academy of Sciences
Research Centre for Natural Sciences
Plasma Chemistry Research Group
Magyar tudósok krt. 2
1117 Budapest
Hungary

and

Szent István University
Department of Applied Chemistry
Villányi út 29-43
1118 Budapest
Hungary

Matthias Rarey
University of Hamburg
ZBG - Center for Bioinformatics
Bundesstraße 43
20146 Hamburg
Germany

James F. Rathman
Molecular Networks GmbH
Neumeyerstr. 28
90411 Nürnberg
Germany

and

Altamira LLC
1455 Candlewood Dr.
Columbus OH 43235
USA

and

The Ohio State University
Department of Chemical and
Biomolecular Engineering
Columbus, OH 43210
USA

Dina Robaa
Martin-Luther-Universität
Halle-Wittenberg
Institute of Pharmacy
Wolfgang-Langenbeck-Str. 4
06120 Halle (Saale)
Germany

Judith M. Rollinger
University of Vienna
Department of Pharmacognosy
Althanstraße 14
1090 Vienna
Austria

Oliver Sacher
Molecular Networks GmbH
Neumeyerstr. 28
90411 Nürnberg
Germany

Klaus-Jürgen Schleifer
BASF SE, Computational Chemistry
and Bioinformatics
A30, 67056 Ludwigshafen
Germany

Gisbert Schneider
Swiss Federal Institute of Technology
(ETH)
Department of Chemistry and
Applied Biosciences
Vladimir-Prelog-Weg 4
CH-8093 Zurich
Switzerland

Nadine Schneider
University of Hamburg
ZBG - Center for Bioinformatics
Bundesstraße 43
20146 Hamburg
Germany

Daniela Schuster
University of Innsbruck
Center for Molecular Biosciences
Innsbruck
Institute of Pharmacy/Pharmaceutical
Chemistry
Computer-Aided Molecular Design
Group
Innrain 80-82
6020 Innsbruck
Austria

Christof Schwab
Molecular Networks GmbH
Neumeyerstr. 28
90411 Nürnberg
Germany

Thomas Seidel
University of Vienna
Department of Pharmaceutical
Chemistry
Althanstraße 14
1090 Vienna
Austria

Wolfgang Sippl
Martin-Luther-Universität
Halle-Wittenberg
Institute of Pharmacy
Wolfgang-Langenbeck-Str. 4
06120 Halle (Saale)
Germany

Simon Spycher
Eawag, Environmental Chemistry
Überlandstrasse 133
8600 Dübendorf
Switzerland

Andreas Steffen
Bayer Pharma Aktiengesellschaft
PH-DD-TRG-CIPL-Bioinformatics
Müllerstr. 178
13342 Berlin
Germany

Aleksey Tarkhov
Molecular Networks GmbH
Neumeyerstr. 28
90411 Nürnberg
Germany

Lothar Terfloth
Insilico Biotechnology AG
Meitnerstrasse 9
70563 Stuttgart
Germany

Igor V. Tetko
Helmholtz Zentrum München
Deutsches Forschungszentrum für
Gesundheit und Umwelt (GmbH)
Ingolstädter Landstr. 1, 60w
85764 Neuherberg
Germany

Alexander Tropsha
University of North Carolina
UNC Eshelman School of Pharmacy
100K Beard Hall
Chapel Hill
NC 27599
USA

Andrea Volkamer
Charité - Universitätsmedizin Berlin
Institute of Physiology
In-silico Toxicology Group
Virchowweg 6
10117 Berlin
Germany

Mathias M. von Behren
University of Hamburg
Center for Bioinformatics
Bundesstraße 43
20146 Hamburg
Germany

Bertram Weiss
Bayer Pharma Aktiengesellschaft
PH-DD-TRG-CIPL-Bioinformatics
Müllerstr. 178
13342 Berlin
Germany

David A. Winkler
Latrobe University
Latrobe Institute for Molecular Science
Bundoora 3082
Australia

and

Monash University
Monash Institute of Pharmaceutical Sciences
392 Royal Parade
Parkville 3052
Australia

and

Flinders University
School of Chemical and Physical Sciences
Bedford Park 5042
Australia

and

CSIRO Manufacturing
Bag 10
Clayton South MDC 3169
Australia

Gerhard Wolber
Freie Universität Berlin
Institute of Pharmacy,
Computer-Aided Drug Design,
Pharmaceutical and Medicinal Chemistry
Königin-Luisestr. 2+4
14195 Berlin
Germany

Jun Xu
Sun Yat-Sen University
School of Pharmaceutical Sciences
132 East Circle at University City
Guangzhou 510006
P. R. China

Aixia Yan
Beijing University of Chemical Technology
Department of Pharmaceutical Engineering
State Key Laboratory of Chemical Resource Engineering
15 BeiSanHuan East Road
Beijing 100029
P. R. China

Chihae Yang
Molecular Networks GmbH
Neumeyerstr. 28
90411 Nürnberg
Germany

and

Altamira LLC
1455 Candlewood Dr.
Columbus, OH 43235
USA

and

The Ohio State University
Department of Chemical and Biomolecular Engineering
Columbus, OH 43210
USA

1 Introduction

Thomas Engel[1] and Johann Gasteiger[2]

[1] Ludwig-Maximilians-University Munich, Department of Chemistry, Butenandtstr. 5 - 13, 81377 Munich, Germany
[2] Computer-Chemie-Centrum, Universität Erlangen-Nürnberg, Nägelsbachstr. 25, 91052 Erlangen, Germany

> **Outline**
>
> 1.1 The Rationale for the Books, 1
> 1.2 Development of the Field, 2
> 1.3 The Basis
>   of Chemoinformatics and the
>   Diversity of Applications, 3

1.1 The Rationale for the Books

In 2003 we issued the book

> **Chemoinformatics: A Textbook**
> (J. Gasteiger, T. Engel, Editors, Wiley-VCH Verlag GmbH, Weinheim, Germany, ISBN 13: 978-3-527-30681-7)

which was well accepted and contributed to the development of the field of chemoinformatics. However, with the enormous progress in chemoinformatics, it is now time for an update. As we started out on this endeavor, it became rapidly clear that all the developments require presenting the field in more than a single book. We have therefore edited two volumes:

- Chemoinformatics – Basic Concept and Methods [1]
- Applied Chemoinformatics – Achievements and Future Opportunities

In the first volume, "Basic Concept and Methods," the essential foundations and methods that comprise the technology of chemoinformatics are presented. The links to this first volume are referenced in the present volume by "**Methods Volume.**"

The *Application Volume* – tagged as shortcut in the first volume – emerged from the single "Applications" chapter of the 2003 textbook. The fact that applications now merit a book of their own clearly demonstrates how enormously the field has grown. Chemoinformatics has certainly matured to a scientific discipline of its own with many applications in all areas of chemistry and in related fields.

Both volumes consist of chapters written by different authors. In order to somehow ensure that the material is not too heterogeneous, we have striven to adapt the contributions to an overall picture and inserted cross-references as mentioned above. We hope that this helps the reader to realize the interdependences of many of the methods and how they can work together in solving chemical problems.

Both volumes are conceived as textbooks for being used in teaching and self-learning of chemoinformatics. In particular, the first, "Methods Volume," is addressed to students, explaining the basic approaches and supporting this with exercises. Altogether, we wanted to present with both books a comprehensive overview of the field of chemoinformatics for students, teachers, and scientists from all areas of chemistry, from biology, informatics, and medicine.

1.2 Development of the Field

We are happy to see – and demonstrate within this book – that chemoinformatics has ventured into and found applications in many fields other than drug discovery. Drug discovery is still the most important area of application of chemoinformatics methods (Chapter 6), but we expect that the other fields of applications will continue to grow.

Some comments on terminology are appropriate. The varying use of the terms *chemoinformatics* and *cheminformatics* seems to indicate a geographical (or perhaps cultural) divide, with "cheminformatics" mainly used in the United States and "chemoinformatics" originating and more widely used in Europe and the rest of the world.

Chemometrics is a field that originated in the early 1970s and is mainly associated with analytical chemistry, as is shown in Chapter 9. The two fields, chemometrics and chemoinformatics, have borrowed heavily from each other and use many of the same methods. As chemoinformatics has a much broader focus, it is fair to consider the data analysis methods in chemometrics as a part of chemoinformatics.

Molecular informatics has appeared as a term, but this is certainly too narrow a name to cover all potential chemoinformatics applications, as many tasks in chemistry deal with compounds and materials that are not limited to or cannot be identified with single molecular structures. The applications in materials science, presented in Chapter 12, are a case in point, as are the applications in process control illustrated in Chapter 13.

More fortunate developments have brought chemoinformatics closer together with other disciplines. The borders between chemoinformatics and *bioinformatics* are becoming quite transparent. Some important and challenging tasks in drug design, and cosmetics development, and in trying to understand living systems need approaches from both fields. We have taken account of this with Section 4.3 on biochemical pathways; Section 6.2 on drugs, targets, and diseases; and Chapter 11 on computational approaches in cosmetics products discovery.

In a similar manner, many tasks faced in chemistry and related fields are tackled by methods of both chemoinformatics and *computational chemistry*, thus blurring the borders between the two disciplines. We indicate the combined utilization of chemoinformatics and computational chemistry in Chapter 5 on structure–spectra correlations and *computer-assisted structure elucidation* (CASE), in Chapter 7 on computational approaches in agricultural research, and in Chapter 11 on computational approaches in cosmetics products discovery. Furthermore, the basic concepts of *computational chemistry* are described in Chapter 8 of the *Methods Volume* [1].

It is exciting to see that the use of computers in chemistry, by methods both of chemoinformatics and of computational chemistry, has gained more widespread recognition. This culminated in the awarding of the *Nobel Prize in Chemistry in 2013* to Martin Karplus, Michael Levitt, and Arieh Warshel. The Swedish Academy of Sciences motivated its decision by stating:

> "Today the computer is just as important a tool for chemists as the test tube."
> (https://www.nobelprize.org/nobel_prizes/chemistry/laureates/2013/press.html)

1.3 The Basis of Chemoinformatics and the Diversity of Applications

We were fortunate to gain many excellent scientists to demonstrate the various applications of chemoinformatics in chemistry and related fields by writing chapters for this book, but it should be emphasized from the very beginning that these applications are, by no means, all of the established or possible applications: the field is still expanding and growing.

1.3.1 Databases

Probably the most important achievement of chemoinformatics is the building of databases on chemical information. The development of chemoinformatics technologies made it possible to store and retrieve chemical information in a variety of databases. Interaction with these databases can be made by the international language of the chemist that is graphical in nature: structure diagrams and reaction equations. Databases are so routinely used in chemical research that it seems to have been forgotten that their construction was only possible by the work and developments by chemists, mathematicians, and computer scientists in the decades from 1960 to 1990. The building of databases on chemical information is presented in Chapter 6 of the *Methods Volume* [1]. In the present volume the exploitation of these databases will be shown in a variety of applications presented in the chapters.

Without these databases an overview of chemical information, which has grown enormously in recent decades, could not be maintained anymore. It is fair to say that modern chemical research can nowadays not be done without consulting databases on chemical information.

1 Introduction

1.3.2 Fundamental Questions of a Chemist

In structuring the presentation of the applications of chemoinformatics methods, we were motivated by the fundamental questions of a chemist and how they can be supported by chemoinformatics methods (see Figure 1.1).

1.3.2.1 Prediction of Properties

In his Norris Award Lecture of 1968, George S. Hammond said:

> "The most fundamental and lasting objective of synthesis is not production of new compounds but *production of properties*."

With this in mind, the first fundamental task of a chemist is to relate the desired property, be it a drug, a pesticide, a paint, or an antiaging compound, with a chemical structure. This is the domain of *structure–property relationships* (SPR) or *structure–activity relationships* (SAR), or even finding such relationships on a quantitative basis by quantitative structure- property relationships (QSPR) or quantitative structure–activity relationships QSAR.

Chapter 2 presents the methodology of the QSPR and QSAR approach for establishing models that allow the prediction of properties. Quite a few of the steps in this approach have been outlined in more detail in the *Methods Volume* [1] in Chapters 9–12. However, because of the importance of this approach in many areas of chemistry and for the prediction of a wide variety of properties,

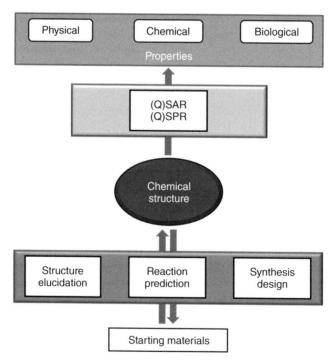

Figure 1.1 Fundamental questions of a chemist and the chemoinformatics methods that can be used in providing support for solving these tasks.

we have decided to outline the QSPR/QSAR approach as an entry to the various chapters in this *Application Volume*.

Many such QSPR and QSAR models for the prediction of physical, chemical, or biological properties have been established. Chapter 3 on the prediction of physicochemical properties presents various approaches to the prediction of some important physical and chemical properties. The prediction of some more properties relevant for drug discovery is indicated in Section 6.9 on the prediction of absorption, distribution, metabolism, and excretion(ADME) properties. The effects of chemicals on living species and on the environment are of much concern to society. The evaluation of these effects is indicated in Chapter 8 on chemoinformatics in modern regulatory science. The prediction of properties of importance in other fields of chemistry is presented in Chapter 7 on computational approaches to agricultural research and in Chapter 9 on chemometrics in analytical chemistry.

Some more recent areas of applications are given in Chapter 10 on chemoinformatics in food chemistry, in Chapter 11 on computational approaches in cosmetics products discovery, and in Chapter 12 on chemoinformatics in material science.

1.3.2.2 Chemical Reaction Prediction and Synthesis Design

Once a chemist has an idea which chemical structure will carry the desired property, he or she is faced with the task of synthesizing this compound and first coming up with a plan for performing the synthesis. This is the domain of *synthesis design* and *planning chemical reactions*. Section 4.2 on chemical reactions and synthesis design addresses these tasks, reflecting on work on computer-assisted synthesis design (CASD) that can be considered as one of the roots of chemoinformatics. Particularly interesting types of reactions are those that keep living organisms alive and thriving. These are discussed in Section 4.3 on biochemical pathways. This section also shows how chemoinformatics and bioinformatics can work together to obtain interesting insights such as discovering the essential pathways of diseases.

1.3.2.3 Structure Elucidation

As our knowledge on the driving forces of chemical reactions is still incomplete and many factors influence the outcome of a reaction, we must first verify whether the reaction that was performed produced the desired product. This is usually established by measuring various spectra to deduce the structure of the reaction product. This requires analysis of the relationships between the spectral data and chemical structure. At an early stage, in the late 1960s, methods for CASE were developed and can be considered as another root of chemoinformatics. In fact, the DENDRAL project at Stanford University is widely considered to be the first application of artificial intelligence to chemical problems. Chapter 5 on structure–spectra correlations and CASE presents the most recent developments in this field.

1.3.3 Drug Discovery

Drug discovery is certainly the most prominent field for the application of chemoinformatics methods. All major drug companies have divisions of chemoinformatics in various guises, and drug companies are the largest employers of chemoinformatics specialists. Furthermore, all drugs developed in the last few years have benefited from the use of chemoinformatics methods in one way or other. Because of its importance, the chapter on drug discovery is the most extensive one in this book. The introductory material found in Section 6.1 on drug discovery provides the framework for a series of the more specialized Sections 6.2–6.13. In addition, it also mentions methods and applications that are not represented in these sections. These 12 specialized sections present a variety of topics such as the use of bioinformatics methods on drugs, targets, and diseases in Section 6.2, the exploitation of natural products for drug discovery in Section 6.3, and in Section 6.4 on chemoinformatics in Chinese herbal medicines. Section 6.3 also shows applications in natural products research that have other focuses beyond drug discovery. In Section 6.5, the efforts made at the US National Institutes of Health are described in providing the general public with data on biological screening results on chemical compounds by PubChem. The PubChem database is an enormous resource for academic and industrial researchers and promises to heavily increase the insights made by scientists working in drug discovery.

The following three sections present methods for analyzing the three-dimensional structure of drug candidates and of proteins as targets for a biological response: Section 6.6 on pharmacophore perception and analysis, Section 6.7 on prediction and analysis of active sites, and Section 6.8 on structure-based virtual screening.

The further development of many drug candidates has to be discontinued in the preclinical phase because of adverse effects such as poor bioavailability, unfavorable pharmacokinetic effects, and undesirable metabolic stability. Many models have been developed for the prediction of the corresponding ADME properties. These are presented in Section 6.9 on the prediction of ADME properties and in Section 6.10 on the prediction of xenobiotic metabolism. The Computer-Aided Drug Design (CADD) Group at the National Cancer Institute has collected and developed a series of chemoinformatics methods for assisting in the drug design process and is making them generally available as detailed in Section 6.11. Many different information resources, such as those found in printed media and various databases, offer interesting information that can be harnessed for drug discovery. The use of these resources is discussed in Section 6.12 on the exploration of new data sources. The last section in Section 6.13 on drug design, status, and future of drug design, offers the personal views of an active academic researcher on where the drug discovery process will develop. His ideas are supported with recent developments from his research group.

1.3.4 Additional Fields of Application

It is gratifying to observe that, as we had hoped in the chapter on future directions in the 2003 textbook, various new fields have benefited from the use of chemoinformatics methods. A number of these will be shown in Chapters 7–13.

The *agrochemical industry* is employing the very same methods used in drug discovery for the development of new plant protection products. These include ligand-based and structure-based methods and are discussed in Chapter 7 on computational approaches in agricultural research.

The effects of chemicals on *human health* and their impact on the *environment* have become of increasing concern to society. At the same time, there is much interest in society to reduce, or even phase out, the use of animals for testing for toxicity of chemicals. For this purpose, computer models for the prediction of toxicity, environmental effects, and bioaccumulation have gained much interest for the registration of new and old chemicals, as indicated in Chapter 8 on chemoinformatics in modern regulatory science.

The analysis of data from *analytical chemistry* by statistical and other mathematical methods, subsumed under the term chemometrics, has a long history. Chapter 9 on chemometrics in analytical chemistry presents the methods used in this field and some typical applications (see also *Methods Volume*, Chapter 11).

It is interesting to see that rather novel areas of applications have been developed in recent years. For example, this is true for *food chemistry* as shown in Chapter 10 on chemoinformatics in food chemistry.

With the pressure on reducing, or even phasing out, the testing of chemicals with animals, the cosmetics industry has started to employ chemoinformatics and other computational chemistry methods. This has meant that the development of new *cosmetics products* is made more efficient, as presented in Chapter 11 on computational approaches in cosmetics products discovery.

One of the potentially largest and most promising new fields of application of chemoinformatics is *materials science*. Here, new ways for representing the objects of study have to be developed, since for many of the materials, no molecular structure can be given. The challenge of properly representing materials in the computer is of crucial influence on the success of any study. This will become clear in Chapter 12 on chemoinformatics in material science.

Many processes in the chemical industry are influenced by a multitude of factors and are governed by these factors in a nonlinear fashion. Indeed, in many cases no explicit mathematical relationship can be specified. In these situations, chemoinformatics methods can be helpful in *controlling* processes, mainly by measuring various control factors through sensors, and using those data for modeling a process. This is shown with a few examples in Chapter 13 on process control and soft sensors.

We hope that the many varied applications in these chapters show the importance of chemoinformatics in supporting chemical research and development in many fields, but it should be emphasized that by far, not all of the problems have been solved. Many challenging problems remain, and chemoinformatics will further evolve in parallel with chemical and biological sciences to provide deeper insights into these and related sciences. Thus, chemoinformatics is in very active development and provides many opportunities for future students. Chapter 14 on future directions collects some ideas on the further development of the field.

Reference

[1] Engel, T. and Gasteiger, J. (eds) (2017) *Chemoinformatics – Basic Concepts and Methods*, Wiley-VCH, Weinheim, 600 pp.

2 QSAR/QSPR

Wolfgang Sippl and Dina Robaa

Martin-Luther-Universität Halle-Wittenberg, Institute of Pharmacy, Department of Pharmaceutical Chemistry, Wolfgang-Langenbeck-Str. 4, 06120 Halle (Saale), Germany

Learning Objectives

- Assess the basic concepts, applications, and limitations of quantitative structure–activity relationships (QSAR) and quantitative structure–property relationships (QSPR) models.
- Describe the chemometric methods of grouping compounds according to their structural similarity, activity, and physicochemical properties (principal component analysis (PCA), hierarchical cluster analysis).
- Explain the methods of calibration models: multivariate linear regression (MLR), principal component regression (PCR), and partial least squares (PLS) regression.
- Discuss the basic concepts of nonlinear models.
- Distinguish internal and external validation: validation strategies; OECD recommendations on validation of QSARs; cross-validation; methods of data splitting into training and validation sets; and validation parameters.

Outline

2.1 Introduction, 9
2.2 Data Handling and Curation, 13
2.3 Molecular Descriptors, 14
2.4 Methods for Data Analysis, 17
2.5 Classification Methods, 19
2.6 Methods for Data Modeling, 20
2.7 Summary on Data Analysis Methods, 30
2.8 Model Validation, 30
2.9 Regulatory Use of QSARs, 46

2.1 Introduction

The fundamental physicochemical properties and the structural features of a molecule determine its behavior in physical, chemical, biological, or environmental processes and are therefore of interest for understanding and modeling the action of compounds.

Applied Chemoinformatics: Achievements and Future Opportunities, First Edition.
Edited by Thomas Engel and Johann Gasteiger.
© 2018 Wiley-VCH Verlag GmbH & Co. KGaA. Published 2018 by Wiley-VCH Verlag GmbH & Co. KGaA.

2 QSAR/QSPR

The methodology of quantitative structure–activity relationship (QSAR) tries to establish a quantitative connection between the molecular structure of a compound and its biological activity. In a similar manner, quantitative structure–property relationship (QSPR) tries to model the relationship between a chemical structure and a variety of physical or chemical properties. This chapter gives an overview of the QSAR and QSPR methodology. Some of the steps involved in this process are explained in more detail in other chapters of this book and in the *Methods Volume*.

Although the amount of experimental data is continuously growing, the number of newly synthesized or *in silico* designed compounds is increasing even more quickly. Due to virtual library screening, millions of potential compounds for which no experimental data are available can be designed *in silico*. Their properties or biological activities can nevertheless be modeled on the basis of reliable models. The basic approach to the problem of predicting properties or biological activities based on structural features can be written in a simple form where the property or biological activity can be expressed as shown in Eq. (2.1):

$$\text{Activity (property)} = \text{function (structure)} \qquad (2.1)$$

Quite often, direct methods to estimate the activity or many other properties of a chemical compound based on the structure of a molecule are not available. Therefore, indirect methods are applied to generate mathematical models able to describe the relationship (Figure 2.1). Indirect means that various mathematical algorithms are applied for the calculation of molecular descriptors that are then used instead of taking the original chemical structures. Based on the molecular descriptors calculated for a set of compounds and considering the corresponding biological activities or physical or chemical properties, a QSAR/QSPR can be established. Both QSAR and QSPR rely on the assumption that structurally similar molecules have similar activities/properties. However, it is also important to discriminate between QSPR that predict physical and physicochemical properties and QSAR that predict chemical reactivity or biological activities, because there are conceptual differences between the two types that affect the

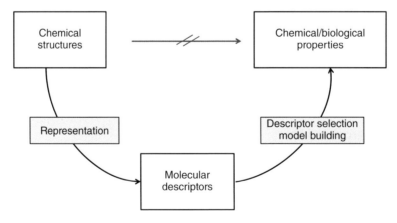

Figure 2.1 The general QSAR/QSPR procedure.

application of such models. Many physical properties (e.g., solubility or log P) are macroscopic properties, in that every atom of the molecule contributes to the measured property. Chemical reactivity data (e.g., pK_a values for acidity) are data specific for a certain site of a molecule and require descriptors that characterize that site (Section 3.3.7). Similarly, biological activities (e.g., receptor binding, enzyme inhibition, metabolism, toxicity, or membrane transport) are also primarily determined by a specific part of the molecule. These properties result from a specific arrangement of special features (called pharmacophore) (Chapter 6.6) present in the molecule that are responsible for addressing the biological effect. Thus, in contrast to models for physicochemical properties, the proper pharmacophoric arrangement of features is fundamental to address a biological effect on the molecular level.

Corwin Hansch and coworkers [1] deserve the credit for having propagated the application of physicochemical properties and statistical methods in structure–activity relationship studies. The original work by Hansch involved the linear combination of several properties using multiple linear regression (MLR) to obtain a quantitative model. The modeled properties were usually equilibrium constants, K, in log units. Since the log K is linearly related with free energy, this approach has also been named linear free energy relationship (LFER). The first QSAR models were based on the observation that partition coefficients (e.g., log P) are correlated with certain biological data. In some cases, the relationship appeared not to be linear. For those cases, parabolic models and bilinear models [2] were developed to better reproduce the experimental data. A further approach that was developed in the early days of QSAR is the Free-Wilson method where the biological activity is correlated with the presence of specific structural features in molecules. For more information the reader is referred to the literature [3, 4].

Usually, a QSAR/QSPR model is derived using a set of already characterized compounds (training set). Mathematical models are generated to correlate the calculated molecular descriptors with the biological activities or chemical properties. To this end, the molecular descriptors, which represent the structural features determining the property or activity of a set of molecules, have to be identified. With these descriptors, various modeling and learning methods (see *Methods Volume*, Chapter 11) are used to build models that describe the relationship between the molecular descriptors and the activity or property of interest. The final model built from the optimal parameters will then undergo validation using test sets to ensure that the model is appropriate and useful (see Section 2.6 for details). The complete process of building a QSAR/QSPR model is depicted in Figure 2.2 as a flow chart.

There are two basic models to be built: *classification* models that classify an object into a category (often called structure–activity/property relationships (SAR or SPR)) and *regression* or *correlation* models that predict an activity or property in a *quantitative* manner (QSAR or QSPR). The last approach is often referred to as *modeling* or, in chemometrics, as *calibration*.

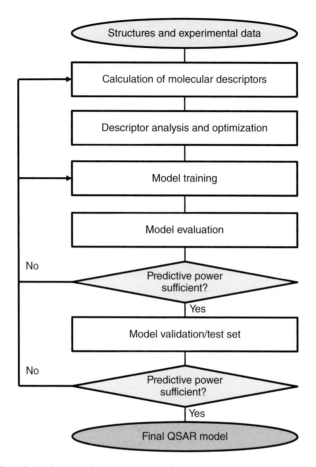

Figure 2.2 Flow chart showing the general steps for generating QSAR models.

The various steps implicated in the generation of QSAR models are presented here. More details on some of the steps are given in separate chapters of the *Methods Volume*:

- Data types
- The data science of QSAR modeling
- Calculation of structure descriptors
- Methods for multivariate data analysis
- Artificial neural networks
- Best practices in QSAR

QSAR/QSPR models can be found in all disciplines where chemicals are used or studied. The relationships established in the models allow the prediction of the properties or activities of novel compounds. Furthermore, appropriate structure descriptor and modeling techniques can contribute to an understanding and rationalization of the underlying relationships between molecular structure and a property or the mechanism of action within a series of molecules.

A few typical scenarios where QSAR/QSPR models are used are listed here [5]:

- Optimization of the biological activity for lead compounds
- Rational identification of novel hits
- Identification of hazardous compounds at early stages of product development
- Predicting off-target effects or toxicity for novel compounds
- Prediction of toxicity to humans and environmental species
- Optimizing pharmacokinetic (ADME) properties of lead compounds
- Prediction of physicochemical properties of compounds
- Prediction of the fate of compounds that are released into the environment

In this chapter, we will focus on QSAR models for the prediction of biological activities. More information considering QSPR for the prediction of physicochemical properties can be found in Chapters 3 and 6.9.

An extensive overview of the evolution of QSAR, current trends, novel applications, and future trends with 300 references has been published [6].

2.2 Data Handling and Curation

2.2.1 Structural Data

The major task of any QSAR analysis is to find a relationship between the biological data and the molecular structures of compounds. To setup a QSAR study, the molecular data of all compounds included in the analysis are needed. The last decade has seen several initiatives where chemical and biological data have been collected for the general public. These databases are quite useful for validating novel QSAR models and methods. The two largest databases that can be freely accessed are ChEMBL [7] and PubChem (Chapter 6.5) [8]. ChEMBL contains structural data of chemicals and bioactivity data for thousands of drug targets. The current version (ChEMBL22) includes over 14.4 million bioassay data covering around 2.0 million compounds and about 11,200 targets. Another open repository for chemical and biological data is the PubChem database that is maintained by the National Institutes of Health of the United States. PubChem contains bioactivity results from 1.1 million high-throughput screening programs with several million values and about 94 million compounds compiled from over 70 depositing organizations.

A fundamental requirement of any modeling study is that the input data in a dataset are correct. Data curation is crucial for any QSAR analysis (see *Methods Volume*, Chapter 12) especially nowadays when the availability of compound databases in public domain has skyrocketed in the recent years. Recent investigations by several groups [9] have clearly demonstrated that the type of chemical descriptors has a much greater influence on the prediction performance of QSAR models than the nature of model optimization techniques. These results emphasize that having erroneous structures in a dataset have a strong effect on the model performance. As a consequence, models developed using incorrect data will result in incorrect models and will be unreliable for prediction. Thus, it is important, whenever possible, to verify the accuracy of primary data before developing any model.

Chapter 12.2 in the *Methods Volume* and Ref. [10] present detailed recommendations on the curation of structural data before setting up a QSAR study.

Several approaches and programs have been developed in the last few years to help identify errors in datasets. This is getting more and more important, since the size of the studied datasets is continuously increasing (e.g., in PubChem and ChEMBL). Moreover, several efforts have been undertaken to cure the publicly available chemical data, for example, in the ChemSpider project [11]. To successfully and rapidly achieve the removal of duplicates, for example, the ISIDA/Duplicates [12] and HiT QSAR [13] programs (free for academic users) can be used. Protocols describing good practices for dataset curation that might be extremely helpful can be found in the literature [10, 14].

2.2.2 Biological Data

Biological data are often expressed in terms that cannot be directly used for QSAR modeling. Since QSAR is based on the relationship between the free energy and equilibrium constants (such as K_i values), the data for a QSAR model have to be correlated with free energy changes of a biological signal. When examining the activity of a biologically active compound, the change in free energy can be derived by the following equation (Eq. 2.2):

$$\Delta G = -2.3\,RT\,\log K_i \tag{2.2}$$

Biological data are generally asymmetrically distributed and the logarithmic transformation moves the data to a nearly normal distribution. Thus, usually log [C] or log 1/[C] are used to express the concentrations derived from the biological experiment. It should be clearly understood that biological data can usually not be determined to a high degree of accuracy but have an experimental error range. Thus, any prediction can only be made to a limited accuracy. Furthermore, biological data also need curation as detailed in Chapter 12.2 in the *Methods Volume*.

2.3 Molecular Descriptors

Molecular descriptors are mathematical representations of molecules resulting from a procedure that transforms the structural features into a symbolic representation. Molecular descriptors can be derived by applying principles from different theories, such as quantum chemistry, computer science, organic chemistry, and graph theory. They are used to model a wide range of properties in all fields of chemistry and biology, for example, environmental sciences, toxicology, analytical chemistry, physical chemistry, medicinal chemistry, and life sciences. In addition to the classical theoretical (calculated) descriptors, there are also experimentally derived descriptors (e.g., log P, dipole moment) that are sometimes used in QSAR studies. In the following, we will solely focus on theoretical molecular descriptors. More details on molecular descriptors may be found in Chapter 10 in the *Methods Volume*.

The growing interest of the scientific community in novel descriptors can be recognized by the existence of more than 5000 reported molecular descriptors in literature [15]. According to Todeschini and Consonni, a molecular descriptor has to fulfill the following mathematical requirements [16]:

1. Invariance with respect to atom labeling and numbering
2. Invariance with respect to rotation and translation
3. An unambiguous algorithmically computable definition
4. Values in a suitable numerical range for the set of studied molecules

The information content of any descriptor depends on two factors: the molecular representation of the compound and the mathematical algorithm that is used for the calculation. Molecular descriptors can be classified by the data type of the descriptors (Table 2.1).

Table 2.1 Classification of molecular descriptors.

Data type	Example
Boolean	Compound has at least one ring
Integer number	Number of carbon atoms
Floating number	Log P
Vector	Dipole moment
Tensor (3×3 matrix)	Electronic polarizability
Scalar field	Electrostatic potential
Vector field	Gradient of the electrostatic potential, that is, force

Molecular descriptors can also be classified on the basis of the dimensionality of the structural representation. Descriptors that are based on single numbers (integer or real numbers) such as molecular weight or log P values are referred to as 0D. These descriptors are fast to generate, but they are very general and not likely to discriminate sufficiently when used alone.

2.3.1 Structural Keys (1D)

Structural keys can be considered as a one-dimensional (1D) representation of compounds and provide 1D descriptors. A structural key describes the composition and structural features of a molecule represented as a Boolean array that can easily be calculated and handled. If a certain fragment or a structural feature is present in a molecule, a particular bit in the string is set to 1 (true), otherwise to 0 (false). Each bit in this array encodes a particular fragment. A fingerprint is a Boolean array, where a list of patterns is generated for each atom and for each pair of adjacent atoms and all bonds, as well as group of atoms joined by longer pathways. Each pattern of a molecule is assigned a unique set of bits along the fingerprint by a hash-coding algorithm. The set of generated bits is added to the fingerprint with a logical OR.

2.3.2 Topological Descriptors (2D)

Topological descriptors represent the constitution of molecules and can be calculated from their molecular graph (for details see Section 2.4.1). By considering organic molecules as graphs, theorems of graph theory allow the calculation of graph invariants that are called topological indices.

Although topological indices encode the same properties as fingerprints do, they are harder to interpret but can be calculated numerically in an easy way.

Another interesting class of descriptors is represented by autocorrelation vectors that were first published by Moreau and Broto [17]. The autocorrelation vectors can be obtained by taking the atoms as a set of points in space and an atomic property as a function evaluated at the points. The autocorrelation vectors are then obtained by summation of the products of the function calculated at atom x and atom $x + k$, where k is the topological distance.

2.3.3 Geometric Descriptors (3D)

Biological activity is often the result of the shape and the electrostatic complementarity of a small molecule and a protein target structure. Therefore, three-dimensional (3D) descriptors that are able to describe these interactions have been developed. 3D descriptors are usually conformation dependent and usually provide more information than the simpler topological descriptors. However, geometric descriptors also show some drawbacks. It is often not known which conformation of a molecule represents the bioactive one. Therefore, either multi-conformation approaches have to be used where statistical averages are considered for model building or pharmacophore modeling or docking methods can be applied to derive the bioactive conformation for all compounds under study.

Further geometric descriptors are the 3D autocorrelation vectors and radial distribution function (RDF) codes. The RDF codes as well as the related 3D-Molecule Representation of Structures based on Electron diffraction (3D-MoRSE) codes can be interpreted as a probability distribution of individual interatomic distances in a molecule [18]. RDF and 3D-MoRSE descriptors have been shown to have good modeling power for different biological and physicochemical properties [19].

Another class of 3D descriptors is derived from force field or quantum mechanical calculations. Examples are the individual potential energy terms, such as van der Waals energy, electrostatic energy, internal strain, HOMO (highest occupied molecular orbital), and LUMO (lowest unoccupied molecular orbital) energies.

A further group of geometric descriptors widely used in QSAR studies is derived from molecular surface or volume calculations. Different properties (e.g., partial charges) can be projected onto the molecular surface and can be used to derive various surface descriptors (e.g., the charged partial surface area (CPSA) descriptors) [20].

Other geometric 3D descriptors are the VolSurf and GRIND descriptors [21]. The idea behind these descriptors is to use specific probes to calculate the interaction energies with a given molecule, for example, hydrophobic interactions or hydrogen-bond acceptor or donor groups. The concept is similar to the well-known GRID approach [22]. A comprehensive and detailed overview of all kinds of molecular descriptors is given by Todeschini and Consonni in the textbook *Molecular Descriptors for Chemoinformatics* [15].

A general question is how many descriptors are necessary to describe the properties or biological activities of a given set of molecules. One approach is to compute as many descriptors as possible and then to select an optimal subset by applying mathematical algorithms. Another general problem is to determine the optimum number of descriptors for a given dataset. For linear regression-based methods one has to keep the number of descriptors as low as 20% of the number of compounds in the dataset.

Another important step is the removal of highly correlated and constant descriptors. A variety of computational methods has been applied for that purpose, including stepwise regression, forward-selection, backward-elimination, genetic algorithm, and trend-vector methods, simulated annealing, and other evolutionary variable selection approaches. A comparison of these selection methods can be found in the following comprehensive reviews [23, 24].

2.4 Methods for Data Analysis

2.4.1 Overview

Chemical data contain information about various characteristics and properties of molecules, and a wide spectrum of methods is available for extracting the relevant information from datasets. These methods come with various names such as pattern recognition, machine learning, data mining, chemometrics, and artificial neural networks (ANN). A brief introduction into these methods is given here. A deeper discussion is contained in the *Methods Volume*, Chapters 11.1 and 11.2.

The learning process usually starts with the selection of a dataset, which is then divided into two or more subsets: a training set that is used to train the system and test sets that provide a means to evaluate the results (see also Section 2.8). The applied algorithm that can be linear or nonlinear uses the training set and tries to learn from these examples. In other words, a function has to be discovered that maps the molecular descriptors onto the biological activity/property. The quality of the learning/mapping is estimated by appraising the ability of the model to predict the outputs of test sets. The two major learning strategies are supervised learning and unsupervised learning.

2.4.2 Unsupervised Learning

The goal of unsupervised learning is to build a representation of the data without considering biological activities. This representation can then be used to detect clusters within the data or to reduce the dimensionality of the data. Common tasks of unsupervised learning are clustering, data compression, or outlier detection. Among the methods used for unsupervised learning, are principal component analysis (PCA) as well as some ANNs (e.g., Kohonen network, also called self-organizing map (SOM)) and conceptual clustering. In addition, several algorithms discussed for supervised learning (e.g., random forest (RF) and support vector machines (SVMs)) can also be used for unsupervised learning if the biological activity is not included.

2.4.3 Supervised Learning

The aim of supervised learning is to make a system learn to associate input data (descriptors) with the corresponding output data (biological activities). In addition to a set of input data, in supervised learning the model is also given a set of target outputs, and its task is to learn to generate the correct output for a newly given input. The output of the model is compared with the experimental data, and thus an error is obtained. Supervised learning methods try to minimize this error. Supervised learning can be used both for classification problems and for modeling or prediction of experimental data. Common methods that use a supervised learning approach are decision trees (DTs), random forest (RF), partial least squares (PLS) and principal component regression (PCR), multiple linear regression analysis (MLRA), support vector machines (SVMs), ANN, and several evolutionary algorithms (e.g., genetic algorithm). These methods are discussed in details in Sections 2.6.2 and 2.6.3.

The following sections give a more detailed description of the methods mentioned earlier and are summarized in Table 2.2.

Table 2.2 Commonly used data analysis methods in cheminformatics.

Unsupervised	Supervised
	Multiple linear regression analysis (MLRA)
Principal component analysis (PCA)	Partial least squares (PLS), principal component regression (PCR)
Conceptual clustering	
Support vector machines (SVMs)	Counterpropagation networks
Random forest (RF)	Back-propagation neural networks
Kohonen network (SOM)	Support vector machines (SVMs)
	Decision trees (DT)
	Random forest (RF), rotation forest
	Evolutionary algorithms (e.g., genetic algorithm, particle swarm optimizer (PSO))
	Linear discrimination analysis (LDA)
	Naive Bayes classifier (NB)
	k-Nearest neighbor (kNN)

There are cases where biological activity values cannot be determined accurately for a variety of reasons, for example, lack of sensitivity of a particular test system. Thus, regression-based analysis cannot be applied. Alternative statistical techniques can be used in these cases, where the problem is simplified to a classification scheme in which compounds are labeled as active, partially active, and inactive. The resulting dataset is then searched for patterns that help to classify the compounds according to these categories.

2.5 Classification Methods

Another important division of the mapping methods is based on the nature of the activity variable. When predicting a continuous value, one is dealing with a regression problem. Meanwhile, when only some categories of classes of the activity need to be predicted, for example, discriminating active and inactive compounds, it is a classification problem. In classification problems, the resulting model is defined by decision boundary separating the classes in the descriptor space. Many of the mapping methods can be applied either for predicting continuous values or for classifying compounds into groups.

2.5.1 Principal Component Analysis

PCA allows the representation of multivariate data in a lower-dimensional space spanned by new orthogonal variables. These variables are generated as a linear combination of the original descriptors by maximizing the description of data variance. PCA gives an overview of dominant patterns and major trends in the data. In mathematical terms, it transforms a number of correlated variables into a smaller number of uncorrelated variables, the so-called principal components (or latent variables). Prior to PCA, the data are often preprocessed to convert them into a form most suitable for the application of PCA. Commonly used preprocessing methods for PCA are scaling and mean centering of the data.

In matrix notation, PCA approximates the data matrix X, which has n objects and m variables, by two smaller matrices: the score matrix T (n objects and d variables) and the loading matrix P (d objects and m variables), where $X = TP^T$.

Related approaches to reduce a data matrix to a smaller number of new variables are factor analysis (FA), correspondence factor analysis (CFA), and nonlinear mapping (NLM).

2.5.2 Linear Discriminant Analysis

Linear discriminant analysis (LDA) is commonly applied to classify compounds and to reduce the data dimensionality. LDA uses a linear combination of features to separate molecules belonging to different classes. In LDA, a linear transformation of the original feature space into a reduced space is carried out, which maximizes the interclass separation and minimizes the within-class variance [25]. LDA tries to generate a hyperplane that is able to separate different classes of compounds (e.g., actives and inactives). The hyperplane is defined by a linear discrimination function derived from linear combination of selected molecular descriptors.

2.5.3 Kohonen Neural Network

A Kohonen network, also known as SOM, is an example for an unsupervised learning by a neural network. It is a method for mapping multidimensional data onto a two-dimensional (2D) surface. An SOM tries to group the input data on the basis of their similarities. Those data points that are similar to each other are allocated to the same neuron or to closely adjacent neurons. It is used to detect patterns in large datasets and to classify compounds into distinct families that can easily be visualized. Further details can be found in the *Methods Volume*, Section 11.2 and in the literature [26–28].

2.5.4 Other Classification Methods

Section 2.7 lists methods that can typically be used both for classification and modeling purposes.

2.6 Methods for Data Modeling

A first step in building a QSAR model should always be an unsupervised learning process to see whether the chosen descriptors are good for classifying the objects into the desired category (e.g., actives and inactives). Only when a good descriptor set is found, a functional relationship between the calculated descriptors and the activity/property should be sought. The first QSAR models were derived assuming a linear relationship between biological data and descriptors (e.g., MLRA). In general, linear models are easily interpretable and sufficiently accurate for small datasets of congeneric series of molecules especially when the descriptors are carefully selected for a given activity. Modern nonlinear modeling methods extend this approach to more complex relationships. In a regression-based analysis, the dependent variable is modeled as a continuous function of the descriptors.

2.6.1 Regression-Based QSAR Approaches

Regression analysis represents a statistical process for estimating the relationships among variables. It includes many techniques for modeling and analyzing several variables, when the focus is on the relationship between a dependent variable and one or several independent variables. Historically, the first QSAR models developed to predict molecular properties were derived from regression analysis. One such example is the LFER (see *Methods Volume*, Section 8.1). The LFER concept is based on the pioneering work of Hammett, who used this approach for estimating the chemical reactivity of a congeneric series of molecules [29]. The basic assumption is that the influence of the structural feature on the free energy change of a chemical process is constant for a congeneric series of compounds.

2.6.1.1 Multiple Linear Regression Analysis (MLRA)

Linear regression models a linear relationship between two variables or vectors, x and y. Thus, in two dimensions this relationship can be described by a straight line given by the equation $y = ax + b$ (a gives the slope of the line and b represents the intercept of the line on the y-axis). The goal of linear regression is to adapt the values of the slope and of the intercept so that the line gives the best prediction of y from x. This is achieved by minimizing the sum of the squares of the vertical distances of the points from the line. The quality of the regression is measured by the correlation coefficient r^2.

While simple linear regression uses only one independent variable for modeling, MLR uses more variables. Given n input variables x_i, the variable y is modeled in a similar way as in the case of simple linear regression with one input variable (Eq. 2.3):

$$y = a_i + a_1 x_1 + a_2 x_2 + \cdots + a_n x_n \tag{2.3}$$

The x_i should be uncorrelated. If they are correlated, however, one way of finding a solution is a stepwise MLR where only those x_i that are not correlated with already used x_i are chosen for the model. There are important caveats in applying multiple regression analysis. The first is based on the fact that, given enough parameters, any dataset can be fitted to a regression line. The consequence of this is that regression analysis generally requires significantly more compounds than descriptors; a useful rule of thumb is at least five times the number of descriptors under consideration. In addition, it has to be checked that the descriptor values are well distributed and not clustered.

MLRA attempts to maximize the fit of the data to a regression equation (minimize the squared deviations from the regression equation) for the activity (maximize the r^2 value) by adjusting each of the available parameters up or down. Regression programs often approach this task in a stepwise fashion, either using a forward-selection or a backward-elimination regression. As the names imply, forward selection starts with an equation involving just one descriptor usually with the one that makes the most contribution. The second and subsequent terms are then added, again choosing the most contributing descriptors. Backward-elimination regression works in the reverse sense; initially, an equation is derived using all calculated descriptors, and then terms with less contribution are removed stepwise. For both forward-selection and backward-elimination regression, the final model may be the one with the best fit to the experimental data. There are some essential problems with stepwise methods that are summarized in Ref. [30].

Other variable selection methods are often used in combination with MLRA: genetic algorithms, particle swarm optimizer (PSO), and simulated annealing [31].

Genetic algorithms often show the best performance. These evolutionary optimizers are widely applied in various fields of chemistry. The natural principles of the evolution of biological species are applied that include the survival of the fittest concept and that improvement can be obtained by different kinds of recombination of independent variables (reproduction, mutation, crossover). The goodness of fit of a derived solution is scored by a function that has to be optimized.

A related method is the so-called logistic regression (LR). LR can be used to model the probability of a compound belonging to a certain target property; thus it is used for classification problems [31].

2.6.1.2 Partial Least Squares

Multivariate linear modeling algorithms, such as PLS, are well established within the QSAR community. PLS, also referred to as projection of latent structures, can be used to generate quantitative models even when the descriptor values are highly correlated and when the number of values greatly exceed the number of samples. In PLS, the so-called latent factors or latent variables [32] have to be calculated using the descriptor space (X-matrix) and the compound property space (Y-matrix). First, the latent variables (also named principal components) for the two matrices X and Y are calculated separately. The scores of the X-matrix are then used for a regression model to predict the scores of Y, which can then be used to predict the compound properties. A commonly used algorithm for calculating PLS is SIMPLS. The optimal number of principal components used for the final model can be obtained by applying a variety of validation methods (described in Section 2.8).

Advanced PLS methods have been developed, like quadratic-PLS and kernel-PLS, kernel-ridge regression-PLS, multiway-PLS, unfolding-PLS, hierarchical-PLS, three-block bifocal PLS, and related approaches. For these advanced regression techniques, the interested reader is referred to the review by Hasegawa and Funatsu [33].

2.6.1.3 Principal Component Regression

PCR can be regarded as a combination of PCA and MLRA, which have been described before. First, a PCA is carried out that yields a loading matrix P and a scores matrix T. For the ensuing MLR only PCA scores are used for modeling Y. Since the PCA scores are inherently uncorrelated, they can be employed directly for MLRA. A more detailed description and applications can be found in Ref. [34].

The selection of relevant effects for the MLR in PCR can be quite complicated. A straightforward approach is to take those PCA scores that have a variance above a certain threshold. By varying the number of the final PCA components, regression models can be optimized. However, if the relevant effects are rather small compared with the irrelevant effects, they will not be included in the first few principal components. A solution to this problem is the application of PLS.

2.6.2 3D QSAR

The number of 3D QSAR studies has exponentially increased over the last few decades, since a variety of methods is commercially available in user-friendly, graphically guided software [35].

2.6.2.1 CoMFA

Comparative molecular field analysis (CoMFA) [35] was developed as a tool for finding 3D QSAR. The underlying idea of CoMFA is that differences in a target property, for example, biological activity, are often closely related to equivalent changes in the shapes and strengths of the non-covalent interaction fields surrounding the molecules. Stated in a different way, the steric and electrostatic fields provide all the information necessary for understanding the biological properties for a set of compounds. The aligned molecules are located in a cubic grid, and the interaction energies between the molecule and a defined probe are calculated for each grid point. Normally only two potentials, namely, the steric potential as a Lennard-Jones function and the electrostatic potential as a simple Coulomb function, are used within a CoMFA study. It is obvious that the description of molecular similarity is not a trivial task, nor is the description of the process of interaction between ligands and the corresponding biological target. In the standard application of CoMFA, only enthalpic contributions of the free energy of binding are provided by the used potentials.

The relationships between the biological activities and the generated interaction fields are evaluated by PLS regression [32]. The PLS method is able to build a statistical model even though there are more numbers of energy values than compounds because the various energy values are correlated with each other and many are unrelated to biological activity. These assumptions give PLS the power to extract a weak signal dispersed over many variables.

One of the biggest advantages of 3D QSAR over classical QSAR is the graphical interpretability of the statistical results. Equation coefficients can be visualized in the region around the ligands. Upon visual inspection, regions of space contributing most to the activity can be easily recognized. The interpretation of the graphical results allows on one hand the easy check of the reliability of the models and on the other hand the design of modified compounds with improved activity or selectivity. In this respect, 3D QSAR methods like CoMFA and GRID/GOLPE [35, 36] have proven to be very useful. The contours that represent 3D locations of fields with significant contribution to the model are displayed. Steric and electrostatic contributions are contoured separately and shown in different colors. The steric contours are relatively easy to interpret. Positive contoured maps show regions in space that, if occupied, increase potency, while negative contours decrease potency. The interpretation of the electrostatic maps is more complicated because of the electroneutrality requirement and because either positive or negative charges in electrostatics can increase potency. If a CoMFA analysis shows significant electrostatic effects, the user must examine the underlying electronic effects of corresponding functional groups to establish which is the true effect and which is an artificial correlation.

A related approach was developed by Clementi and Cruciani who used the GRID interaction fields as descriptors for building PLS models [36]. The program named GOLPE also offers the possibility to reduce the number of variables by a variety of chemometric tools such as fractional factorial design, and D-optimal design. Variable selection methods have also been included for optimal region selection, and it was shown that improved QSAR models can be derived as compared with the original CoMFA technique.

To optimize CoMFA and improve its reproducibility, the cross-validated R^2-guided region selection (q^2-GRS) routine was developed. The q^2-GRS method was developed on the basis of independent analyses of small areas (or regions) of near-molecular space to address the issue of optimal region selection in molecular interaction field (MIF)-derived 3D QSAR. An important aspect in the q^2-GRS routine is that it eliminates areas of 3D space where changes in steric and electronic fields do not correlate with changes in biological activity (low q^2 values) from the analysis, hence affording an optimization of the region selection for the final PLS analysis.

2.6.2.2 CoMSIA

Due to the problems associated with the form of the Lennard-Jones potential applied in most CoMFA models [35], Klebe *et al.* [37] have developed a similarity indices-based CoMFA method, which is named comparative molecular similarity indices analysis (CoMSIA). The method uses Gaussian-type functions instead of the traditional CoMFA potentials. Three different indices related to steric, electrostatic, and hydrophobic potentials are used. The clear advantage of this method lies in the functions used to describe the molecules as well as the resulting contour maps, which are easier to interpret compared to the CoMFA plots. The CoMSIA procedure also avoids the cutoff values used in CoMFA to restrict the potential functions from assuming extremely large values. For a detailed description of the method as well as its application, the reader is referred to the literature [37].

2.6.2.3 Open3DQSAR and 3-D QSAutogrid

In June 2011, all patent restrictions of CoMFA were dropped, and several open-source projects for generating PLS models based on interaction fields were launched, for example, Open3DQSAR [38] and 3-D QSAutogrid [39]. Open3DQSAR aims at pharmacophore exploration by applying PLS analysis of MIFs. The program can be used to generate steric potential, electron density, and MM/QM electrostatic potential fields and can also import MIFs generated with other programs such as GRID and CoMFA. 3-D QSAutogrid uses AutoGrid to generate MIFs and R-based scripts for statistical treatment of the merged biological activity and MIFs.

2.6.2.4 Alignment-Independent 3D QSAR Methods

The most crucial and difficult step in a CoMFA-related analysis is how to align the studied molecules in a realistic manner. A recent development of the CoMFA method allowing to avoid the alignment problem has been described by several groups [40–42]. Silverman and Platt [40] used in their comparative molecular moment analysis (CoMMA) method descriptors that characterize shape and charge distribution such as the principal moments of inertia and properties derived from dipole and quadrupole moments, respectively. The authors investigated a number of datasets and obtained models with good consistency and predictive power. A comparable approach using the GRID force field for the generation of principal moments has been reported by Cruciani *et al*. They have integrated their methods in the programs VolSurf and Almond [41, 42].

Several other 3D QSAR approaches have been developed during the last few years including 4D QSAR and 5D QSAR. In these approaches, the conformational space of the molecules is considered, or the interaction with a receptor is modeled and included as a further descriptor. The GERM [43], COMPASS [44], receptor surface [45], and QASAR [46] methods rely on properties calculated at discrete locations in the space at or near the union surface of the active ligands. The so generated "receptor surface" should simulate the macromolecular binding site. If all molecules of the dataset bind in a manner that does not distort the residues at the binding site too much, this can be a reasonable approach.

2.6.2.5 Template and Topomer CoMFA

Two CoMFA-related QSAR approaches developed by Cramer *et al.* have shown some advantages over the classical CoMFA method. Template and Topomer CoMFA represent in principle conventional CoMFAs using the topomer methodology for deriving the alignments [47]. A topomer is defined as a structural fragment having a single pose (conformation plus position, at least one open valence) that can be positioned in space by superimposing its open valence onto a fixed Cartesian vector. Only the topology of the topomer is considered. Similar fragment topologies should afford similarly shaped topomers. The 3D geometries of the topomers are generated, followed by canonically determined adjustments to acyclic single bond torsions, stereochemistry, and ring puckering. Topomer CoMFA simply uses topomers from the fragmented training set as the 3D QSAR-requisite aligned input structures and otherwise differs from a normal CoMFA only in its use of multiple CoMFA columns one for each set of fragments (R-groups). The main advantage compared with traditional CoMFA is the fully automated model generation. Thus, a user-biased alignment procedure that is a prerequisite for a traditional CoMFA is not needed.

2.6.3 Nonlinear Models

The following methods all generate nonlinear models, thus extending the traditional linear QSARs to nonlinear functions of the input descriptors. Nonlinear methods might be the better choice for large and diverse datasets and complex biological data. However, usually they are harder to interpret. Complex nonlinear models may also fall prey to overfitting, and therefore extensive validation has to be carried out.

2.6.3.1 Decision Tree (DT)

DTs, also known as recursive partitioning, constitute a machine learning technique that gives a graphical representation of a procedure for classification. A DT is built from a training set, and the tree can then be used for predicting properties or activities of new compounds. A DT consists of nodes and branches. An instance is classified by sorting it down the tree from the root to a leaf node. Depending on the value of a property on each node, one of the branches is taken. The leaf node finally gives the classification of the instance. DTs have the advantage of easy interpretation especially if they are not too big. In addition, unnecessary descriptors do not affect the performance of a tree. DTs are usually applied for classification purposes but can also be used for regression problems. This is done by associating each leaf with a numerical value instead of the categorical class. An example of a simple tree is shown in Figure 2.3. For further reading see [48].

Ensemble learning is applied in several machine learning approaches. Multiple models, called learners, are used and trained to solve a problem instead of using a single model. By constructing a set of hypotheses and combining them, a better correlation or classification might be observed. An ensemble contains a number of learners that are the so-called base learners. An important feature of ensemble models is that they usually have a stronger generalization ability compared with a

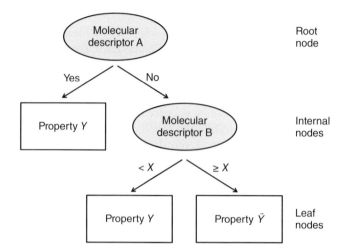

Figure 2.3 Example of a simple decision tree.

single model. The base learners are obtained using a training set by a base learning algorithm. All kinds of machine learning methods (e.g., DT, neural network, or others) can be taken as algorithm. The ensemble model is constructed in two steps. First, individual base learners are generated. Then, the base learners are combined for use. The most popular combination schemes are "majority voting" for classification problems and "weighted averaging" for regression models. There are many effective ensemble methods including Boosting [49] (implemented in the program ADABoost), Bagging [50], and Stacking [51]. In Bagging, bootstrap samples of objects are used to train a classifier on each sample. Then, the classifier votes are combined by majority voting. In general, the classifiers in such an ensemble model have relatively high classification accuracy. The factor encouraging diversity between the classifiers is the proportion of different molecules in the training samples that has to be carefully checked.

Bagging is, for example, implemented as a learning algorithm in the RF method that is described in the next section. Boosting in QSAR models has been applied and described for example in the study by Agrafiotis et al. [52]. A comparison of different ensemble learning methods is given in Ref. [53].

2.6.3.2 Random Forest (RF)

RF, first introduced by Breiman in 2001 [54], uses a large number of different DTs (named forest) obtained from bootstrap samples generated for a training set of objects. It applies the concept of "wisdom of crowds," which means that utilizing multiple independent models is more beneficial compared with a single model.

RF makes a final prediction based on the majority prediction from each of the trees. To construct each DT, a training sample is randomly selected with replacement from an original dataset. Using the new training sample, a DT is grown with randomly selected descriptors and it is not pruned. RF is easy to use, and only two parameters have to be defined: the number of trees in the forest and the number of descriptors in each tree. It is recommended that a large number of trees should be grown and the number of descriptors to be taken from the square root of the total number of descriptors. For regression models, the final output is calculated as the arithmetic mean of the prediction from all individual trees in the RF.

RF can handle large numbers of training data and descriptors. Besides classifying, it can be also used for unsupervised clustering. It was found that RF is less affected by noisy data. With RF methods the problem of overfitting should be reduced while improving the accuracy. As with any classification method, problems occur with unbalanced datasets.

A variety of software packages and websites offer RF for classification and regression, for example, R statistical package [55], Orange [56], WEKA [57], Chamming [58]. For a comprehensive overview, the following reference is suggested [59].

A novel method for generating classifier ensembles based on feature extraction has been reported and named "rotation forest" [53]. To select the training set for a base classifier, the feature set is split into k random subsets, and a PCA is carried out for all subsets separately. Then, reassembling a new extracted feature set while keeping all the components is done. Thus, the data is transformed linearly into the new features. A decision tree is trained with the resulting data and diverse classifiers are derived. The general idea of the rotation approach is to increase simultaneously the classifier accuracy and the diversity within the ensemble of models. The diversity of the classifiers is promoted through the feature extraction for each base classifier.

2.6.3.3 k-Nearest Neighbors

The k-nearest neighbors (kNN) method uses simple distance learning approaches whereby an unknown molecule is classified according to the majority of its kNN in a studied training set. Thus, kNN, which is a nonlinear, nonparametric method, can be used to predict the activity of a compound as distance weighted average of the bioactivities of its kNN. The method represents a simple decision scheme that requires practically no training and is asymptotically optimal, that is, with an increase in training data it converges to the optimal prediction error. Whether a molecule is near to others is measured by an appropriate distance metric (e.g., the Euclidean distance). The standard kNN method first calculates distances between an unknown molecule u and all the molecules in the training set. Then, k molecules most similar to the unknown molecule u are selected from the training set, according to the calculated distances. Next, molecule u is classified with the group to which the majority of the k compounds belong. The number of neighbors, k, is a user-defined value that needs to be optimized as it will affect the performance of the model. Since kNN relies on measuring distances between compounds, the used descriptors should be scaled to avoid descriptors with a large magnitude from adversely influencing the kNN model.

2.6.3.4 Naive Bayes Classifier

In machine learning, naive Bayes (NB) classifiers are a family of simple probabilistic classifiers. They are also named simple Bayes or independence Bayes models.

NB can handle multiple classes. An NB classifier makes use of Bayes' theorem to predict the probability $P(x|y)$ of an instance x to belong to class y (Eq. 2.4):

$$P(x|y) = \frac{P(x|y)(Py)}{P(x)} \tag{2.4}$$

NB classifiers are called naïve because they assume all features to be totally independent of each other. To generate an NB model, a training set of labeled instances with different class labels is utilized for supervised learning. The classification works surprisingly well given its simplicity, one suggestion being that the dependence between features at least partly cancels out. In addition, the method is very efficient and computationally not demanding. NB has been found to work better than some more complex methods in the presence of noisy data [59].

2.6.3.5 Artificial Neural Networks

There is a variety of ANN methods that try to model the information processing in the human brain (*Methods Volume*, Section 11.2) [29]. An SOM models the mapping of sensory signals in the brain and can be used for classification (Section 2.5.3). Other methods can be used both for classification and for QSAR modeling.

An ANN uses a simplified mathematical model representing a biological neuron, defined by Eq. (2.5):

$$y(x_1, \ldots, x_n) = f \sum_{i=1}^{N} (w_i \times x_i) \tag{2.5}$$

Where $x_1, \ldots, x_n$ are inputs, w_i are weights of a neuron, and f is a nonlinear response function (Figure 2.4, right).

A number of neurons are organized in layers, where the input to each neuron of a layer is an output of a neuron from the previous layer usually modified by a sigmoidal transfer function.

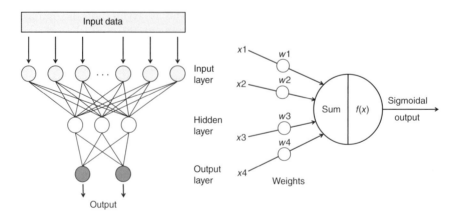

Figure 2.4 ANN, example of a two-layered ANN (there are only two layers of weights!).

A neural network is completely defined by the set of neuron weights $W = \{w_{ij}, i = 1, \ldots, L, j = 1, \ldots, N_i\}$, where L is the number of layers and N_i is the number of neurons in ith layer. During the training of a network, the weights are optimized to minimize the sum of squares of errors.

The most commonly used method for the determination of the weights is the error back-propagation algorithm, which is a supervised learning strategy. Usually, such a network consists of several layers of neurons (input layer, hidden layers, and output layer). Figure 2.4, left, gives a typical ANN architecture. The input layer of the network represents molecular descriptors, and the output layer represents the target properties of compounds.

The general principle of back-propagation learning is to adjust the weight of each neuron depending on the error of its output signal, with the goal of minimizing the output error [60]. In each cycle of the training process, the final output is compared with the target property. This yields the error signal that is back-propagated through the layers from the output to the input layer, and the weights of each layer are adapted correspondingly. The approximation improves as the errors converge after numerous training cycles. After a certain number of training cycles, the network will begin to "memorize" the training data. That means it will accurately model the training data but the ability to predict other data will decrease. Therefore, there are various methods such as the use of a validation set or techniques like early stopping, pruning, and weight decay to minimize the risk of overfitting an ANN [61].

Further network concepts are counterpropagation (CPG) network, Bayesian regularized artificial neural network (BRANN), probabilistic neural network (PNN), radial basis function neural networks (RBF), and general regression neural network (GRNN). A comprehensive review of available ANN concepts applied in drug design can be found in the literature [62].

2.6.3.6 Support Vector Machine

An SVM is based on the so-called structural risk minimization principle from statistical learning [63]. It constructs a descriptor space hyperplane that separates the training set samples into two classes. It represents one of the most popular machine learning methods in QSAR, which is indicated by the increasing number of publications in the field. For classification, SVM aims at creating a decision hyperplane that maximizes the margin between the closest points, known as support vectors, and the hyperplane. For regression purposes, SVM maps the data into a high-dimensional space, using a so-called kernel function that is typically nonlinear. In brief, an SVM is performed not in the original space but in the feature space obtained via a nonlinear kernel transformation of the original space. The feature space has a higher dimensionality (often, it has an infinite number of dimensions), which makes it possible to separate classes that were not linearly separable in the original non-transformed space. If an SVM is used for regression purposes, it is also referred to as support vector regression (SVR).

The SVM optimization tends to a single, global minimum, and, in the majority of cases, the solution is unique for a given set of free parameters. This is not the case for stochastic methods, such as ANN and evolutionary algorithms. The SVM method has been shown to exhibit low overfitting and thus allows for good generalization to the previously unseen compounds. It is also relatively robust when only a small number of examples of each class are available. The interested reader is referred to two comprehensive reviews of SVMs [64, 65].

2.7 Summary on Data Analysis Methods

Although a continuously growing number of articles applying data analysis methods in the field of QSAR are published, there is no general rule which method performs best for a given dataset. Thus, there is no single best method for all problems. The relative abilities of methods depend on many aspects, including the number and distribution of the studied compounds in chemical space, the linearity or complexity of the studied biological data, and the used descriptors and observed intercorrelations. Several studies have been published where the performance of various machine learning methods was analyzed and critically discussed [58, 64, 66, 67].

2.8 Model Validation

The process of validation tries to identify "useful" models from the pool of ill-conceived ones. If the entire model development procedure is performed in a correct way, according to state-of-the-art methods, there is a considerably high probability that the model is not only useful but also correct. This is where model validation sets in. The term "validation" encompasses numerous methods that might help to distinguish bad models from good ones [68].

The development of good QSAR models is considered to be so important that, in addition to this section here, an entire chapter of its own is devoted to this topic treating it in more detail (see *Methods Volume*, Chapter 12).

Generally, the validation of QSAR/QSPR models is performed by using the following approaches:

- Internal validation or cross-validation
- External validation by splitting the available dataset into a training set for model development and a test set for testing the predictive power of the final model
- Blind external validation by application of the model to new external data
- Data randomization or Y-scrambling for verifying the absence of chance correlation between activity and modeling descriptors

What makes QSAR/QSPR and similar methodologies different from general statistical modeling is the extreme complexity of the nature of dependent variables (physiochemical and, in particular, biological data) and the problems with the adequate representation of the objects (structural representation–molecular descriptors).

In this section, we will discuss the most useful and widely used validation methods in the field of QSAR/QSPR modeling. Through this section, we hope to encourage readers to apply rigorous validation routines and to use as many diverse methods as possible. Again, a more comprehensive treatise is contained in the *Methods Volume*, Section 12.2.

2.8.1 Proper Use of Validation Routines

There are many ways for categorizing validation methods. From the QSAR/QSPR point of view, the distinction between external and internal validation seems to be the most important one. *Internal validation* estimates model performance either analytically, by calculating appropriate criteria and statistics, or on the basis of explicitly reusing the training set. The most prominent methods are cross-validation, bootstrapping, and Y-randomization.

In contrast, the *external validation* methods depend upon additional data that were never used in the process of model training. This additional data – the external data – are used in a direct way to estimate the generalization ability of the model by calculating a prediction error over any test set. Rigorous external validation must be considered as an integral part of any model development. The situation when new experimental data for new compounds become available after a model has been generated is the ultimate test for any model.

The prediction error can simply be calculated by comparing the true values with the predicted ones. For example, the mean absolute error (MAE) is calculated according to Eq. (2.6):

$$\text{MAE} = \frac{1}{N} \sum_{i=1}^{N} |y_i - \hat{y}_i| \qquad (2.6)$$

where N is the size of the dataset, y_i is the true value, and $\hat{y}_i$ is the predicted value of the property for compound i. The MAE is a direct measure of the prediction error over the dataset. The absolute error is not necessarily the best measure and is used here only as the simplest example. The more sophisticated and accurate measures are discussed in Section 2.8.8.

Although the common consensus among QSAR practitioners is that external validation is superior to internal validation in terms of reliability of error estimates, it is also agreed that internal validation should always be carried out in parallel with external validation.

The ultimate goal of validation is to assure that the final model is of high quality and capable of yielding highly predictive results. This is quite often achieved in a two-step process that involves fulfillment of two slightly different goals [69]:

1. Model selection – Selection of the most promising model from the collection of available ones
2. Model assessment – Estimation of the generalization ability of the final model

The first scenario is usually applied in cases where the process of model development produces a collection of models by varying some parameters, which usually influence the complexity of models, and/or using different sets of descriptors and different data analysis methods. Validation methods are then applied to find the best model, that is, the set of parameters that offers the optimum trade-off between model complexity and its potential generalization ability.

The second scenario is quite different. Here we have the "final" model and its generalization ability has to be estimated. In other words, a statement concerning model performance over new data has to be made.

It is common practice to put greater emphasis on internal validation in model selection and to give higher priority to external validation in model assessment [14]. Data are often scarce in QSAR practice; usage of internal validation at the stage of model selection can spare it substantially. Nevertheless, it is advised to use both types of validation whenever possible.

However, it should be realized that the estimated prediction ability essentially holds true only for the particular external data used to estimate it. To circumvent this, probably unexpected difficulty, several methods are in place.

External data can be carefully collected to ensure they belong to the chemical space of the training set. In this approach, external data can be either collected *a priori* by splitting available data using one of the methods of representative subset selection or *a posteriori* by compiling data from other external sources. The posterior compilation of data should be supported by consideration of the applicability domain (AD) of the model – for more details see also Section 2.8.9 [70, 71]. Methods of representative subset selection, in contrast, are usually designed to produce, in a systematic manner, representative datasets that are supposed to evenly cover the chemical space in question [72]. This reduces the risk of extrapolation and makes error estimates more self-consistent [73].

2.8.2 Modeling/Validation Workflow

Before proceeding any further, it is desirable to define a general modeling/validation workflow. Several attempts to define such a workflow have already been described in the literature [9, 74]. The workflow is shown in Figure 2.5, indicating the flow of the data. Indeed, data is the most important part of any QSAR undertaking. The quality of data, the amount of data, and its proper use are the most influential factors that allow building predictive models (see Section 2.6.2 for further details). The workflow starts with the splitting of the complete dataset into modeling and test subsets, and then the modeling subset is split into training and validation sets. Eventually, the training set is used for model building and is also submitted to various internal validation routines like cross-validation, bootstrapping, and Y-randomization. Models that do not pass quality criteria are rejected. Eventually chosen, the so-called final model is submitted to the external validation over the test set that was kept aside during modeling.

2.8.3 Splitting of Datasets

The modeling/validation workflow shown on Figure 2.5 gives a rough outline of the general data splitting scheme. It starts with the complete dataset and splits it accordingly. Eventually, the scheme may contain additional sets that are not part of the starting dataset. The naming convention used is partially compliant with conventions present in the literature. Nevertheless, some discrepancies with commonly agreed names may occur. This scheme should be regarded as

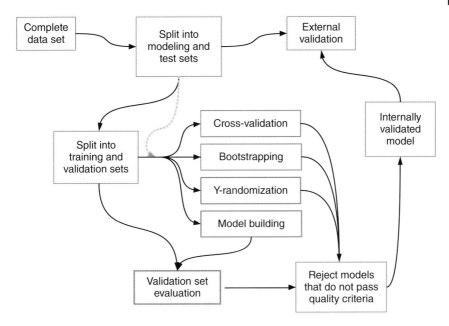

Figure 2.5 General modeling/validation workflow.

a general suggestion and should be adapted to the needs of a particular QSAR exercise.

- Complete dataset – The entire data, available prior to the modeling exercise
  - Modeling set – Part of the complete dataset used for the model building
    o Training set – Part of the modeling set used to fit models
    o Control set – Part of the training or modeling set used sometimes to optimize model parameters
    o (Internal) validation set – Part of the modeling set used to select the best model
  - (External) test set – Part of the complete dataset used for evaluation of the final model only
- Additional external sets – The data available *posterior* to the modeling, used for additional tests of the final model
- Additional modeling sets – The data available *posterior* to the modeling, used to extend the training or modeling set (update of the final model)

The question concerning the proportions of the splits emerges. Unfortunately, there is no general rule here. It might be advisable to start from 4/1, 3/1, or 2/1, where 4, 3, and 2 reflect the size of modeling or training sets and 1 the size of validation or test sets. If the above scheme is followed with a 4/1 proportion split, the modeling and test sets are 80% and 20% of the complete dataset, respectively. Consequently, the training and validation sets are 64% (80% of 80%) and 16% (20% of 80%) of the complete dataset. In general, QSAR analysis is to a great extent data driven. This means that the appropriate size (and quality) of the training sets is crucial to obtain highly predictive models. Therefore, it is recommended that the training set is at least 50% of the complete dataset.

2.8.4 Compilation of Modeling, Training, Validation, Test, and External Sets

We have already discussed possible approaches to partition data into separate sets. Between two extreme possibilities of systematic sampling and random splits, there are many intermediate methods. It seems almost certain that the fully random approach should deliver the most trustful error estimates. It is because at the very heart of almost all statistical methods, there is an assumption of random sampling [74].

However, as we have already mentioned, to make error estimates calculated for random splits truly reliable, it is necessary to repeat splitting several times (at least 10 times) preferably with completely separate test sets. In some cases, this might be considerably expensive in terms of time and computational resources to repeat the entire experiment several times. Probably, the more severe limitation here is the scarcity of data. Quite often there are not enough data to repeat the validation with 10 (or more) mutually independent test sets being also independent with respect to the modeling set.

Hence, random splitting can lead to an unjust error estimate if it is performed only once. This explains the effort spent in developing methods to make data splits more efficient.

Another method of reducing possible incompatibilities in randomly selected sets that deserve particular attention is stratification – a method of random sampling that ensures that the distribution of a particular property in the sets is the same as in the initial dataset (for more information the reader is referred to [75]).

The spectrum of methods for the compilation of test sets is not limited to random sampling methods though. There are numerous methods of representative or rational subset selection available. Rational selection methods are commonly used in QSAR modeling, and some of them deliver results very similar to those of random splits [71]. There are also some other important applications in which rational methods are if not indispensable then at least useful. These are, for instance, the analysis of large databases and model selection/model optimization. In the former case, they are used simply to reduce the amount of data by taking a subset of representative objects from the database. Reduction of data is often necessary to reduce computational costs of the analysis, and it is obvious that it is desirable to conduct statistical modeling on the representative subset rather than on the random sample. Reduction of the dataset size can also be applied in order to balance datasets. In the case of model selection/optimization, the usage of representative subsets produced by systematic split methods is often desirable and may significantly ease the development of predictive models. It is also useful to compare error estimates resulting from random and systematic subset selection methods. It gives better insight into the behavior of the modeling procedure and helps to improve it.

On the next pages, there is a short summary of the most popular methods of representative subset selection. The selection of methods is probably biased toward the so-called uniform design methods that aim to cover the chemical space evenly. The cluster-based methods are nonetheless important as well as efficient.

To better illustrate differences and similarities of the presented methods, we have used the well-known Iris dataset [76]. Figure 2.6 shows a PCA projection on the two first principal components of all 150 points in the dataset. This projection is also used to show the distribution of objects selected by the Kennard–Stone algorithm.

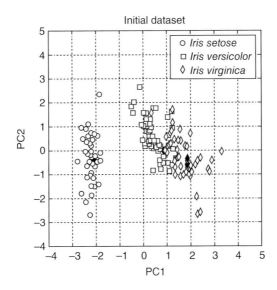

Figure 2.6 PCA score projection of Iris dataset objects on the plane defined by principal components PC1 and PC2; (○) Iris setosa, (□) Iris versicolor, (◊) Iris virginica.

2.8.4.1 Kennard–Stone Algorithm

Probably, the best known method of uniform design among QSAR practitioners is the Kennard–Stone algorithm. It was included in the program CADEX [77]. The algorithm selects a representative training subset according to relatively simple rules that can be summarized in following steps:

1. Select object closest to the mean (the object closest to the mean is supposed to be the most representative).
2. Select object that is the most dissimilar to the first.
3. Select object that is the most dissimilar to its nearest object already belonging to the subset.
4. Stop if the subset contains the desired number of objects.

Figure 2.7 shows the distribution of objects selected by the CADEX method on the example of the Iris dataset. Objects assigned to the training set are shown as open symbols; objects assigned to the test set are shown as filled symbols. The training set spans the entire space whereas the test set occupies the central or inner part of the space.

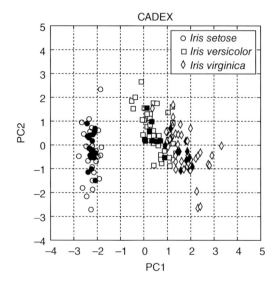

Figure 2.7 Example of the distribution of objects selected by the CADEX method. A PCA score projection of the Iris dataset objects on the plane defined by PC1 and PC2; open and filled symbols represent training and test sets, respectively.

The detailed implementations often differ from the general algorithm described previously. The initial object can be selected randomly or as the most distant one from the centroid. Different measures of dissimilarity can be used, ranging from the Euclidean distance to the Tanimoto coefficient [78].

An interesting modification or extension of this algorithm was published by Snee as the DUPLEX algorithm [79]. The DUPLEX algorithm aims at creating two subsets (training and test) that have similar statistical properties.

Some further often applied subset selection approaches are Sphere exclusion [80], OptiSim [81], and D-optimal design [82].

2.8.4.2 Cluster-Based Methods

Subset selection can also be done on the basis of clustering methods. The idea is to select objects from the clusters in a way that ensures optimal diversity and representativeness of selection. There are many clustering methods that can be used for this purpose including hierarchical, nonhierarchical, and density-based methods.

2.8.5 Cross-Validation

Cross-validation is probably one of the most widely used internal validation techniques. Moreover, it is used not only for model validation itself but also in other contexts such as variable selection or optimization of model parameters.

In a k-fold cross-validation, the initial set is split into k subsets in a random or semi-random way. One of the k subsets is held out and the remaining $k-1$ subsets are used as a training set. A subset not being part of the training set (a hold-out set) is used as a test set. This procedure is repeated k times in a way that each of the k subsets is used once as a test set – a hold-out set. When the procedure is completed and the predictions for all hold-out sets are collected, the cross-validated estimators of errors can be calculated.

The sizes of the training and hold-out sets in a k-fold cross-validation depend on the value of k. In each fold, the test set is approximately $1/k$ of the initial set, and the rest of it $(1-1/k)$ becomes the training set. The lowest possible value of k is 2. In that extreme case, the initial set is divided only into two subsets, and the training and test subsets are of approximately of equal size – half of the initial dataset. The other extreme is when k is equal to N, that is, the number of cases in the initial dataset. In this setting, the hold-out sets are smallest and comprise exactly one compound. This type of cross-validation is known as leave one out (LOO). Table 2.3 summarizes the types of cross-validation depending on the value of k.

Table 2.3 Cross-validation types depending on the value of k.

Value of k	Percent of hold-out set (%)	Cross-validation type
2–7	50–15	LMO – leave many out
5–20	20–5	LSO – leave several out
N	$100/N$	LOO – leave one out

There is no clear threshold between LMO and LSO.

In the prevailing number of studies, LOO or LSO cross-validation is used where values of k usually range from 5 to 10 [74]. Generally, LOO should not be used as the sole cross-validation method. In the majority of cases, it is better to use k-fold cross-validation with k in the aforementioned range. However, the best practice is to use several types of cross-validation together with LOO. Due to the nondeterministic nature of k-fold cross-validation, it is also advisable to repeat the cross-validation several times. It should be stressed that we do not recommend it in order to improve results; on the contrary, it should be done only to exclude the possibility that the estimates yielded by a single run are, by chance, overoptimistic; an error estimate obtained by a single k-fold cross-validation is not reliable.

In many cases, the dataset is too small, and splitting it into modeling and test sets of sizes appropriate for efficient model building and validation is not possible. To overcome this problem, LOO or LSO (with a relatively high value of k) cross-validation could be used as a replacement of the external validation set. It is important to stress that although small datasets are quite frequent in some types of QSAR studies, these kinds of cases should be always considered as special and special attention should be paid to the analysis of their results.

2.8.6 Bootstrapping

In QSAR studies, one is interested in a particular bootstrapping procedure that can be used to estimate the generalization error by specific reusing of the training set. Bootstrapping is a quite popular method [72]. It is especially efficient for very small datasets but can potentially be applied to any data.

The idea behind it is to fit the model with a set of the size of the training set but still hold out some part of it for further validation. This is achieved by drawing a random sample with replacement from the training set. If the size of the set is N, the probability of compound i being in the bootstrap sample is defined as follows by Eq. (2.7):

$$P(i \in \text{bootstrap sample}) = 1 - \left(1 - \frac{1}{N}\right)^2 \qquad (2.7)$$

This can be approximated by $1 - e^{-1}$ and is usually rounded to 0.632. Hence, there are a 63.2% chance of an object being in the bootstrap sample and a 36.8% of it being in the hold-out sample. It also means that, on average, the new training set (bootstrap sample) contains 63.2% of the original training set, and the hold-out sample contains 36.8% of it. The sampling is repeated several times (100 times is a good standard). Error estimates are calculated in each iteration and at the end results are averaged.

Each time a new model is fitted with a bootstrap sample, the initial training set can be used to calculate the error of prediction; however this estimate is far too optimistic. On average, 63.2% of the training set, over which the error is calculated, was used to fit the models. A straightforward alternative is to use the hold-out sets for error estimation. In this case, estimates are similar to those of a two- or threefold cross-validation and are usually too pessimistic.

It is common practice to combine both types of error estimates into one measure, the so-called 0.632 bootstrap error estimate (Eq. 2.8):

$$\text{err}^{0.632} = 0.632 \times \text{err}_{\text{holdout}} + 0.368 \times \text{err}_{\text{training}} \qquad (2.8)$$

This gives a reasonably good measure of the prediction error that can safely be used in most of the cases. It could be further improved by taking into account the degree of overfitting and possibly other aspects as well. More details can be found in the literature [14].

2.8.7 Y-Randomization (Y-Scrambling)

A powerful validation method that gained much attention is the so-called Y-randomization. This technique is also known as Y-scrambling, response randomization, or Y-permutation [14, 71]. In contrast to cross-validation or bootstrapping, it does not aim at estimating errors of prediction, but its goal is to estimate the reliability of these kinds of errors [83]. In its simplest form, it randomly shuffles the response-dependent variable (traditionally denoted as Y, thus the name Y-randomization) and builds a shuffled model with the original descriptors and the shuffled dependent variable. This simple procedure is repeated several times (again, 100 times is a good standard), and the error estimate of the normal model is compared with the distribution of error estimates of the shuffled models. If the error estimate of the normal model is in a very tail of this distribution, then it is likely that the estimate is statistically significant.

How can "being in a very tail" be quantified? Shen et al. proposed to use the Z-test for that purpose [84]. Suppose that μ is an average value of the error estimates of the shuffled models and σ is its standard deviation. To test whether the error estimate of the normal model *err* is significant, one has to calculate the Z score by Eq. (2.9)

$$Z = \frac{\text{err} - \mu}{\sigma} \quad (2.9)$$

and compare it with the critical values for a desired significance level. The critical values can be easily calculated by submitting the appropriate probabilities to the inverse of the cumulative distribution function of the normal distribution. Table 2.4 contains critical values calculated for the most commonly used significance levels.

Table 2.4 Critical values for the Z-test.

Significance level α	Critical value for the Z-test
0.1	1.282
0.05	1.645
0.01	2.326
0.005	2.576
0.001	3.090

2.8.8 Goodness of Prediction and Quality Criteria

Undoubtedly, the most used (and abused) quality criterion in QSAR studies is the so-called cross-validated predictive coefficient of determination, commonly denoted as q^2 (Eq. 2.13). It is based on the predictive coefficient of determination, R^2, which can be calculated by Eq. (2.10):

$$R^2 = 1 - \frac{\sum_{i=1}^{N}(y_i - \hat{y}_i)^2}{\sum_{i=1}^{N}(y_i - \bar{y})^2} \quad (2.10)$$

The numerator of the second term in this equation is called the residual sum of squares (RSS), and the denominator is called the total sum of squares (TSS). It can be, therefore, defined in the following way by Eq. (2.11):

$$R^2 = 1 - \frac{\text{RSS}}{\text{TSS}} \quad (2.11)$$

Now, RSS can be replaced by the predicted residual sum of squares (PRESS), a statistical value frequently calculated for cross-validated predictions (Eq. 2.12):

$$\text{PRESS} = \sum_{i=1}^{N}(y_i - \hat{y}_{i,-i})^2 \quad (2.12)$$

where $\hat{y}_{i,-i}$ is the value predicted when i is in a hold-out set. q^2 can now easily be defined by Eq. (2.13):

$$q^2 = 1 - \frac{\text{PRESS}}{\text{TSS}} = 1 - \frac{\sum_{i=1}^{N}(y_i - \hat{y}_{i,-i})^2}{\sum_{i=1}^{N}(y_i - \bar{y})^2} \quad (2.13)$$

It is good practice to use subscripts to indicate the type of cross-validation. Hence q^2_{LOO} or $q^2_{10\text{-fold}}$ refer to LOO and to 10-fold cross-validation, respectively.

A common practice is to consider models as predictive if $q^2 > 0.5$ (or even $q^2 > 0.6$) and $R^2 > 0.6$ (or even $R^2 > 0.7$) [14, 85]. However, as reported in the literature and according to our own experience, one has to be very careful when q^2 is too high as it may indicate overfitted models. In particular, one should never use q^2 as the sole measure of prediction quality. Generally, the desired range of q^2 is 0.65–0.85.

Since cross-validation is sometimes prone to deliver overoptimistic estimates of prediction ability, it should be accompanied by estimates calculated over an external test set.

It is straightforward to use a parameter similar to R^2 for an external test set. Replacing RSS and TSS by the appropriate values, R^2_{ext} can be defined by mimicking the R^2 definition in Eq. (2.14):

$$R^2_{ext} = 1 - \frac{\sum_{i=1}^{N}(y_i - \hat{y}_i)^2}{\sum_{i=1}^{N}(y_i - \bar{y})^2} \tag{2.14}$$

Here, summations are done over the external test set. N is the size of the external set, and $\bar{y}$ is the mean of the experimental value in the external set. Another similar parameter for assessing external predictability, namely, Q^2_{ext}, has been proposed (Eq. 2.15) [70]:

$$Q^2_{ext} = 1 - \frac{\sum_{i=1}^{N}(y_i - \hat{y}_i)^2}{\sum_{i=1}^{N}(y_i - \bar{y}_{TR})^2} \tag{2.15}$$

The equation is very similar to Eq. (2.14), but $\bar{y}_{TR}$ is the mean value of the dependent variable in the training set. There is ongoing discussion in the QSAR community concerning the potential superiority of Q^2_{ext} (denoted also as Q^2_{F1}) over R^2_{ext} (denoted also as Q^2_{F2}) and vice versa. In the course of this discussion, another parameter was proposed by Consonni et al. that combines R^2 and R^2_{ext} in one measure denoted as Q^2_{F3} (Eq. 2.16) [70]:

$$Q^2_{F3} = 1 - \frac{\sum_{i=1}^{N}(y_i - \hat{y}_i)^2}{\text{TSS}} \times \frac{N_{TR}}{N} = 1 - \frac{\sum_{i=1}^{N}(y_i - \hat{y}_i)^2}{\sum_{i=1}^{N_{TR}}(y_i - \bar{y}_{TR})^2} \times \frac{N_{TR}}{N} \tag{2.16}$$

Here N is the number of objects in the external test set, N_{TR} is the number of objects in the training set, and the sum in the denominator runs over the training set.

Another interesting measure of performance is Lin's concordance correlation coefficient $\hat{\rho}_C$ often denoted as CCC (Eq. 2.17) [85]:

$$\hat{\rho}_C = \frac{2\sum_{i=1}^{N}(y_i - \bar{y})(\hat{y}_i - \bar{\hat{y}})}{\sum_{i=1}^{N}(y_i - \bar{y})^2 + \sum_{i=1}^{N}(\hat{y}_i - \bar{\hat{y}})^2 + N(\bar{y} - \bar{\hat{y}})^2} \tag{2.17}$$

where $\bar{\hat{y}}$ is the mean value of the predicted values.

In the case of R^2_{ext}, Q^2_{ext}, and Q^2_{F3}, the threshold of predictive models is agreed to be 0.7; for CCC usually a value of 0.85 is suggested.

The goodness of prediction can also be efficiently measured by different error estimates. The simplest one, from the conceptual point of view, is probably MAE. It was already mentioned in Section 2.8.1 as Eq. (2.6); it is presented here again for convenience (Eq. 2.18):

$$\text{MAE} = \frac{1}{N}\sum_{i=1}^{N}|y_i - \hat{y}_i| \tag{2.18}$$

In the QSAR reality, MAE is rarely used; much more often the root-mean-square error (RMSE) is reported (Eq. 2.19):

$$\text{RMSE} = \sqrt{\frac{1}{N}\sum_{i=1}^{N}(y_i - \hat{y}_i)^2} \tag{2.19}$$

RMSE can be easily calculated for (a test set) prediction (RMSEP) or cross-validation (RMSECV). It could also directly be used to calculate the 0.632 bootstrap error estimate (see Eq. 2.8), for example, by Eq. (2.20):

$$\text{RMSE}_{boot}^{0.632} = 0.632 \times \text{RMSEP}_{holdout} + 0.368 \times \text{RMSE}_{training} \tag{2.20}$$

According to the OECD guidelines (see Section 2.9), the standard deviation error of prediction (SDEP) is identical to the aforementioned RMSECV and can be effectively defined by Eq. (2.21):

$$\text{SDEP} = \sqrt{\frac{\text{PRESS}}{N}} = \sqrt{\frac{1}{N}\sum_{i=1}^{N}(y_i - \hat{y}_{i,-i})^2} \tag{2.21}$$

2.8.9 Applicability Domain and Model Acceptability Criteria

In the QSAR field, the applicability domain (AD) is calculated for testing the limitations of a model, that is, the range of molecular properties or structures for which the model is considered to be applicable [86]. This is an important issue since many QSAR models that are developed can be regarded as local models, that is, a model with a narrow applicability for a specific class of compounds but with a good predictive power. Global models with a broad applicability but decreased predictive power, on the other hand, are based on larger and more diverse datasets. AD is important for validating QSAR models. In particular, if an external validation is being performed, the predictions must be made for compounds not included in the training set. However, to ensure that the external "validation" set is appropriate, the test set compounds should fall within the AD deduced from the training set. Ideally, a QSAR model should only be used to make predictions for compounds that are within the AD by interpolation, not extrapolation.

Various approaches for defining the AD have been based on a similarity analysis [75, 87]. All of these approaches are based on the premise that a QSAR prediction is reliable if the compound for which a prediction is being made is similar to the ones in the training set.

Four major approaches have been recognized to estimate the AD for QSAR models [75]:

- Range-based methods
  The simplest method for describing the AD is to consider the ranges of individual descriptors. This defines an n-dimensional hyper-rectangle with sides parallel to the coordinate axes. The data distribution is assumed to be uniform. An often applied range-based method is PCA (Figure 2.8).
- Geometric methods
  Using geometric approaches, the AD can be estimated by calculating the smallest convex area containing the entire training set. The most straightforward empirical method for defining the coverage of an n-dimensional set is the convex hull, which is the smallest convex area that contains the original set. The calculation of the convex hull represents a computational geometry problem. More details can be found in Ref. [87].
- Distance-based methods
  These approaches define AD by calculating distances of a query data point within the descriptor space of the training data. The distance between the defined point and the dataset points is then compared with a predefined threshold. Three distance-based approaches have been found to be most useful in QSAR research, namely, the Euclidean, Mahalanobis, and city-block distances.
- Probability density distribution-based methods
  The probability density function of a dataset can be estimated by parametric or nonparametric methods. Parametric methods assume that the density function has the shape of a standard distribution (e.g., Gaussian distribution). Alternatively, a number of nonparametric techniques are available that do not make any assumptions about the data distribution. Nonparametric techniques

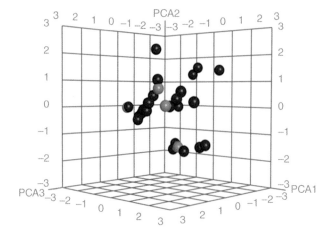

Figure 2.8 Schematic representation of the chemical space for a set of compounds described by the first three principal components (PC1-3). The test set molecules (grey balls) are located within the applicability domain covered by the molecules of the training set (black balls).

allow the probability density to be estimated solely from data by kernel density estimation or mixture density methods.

Further methods have been developed and implemented in some commercial software packages. A comprehensive survey can be found in the "Report and Recommendations of an ECVAM Workshop" [88].

2.8.10 Scope of External and Internal Validation

Let us look again at the workflow presented in Figure 2.5 in Section 2.8.2. It is shown again on Figure 2.9 for convenience. Let us analyze it one more time. First, the initial (complete) dataset is split into the modeling and the external test sets. The modeling set is then used for model building, and the test set is used afterward to perform an external validation. Although the methodology of splitting might raise some concerns on the full independence of the resulting sets (see discussion in Section 2.8.1), it can be assumed that as long as the test set is not used in the model building process, it can deliver proper external validation error estimates. The splitting of the modeling set into training and validation sets is optional. If it is done and if appropriate requirements are met, it might be stated that the validation set delivers proper validation error estimates as well. Moreover, in the central part of the workflow, there are several blocks of internal validation. These also give appropriate error estimates.

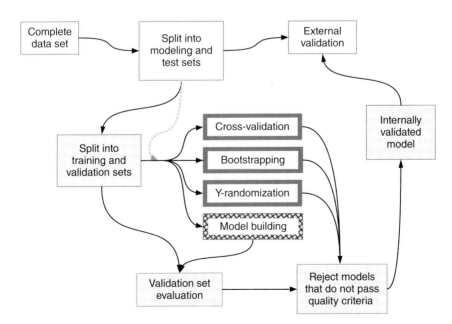

Figure 2.9 Modeling/validation workflow. The scope of the external validation is shown in gray color; the scope of the inner validation over the validation set is shown in cross-hatching. The internal validation methods are shown with a bold frame.

How are these validations (error estimates) related to each other? It is rather obvious that the superior one is the external validation calculated over the external test set. But what is its scope? What conclusions can be drawn from it? Similar questions can be asked, in principal, for any kind of validation.

In the case of external validation performed as proposed by the workflow, it simply validates the final model that resulted from this part of the workflow that involves the modeling set (the outcome of validation might be either positive or negative). In Figure 2.9, this part of the workflow is shown in gray color.

This leads to the very important rule: validation, either external or internal, holds true for the sets used to perform it, and any generalization, although anticipated and possible, must be considered with extreme care. The decision whether the particular set is "covered" by the validation can be alleviated by consideration of the AD of the model. In fact, the AD should always be considered (see Section 2.8.9 for a detailed discussion).

The assumption that both internal validation and model building use exactly the same modeling procedure is crucial here. Figure 2.10 shows two possible modeling workflows that employ cross-validation. In the top workflow, the training set is submitted to some preprocessing routines like variable selection, normalization, transformation, etc. After that it is submitted to model building and cross-validation. The scope of this cross-validation is shown by the gray color. It is limited to the model built with the preprocessed data. It cannot go beyond the preprocessing and should be considered as local. It is a common mistake to use the results of such cross-validation to make conclusions that go far beyond the limiting preprocessing "box." In fact, error estimates obtained as a result of this procedure are very likely to be far too optimistic and should not be used per se.

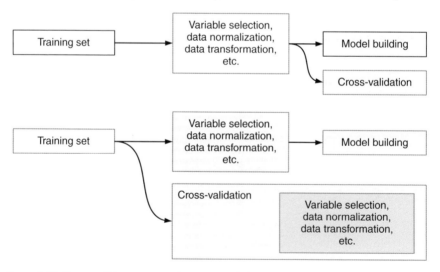

Figure 2.10 Two possible ways to perform cross-validation.

How should the cross-validation be done to broaden its scope? The preferred way is shown in Figure 2.10 at the bottom workflow. Here, the entire preprocessing procedure is repeated in every fold of cross-validation (this is schematically shown by nesting the preprocessing box inside the cross-validation box). The scope of this cross-validation is shown by the gray color and is evidently much broader. This discussion is obviously relevant not only for cross-validation but is true for any type of validation including Y-randomization and bootstrapping [75].

2.8.11 Validation of Classification Models

Measuring the success of a classification model is not as straightforward as it might initially seem. For a binary classification exercise, the predictions can be classed as true positives (TP), false positives (FP), true negatives (TN), and false negatives (FN). These are combined into a confusion matrix of actual against predicted classes, the diagonal elements being the numbers of TP and TN, and the off-diagonal ones the numbers of FP and FN.

A commonly used measure for binary classification is the so-called Matthews correlation coefficient (MCC). The MCC can be used even if the classes are of different sizes. MCC represents a correlation coefficient between the observed and the predicted binary classification; values between −1 and +1 are obtained. A coefficient of +1 means a perfect prediction, 0 means no better than random prediction, and −1 indicates total disagreement between prediction and observed data. MCC is derived from the confusion matrix using Eq. (2.22):

$$\text{MCC} = \frac{\text{TP} \times \text{TN} - \text{FP} \times \text{FN}}{\sqrt{(\text{TP}+\text{FP})(\text{TP}+\text{FN})(\text{TN}+\text{FP})(\text{TN}+\text{FN})}} \quad (2.22)$$

A graphical representation is also commonly used to measure the performance of classification and ranking. A so-called receiver operating characteristic (ROC) curve can be plotted using the true positive rate (TPR = TP/(TP + FN), also known as sensitivity) versus the false positive rate (FPR = FP/(FP + TN), also known as 1 − specificity). The line in Figure 2.11 indicates the performance of the

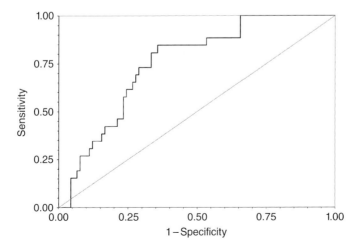

Figure 2.11 Example ROC curve (bold line) for a binary classification problem. The straight line is obtained by guessing the class membership.

classification. As an additional parameter the area under the curve (AUC) can be used. Random guessing would result in an AUC of 0.5. A classifier performs well if the curve climbs rapidly toward the upper left-hand corner, resulting in AUC values above 0.5.

2.9 Regulatory Use of QSARs

The international cooperation among the OECD member countries on QSARs started in 1990 with a review of the availability and the predictive power of QSAR models, and a number of reports were produced. Following a workshop on "Regulatory Use of QSARs for Human Health and Environmental Endpoints" in Setubal, Portugal, the OECD member countries agreed to define an accepted set of criteria, as well as procedures for the evaluation of QSAR models. The OECD principles for the validation of QSAR models were agreed upon in 2004, and a guidance document was made available in 2007 [89]. In 2004, the OECD member countries recognized that the focus of the work should shift to the regulatory use and application of QSARs. The requirement for technical guidance concerning QSAR modeling and validation has also been recognized by OECD expert groups as a crucial and urgent point in recent years. This has led to the development, for regulatory purposes, of the OECD principles for the validation of QSAR models. The principles can be summarized as follows:

1. The used experimental data should have a clearly defined endpoint.
2. The models have to be generated applying an unambiguous algorithm.
3. For prediction purposes an applicability domain has to be defined.
4. An appropriate measure of goodness of fit, robustness, and predictive power should be given.
5. A mechanistic interpretation is desirable.

More details on the use of QSAR methods for regulatory purposes can be found in Chapter 8.

Essentials

- The abbreviation QSAR stands for quantitative structure–activity relationships. QSPR means quantitative structure–property relationships.
- As many properties of molecules cannot directly be predicted from its structure, indirect approaches are used to solve this problem. In the first step, molecular descriptors are calculated for a set of molecules. Secondly, linear and nonlinear data analysis algorithms are applied to predict the activity (or property) of interest using the calculated descriptors.
- Molecules can be represented by structure descriptors in a hierarchical manner with respect to (i) the type of descriptor and (ii) the molecular representation of the compound.
- Besides traditional methods, such as MLRA and PLS, machine learning techniques are nowadays increasingly applied for QSAR modeling. Methods such as

random forest, artificial neural networks, and support vector machines can be applied for regression, classification, and ranking problems.
- The validation of QSAR models is of crucial importance.
- The applicability domain of a model is defined to be the set of structures as represented by molecular descriptors.
- QSAR predictions should only be made within the applicability domain of a model.
- QSAR model validation has also been recognized by specific OECD expert groups as a crucial and urgent point in recent years, and this has led to the development of the OECD principles for the validation of QSAR models for regulatory purposes.

Available Software and Web Services (accessed January 2018)

- Open3DQSAR – An open-source tool aimed at pharmacophore exploration by applying PLS analysis of molecular interaction fields (MIFs) (http://open3dqsar.sourceforge.net).
- 3-D QSAutogrid – 3-D QSAR modeling for virtual screening (http://www.3d-qsar.com); software for the development and validation of multiple linear regression (MLR) models by ordinary least squares (OLS) (http://www.qsar.it).
- CORALSEA – Freeware to build quantitative structure–property/activity relationships (QSPR/QSAR) (http://www.insilico.eu/coral/CORALSEA.html).
- McQSAR – Generates quantitative structure–activity relationships (QSAR equations) using the genetic function approximation paradigm (http://users.abo.fi/mivainio/mcqsar).
- BlueDesc – Open-source descriptor calculator (http://www.ra.cs.uni-tuebingen.de/software/bluedesc/welcome_e.html).
- E-DRAGON – Online descriptor calculator (http://146.107.217.178/lab/edragon/).
- Mold2 – Free descriptors generator software (http://www.fda.gov/ScienceResearch/BioinformaticsTools/Mold2/default.htm).
- CODESSA – COmprehensive DEscriptors for Structural and Statistical Analysis (http://www.codessa-pro.com).
- Molecular descriptors – Collection of free online resources (http://www.moleculardescriptors.eu/resources/resources.htm).
- PaDEL-Descriptor – Open-source Java-based software for the calculation of molecular descriptors (http://www.yapcwsoft.com/dd/padeldescriptor/).
- CORINA Symphony – Software for computing molecular descriptors (https://www.mn-am.com/products/corinasymphony).
- RapidMiner – An open-source system with a large collection of algorithms for data analysis and model development (https://rapidminer.com/).
- R – A language and environment for statistical computing and graphics (http://www.r-project.org).

- WEKA – Open-source compilation of modeling methods and tools for data preprocessing, classification, regression, clustering, and visualization (http://www.cs.waikato.ac.nz/ml/weka).
- AutoWEKA – User-friendly graphical interface for WEKA, open source (http://www.mt.mahidol.ac.th/autoweka).
- SNNS – (Stuttgart Neural Network Simulator) is a neural network simulator (http://www.ra.cs.uni-tuebingen.de/SNNS/).
- SONNIA – Self-Organizing Neural Network Package (https://www.mn-am.com/products/sonnia).
- Orange/AZOrange – Offers tools for data preparation, evaluation, visualization, classification, regression, and clustering (http://www.ailab.si/orange).
- CHARMMing – Web-based QSAR and data mining tools (www.charmming.org).
- OChEM – Online Chemical Modeling Environment is a web-based platform for QSAR modeling (http://www.ochem.eu).
- Chembench – A free portal that enables researchers to mine available chemical and biological data (https://chembench.mml.unc.edu/home).
- PubChem – PubChem is a free chemistry database, serving the research community by providing information on millions of chemicals, their biological activities, patents, and targets (https://pubchem.ncbi.nlm.nih.gov).
- ChEMBL – An open data resource of binding, functional, and ADMET bioactivity data (https://www.ebi.ac.uk/chembl).
- ChemSpider – A free chemical structure database providing fast text and structure search access to over 32 million structures (http://www.chemspider.com).

Selected Reading

- Alpaydin, E. (2010) *Introduction to Machine Learning*, 2nd edn, MIT Press, Cambridge, MA, 584 pp.
- Cherkasov, E., Muratov, N., Fourches, D., Varnek, A., Baskin, I.I., Cronin, M., Dearden, J., Gramatica, P., Martin, Y.C., Todeschini, R., Consonni, V., Kuz'min, V.E., Cramer, R., Benigni, R., Yang, C., Rathman, J., Terfloth, L., Gasteiger, J., Richard, A., and Tropsha, A. (2014) *J. Med. Chem.*, **57**, 4977–5010.
- Cumming, J.G., Davis, A.M., Muresan, S., Haeberlein, M., and Chen, H. (2013) *Nat. Rev. Drug Discov*, **12**, 948–962.
- Devillers, J. (2007) *Neural Networks in QSAR and Drug Design*, Elsevier Ltd., 284 pp.
- Todeschini, R. and Consonni, V. (2009) *Molecular Descriptors for Chemoinformatics*, Wiley-VCH, Weinheim, Germany, 1257 pp.
- Tropsha, A. (2010) *Mol. Inform.*, **29**, 476–488.
- Varnek, A. and Baskin, I. (2012) *J. Chem. Inf. Model.*, **52**, 1413–1437.
- Zupan, J. and Gasteiger, J. (1999) *Neural Networks in Chemistry and Drug Design*, 2nd edn, Wiley-VCH, Weinheim, Germany, 400 pp.

References

[1] Hansch, C. and Fujita, T. (1964) *J. Am. Chem. Soc.*, **86**, 1616–1626.
[2] Kubinyi, H. (1976) *Arzneimittelforschung*, **26**, 1991–1997.
[3] Free, S.M. and Wilson, J.W. (1964) *J. Med. Chem.*, **7**, 395–399.
[4] Kubinyi, H. (1976) *J. Med. Chem.*, **19**, 587–600.
[5] Dearden, J.C., Cronin, M.T., and Kaiser, K.L. (2009) *SAR QSAR Environ. Res.*, **20**, 241–266.
[6] Cherkasov, A., Muratov, E.N., Fourches, D., Varnek, A., Baskin, I.I., Cronin, M., Dearden, J., Gramatica, P., Martin, Y.C., Todeschini, R., Consonni, V., Kuz'min, V.E., Cramer, R., Benigni, R., Yang, C., Rathman, J., Terfloth, L., Gasteiger, J., Richard, A., and Tropsha, A. (2014) *J. Med. Chem.*, **57**, 4977–5010.
[7] Overington, J. (2009) *J. Comput. Aided Mol. Des.*, **23**, 195–198.
[8] Kaiser, J. (2005) *Science*, **308**, 774.
[9] Tetko, I.V., Sushko, I., Pandey, A.K., Zhu, H., Tropsha, A., Papa, E., Oberg, T., Todeschini, R., Fourches, D., and Varnek, A. (2008) *J. Chem. Inf. Model.*, **48**, 1733–1746.
[10] Fourches, D., Muratov, E., and Tropsha, A. (2010) *J. Chem. Inf. Model.*, **50**, 1189–1204.
[11] Chemspider, RSC, http://www.chemspider.com (accessed January 2018).
[12] ISIDA Software, University of Strasbourg, France, http://infochim.u-strasbg.fr (accessed March 2017).
[13] Kuz'min, V.E., Artemenko, A.G., and Muratov, E.N. (2008) *J. Comput. Aided Mol. Des.*, **22**, 403–421.
[14] Tropsha, A. (2010) *Mol. Inform.*, **29**, 476–488.
[15] Todeschini, R. and Consonni, V. (2009) *Molecular Descriptors for Chemoinformatics*, Wiley-VCH, Weinheim, 1257 pp.
[16] Todeschini, R. and Consonni, V. (2010) Molecular descriptors, in *Recent Advances in QSAR Studies* (eds T. Puzyn, J. Leszczyynski, and M.T.D. Cronin), Springer, New York, 29–102.
[17] Moreau, G. and Broto, P. (1980) *Nouv. J. Chim.*, **4**, 757–776.
[18] Gasteiger, J., Sadowski, J., Schuur, J., Selzer, P., Steinhauer, L., and Steinhauer, V. (1996) *J. Chem. Inf. Comput. Sci.*, **36**, 1030–1037.
[19] Schuur, J. and Gasteiger, J. (1997) *Anal. Chem.*, **69**, 2398–2405.
[20] Stanton, D.T., Mattioni, B.E., Knittel, J.J., and Jurs, P.C. (2004) *J. Chem. Inf. Comput. Sci.*, **44**, 1010–1023.
[21] Cruciani, G., Pastor, M., and Mannhold, R. (2002) *J. Med. Chem.*, **45**, 2685–2694.
[22] Goodford, P.J. (1985) *J. Med. Chem.*, **28**, 849–857.
[23] Gramatica, P. (2013) *Methods Mol. Biol.*, **930**, 499–526.
[24] Eriksson, L., Jaworska, J., Worth, A.P., Cronin, M.T., McDowell, R.M., and Gramatica, P. (2003) *Environ. Health Perspect.*, **111**, 1361–1375.
[25] Guha, R. and Jurs, P.C. (2005) *J. Chem. Inf. Model.*, **45**, 65–73.
[26] Ivanciuc, O. (2009) *Encyclopedia of Complexity and Systems Science*, Springer, pp. 2139–2159.

[27] Devillers, J. (2007) *Neural Networks in QSAR and Drug Design*, Elsevier Ltd., 284 pp.
[28] Zupan, J. and Gasteiger, J. (1999) *Neural Networks in Chemistry and Drug Design: An Introduction*, 2nd edn, Wiley-VCH, Weinheim, 400 pp.
[29] Hammett, L.P. (1937) *J. Am. Chem. Soc.*, **59**, 96–103.
[30] Efron, B., Hastie, T., Johnston, I., and Tibshirani, R. (2004) *Ann. Stat.*, **32**, 407–499.
[31] Goodarzi, M., Dejaegher, B., and Vander Heyden, Y. (2012) *J. AOAC Int.*, **95**, 636–651.
[32] Wold, S. (1991) *Quant. Struct. Act. Relat.*, **10**, 191–193.
[33] Hasegawa, K. and Funatsu, K. (2010) *Curr. Comput. Aided Drug Des.*, **6**, 103–127.
[34] Hemmateenejad, B. and Elyasi, M. (2009) *Anal. Chim. Acta*, **646**, 30–88.
[35] Cramer, R. (1992) *J. Comput. Aided Mol. Des.*, **6**, 475–486.
[36] Baroni, M., Constantino, G., Cruciani, G., Riganelli, D., Valigi, R., and Clementi, S. (1993) *Quant. Struct. Act. Relat.*, **12**, 9–20.
[37] Klebe, G., Abraham, U., and Mietzner, T. (1994) *J. Med. Chem.*, **37**, 4130–4146.
[38] Tosco, P. and Balle, T. (2011) *J. Mol. Model.*, **17**, 201–208.
[39] Ballant, F., Caroli, A., Wickersham, R.B., and Ragno, R. (2014) *J. Chem. Inf. Model.*, **54**, 956–969.
[40] Silverman, B.D. and Platt, D.E. (1996) *J. Med. Chem.*, **39**, 2129–2140.
[41] Cruciani, G., Crivori, P., Carupt, P.A., and Testa, B. (2000) *J. Mol. Struct. (THEOCHEM)*, **503**, 17–31.
[42] Pastor, M., Cruciani, G., McLay, I., Pickett, S., and Clementi, S. (2000) *J. Med. Chem.*, **43**, 3233–3243.
[43] Walters, E. (1998) *Perspect. Drug Discov. Des.*, **12**, 159–166.
[44] Jain, A.N., Koile, K., and Chapman, D. (1994) *J. Med. Chem.*, **37**, 2315–2327.
[45] Hahn, M. and Rogers, D. (1998) *Perspect. Drug Discov. Des.*, **12**, 117–134.
[46] Vedani, A. and Zbinden, P. (1998) *Pharm. Acta Helv.*, **73**, 11–18.
[47] Cramer, R.D. and Wendt, B. (2014) *J. Chem. Inf. Model.*, **54**, 660–671.
[48] Schwaighofer, A., Schroeter, T., Mika, S., and Blanchard, G. (2009) *Comb. Chem. High Throughput Screen.*, **12**, 453–468.
[49] Svetnik, V., Wang, T., Tong, C., Liaw, A., Sheridan, R.P., and Song, Q. (2005) *J. Chem. Inf. Model.*, **45**, 786–799.
[50] Breiman, L. (1996) *Mach. Learn.*, **24**, 123–140.
[51] Wolpert, D.H. (1992) *Neural Netw.*, **5**, 241–260.
[52] Agrafiotis, D.K., Cedeno, W., and Lobanov, V.S. (2002) *J. Chem. Inf. Comput. Sci.*, **42**, 903–911.
[53] Rodríguez, J.J., Kuncheva, L.I., and Alonso, C.J. (2006) *IEEE Trans. Pattern Anal. Mach. Intell.*, **28**, 1619–1630.
[54] Breiman, L. (2001) *Mach. Learn.*, **45**, 5–32.
[55] R Statistical Package, http://www.r-project.org/ (accessed January 2018).
[56] Orange Data Mining, http://orange.biolab.si/ (accessed January 2018).
[57] WEKA, http://www.cs.waikato.ac.nz/ml/weka (accessed January 2018).

[58] Weidlich, I.E., Pevzner, Y., Miller, B.T., Filippov, I.V., Woodcock, H.L., and Brooks, B.R. (2015) *J. Comput. Chem.*, **36**, 62–67.
[59] Varnek, A. and Baskin, I. (2012) *J. Chem. Inf. Model.*, **52**, 1413–1437.
[60] Rumelhart, D.E., Hinton, G.E., and Williams, R.J. (1986) *Nature*, **323**, 533–536.
[61] Livingstone, D.J. and Mannalack, A. (2003) *QSAR Comb. Sci.*, **22**, 510–518.
[62] Winkler, D. (2004) *Mol. Biotechnol.*, **27**, 139–167.
[63] Cortes, C. and Vapnik, V. (1995) *Mach. Learn.*, **20**, 273–297.
[64] Trotter, M.W.B. and Holden, S.B. (2003) *QSAR Comb. Sci.*, **22**, 536–548.
[65] Noble, W.S. (2006) *Nat. Biotechnol.*, **24**, 1565–1567.
[66] Novotarskyi, S., Sushko, I., Körner, R., Pandey, A.K., and Tetko, I.V. (2011) *J. Chem. Inf. Model.*, **51**, 1271–1280.
[67] Bruce, C.L., Melville, J.L., Pickett, S.D., and Hirst, J.D. (2007) *J. Chem. Inf. Model.*, **47**, 219–227.
[68] Golbraikh, A. and Tropsha, A. (2002) *J. Mol. Graph. Model.*, **20**, 269–276.
[69] Hastie, T., Tibshirani, R., and Friedman, J. (2009) *The Elements of Statistical Learning*, Springer, New York, NY, p. 745.
[70] Consonni, V., Ballabio, D., and Todeschini, R. (2009) *J. Chem. Inf. Model.*, **49**, 1669–1678.
[71] Martin, T.M., Harten, P., Young, D.M., Muratov, E.N., Golbraikh, A., Zhu, H., and Tropsha, A. (2012) *J. Chem. Inf. Model.*, **52**, 2570–2578.
[72] Golbraikh, A. and Tropsha, A. (2002) *J. Comput. Aided Mol. Des.*, **16**, 357–369.
[73] Polanski, J., Bak, A., Gieleciak, R., and Magdziarz, T. (2006) *J. Chem. Inf. Model.*, **46**, 2310–2318.
[74] Tropsha, A., Gramatica, P., and Gombar, V. (2003) *QSAR Comb. Sci.*, **22**, 69–77.
[75] Tetko, I.V., Sushko, I., Pandey, A.K., Zhu, H., Tropsha, A., Papa, E., Oberg, T., Todeschini, R., Fourches, D., and Varnek, A. (2008) *J. Chem. Inf. Model.*, **48**, 1733–1746.
[76] Fisher, R.A. (1936) *Ann. Eugen.*, **7**, 179–188.
[77] Kennard, R.W. and Stone, L.A. (1969) *Technometrics*, **11**, 137–148.
[78] Snarey, M., Terrett, N.K., Willett, P., and Wilton, D.J. (1997) *J. Mol. Graph. Model.*, **15**, 372–385.
[79] Snee, R.D. (1977) *Technometrics*, **19**, 415–428.
[80] Golbraikh, A. (2000) *J. Chem. Inf. Model.*, **40**, 414–425.
[81] Clark, R. (1997) *J. Chem. Inf. Model.*, **37**, 1181–1188.
[82] Rodionova, O.Y. and Pomerantsev, A.L. (2008) *J. Chemom.*, **22**, 674–685.
[83] Klopman, G. and Kalos, A.N. (1985) *J. Comput. Chem.*, **6**, 492–506.
[84] Shen, M., LeTiran, A., Xiao, Y., Golbraikh, A., Kohn, H., and Tropsha, A. (2002) *J. Med. Chem.*, **45**, 2811–2823.
[85] Chirico, N. and Gramatica, P. (2011) *J. Chem. Inf. Model.*, **51**, 2320–2335.
[86] Golbraikh, A., Muratov, E., Fourches, D., and Tropsha, A. (2014) *J. Chem. Inf. Model.*, **54**, 1–4.
[87] Horvath, D., Marcou, G., and Varnek, A. (2009) *J. Chem. Inf. Model.*, **49**, 1762–1776.

[88] Netzeva, T.I., Worth, A.P., Aldenberg, T., Benigni, R., Cronin, M.T.D., Gramatica, P., Jaworska, J.S., Kahn, S., Klopman, G., Marchant, C.A., Myatt, G., Nikolova-Jeliazkova, N., Patlewicz, G.Y., Perkins, R., Roberts, D.W., Schultz, T.W., Stanton, D.T., van de Sandt, J.M., Tong, W., Veith, G., and Yang, C. (2005) *Altern. Lab Anim.*, **33**, 155–173.

[89] OECD (2007) *Guidance Document on the Validation of (Q)SAR Models*, OECD, Paris.

3 Prediction of Physicochemical Properties of Compounds

Igor V. Tetko[1], Aixia Yan[2], and Johann Gasteiger[3]

[1] Institute of Structural Biology, Helmholtz Zentrum München, Deutsches Forschungszentrum für Gesundheit und Umwelt (GmbH), Ingolstädter Landstr. 1, 60w, Neuherberg 85764, Germany
[2] Beijing University of Chemical Technology, State Key Laboratory of Chemical Resource Engineering, Department of Pharmaceutical Engineering, 15 BeiSanHuan East Road, Beijing 100029, P. R. China
[3] Computer-Chemie-Centrum, Universität Erlangen-Nürnberg, Nägelsbachstr. 25, 91052 Erlangen, Germany

Learning Objectives

- To derive quantitative relationships between a property and a structure.
- To acquire knowledge about different types of methods available in the field.
- To acquire an overview of methods for modeling several important properties.
- To identify and understand the limitations of different modeling techniques.

Outline

3.1 Introduction, 53
3.2 Overview of Modeling Approaches to Predict Physicochemical Properties, 54
3.3 Methods for the Prediction of Individual Properties, 59
3.4 Limitations of Statistical Methods, 76
3.5 Outlook and Perspectives, 76

3.1 Introduction

Fundamental physical and chemical properties of a compound such as its solubility, its lipophilicity, or its acidity determine its behavior in chemical, biochemical, or environmental processes. Therefore their knowledge is required for understanding and modeling the action of a compound in drug discovery, environmental chemistry, and other chemical industries. Although the amount of experimental data is growing, the number of newly synthesized or designed compounds is increasing more quickly, especially through high-throughput methods such as parallel synthesis and combinatorial chemistry. Many compounds can be designed virtually and then used in computer screening against protein targets. The prediction of their physicochemical properties as well as the simplicity of their chemical synthesis is very important to guide medicinal chemists to select the most appropriate hits for synthesis and experimental validation. Thus, there is a strong need for accurate methods for the prediction of physicochemical properties. The modeling of several properties, that is,

Applied Chemoinformatics: Achievements and Future Opportunities, First Edition.
Edited by Thomas Engel and Johann Gasteiger.
© 2018 Wiley-VCH Verlag GmbH & Co. KGaA. Published 2018 by Wiley-VCH Verlag GmbH & Co. KGaA.

octanol/water partition and distribution coefficients, water solubility, melting point (MP), pK_a value, and thermochemical properties, will be exemplified.

3.2 Overview of Modeling Approaches to Predict Physicochemical Properties

The basic approach to the problem of estimating properties can be written in a very simple form that states that a molecular property P can be expressed as a function of the molecular structure C (Eq. 3.1):

$$P = f(C) \tag{3.1}$$

The function $f(C)$ may have a very simple form, as is the case for the calculation of the molecular weight from the relative atomic masses. In most cases, however, $f(C)$ will be complicated when it comes, for example, to describing the structure by quantum mechanics, when the property is derived directly from the wave function such as the calculation of the dipole moment by applying the dipole operator.

Most physicochemical properties are related to interactions between a molecule and its environment. For instance, the partitioning between two phases is a temperature-dependent constant of a substance with respect to the solvent system. Eq. 3.1 therefore has to be rewritten as a function of the molecular structure, C, and of the experimental conditions of the solvent, S; the temperature, T; and so on (Eq. 3.2):

$$P = f(C, S, T, K) \tag{3.2}$$

A collection of comprehensive data to cover all conditions is unfeasible and can also be redundant. Indeed, the changes of properties as a function of certain conditions are frequently well approximated using theoretical or empirical models. Therefore, for many properties it is important to have measurements under some predefined standard experimental conditions, which are kept the same, in order to allow the merging of data from different experiments. The Organization for Economic Co-operation and Development (OECD) has developed a number of such standard protocols for different properties (OECD Guidelines for the testing of chemicals [1]). The use of the same experimental conditions allows the treatment of the effects, which are not directly related to the molecular structure as constant. This significantly simplifies the modeling of physicochemical properties since it allows the collection and merging of data measured in different experiments. However, as an important consequence, the prediction models are valid only for the system under investigation. A model for the prediction of the acidity constant pK_a in aqueous solutions cannot be applied to the prediction of pK_a values in DMSO solutions.

It should be mentioned that even the most precise experimental measurements could still have high variability due to the analyzed compound itself. In practice, we almost never deal with properties of pure (ideal) chemical compounds. For example, the MP can be measured very precisely and does not strongly depend on the conditions of experiment. MP values in a set of 225,000 compounds collected from patents [2] have an average experimental error of 3 °C (intralaboratory reproducibility) according to the ranges provided for individual compounds. However, for the same data the duplicated measurements in different experiments contributed to an error at >30 °C [2]. The large interlaboratory variations could possibly be due to different purities of compounds and the presence of different polymorphic and/or amorphic forms. Indeed, even if a compound was synthesized using exactly the same protocol but was stored under different conditions, the measurement results could be different due to different degrees of decomposition.

There can also be an uncertainty in the representation of chemical structures for modeling. Problems such as tautomers, definition of aromaticity, and 3D conformations of compounds (which are required for 3D descriptors), among others, could significantly contribute to the differences in calculated descriptors and as a result to variations in the results obtained by the equations and in the lack of reproducibility of the models. Thus, the use of a strict experimental protocol as well as the use of a detailed and robust workflow to process a chemical compound is a prerogative for the creation of reliable models.

In the following sections we will first provide an overview of typical methods for the prediction of physicochemical properties, which will be followed by a more detailed analysis of several physicochemical properties important for drug discovery and chemical industry in general.

3.2.1 Prediction of Properties Based on Other Properties

Many physicochemical properties of compounds are strongly interconnected. For example, compounds with a higher MP are often less soluble. The relationships between physicochemical properties can be derived based on a theoretical analysis or/and found empirically.

3.2.2 Prediction of Properties Based on Theoretical Calculations

The development of quantum chemistry methods has contributed new powerful approaches to the prediction of physicochemical properties. An example for such an approach is the continuum solvation model for real solvents (COSMO-RS) [3], which considers interactions in a liquid system as contact interactions of the molecular surfaces. The density functional theory (DFT)/conductor-like screening model (COSMO) is used to calculate the total energy and the polarization (or screening) of charge density σ on the surface of a molecule. This is a time-consuming step, which is done once for each molecule and stored in a file for further analysis. The interaction energies of, for example, hydrogen bonding and electrostatic terms are written as pairwise interactions of the respective polarization charge densities. This approach was successfully used to predict different physicochemical properties including partition coefficients between different media and also water solubility [4, 5].

A complete theoretical *ab initio* modeling of many physicochemical properties using quantum-chemical calculations is currently not possible due to the complexity of the considered systems. The methods based on theoretical calculations simplify these calculations by capturing the most essential interactions. Like with approaches analyzed in the previous section, their power is in a clear mechanistic interpretation and in the identification of violations of the assumptions used for the simplification of the considered systems.

3.2.3 Additivity Schemes for Property Prediction

Molecular structures consist of atoms that are held together by bonds. The forces between molecules are typically one or two orders of magnitude lower than the forces that hold atoms together in molecules. The picture of an atom in a molecule is intuitively quite appealing, and it raises the question if atoms bring with them properties that still can be discerned in the bound state of an atom in a molecule. Or, conversely, can we break up a molecular property into contributions of its constituent atoms and calculate molecular properties by adding these contributions from its atoms? It will be shown that this can be done with satisfactory accuracy only for a few molecular properties. However, the idea of calculating molecular properties by contributions from its atoms can be extended to considering contributions from structural units larger than atoms, such as of bonds or of groups.

In fact, there is a hierarchy in calculating molecular properties by additivity of atomic, bond, or group properties as pointed out some time ago by Benson [6, 7]. The larger the substructures that have to be considered, the larger is the number of increments that can be derived, and the higher the accuracy in the values obtained for a molecular property. A basic assumption in such additivity schemes is that the interactions between the atoms of a molecule are of a rather short-range nature. This fact can be expressed in a more precise manner. The law of additivity can be expressed in a chemical equation [7]. Let us consider the atoms (or groups) X and Y attached to a common skeleton, S, and consider the redistribution of these atoms on that skeleton as expressed by Eq. (3.3):

$$X-S-X + Y-S-Y \leftrightarrows 2X-S-Y \tag{3.3}$$

The law of additivity then says that the sum of the properties of the molecules on the right-hand side is the same as the sum of the properties on the left-hand side of Eq. (3.3). When additivity of atomic properties is valid, then the skeleton S disappears and Eq. (3.3) can be rewritten as Eq. (3.4):

$$X-X + Y-Y \leftrightarrows 2X-Y \tag{3.4}$$

The sum of the properties of the diatomic species X_2 and Y_2 is the same as twice the property of XY. This is the zero-order approximation to additivity rules. If S is a single atom or a group of atoms with the bonds attached to the same atom (such as a CH_2 group), then we have the additivity of bond properties, the first-order approximation, as given by Eq. (3.5):

$$X-CH_2-X + Y-CH_2-Y \leftrightarrows 2X-CH_2-Y \tag{3.5}$$

When group additivity is valid, S consists of a group with the bonds of X and Y attached to two adjacent atoms such as in Eq. (3.6):

$$X-CH_2-CH_2-X + Y-CH_2-CH_2-Y \leftrightarrows 2X-CH_2-CH_2-Y \tag{3.6}$$

This is the second-order approximation to additivity rules.

Equations (3.3)–(3.6) clearly illustrate the increase in distance in the interactions between atoms X and Y in going from the additivity of atomic to bond and further to group contributions.

3.2.3.1 Atom-Based Contribution Methods

Physicochemical properties can be considered in a first approximation as additive with respect to its atoms and thus calculated by Eq. (3.7):

$$P = \sum_{i=1}^{N} a_i n_i \tag{3.7}$$

where a_i is the contribution of an atom i and n_i is the count of the atoms of type i. This approximation is perfectly working for the molecular weight of a molecule, where the exact mass is calculated as the mass of the individual atoms. The same approximation also works rather well for other molecular properties, such as molecular refraction or parachor. However, its accuracy decreases for more complex properties, for example, log P or water solubility, where interactions between atoms start to play a significant role. In order to address this problem, one possibility is to extend the equation by introducing correction factors, b_j, where n_j is their frequency or/and to distinguish different types of atoms depending on their environment, for example, aromatic and aliphatic carbon atoms, nitrogen atoms in nitro, nitroso, *etc.* groups, and so on (see Eq. 3.8):

$$P = \sum_{i=1}^{N} a_i n_i + \sum_{j=1}^{K} b_j n_j \tag{3.8}$$

In this case, different types of the same atom will provide different contributions to Eq. (3.7), while the correction factors can account for additional interactions not captured by the main term. Such approaches were used to develop models for more complex properties, such as in the ALOGP [8] and XLOGP [9] methods to predict log P.

3.2.3.2 Structural Groups

Instead of subdividing a molecule into atoms, one can subdivide it into a number of structural groups. Thus, each group will accumulate interactions between its atoms and consequently will provide a better representation of the chemical structure. The equation for the prediction of properties based on structural groups is the same as in Eq. (3.7) or (3.8), with the only difference being that the coefficients and the frequencies correspond to those of groups and not to individual atoms.

One of the most extensively developed approach in this field, UNIFAC (UNIQUAC Functional-group Activity Coefficients) [10], is widely used to predict nonelectrolyte activity in nonideal mixtures. This method includes binary interaction parameters between the functional groups, which are usually estimated experimentally or can be predicted.

The development of structural groups is frequently oriented towards a target property. For example, the presence of a particular structural group is used in drug discovery to flag unstable, reactive, or toxic compounds, a comprehensive compilation of which is available online [11]. The flagging is actually a simple model with a binary output yes/no. Structural groups used to predict the toxicity of chemical compounds are frequently named as "toxalerts." A set of toxalerts, called ToxPrint, is publicly available [12]. It can be processed by the freely available software Chemotyper [13, 14]. There are also structural groups developed to flag potential frequent hitters [15, 16]. Such structural groups are developed following a detailed analysis and comparison of the activity of compounds with and without the group.

3.2.3.3 Linear Free Energy Relationships (LFERs)

What is the basis for the use of structural groups for modeling properties? In fact, a group of so-called linear free energy relationship (LFER) methods provides their physical foundation. This approach is based on the pioneering work of Hammett, who introduced this method for the prediction of chemical reactivity. The basic assumption is that the influence of a structural feature on the free energy change of a chemical process is constant for a congeneric series of compounds. A property Φ that is linearly dependent on a free energy change can then be calculated by the property of the basic element of this series, the so-called parent element, and the constant Φ_x for the structural feature X (see Eqs. 3.9–3.12) [17]:

$$\Delta G = -2.3 RT \; \log \Phi \tag{3.9}$$

$$\Delta\Delta G = \Delta G_{R-X} - \Delta G_{R-H} = -2.3 RT \; \log \Phi_{R-X} + 2.3 RT \; \log \Phi_{R-H} \tag{3.10}$$

$$\log \Phi_{R-X} - \log \Phi_{R-H} = -\frac{1}{2.3RT} \Delta\Delta G = k\Delta\Delta G = \varphi_X \tag{3.11}$$

$$\log \Phi_{R-X} = \log \Phi_{R-H} + \varphi_X \tag{3.12}$$

The LFER method can be extended by the further separation of the parent structure into non-overlapping predefined fragments. However, the use of smaller fragments decreases the accuracy of the approach. The promise of strict linearity does not hold true in most cases, so corrections have to be applied in the majority of methods. Correction terms are often related to long-range interactions such as resonance or steric effects.

3.2.3.4 General Fragment Approaches

The separation of a molecule into non-overlapping groups needs careful procedures. Instead, it is also possible to separate molecules into a number of fragments by using a certain rule for their generation. Such an approach, for example, is used in the *in silico* design and data analysis (ISIDA) fragmentor [18], which subdivides molecules into a number of fragments and uses their counts as descriptors. The main difference of these methods to the structural groups is that they generate a much larger number of (overlapping) fragments. The increase in the number of generated fragments and their redundancy due to overlap invalidates the purely additive formalism of LFER approaches to identify independent contributions for each fragment. Therefore, the methods based on such fragmentation schemes frequently use statistical methods to correlate fragments and properties of molecules as described in the section below. The use of LFER-based approaches as well as other additivity schemes provides interpretable models based on the assumption of additivity of properties as a function of the respective contributing groups.

3.2.4 Statistical Quantitative Structure–Property Relationships (QSPRs)

The calculation of a set of descriptors (see *Methods Volume*, Chapter 10) can provide a more universal representation of molecules. The descriptors can be used as input into general-purpose machine learning methods (see *Methods Volume*, Sections 11.1 and 11.2) in the so-called quantitative structure–property relationship (QSPR) studies. Powerful methods, such as neural networks, support vector machines (SVMs), and Gaussian processes, can identify statistical relationships between molecular descriptors and the properties and use them to predict new molecules. Linear regression methods, of course, can also be used. However, in comparison with the additive methods mentioned in the previous section, the resulting equations will be based on statistical rather than on mechanistic assumptions and can be more difficult to interpret.

In order to successfully model one or another property with such methods, it is important to select descriptors that properly describe the property considered. This is one of the most important steps in QSPR, and the development of powerful descriptors is of central interest in this respect. Descriptors can range from simple atom or structural group counts, considered in the previous section, to quantum-chemical descriptors. Which kind of descriptors should or can be used is primarily dependent on the size of the dataset to be studied and the required accuracy. For example, for highly noisy data, one may need to use a more general representation of a chemical structure as compared with cleaner data, where one could better distinguish fine effects due to variations in chemical structures. Chapter 10 in the *Methods Volume* gives a detailed introduction to the calculation methods for molecular descriptors, while Sections 11.1 and 11.2 in the *Methods Volume* provide details on model development and validation. Methods for the prediction of several individual properties are analyzed in the next section.

3.3 Methods for the Prediction of Individual Properties

In this section we will overview several representative models to predict individual properties of compounds. The selection of methods was done to demonstrate diverse approaches rather than to provide a comprehensive overview of existing methods, which can be found in specialized reviews.

3.3.1 Mean Molecular Polarizability

The estimation of mean molecular polarizabilities from atomic refractions has a long history, dating back more than one hundred years [19]. Miller and Savchik were the first to propose a method that considered atom hybridization in which each atom is characterized by its state of atomic hybridization [20]. Kang and Jhon [21] showed that mean molecular polarizabilities, $\bar{\alpha}$, can be estimated from atomic hybrid polarizabilities, α_i, by a simple additivity scheme summing over all N atoms (Eq. 3.13):

$$\bar{\alpha} = \sum_{i=1}^{N} \alpha_i \qquad (3.13)$$

Miller [22] later somehow revised these atomic contributions, α_i, based on new experimental data.

Table 3.1 lists some comparisons between experimental mean molecular polarizabilities and those estimated by Eq. (3.13). In this scheme, the estimation of mean molecular polarizability for acetic acid needs five values, values for sp^3-C, for sp^2-C, for sp^3-O, for sp^2-O, and for a hydrogen atom.

Table 3.1 Experimental mean molecular polarizabilities and values calculated by Eq. (3.13).

Molecule	$\bar{\alpha}$ (Å^3) Exp.	$\bar{\alpha}$ (Å^3) Calc.
H_2	0.79	0.77
CH_4	2.60	2.61
n-C_5H_{12}	9.95	9.95
neo-C_5H_{12}	10.20	9.95
$cyclo$-C_6H_{12}	10.99	11.01
$CH_2=CH_2$	4.26	4.25
C_6H_6	10.39	10.43
CH_3F	2.62	2.52
CF_4	2.92	2.25
CCl_4	10.47	10.32
NH_3	2.26	2.13
Aniline	11.58–12.12	11.91
Acetic acid	5.05–5.15	5.17
Pyridine	9.14–9.47	9.72

3.3.2 Thermodynamic Properties

3.3.2.1 Additivity of Atomic Contributions

Investigations to find additive constituent properties of molecules go back to the 1920s and 1930s with work by Fajans [23–25] and others. In the 1940s and 1950s, the focus had shifted to the estimation of thermodynamic properties of molecules such as heat of formation, $\Delta H_f°$; entropy, $S°$; and heat capacity, $C_p°$.

As Benson [7] pointed out, the additivity of atomic contributions is an insufficient approximation for estimating enthalpy, $\Delta H_f°$, for it would lead to unacceptably large errors. However, the error in the estimation of the molar heat capacity $\Delta C_p°$ and of entropy $\Delta S°$ by Eq. (3.13) is usually quite acceptable.

3.3.2.2 Additivity of Bond Contributions
The next higher order of approximation, the first-order approximation, is obtained by estimating molecular properties by additivity of bond contributions. In the following, we will concentrate on thermochemical properties only. Such a scheme seems to be sufficient for the estimation of $C_p°$ and $S°$, but does not give sufficiently accurate values for the heat of formation, $\Delta H_f°$. Furthermore, this scheme cannot distinguish between the values for isomeric hydrocarbons because these compounds have the same number and types of bonds.

3.3.2.3 Additivity of Group Contributions
The second-order approximation for the estimation of molecular properties consists of summing up contributions of groups. A group consists of a central atom and its directly bonded neighbor atoms. A molecule is fragmented into all such monocentric groups, a value is taken for each of these groups, and these contributions are summed to obtain the molecular property. The values for these group contributions can be obtained from a multilinear regression analysis of the properties of a series of molecules.

Figure 3.1 shows the groups that are obtained for alkanes and the corresponding notation of these groups as introduced by Benson [7].

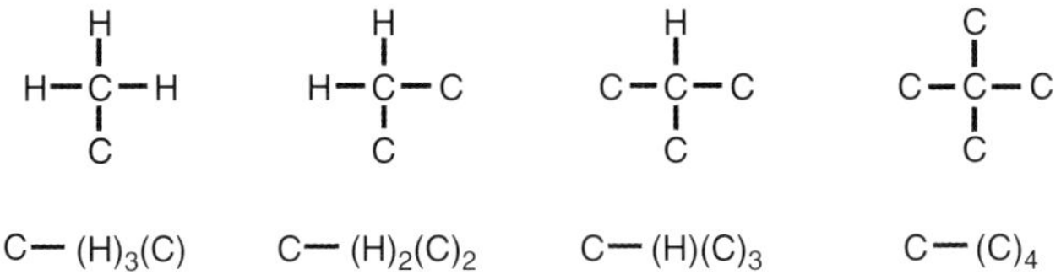

$C-(H)_3(C)$ $C-(H)_2(C)_2$ $C-(H)(C)_3$ $C-(C)_4$

Figure 3.1 Functional groups for alkanes according to Benson's [7] notation.

Table 3.2 contains the group contributions to important thermochemical properties of alkanes. Results obtained with these increments and more extensive tables can be obtained from Refs. [6] and [7].

Table 3.2 Group contributions to $C_p°$, $S°$, and $\Delta H_f°$ for ideal gases at 25 °C, 1 atm. for alkanes.

Group	Contribution to			
	$C_p°$ [a]	$S°$ [a]	$\Delta H_f°$ [b]	ΔH_a [b]
$C-(H)_3(C)$	25.95	127.30	−42.19	1412.31
$C-(H)_2(C)_2$	22.81	39.43	−20.72	1172.08
$C-(H)(C)_3$	18.71	−50.53	−6.20	936.20
$C-(C)_4$	18.21	−146.93	8.16	704.42

a) Values in J/mol K.
b) Values in kJ/mol.

With group additivity, the estimated values for $C_p°$ and $S°$ now agree on the average to within ±1.2 J/mol K, which is well within the experimental error. The agreement for $\Delta H_f°$ is now much improved against the results from bond additivity and is on the average within ±1.7 kJ/mol of the experimental value with only exceptional cases going as high as ±12.0 kJ/mol.

Table 3.2 also contains beyond the values for heats of formation, $\Delta H_f°$, values for the heats of atomization, ΔH_a, as these are easier to interpret. Heats of atomization can be obtained from heats of formation by addition of values for converting elements in their standard state into gaseous atoms. Heats of atomization can be better interpreted because they refer to the enthalpy required to transform a molecule into its constituent atoms in the gas phase.

In order to develop a quantitative interpretation of the effects contributing to heats of atomization, we will introduce other schemes that have been advocated for estimating heats of formation and heats of atomization. In the following we will discuss two schemes and illustrate them with the example of alkanes. Laidler [26] modified a bond additivity scheme by using different bond contributions for C—H bonds, depending on whether hydrogen is bonded to a primary $(E(C—H)_p)$, secondary $(E(C—H)_s)$, or tertiary carbon $(E(C—H)_t)$ atom. Thus, in effect, Laidler also used four different types of structure elements to estimate heats of formation of alkanes, in agreement with the four different groups used by Benson.

Another scheme for estimating thermochemical data, introduced by Allen [27], accumulated the deviations from simple bond additivity in the carbon skeleton. To achieve that, he introduced, over and beyond a contribution from a C—C and a C—H bond, a contribution G(CCC) every time a consecutive arrangement of three carbon atoms was met and a contribution D(CCC) whenever three carbon atoms were bonded to a central carbon atom. Table 3.3 shows the substructures, the symbols, and the contributions to the heats of formation and to the heats of atomization.

Table 3.3 Allen's scheme: Substructures, notations, and contributions to heats of formation and heats of atomization (values in kJ/mol).

	C—C	C—H	C—C—C	C—C(C)$_2$
	B(CC)	B(CH)	G(CCC)	D(CCC)
$\Delta H_f°$	18.80	−17.33	−4.6	0.25
ΔH_a	338.90	414.29	4.6	−0.25

It can be shown that all three schemes, the Benson, the Laidler, and the Allen schemes, are numerically equivalent and thus provide the same accuracy.

Any one of these additivity schemes can be used for the estimation of a variety of thermochemical molecular data, most prominently for heats of formation, with high accuracy [28]. A variety of compilations of thermochemical data is available [29, 30]. A computer program, based on Allen's scheme, has been developed [31].

The use of group contribution methods for the estimation of properties of pure gases and liquids [32, 33] and of phase equilibrium [34] has also a long history in chemical engineering.

3.3.2.4 Effects of Rings

The entire discussion on atom, bond, and group additivity schemes has tacitly assumed that there are no rings in organic structures. Rings can, however, exert drastic deviations from atom, bond, or group additivity for physical or chemical data in general and thermochemical data in particular. Rings can either stabilize or destabilize molecules beyond what is to be expected from a simple additivity scheme. Stabilization comes from aromatic ring systems such as in benzene, or in naphthalene derivatives, or in heterocyclic systems as in furane, pyrrole, thiophene, or pyridine compounds. Destabilization is observed in small rings such as in three- and four-membered ring systems due to bond angle strain. Medium-sized rings such as nine- and ten-membered ring systems are destabilized due to eclipsed conformations. Combinations of rings, particularly if small rings are involved, can introduce additional strain. All these effects can be accounted for by extensions of an additivity scheme, when special increments are attributed to monocyclic structures and the combination of two ring systems having one, two, or three atoms in common [35]. Combination of a table containing values for these ring fragments with an algorithm for the determination of the smallest set of smallest rings (SSSR) [36] (see also *Methods Volume*, Chapter 3) allows such a procedure to be automatically performed.

In recent years the estimation of the heats of formation of the compounds involved in biochemical pathways and other metabolism reactions has become of great interest [37]. Their heats of formation are needed in the quantitative modeling of metabolic networks.

3.3.3 Octanol/Water Partition Coefficient (Log P)

3.3.3.1 Historical Consideration

The lipophilicity of chemical compounds is an important physicochemical property particularly so in drug design. It governs the distribution of a compound between water and organic phases and is frequently measured through its partition between octanol and water for neutral species as in Eq. (3.14):

$$P_{oct} = \frac{[Compound^N]_{oct}}{[Compound^N]_{w}} \tag{3.14}$$

where $Compound^N$ represents its neutral form. Usually not P_{oct} but $\log P_{oct}$ is used. According to Lipinski *et al.* [37], an optimal lipophilicity is required for a good oral bioavailability of drugs. The wide recognition of the octanol/water partition coefficient to be taken as a model property for lipophilicity is due to the work of Corwin Hansch in the 1960s. He selected it considering a number of practical reasons [38]:

- Octanol is a good mimic for the long hydrocarbon chains with a polar headgroup found in membranes.

- Octanol dissolves water, thus emulating the aqueous component of biological hydrophobic regions such as membranes.
- Octanol is cheap, easy to purify, and lacks a chromophore, which would interfere with the spectroscopic determination of compound concentrations.

The success and widespread use of this coefficient in different fields of chemistry indicates the strong vision of Hansch. Here we give an overview of some methods for the calculation of log P. A more comprehensive analysis and coverage of different methods to predict log P can be found in a dedicated review [39].

3.3.3.2 First Approaches to the Prediction of Log P

Fujita *et al.* were the first to develop a calculation method that was based, analogously to the Hammett approach, on substituent constants π_X, as shown in Eq. (3.15):

$$\pi_X = \log P_{R-X} - \log P_{R-H} \tag{3.15}$$

The hydrophobic constant π_X is a measure of the contribution of a substituent X to the lipophilicity of compound R–X compared with that of the compound R–H. This approach is a direct implementation of an LFER where Φ is replaced by P (see Eq. 3.12). The lipophilicity constants π_X allow the estimation of log P values for congeneric series of compounds having various substituents according to Eq. (3.16):

$$\log P_{R-X,Y,Z} = \log P_{R-H} + \sum \pi_{X,Y,Z} \tag{3.16}$$

The drawback of this method is that the parent solute, at least, has to be available or must be synthesized and its log P value has to be determined experimentally. This problem was particularly challenging for complex and diverse drug-like molecules.

Nys and Rekker therefore developed a fragment constant approach, which is based on the additivity of fragment contributions to molecular lipophilicity (Eq. 3.17) [40]:

$$\log P = \sum_{i=1}^{n} a_i f_i + \sum_{i=1}^{n} k_i C_M \tag{3.17}$$

where a_i is the incidence of fragment i, f_i the lipophilic fragment constant, C_M a correction factor, and k_i the frequency of C_M, which is equivalent to Eq. (3.8) described earlier. In order to obtain the fragment constants, sets of molecules for which experimental log P values are available were cut into the predefined fragments, and the numerical contribution of each fragment was determined by multilinear regression analysis. The correction factors were explained by structural features such as resonance interactions, condensation in aromatics, or even hydrogen atoms bound to electronegative groups. Amazingly, the C_M values were all the same and equal to 0.289 in the original method and were refined to 0.219 in the revised version of their system [41]. Therefore, they were also called "magic constants."

The first computational approaches to predict log P were essential to establish this coefficient as an important one for different applications in drug discovery. Moreover, they initiated the development of methodologies, for example, LFER, functional groups, and so on, which are widely used in contemporary chemoinformatics applications.

3.3.3.3 Other Substructure-Based Methods

In recent decades various other substructure-based approaches were developed, which are available as computer programs [42]. Hansch and Leo developed the CLOGP program [39], which is based on fragment constants and correction terms. Fragment constants were derived from solutes where the fragment occurs in isolation. Furthermore, the bonding environment is taken into account. The correction factors were calculated to account for specific interactions, which are not covered by the fragments.

Several models have been published based on atom contributions. One of the most frequently used atomic increment systems, ALOGP, was developed by Ghose and Crippen [43]. Atoms were classified by their neighboring environment, and carbon atoms were differentiated by their hybridization state. Despite its broad application, the method exhibits larger deviations for more complex compounds and shows a bias toward underestimating log P. The method, ALOG98, was therefore revised using a training set of 9920 compounds, yielding a standard deviation of 0.67 log units [44].

XlogP is another popular atom-based approach that was originally using 76 atom types and 4 correction factors [9]. The model calculated a standard deviation of 0.37 for a training set of $N = 1831$ molecules. The latest version of this approach, XlogP3, was redeveloped for $N = 8119$ compounds using 87 atom types and two correction factors [45]. An interesting feature of XlogP3 is that for each query molecule, the method finds a nearest neighbor in the training dataset and uses its log P value as a basis for the prediction of the value of the new compound and corrects this value based on differences in atom-type compositions of both molecules. Thus XlogP3 resembles the method of Fujita with the exception that substituent constants are calculated "on the fly."

Because of their simplicity, the atom-based methods can easily be implemented, and, for example, both the original and the revised version of ALOGP are widely used in various packages [39].

3.3.3.4 QSPR Models

Besides these LFER-type models, approaches have been developed using whole-molecule descriptors and learning algorithms other than multilinear regression. Two methods will be mentioned here as an example.

ALOGPS program was developed using 75 atom and bond-type E-state descriptors [46] for $N = 12{,}908$ molecules [47]. Some of these descriptors were a combination of several atom types, which had similar linear regression coefficients, or were expected to describe similar atom or bond types, but were underrepresented in the training set. A backpropagation neural network model with 10 neurons in the hidden layer was produced, giving significantly better results ($R^2 = 0.95$, RMSE $= 0.42$) compared with a linear regression model ($R^2 = 0.89$, RMSE = 0.61) using the same descriptors. The accuracy of predictions decreased with the size of the molecules; therefore, predictions for large molecules could become unreliable.

At about the same time, Beck *et al.* published a model based on semiempirical quantum chemistry descriptors and a backpropagation neural network [48]. The training dataset consisted of 1085 compounds, and 36 descriptors were derived from AM1 and PM3 calculations describing electronic and spatial effects. Following descriptor selection the best results with a standard deviation of 0.56 were obtained with network architecture of 16–10–1. For a test dataset, a standard deviation of 0.39 was reported, which is similar to the error calculated by the ALOGPS method.

Both models were developed using neural networks having a similar architecture (10 hidden neurons). However, the first model has the advantage that it does not need the 3D structure and therefore does not require a choice on the conformation of molecules and does not need time-consuming quantum-chemical calculations. Moreover, ALOGPS was developed with a much larger set of compounds. A practical suggestion for the development of a log P model would be to try first simpler descriptors (e.g., 2D) and approaches (e.g., linear regression) and increase the complexity of the model only when it is justified by an improved performance of more complex approaches.

3.3.3.5 Factors Limiting the Prediction Power of log *P* Methods

Considering that the log P property has been actively used in R&D for more than half a century, one can assume that the existing approaches can reliably predict it. The benchmarking study by Mannhold *et al.* [39] was done using public and in-house data for models contributed by leading groups and chemical software. The performance of the best methods for a public set of $N = 266$ molecules was similar to the estimated experimental errors of 0.3 log P units [39]. However, the models provided a much lower prediction ability of an RMSE > 1 for 95,000 in-house compounds measured in the Pfizer company.

The models showed a parabolic dependency of the prediction errors on the number of non-hydrogen atoms (NHA) in the molecules, that is, the models failed to accurately predict both small and large molecules (Figure 3.2). The best accuracy was achieved for molecules with about 17 NHA, which was the median value of the number of atoms in the public dataset used by ALOGPS and possibly by many other programs. Thus, the developed approaches provided a rather poor extrapolation performance out of the chemical space used for developing the models.

This conclusion was confirmed in a later study by the same authors [49]. The local correction of predictions with ALOGPS (the so-called LIBRARY mode) decreased the errors from an RMSE = 1.02 to 0.59 log units for the Pfizer in-house set. The LIBRARY mode uses new experimental values to extend the training set. After prediction of the log P value for a new molecule, the LIBRARY mode finds analogs (nearest neighbors) of the molecule in the extended training set and also predicts their log P values. Then, the model calculates the average prediction error for the analogs and uses this error to correct predictions of the new molecule. Importantly, the accounting for the applicability domain allowed identification of the top 60% most accurate predictions, which had errors

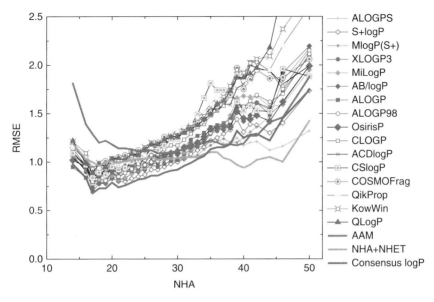

Figure 3.2 RMSE of different models as a function of the number of non-hydrogen atoms (NHA) in molecules. (Reprinted with permission [9].)

similar to the estimated experimental accuracy of 0.33 log units [49]. For these compounds one could just use the predicted values, thus making their measurements redundant.

Low predictions of models could be due to differences in chemical diversity of the molecules in the training and test sets. The extension of the training set with new molecules can dramatically improve the accuracy of models. Thus, a comparison of different computational approaches for the prediction of new data could be biased due to the differences in their training set composition. Last but not least, the use of the applicability domain of models can identify reliable predictions, which do not need to be measured.

3.3.4 Octanol/Water Distribution Coefficient (log D)

The log P value refers to the partition of non-ionized species, which is usually achieved for ionizable compounds by titration to a pH at which point a compound is not ionized. Apparently, for some molecules, such as zwitterions (also known as inner salts), which always have charged groups, such conditions cannot be achieved, and thus log P cannot be determined. For such molecules as well as for (partially) ionized molecules, the log D value, the octanol/water distribution coefficient is measured. Frequently, molecules have several ionized groups and thus can exist in several ionized states. In this case, the log D coefficient can be calculated as a ratio of contributions of fractions of the different ionized forms of the compound in the water and octanol phases (Eq. 3.18) [50]:

$$\log D = \log \left(\sum_{i=1}^{N} f^i D^i_{\text{oct}} \right) \quad (3.18)$$

where f_i is the mole fraction of the ith form and D_{oct} is the distribution coefficient of this form at the given pH value. If the pK_a constants of the groups are known, the ionized fraction of a monoprotic base for the given pH is calculated as given by Eq. (3.19):

$$f = \frac{10^{pK_a-pH}}{(1+10^{pK_a-pH})} \tag{3.19}$$

while for a monoprotic acid, there is (Eq. 3.20)

$$f = \frac{10^{pH-pK_a}}{(1+10^{pH-pK_a})} \tag{3.20}$$

These formulas indicate that for the analytical calculation of log D at each pH, one should know a number of constants, either measured or predicted, which can be difficult to get in practice. Moreover, considering that even the prediction of a neutral form, log P, is a difficult task, the predictions of log D at arbitrary pH could be an even more challenging task.

Therefore, in pharmaceutical companies the log D value is measured at several fixed pH values, for example, at pH 7.4 and pH 6.5, which correspond to the pH of blood and of the rectum. The measurements at a fixed pH are easier to perform compared with the determination of the ionization constants at multiple pH values. Therefore, a large number of experimental measurements, in the order of tens of thousands of compounds, are available for these properties. For example, more than 100,000 data points were measured and modeled in the work of Sheridan, who used a random forest to obtain an RMSE = 0.74 for these data [51]. The authors used simple descriptors such as atom pairs, which were of the form (Eq. 3.21)

$$\text{Atom-type } i - (\text{distance in bonds}) - \text{Atom-type } j \tag{3.21}$$

One type of descriptor included atoms, which were characterized by the element, the number of non-hydrogen neighbors, and the number of π electrons. The second type of descriptors included an atom type in one of seven forms (cation, anion, neutral donor, neutral acceptor, polar, hydrophobic, etc.). They reported that these types of descriptors provided the most robust predictions in their studies. Thus, for very large datasets, a simpler representation of chemical structures could be sufficient.

The log D property strongly correlates with log P. A LIBRARY correction was proposed to convert log P model to log D by exploiting this dependency. The training log P set was substituted by molecules with measured log D values. The other calculations were performed as described in the previous section. The LIBRARY mode provided highly predictive models for the Pfizer dataset [52] (RMSE = 0.69, N = 17,861). Thus this approach allows a fast and accurate adaptation of models to a new related property. The same method could be useful for developing focused models for new classes of compounds.

Log D values at fixed pH values can easily be measured, and thus they can be well modeled due to availability of experimental data. The accuracy of prediction at arbitrary pH can be lower due to an accumulation of errors in the pK_a and log P predictions.

3.3.5 Estimation of Water Solubility (log S)

Solubility in water is an important property contributing to the biological activity of organic molecules [53]. The binding affinity of drug-like compounds increases with their log P, which, on average, decreases their solubility, thus creating significant difficulties for the development of orally available drugs. The prediction of solubility of organic compounds in water is a major challenge in drug development. This topic is treated here in a somehow cursory manner as a more extensive discussion is contained in Section 6.9 (Prediction of ADME Properties).

3.3.5.1 Different Types of Water Solubility

The difficulties with solubility already start with its definition. Depending on how solubility is measured, one can get different values, which are generally incompatible with one another. One of the most scientifically sound definitions of solubility was provided by Comber [54]:

- Kinetic solubility is the concentration of a compound in solution at the time when an induced precipitate first appears.
- Equilibrium solubility (thermodynamic solubility) is the concentration of a compound in a saturated solution when an excess of solid is present, and the solution and the solid are at their equilibrium.
- Intrinsic solubility is the equilibrium solubility of the free acid or base form of an ionizable compound at a pH where it is fully unionized.

The latter definition traces back to Hörter and Dressman [55] who originally stated, "The intrinsic solubility can be defined as the solubility of a compound in its free acid or base form." The first two definitions implicitly depend on the pH used to perform the measurements and thus can be different for ionizable compounds. The third definition fixes a unique value for a compound, and this value is advantageous for computational modeling. If a molecule is neutral, then equilibrium solubility = intrinsic solubility. In the case that a compound does not form an oversaturated solution, kinetic solubility = equilibrium solubility. In general, kinetic ≥ equilibrium ≥ intrinsic solubility. Thus, intrinsic solubility corresponds to the minimum solubility of a compound. Moreover, water solubility is temperature dependent. The models analyzed below correspond to intrinsic solubility measured at room temperature.

3.3.5.2 Factor Contributing to Difficulties with Solubility Modeling

From the standpoint of thermodynamics, the dissolving process is the establishment of equilibrium between the phase of the solute and its saturated aqueous solution. The first factor, which contributes to solubility, is the crystal form of the compound. Different crystal forms result in different solubilities of a compound. A recent review by Pudipeddi and Serajuddin [56] concluded that the ratio of polymorph solubility is typically less than two. However, this ratio can be as high as 10–1600 fold between the amorphous and crystalline forms of drugs [57]. Thus, the presence of even small amounts of an amorphous form in the crystalline form can dramatically affect the solubility of a compound.

Water solubility also strongly depends on the intermolecular forces that exist between the solute molecules and the water molecules. The solute–solute, solute–water, and water–water adhesive interactions determine the amount of compounds dissolving in water. Additional solute–solute interactions are associated with the lattice energy in the crystalline state. Thus, the solubility depends on many factors, which need to be captured by descriptors to correctly predict it.

We will list here the major approaches to solubility prediction and illustrate them with a simple example each. A more detailed presentation of solubility prediction can be found in Section 6.9.3.

3.3.5.3 Yalkowsky's General Solvation Equation (GSE)

There are several approaches to log S prediction, which are based on other properties. The solubility of a compound is strongly correlated with its MP. Already in 1965 Irmann [58] used the MP to correlate water solubility with the structure of hydrocarbons and halohydrocarbons using Eq. (3.22):

$$\log S = c + \sum a_i n_i + \sum f_i n_i + 0.0095(\text{MP} - 25) \tag{3.22}$$

where a_i and f_i are atomic and fragmental contributions and c is a constant dependent on the type of compound.

Yalkowsky related MP and solubility based on three hypothetical steps [59]:

- The analyzed crystal was heated until it melted.
- The melted liquid was cooled to the water temperature.
- The compound was dissolved in water.

This analysis allowed him to propose a general solubility equation (GSE, Eq. 3.23):

$$\log S = 0.5 - \log P - 0.01(\text{MP} - 25) \tag{3.23}$$

where $\log P$ is the octanol/water partition coefficient of the compound and the solubility is measured as the decimal logarithm of molar concentration. For compounds that are liquid at room temperature (MP < 25 °C), this equation simplifies to Eq. (3.24):

$$\log S = 0.5 - \log P \tag{3.24}$$

The application of Yalkowsky's equation based on predicted MP [2] (see also Section 3.3.6) and log P values provides an RMSE of 0.84 log S units for $N = 1311$ molecules [2] of Huuskonen dataset [60]. This error is higher than that of several other methods, which are reviewed below.

3.3.5.4 Additivity Schemes

Despite the complexity of the contributions to solubility, some researchers also modeled this property by using additivity schemes. Kühne et al. developed a solubility model using experimental data on 351 liquids and 343 solids [61]. The number of fragments was 49, but the model also required additional eight correction factors to calculate an $R^2 = 0.95$ and an AAE = 0.38 log units (AAE: average absolute errors).

3.3.5.5 QSPR Models

A number of models were developed to correlate log S with theoretical descriptors.

Huuskonen [60] compiled a set of 1297 compounds, which was used in multiple studies by different authors [2, 62–64]. The dataset was revised by Tetko et al. [63] and Yan and Gasteiger [64] and is available at the OCHEM website [65]. In his original work, Huuskonen [60] used E-state indices and several other topological indices (a total of 30 indices) to develop a model using $N = 884$ molecules as the training set, while the remaining molecules were used as the test set. The predicted results for the 413 molecules in the test set, $SE = 0.71$, as calculated by linear regression were significantly improved with a neural network model, resulting in an $SE = 0.6$. The use of nonlinear approaches to predict log S is logical considering the complexity of the process of dissolving.

The development of models for the prediction of solubility in water remains a challenging task due to the complexity of the property and the problems in its experimental measurements. Nonlinear methods as well as methods based on a correlation with other properties are the most promising approaches in this direction.

3.3.6 Melting Point (MP)

The MP is an important physicochemical property for drug discovery, as already mentioned, as part of the models analyzed in previous sections. Of particular concerns are the MPs of different polymorphs and the influence of traces of amorphous parts. The recent increase in interest in MP prediction is also connected with the development of green chemistry and ionic liquids [66]. This property depends on the crystal packing and the energies that contribute to it. The MP is considered as one of the most difficult properties to predict although it is relatively easy to measure [67].

A lot of work to predict MP was done for homologous series as reviewed by Dearden [68]. Mills already in 1881 [69] derived Eq. (3.25):

$$\mathrm{MP}(^\circ\mathrm{C}) = \frac{\beta x}{1 + \gamma x} \tag{3.25}$$

where β and γ are constants for a given class of hydrocarbon chain compounds and $x = n - c$, with n equal to the number of carbon atoms in the chain and c a constant. These models achieved an excellent accuracy of prediction with less than 1 °C.

Unfortunately, the predictions of the MP for diverse sets of molecules have a lower accuracy. For example, Karthikeyan et al. [70] found that 2D descriptors provided a better prediction accuracy (RMSE = 48–50 °C) compared with models developed using 3D indices (RMSE = 55–56 °C) using a dataset of 4173 compounds. A combination of 2D and 3D descriptors did not improve the models. Another study to predict the MP of ionic liquids of 717 bromides of nitrogen-containing organic cations was performed by Varnek et al. [66]. The authors calculated an RMSE in the ranges of 26 to 49 °C, depending on the used descriptors and the machine learning methods with higher errors corresponding to more diverse sets of molecules.

Both these examples show the difficulty to correctly predict this property and indicate that errors in the range of tens of degree Celsius can be expected for this property.

The largest models to predict this property were based on more than 275,000 data points [2], most of which were collected from patents. The final consensus model had an RMSE = 37 °C. However, the errors were not the same in different temperature regions (Figure 3.3). The analysis of the MP data showed that about 90% of the compounds, which are of relevance for drug discovery (e.g., drugs, compounds from patents, or molecules available from the chemical provider Enamine have MPs within the 50–250 °C interval). Thus, this interval of MP temperatures provides the most interesting range of temperatures for the prediction of the MP of drug-like compounds. Compounds from this temperature region were predicted with an RMSE of 33 °C.

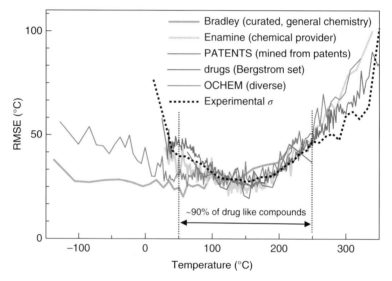

Figure 3.3 The model prediction errors for five sets of compounds are shown as a function of temperature. The model errors for the PATENTS sets match the experimental accuracy of the data in this set. (Adapted from Tetko 2016 [2].)

The 33 °C error may look large compared with the MP model for hydrocarbons that was derived almost 150 years ago and that had an error of less than 1 °C. However, as it was aforementioned, the analyzed compounds can be in different crystalline and/or amorphous stages. The preferred form depends on the condition of crystallization, impurities, and so on, thus contributing to uncertainties in the measurements as well as in the modeling of this property. Indeed, the MP as well as the solubility of molecules refers to the properties of a material rather than the properties of a chemical compound. All these factors contribute to variations in MP. The analysis of duplicated measurements ($N = 18,058$) available in the PATENTS dataset indicated that the prediction accuracy of the models for this set was limited by the experimental accuracy of the data.

This example shows that the difficulty in the prediction of certain properties can be connected with the uncertainty of the experimental data. Moreover, frequently researchers may only be interested in relevant predictions, for example, of "drug-like" molecules considered above. Thus, statistical parameters measured for these subsets are the most relevant features for the estimation of the prediction accuracy of the models.

3.3.7 Acid Ionization Constants

The acid ionization constant is an equilibrium constant, K_a, which is defined as the ratio of the protonated and the deprotonated form of a compound; it is usually indicated as $pK_a = \log_{10} K_a$. The ionization state of the compound dramatically influences its physical, chemical, and biochemical properties [71]. Frequently, compounds have more than one ionization center. Such compounds are called multiprotic. Methods, and particularly, descriptors, which are developed to model properties of whole compounds, are generally not optimal/applicable to predict the pK_a of multiprotic molecules. The successful modeling of ionization constants requires local descriptors, which can properly describe the ionization centers of molecules.

A problem with the pK_a value is that ionization of different centers for multiprotic compounds happens simultaneously, and thus the observed ionizations constants, macroconstants, are formed by a combination of ionization constants for individual groups, that is, microequilibrium constants (microconstants). Figure 3.4 indicates the example of cetirizine for which all 12 microconstants were experimentally measured.

For the purpose of modeling, the microconstants are the most important ones. The titration curves can be used to identify the microconstants experimentally [72]. In cases where no information about microconstants is known (and, unfortunately, this is the case for the majority of data), a determination of the main path of ionization can be used to treat macroconstants as microconstants [73]. For cetirizine the main path corresponds to the four dominant states ●●●, ●●○, ○●○, ○○○ (Figure 3.4c).

For the majority of pK_a estimation methods, the first step is the identification of the ionizable center. For example, 13 centers are identified in the SPARC approach (SPARC Performs Automatic Reasoning in Chemistry), which uses LFERs combined with molecular orbital theory to describe resonance, electrostatic, solvation, and hydrogen bonding effects influencing pK_a [74].

The description of the ionizable center using atoms connected to it and considering concentric levels of bond distances was used in the work of Xing and Glen [75]. The authors counted several Sybyl atom types of different distances from the ionizable center. They also analyzed other types of descriptors, including atomic charges, atomic polarizabilities, bond types, and others. However, the best statistical results were calculated using hierarchical trees of atom types. The partial least squares method applied to derive a model calculated an SE of 0.76 and 0.86 for 645 acids and 384 bases, respectively.

The work of Zhang et al. [76] also used physicochemical attributes of atoms and topological distances (topological spheres) as a basis for descriptor calculations for modeling the pK_a values of aliphatic carboxylic acids (Figure 3.5). However, instead of using just counts of atoms, the authors derived several new descriptors, which accounted for the physicochemical interactions of the reactive center with the rest of the molecule.

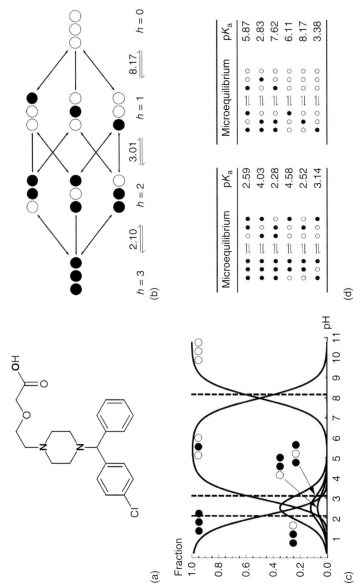

Figure 3.4 Microspecies and constants using the example of cetirizine. The microspecies are represented as triplets, where the first position refers to the hydroxyl group of the carboxylic acid group, the second one refers to the middle nitrogen atom, and the third position refers to the nitrogen atom farthest away from the carboxylic group; for example, ●o● represents the zwitterionic form with one proton bound to the middle nitrogen, the dominant neutral form of cetirizine. (a) Cetirizine. Protonation sites (OH, N, N) in bold face. (b) Protonation scheme. Cetirizine has $n = 3$ protonation sites, and thus $2^3 = 8$ microstates and $3 \times 2^2 = 12$ microequilibria. The $3 + 1 = 4$ macrostates are shown below, with h = number of bound hydrogens. (c) Distribution of microspecies as a function of pH. The microspecies ●oo and oo● are very close to the baseline. (d) Microconstants. All values are experimentally determined. (Reprinted with permission from [71].)

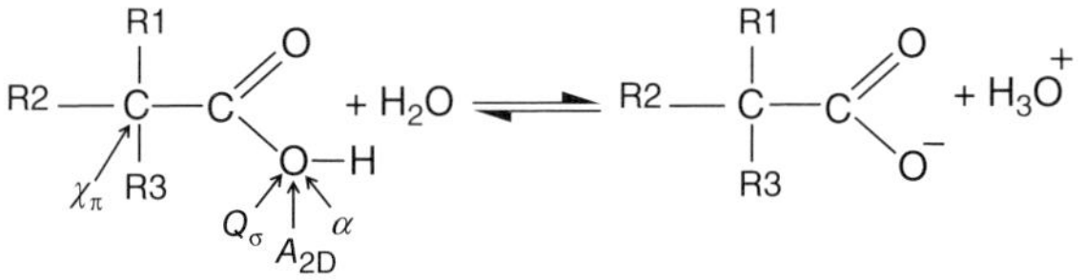

3.3 Methods for the Prediction of Individual Properties

Figure 3.5 Reaction equation for the ionization of aliphatic carboxylic acids with the physicochemical effects is indicated: α_O is effective polarizability, Q_σ is inductive effect on the ionizable atom, A_{2D} is steric hindrance at the ionization site, and χ_π is electronegativity at the π-carbon atom.

The first descriptor accounted for the inductive effect on the ionizable atom (Eq. 3.26):

$$Q_{\sigma,i} = \sum_{d=1}^{7} \sum_{j \in TS_d} \frac{q_{\sigma,j}}{d^2} \quad (3.26)$$

where d is the number of bonds (i.e., spheres) from an atom j (in a topological sphere TS_d) to the central atom i and $q_{\sigma,j}$ is the atomic partial sigma charge on atom j (see *Methods Volume*, Section 8.1) [77]. The weighting by the factor $1/d^2$ was introduced to reflect the decrease of the inductive effect of atoms that are further away from the central atom. The Q_σ descriptor provided a good correlation with Taft σ^* substituent constants ($R^2 = 0.85$, $N = 130$), reflecting its merit for modeling the inductive effect.

Next a descriptor for steric hindrance at the ionization site was defined by Eq. (3.27) to account for differences in the solvation by water molecules:

$$A_{2D,i} = \sum_{d=1}^{5} \sum_{j \in TS_d} V_{rel,j} \frac{1}{N_d} \quad (3.27)$$

In Eq. (3.27) $V_{rel,j}$ is the van der Waals volume for an atom j relative to a carbon atom and N_d is the number of carbon atoms at the topological sphere d (TS_d) in a diamond lattice with $N_d = (4, 12, 32, 88, 240)$. Further effects deemed important for ionization were the effective polarizability, α_O, at the oxygen atom [78] (see *Methods Volume*, Section 8.1) and the π electronegativity, $\chi_{\pi,\alpha-C}$, at the α-carbon atom. Furthermore, as α-amino acids exist as zwitterions, an indicator classifier, I_{amino}, was introduced that had a value of 1 for an amino acid and a value of 0 for all other acids. Equation (3.28) shows the linear model that was thus obtained for the prediction of the pK_a values of 1122 aliphatic carboxylic acids ($R^2 = 0.81$, $s = 0.42$) [76]:

$$pK_a = -37.54 Q_{\sigma,O} + 12.27 A_{2D,O} - 1.02 \alpha_O + 0.11 \chi_{\pi,\alpha-C} - 1.89 I_{amino} + 19.10 \quad (3.28)$$

A linear model using similar descriptors was developed for the pK_a values of 288 aliphatic alcohols ($R^2 = 0.82$ and $s = 0.76$). This is an example where consideration of the chemical nature of ionization allowed the authors to propose new mechanistically interpretable descriptors, which were found as important for the prediction of pK_a values.

The MoKa program extends the idea of circular descriptors to 3D structures by using molecular interaction fields as descriptors [73]. In order to improve the performance, the authors developed 33 pK_a prediction models to cover different ionizable groups. The application of their approach to the in-house library of Roche company ($N = 5581$) calculated an RMSE = 1.09; $R^2 = 0.82$. This result was greatly improved by automated training with an additional set of in-house compounds to RMSE = 0.49, $R^2 = 0.96$ [79].

Thus, the development of descriptors, which capture the basic chemical and physical peculiarities of the analyzed properties, is important to develop highly predictive models. The design of such descriptors was a cornerstone for the development of pK_a models. Moreover, like for previous properties, the pK_a models are not universal, and their training with data, which are similar to the test set compounds, can improve their prediction accuracy.

3.4 Limitations of Statistical Methods

The statistical models derived by learning from data have an applicability domain limited by the structural diversity of the compounds in the training set. Thus, these models cannot reliably predict properties of new compounds, which are very different from those used to develop the model [39]. Frequently, academic users and/or pharma companies are working with compounds, which are rather different from those that are publicly available and were used to develop the existing models. This can happen since the researchers address a new chemical space or/and there is a need to avoid intellectual property rights issues for patenting drugs. In this case, the estimation of the accuracy of prediction and of the applicability domain of models can be very important to avoid spurious predictions. Moreover, as was demonstrated for several examples [39, 52, 79], a development and/or extension of the existing models using several experimental measurements for new chemical series can dramatically improve the accuracy for new chemical series.

3.5 Outlook and Perspectives

Further progress in the development of new powerful physicochemical models can be expected with progress in theoretical approaches, new fast assays for experimental measurements, and the development of better machine learning methods.

The increase in computing power can enable the use of more powerful methods for the prediction of complex properties, for example, the use of *ab initio* methods to model physicochemical properties based on the first principles. These methods can also contribute to the development of new descriptors or improve the calculation of the existing one; for example, calculation of solvent-accessible surface area (SASA) could be based on *ab initio* methods, thus improving the accuracy of models.

The development of new high-throughput screening assays to experimentally measure properties of compounds will be another important contribution to the further development of new methods by providing more and better data. This can be accompanied by the development of better methods for the automatic extraction of information from previously published results in patents and articles as was already done for studies on MPs [2]. These methods contribute to the appearance of "big data," the potential of which still needs to be properly exploited [80].

Another progress can come from the development of new fast machine learning methods. Methods such as deep neural networks appear to be well suited to model complex properties and are expected to significantly contribute to the development of drug discovery [81]. These methods can also be used to develop multi-learning approaches, in which several properties are simultaneously predicted, thus contributing to a better prediction ability for each of the individual properties, as well as making it possible to fully utilize the potentials of large heterogeneous datasets [80].

Essentials

- There is a range of methods available, from rather simple approaches to theoretical calculations, to estimate important physical and chemical properties of compounds.
- Nonlinear methods can provide higher accuracy compared with linear approaches.
- Additivity schemes for estimating molecular properties play an important role in chemical engineering.
- The accuracy of an additivity scheme can be increased by going from atomic contributions through bond contributions to group contributions.
- Heats of formation can be estimated with reasonable accuracy by additivity of group increments and corrections for ring effects.
- A wide range of methods has been developed for estimating octanol/water partition coefficients (log P) and octanol/water distribution coefficients (log D).
- Various approaches to the estimation of water solubility (log S) of organic compounds have been developed.
- The prediction of the melting point of organic compounds has been investigated.
- Different methods for the prediction of acid ionization constants (pK_a values) have been developed.

Available Software and Web Services (accessed January 2018)

- http://www.vcclab.org/lab/alogps – ALOGPS 2.1 program to predict log P and solubility of chemical compounds.
- http://ochem.eu/model/select.do – models to predict solubility in water and in DMSO, partition coefficients, decomposition and ready degradability, melting and boiling points as well as various molecular descriptors.
- http://www.mn-am.com/products – software for prediction of 3D structure of molecules, calculation of descriptors and physicochemical properties.

- http://chm.kode-solutions.net/products_dragon.php – Dragon software for the calculation of chemical descriptors.
- http://toxprint.org
- http://chemotyper.org
- http://www.mn-am.com
- http://www.cosmologic.de

Selected Reading

- Avdeef, A. (2012) *Absorption and Drug Development: Solubility, Permeability, and Charge State*, 2nd edn, John Wiley & Sons, Inc., Hoboken, NJ, 698 pp.
- Shields, G.C. and Seybold, P.G. (2017) *Computational Approaches for the Prediction of pKa Values*, Boca Raton, FL, CRC Press, 175 pp.
- Testa, S., Kramer, D., Wunderli-Allenspach, H., and Folkers, G. (2006) *Pharmacokinetic Profiling in Drug Research: Biological, Physicochemical, and Computational Strategies*, Verlag Helvetica Chimica Acta/Wiley-VCH, Zürich/Weinheim, 500 pp.

References

[1] 10.1787/20745753 (accessed January 2018)
[2] Tetko, I.V., Lowe, D., and Williams, A.J. (2016) *J. Cheminform.*, **8**, 2.
[3] Klamt, A. (1995) *J. Phys. Chem.*, **99**, 2224–2235.
[4] Wittekindt, C. and Klamt, A. (2009) *QSAR Comb. Sci.*, **28**, 874–877.
[5] Klamt, A., Eckert, F., Hornig, M., Beck, M.E., and Burger, T. (2002) *J. Comput. Chem.*, **23**, 275–281.
[6] Benson, S.W. (1976) *Thermochemical Kinetics: Methods for the Estimation of Thermochemical Data and Rate Parameters*, John Wiley & Sons, Inc., New York, 334pp.
[7] Benson, S.W. and Buss, J.H. (1958) *J. Chem. Phys.*, **29**, 546–572.
[8] Ghose, A.K., Viswanadhan, V.N., and Wendoloski, J.J. (1998) *J. Phys. Chem. A*, **102**, 3762–3772.
[9] Wang, R.X., Fu, Y., and Lai, L.H. (1997) *J. Chem. Inf. Comput. Sci.*, **37**, 615–621.
[10] Fredenslund, A., Jones, R.L., and Prausnitz, J.M. (1975) *AIChE J.*, **21**, 1086–1099.
[11] Sushko, I., Salmina, E., Potemkin, V.A., Poda, G., and Tetko, I.V. (2012) *J. Chem. Inf. Model.*, **52**, 2310–2316.
[12] http://toxprint.org (accessed January 2018)
[13] Yang, C., Tarkhov, A., Marusczyk, J., Bienfait, B., Gasteiger, J., Kleinöder, T., Magdziarz, T., Sacher, O., Schwab, C.H., Schwoebel, J., Terfloth, L., Arvidson, K., Richard, A., Worth, A., and Rathman, J. (2015) *J. Chem. Inf. Model.*, **55**, 510–528.
[14] http://chemotyper.org (accessed January 2018)

- [15] Baell, J.B. and Holloway, G.A. (2010) *J. Med. Chem.*, **53**, 2719–2740.
- [16] Schorpp, K., Rothenaigner, I., Salmina, E., Reinshagen, J., Low, T., Brenke, J.K., Gopalakrishnan, J., Tetko, I.V., Gul, S., and Hadian, K. (2014) *J. Biomol. Screen.*, **19**, 715–726.
- [17] Chapman, N.B. and Shorter, J. (1972) *Advances in Linear Free Energy Relationships*, Springer US, Boston, MA, 486pp.
- [18] Varnek, A., Fourches, D., Horvath, D., Klimchuk, O., Gaudin, C., Vayer, P., Solov'ev, V., Hoonakker, F., Tetko, I.V., and Marcou, G. (2008) *Curr. Comput. Aided Drug Des.*, **4**, 191–198.
- [19] Eisenlohr, F. (1911) *Z. Phys. Chem. (Leipzig)*, **75**, 585–607.
- [20] Miller, K.J. and Savchik, J. (1979) *J. Am. Chem. Soc.*, **101**, 7206–7213.
- [21] Kang, Y.K. and Jhon, M.S. (1982) *Theor. Chim. Acta*, **61**, 41–48.
- [22] Miller, K.J. (1990) *J. Am. Chem. Soc.*, **112**, 8533–8542.
- [23] Fajans, K. (1920) *Ber. Deut. Chem. Ges.*, **53B**, 643–665.
- [24] Fajans, K. (1922) *Ber. Deut. Chem. Ges.*, **55B**, 2826–2838.
- [25] Fajans, K. (1921) *Z. Phys. Chem.*, **99**, 395–415.
- [26] Laidler, K.J. (1956) *Can. J. Chem.*, **34**, 626–648.
- [27] Allen, T.L. (1959) *J. Chem. Phys.*, **31**, 1039–1049.
- [28] Gasteiger, J., Jacob, P., and Strauss, U. (1979) *Tetrahedron*, **35**, 139–146.
- [29] Pedley, J.B., Naylor, R.D., Kirby, S.P., and Pedley, J.B. (1986) *Thermochemical Data of Organic Compounds*, Chapman and Hall, London, 792pp.
- [30] Cox, J.D. and Pilcher, G. (1970) *Thermochemistry of Organic and Organometallic Compounds*, Academic Press, London, 643pp.
- [31] Gasteiger, J. (1979) *Tetrahedron*, **35**, 1419–1426.
- [32] Cordes, W. and Rarey, J. (2002) *Fluid Phase Equilib.*, **201**, 409–433.
- [33] Poling, B.E., Prausnitz, J.M., and O'Connell, J.P. (2001) *The Properties of Gases and Liquids*, McGraw-Hill, New York, 768pp.
- [34] Wittig, R., Lohmann, J., and Gmehling, J. (2003) *Ind. Eng. Chem. Res.*, **42**, 183–188.
- [35] Gasteiger, J. and Dammer, O. (1978) *Tetrahedron*, **34**, 2939–2945.
- [36] Gasteiger, J. and Jochum, C. (1979) *J. Chem. Inf. Comput. Sci.*, **19**, 43–48.
- [37] Lipinski, C.A., Lombardo, F., Dominy, B.W., and Feeney, P.J. (1997) *Adv. Drug Deliv. Rev.*, **23**, 3–25.
- [38] Tetko, I.V. and Livingstone, D.J. (2006) Rule-based systems to predict lipophilicity, in *Comprehensive Medicinal Chemistry II: In Silico Tools in ADMET*, (eds B. Testa and H. van de Waterbeemd), Elsevier, Oxford. Vol. 5, pp 649–668.
- [39] Mannhold, R., Poda, G.I., Ostermann, C., and Tetko, I.V. (2009) *J. Pharm. Sci.*, **98**, 861–893.
- [40] Nys, G.G. and Rekker, R.F. (1973) *Chim. Therap.*, **8**, 521–535.
- [41] Rekker, R.F. and Mannhold, R. (1992) *Calculation of Drug Lipophilicity. The Hydrophobic Fragmental Constant Approach*, VCH, Weinheim, 113pp.
- [42] Leo, A.J. (1993) *Chem. Rev.*, **93**, 1281–1306.
- [43] Ghose, A.K. and Crippen, G.M. (1986) *J. Comput. Chem.*, **7**, 565–677.
- [44] Wildman, S.A. and Crippen, G.M. (1999) *J. Chem. Inf. Comput. Sci.*, **39**, 868–873.

[45] Cheng, T., Zhao, Y., Li, X., Lin, F., Xu, Y., Zhang, X., Li, Y., Wang, R., and Lai, L. (2007) *J. Chem. Inf. Model.*, **47**, 2140–2148.
[46] Hall, L.H. and Kier, L.B. (1995) *J. Chem. Inf. Comput. Sci.*, **35**, 1039–1045.
[47] Tetko, I.V., Tanchuk, V.Y., and Villa, A.E. (2001) *J. Chem. Inf. Comput. Sci.*, **41**, 1407–1421.
[48] Beck, B., Breindl, A., and Clark, T. (2000) *J. Chem. Inf. Comput. Sci.*, **40**, 1046–1051.
[49] Tetko, I.V., Poda, G.I., Ostermann, C., and Mannhold, R. (2009) *Chem. Biodivers.*, **6**, 1837–1844.
[50] Lombardo, F., Faller, B., Shalaeva, M., Tetko, I., and Tilton, S. (2007) The good, the bad and the ugly of distribution coefficients: current status, views and outlook, in *Drug Properties: Measurement and Computation* (ed. R. Mannhold), Wiley-VCH, Weinheim, pp 407–437.
[51] Sheridan, R.P. (2012) *J. Chem. Inf. Model.*, **52**, 814–823.
[52] Tetko, I.V. and Poda, G.I. (2004) *J. Med. Chem.*, **47**, 5601–5604.
[53] Balakin, K.V., Savchuk, N.P., and Tetko, I.V. (2006) *Curr. Med. Chem.*, **13**, 223–241.
[54] Comer, J. (2005) The Relationships Between Lipophilicity, Solubility and pKa for Ionizable Molecules. PhysChem forum for Physical Chemists by Physical Chemists, UK.
[55] Horter, D. and Dressman, J.B. (2001) *Adv. Drug Deliv. Rev.*, **46**, 75–87.
[56] Pudipeddi, M. and Serajuddin, A.T. (2005) *J. Pharm. Sci.*, **94**, 929–939.
[57] Hancock, B.C. and Parks, M. (2000) *Pharm. Res.*, **17**, 397–404.
[58] Irmann, F. (1965) *Chem. Ing. Tech.*, **37**, 789–798.
[59] Jain, N. and Yalkowsky, S.H. (2001) *J. Pharm. Sci.*, **90**, 234–252.
[60] Huuskonen, J. (2000) *J. Chem. Inf. Comput. Sci.*, **40**, 773–777.
[61] Kühne, R., Ebert, R.U., Kleint, F., Schmidt, G., and Schuurmann, G. (1995) *Chemosphere*, **30**, 2061–2077.
[62] Hou, T.J., Xia, K., Zhang, W., and Xu, X.J. (2004) *J. Chem. Inf. Comput. Sci.*, **44**, 266–275.
[63] Tetko, I.V., Tanchuk, V.Y., Kasheva, T.N., and Villa, A.E.P. (2001) *J. Chem. Inf. Comput. Sci.*, **41**, 1488–1493.
[64] Yan, A. and Gasteiger, J. (2003) *J. Chem. Inf. Comput. Sci.*, **43**, 429–434.
[65] Sushko, I., Novotarskyi, S., Korner, R., Pandey, A.K., Rupp, M., Teetz, W., Brandmaier, S., Abdelaziz, A., Prokopenko, V.V., Tanchuk, V.Y., Todeschini, R., Varnek, A., Marcou, G., Ertl, P., Potemkin, V., Grishina, M., Gasteiger, J., Schwab, C., Baskin, I.I., Palyulin, V.A., Radchenko, E.V., Welsh, W.J., Kholodovych, V., Chekmarev, D., Cherkasov, A., Aires-de-Sousa, J., Zhang, Q.Y., Bender, A., Nigsch, F., Patiny, L., Williams, A., Tkachenko, V., and Tetko, I.V. (2011) *J. Comput. Aided Mol. Des.*, **25**, 533–554.
[66] Varnek, A., Kireeva, N., Tetko, I.V., Baskin, I.I., and Solov'ev, V.P. (2007) *J. Chem. Inf. Model.*, **47**, 1111–1122.
[67] Gavezzotti, A. (1994) *Acc. Chem. Res.*, **27**, 309–314.
[68] Dearden, J.C. (2003) *Environ. Toxicol. Chem.*, **22**, 1696–1709.
[69] Mills, E.J. (1884) *Philos. Mag.*, **17**, 173–187.
[70] Karthikeyan, M., Glen, R.C., and Bender, A. (2005) *J. Chem. Inf. Model.*, **45**, 581–590.

[71] Rupp, M., Korner, R., and Tetko, I.V. (2011) *Comb. Chem. High Throughput Screen*, **14**, 307–327.
[72] Marosi, A., Kovacs, Z., Beni, S., Kokosi, J., and Noszal, B. (2009) *Eur. J. Pharm. Sci.*, **37**, 321–328.
[73] Milletti, F., Storchi, L., Sforna, G., and Cruciani, G. (2007) *J. Chem. Inf. Model.*, **47**, 2172–2181.
[74] Hilal, S.H., Karickhoff, S.W., and Carreira, L.A. (1995) *Quant. Struct. Act. Relat.*, **14**, 348–355.
[75] Xing, L. and Glen, R.C. (2002) *J. Chem. Inf. Comput. Sci.*, **42**, 796–805.
[76] Zhang, J., Kleinöder, T., and Gasteiger, J. (2006) *J. Chem. Inf. Model.*, **46**, 2256–2266.
[77] Gasteiger, J. and Marsili, M. (1980) *Tetrahedron*, **36**, 3219–3228.
[78] Gasteiger, J. and Hutchings, M.G. (1984) *J. Chem. Soc. Perkin Trans.*, **2**, 559–564.
[79] Milletti, F., Storchi, L., Goracci, L., Bendels, S., Wagner, B., Kansy, M., and Cruciani, G. (2010) *Eur. J. Med. Chem.*, **45**, 4270–4279.
[80] Tetko, I.V., Engkvist, O., Koch, U., Reymond, J.L., and Chen, H. (2016) *Mol. Inform.*, **35**, 615–621.
[81] Baskin, I.I., Winkler, D., and Tetko, I.V. (2016) *Expert Opin. Drug Discov.*, **11**, 785–795.

4 Chemical Reactions

Learning Objectives

- To describe why synthesis is hard.
- To analyze individual reactions and reaction mechanisms.
- To be aware of the data available.
- To figure out issues with data reliability and accessibility.
- To comprehend synthesis design.
- To become familiar with synthetic processes.
- To design synthetic pathways.
- To assess the quality of a synthetic pathway.
- To become familiar with the state of the art of synthesis design systems.
- To obtain an overview of Biochemical Pathways, the central or endogenous metabolism.
- To utilize the information on the Biochemical Pathways Wall Charts by modern chemoinformatics search methods.
- To build a database of biochemical pathways by representing molecules as connection tables and by providing information on the reaction site, the bonds broken and made in the reaction.
- To express essential details of biochemical reactions.
- To evaluate criteria from the classification of enzyme catalyzed reactions.
- To define the EC nomenclature.
- To describe how chemoinformatics and bioinformatics methods can be linked together.
- To utilize the information in Biochemical Pathways for bioengineering applications or chemical synthetic biology.

Outline

4.1 Chemical Reactions–An Introduction, 84
4.2 Reaction Prediction and Synthesis Design, 86
4.2.1 Introduction, 86
4.2.2 Reaction Prediction, 87
4.2.3 Synthesis Design, 94
4.2.4 Conclusion, 102
4.3 Explorations into Biochemical Pathways, 106
4.3.1 Introduction, 106
4.3.2 The BioPath.Database, 110
4.3.3 BioPath.Explore, 111
4.3.4 Search Results, 112
4.3.5 Exploitation of the Information in BioPath.Database, 117
4.3.6 Summary, 129

Applied Chemoinformatics: Achievements and Future Opportunities, First Edition.
Edited by Thomas Engel and Johann Gasteiger.
© 2018 Wiley-VCH Verlag GmbH & Co. KGaA. Published 2018 by Wiley-VCH Verlag GmbH & Co. KGaA.

4.1 Chemical Reactions – An Introduction
Johann Gasteiger

Computer-Chemie-Centrum, Universität Erlangen-Nürnberg, Nägelsbachstr. 25, 91052 Erlangen, Germany

The computer representation and modeling of chemical reactions is by far more difficult than that of chemical structures. Chemical reactions involve the dynamic changes of chemical structures that are influenced by a variety of physicochemical and environmental factors. On the other hand, each day chemists are running chemical reactions to produce new compounds or to make known compounds in a more efficient manner. Thus, any chemoinformatics methods that can support chemists in these endeavors are very much wanted.

As shown in Section 4.2, chemists face three fundamental problems before embarking on running reactions: reaction planning, reaction prediction, and synthesis design. Indeed chemoinformatics has already developed methods for assisting chemists in these three problem areas. Reaction databases can be queried to answer problems of reaction planning and reaction prediction as detailed in the *Methods Volume* (Chapter 6). Quantum chemistry provides the basic theory for predicting the outcome of chemical reactions. However, the calculations are rather time consuming and require careful studies of a variety of transition states, and some environmental influences such as solvents, temperature, or concentrations are still very difficult to model. In such situations, methods of inductive learning as provided by chemoinformatics that can process the huge amount of known reactions are highly desirable.

In Section 4.2 Jonathan Goodman outlines the state of the art of reaction prediction and synthesis design. To set the scene, he emphasizes all the questions and problems synthetic chemists are facing in planning their reactions and syntheses. He briefly mentions some of the more important computer systems that have been developed to tackle these tasks. Details have to be taken from the original publications of the various systems. Furthermore, no mention of the various attempts of knowledge extraction from reaction databases could be made.

Another overview of reaction prediction and synthesis design systems was published by Wendy Warr [1]. Philip Judson detailed the history of early computer-assisted synthesis design systems and some of the reasons for their failure [2]. However, renewed interest in the development of synthesis design systems can be recognized, and some of the systems have matured to a point of being used in industry.

Biochemical reactions are arguably the most important set of reactions because they keep living species alive. Section 4.3 gives an introduction into biochemical pathways and shows how a database of these reactions can be exploited for enhancing our understanding of these reactions. It is also shown how chemoinformatics and bioinformatics can work together to shed light onto the pathways involved in certain diseases and on cheese flavor-forming pathways.

References

[1] Warr, W. (2014) *Mol. Inf.*, **33**, 469–476.
[2] Judson, P. (2009) *Knowledge-Based Expert Systems in Chemistry: Not Counting on Computers*, Royal Society of Chemistry, Cambridge.

4.2 Reaction Prediction and Synthesis Design
Jonathan M. Goodman

Department of Chemistry, University of Cambridge, Lensfield Road, Cambridge, CB2 1EW, UK

4.2.1 Introduction

A central question for chemistry is: What happens when molecules are mixed together? If we can understand individual reactions, we can begin to plan synthetic pathways. Inventing and making new molecules is central to chemistry.

Before operating a chemical reaction, chemists are faced with three fundamental questions:

1. How can I transform a given starting material, A, into a desired product, P?
   This is a question of *reaction planning* (Figure 4.1). It will best be answered by consulting a reaction database.
2. What will be the outcome if I mix two starting materials, A and B?
   This is a question of *reaction prediction*. In order to answer such a question, we need to know the driving forces of chemical reactions.
3. How can I make a desired target compound, P, from available starting materials, A_1, A_2, etc.? This is the task of *synthesis design*.

Designing synthetic pathways requires a knowledge and understanding of a large number of chemical reactions and of the properties of molecules. Some people are very good at this, and there are many extraordinary examples of the synthesis of complex molecules [3–5]. Computers can remember more reports of experiments than people can, are able to search them effectively and quickly, and can calculate quite precisely the relative rates of competing reactions. It seems obvious, therefore, that computers should be better at synthesis design than people. This is not, currently, the case. Computational tools are playing an increasingly prominent role in assisting people to design syntheses, but they show no sign – and even no possibility – of replacing people. It is now unthinkable to design a synthesis without using search engines and databases to help to ensure that relevant literature has been found. The challenge, both for people and computers, is to find the data, to assess its reliability, to work out how it fits into current understanding of synthetic chemistry, and to consider how it might be used to predict the results of new reactions.

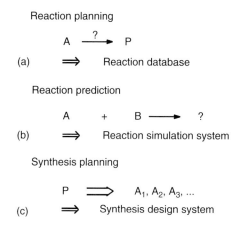

Figure 4.1 Different types of problems encountered in dealing with chemical reactions.

Organic chemists have made huge progress in synthesizing molecules in terms of target complexity, high yields, less and easy separation of by-products, and use of less material and energy.

Maitotoxin is not only an extraordinarily complicated molecule but also a realistic target for total synthesis [6]. It has never been made, but the work that has been done toward making it by synthesizing a precursor (Figure 4.2) is a convincing demonstration that a total synthesis should be achievable. The synthetic design and the execution of many parts of the synthesis were a human endeavor, supported by computerized tools but not controlled nor driven by them.

Eribulin/Halaven (Figure 4.3) has a much simpler structure than maitotoxin but still represents a formidable synthetic challenge [7]. This molecule is manufactured by total synthesis and is sold as a treatment for breast cancer. Structures of this level of complexity could, in principle, be made from a vast number of synthetic routes. Asking an organic chemist or a computer to design a new synthesis of Halaven would be asking for a major task in itself, before any experimentation began.

This chapter investigates the challenges and the opportunities in all of these areas. Why is reaction prediction and synthesis design so hard? The challenge is so high that it is extraordinary that so many exquisite structures have been created, isolated, and characterized.

4.2.2 Reaction Prediction

4.2.2.1 What Is a Molecule?

A chemical reaction can be defined as a chemical change that transforms one or more molecules into one or more other substances. We can only understand reactions if we have a good description of molecular structure.

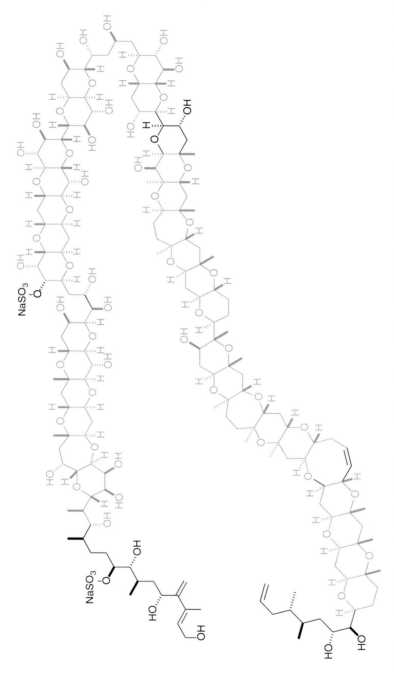

Figure 4.2 Intermediate in the synthesis of maitotoxin, one of the most complex targets for total synthesis. The light grey parts of the molecule show the regions that have been synthesized [6]. Much of the hard work has been completed. However, joining these fragments together will require substantial work, as will completing the remaining fragments of the molecule.

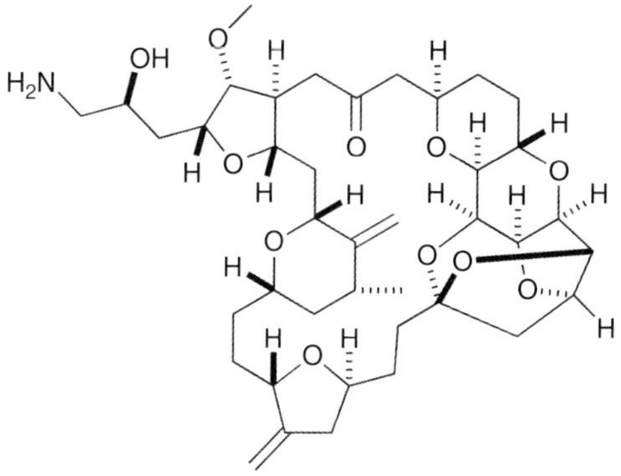

Figure 4.3 Eribulin/Halaven. The most complex molecule synthesized and sold.

There are an enormous number of molecules to consider. Guida et al. estimate 10^{60}, or so, containing 30 atoms or fewer but explain that the estimate is a crude one and that the real number is likely to be much larger, even for molecules with such a low molecular weight [8]. Even this low estimate is dramatically larger than the number of molecules that have ever been characterized, which is of the order of a hundred million (10^8). Thus, we have studied less than one part in 10^{50} of the possible molecules.

Distinct potential molecular structures are not only highly numerous but also highly diverse.

Chemists usually communicate molecular structures with structural diagrams that are unambiguous when drawn carefully. These are hard for computers to interpret, but their computer representation has been solved with connection tables being the standard form (see *Methods Volume*, Chapter 2).

For many applications it would be advantageous to have a string of text that contains all the information required to define a molecule exactly, and no more. Ideally, this identifier would be exactly the same if generated by different people. The InChI (IUPAC-NIST Chemical Identifier) provides an effective way to do this [9, 10].

The first stage of reaction prediction, describing molecules, therefore, is solved for a very large number of important processes. We can describe molecules in a way that is easily used and communicated around the world.

4.2.2.2 What Is a Good Reaction?

A reaction is a process that transforms one or more molecules into another or others. The number of such processes is uncountable.

A good reaction, in the sense of one that is useful for synthesis, has a much more restrictive definition. A reaction must produce a good yield of the desired product and do this without also producing by-products that are difficult to separate, hazardous to handle, or form an expensive waste stream. It must be controllable, so that it begins when required and does not produce excessive heat. It must work fast enough to be useful and must be manageable on an appropriate scale. Most of all, the reaction should be predictable. By thinking about the reaction scheme, it should be possible to determine if it will work for a particular set of starting materials and reagents.

In practice, it is very hard to find good reactions. The number of reactions that are useful in synthesis is tiny, compared with the number of possible reactions. Reaction development has to balance the different desirable outcomes, weighing selectivity against yield and choosing which of the by-products is least undesirable. A product that is hard to separate or hard to distinguish from the desired product may be far more troublesome than accepting a low yield for the reaction.

There is a group of reactions that usually work very well: click chemistry [11]. These are defined by Sharpless as reactions that are modular, wide in scope, and high yielding, generate only inoffensive by-products that can be easily removed, and are stereospecific. In addition, such reactions should require simple reaction conditions, readily available starting materials and reagents, and benign solvent or no solvent and permit the simple isolation of products.

The invention of new reactions is an exciting field, and many new processes are being reported. Most are being discovered by experimentation [12], but there are now a few reports of studies using computational methods to generate new reactivity [13].

Such specific reactions are only one part of the molecular behavior that needs to be understood to predict reactions and to design syntheses. Click chemistry is easy to predict, because it nearly always does just one, very specific, transformation. At the other extreme, adding a simple reagent, for example, an acid, might promote a wide range of different reactions, and a detailed knowledge of the structure and the properties of the substrates will be required in order to predict such reactions.

Reactions where a simple reagent leads to a complex transformation are common. The example of a process to form dolabriferol (Figure 4.4) can reasonably be regarded as a single reaction, although it might also be analyzed as a series of reactions for which none of the intermediates are isolated [14].

If a reaction outcome is to be predicted with confidence, then it must either belong to the very small group of reactions that are very predictable or be so well understood that their behavior can be predicted in novel surroundings. Reaction prediction is critical to the design of syntheses. Reaction prediction must be good if it is to be useful. In order to understand reactions this well, we need to understand their reaction mechanisms in enough detail to know how they work, how they will be affected by their surroundings, and how they might go wrong.

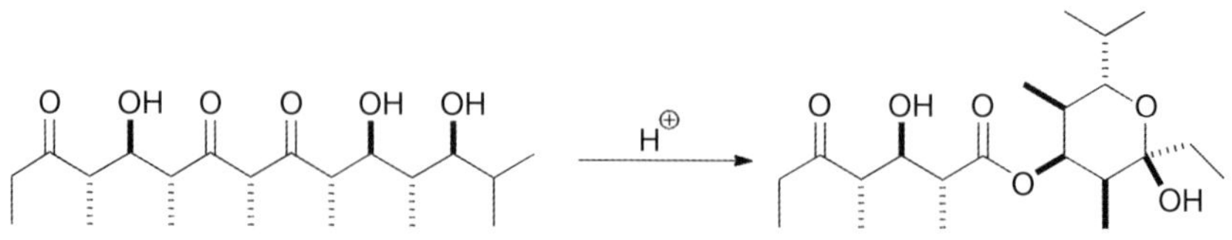

Figure 4.4 Acid-catalyzed rearrangement to dolabriferol.

4.2.2.3 What Is a Mechanism?

For many reactions, it is possible to suggest detailed atomic-scale changes that are likely to correspond to the processes happening in the molecules themselves. Such an analysis is extremely useful, as it can hint how a reaction might change if new factors are introduced, such as other functional groups in the same molecule.

The mechanism might suggest likely by-products and so give a hint of the possible yield. A quantitative measure of the free energy change for a process may also give an estimate of likely yield although the gap between what is being calculated and the property of interest is rather large, as yield is rarely determined completely by a position of equilibrium [15].

4.2.2.4 Can We Trust the Literature?

4.2.2.4.1 Finding Data

In order to predict whether a reaction is likely to occur, and whether it will work well, the obvious approach is to search through records of reactions people have done before. This is not straightforward, because so much chemistry has been recorded and the records are not always consistent.

It is straightforward to search for words. It is more useful, however, to search for chemical structures, and this is possible because we can describe most structures precisely and uniquely (Section 4.2.2.1 and *Methods Volume*, Chapter 2). In addition to searching for specific structures, it is possible to search for all molecules that contain a particular fragment of a molecule and also for structures that are "similar" to a structure of interest. The precise definition of "similar" is a complicated and somewhat non-intuitive issue, about which there is no consensus on the precise meaning. However, there are popular ways of quantifying similarity that provide computationally tractable algorithms for searching for "similar" molecules (see *Methods Volume*, Chapter 7). Despite these concerns, searching for structures is a well-developed field.

Searching for reactions is very much harder. Reaction databases (see *Methods Volume*, Chapter 6) can record reactions, and it is possible to search for a particular starting material that is linked to a particular product. There may be uncertainties because it is not always clear exactly where the atoms of the starting material end up in the product, and this problem becomes much worse if substructures or similar systems are used for searching. The question "Are there any reactions in which a molecule containing a specific fragment is transformed into a molecule containing a different specific fragment?" is well defined, but there is no guarantee that the fragment in the product actually originates from the atoms in the corresponding fragment in the starting material. Such confidence requires knowledge of how the starting material atoms map into the product atoms, and this can be very hard to obtain. It is possible to generate sensible mappings automatically, but they are not guaranteed to be correct in all cases. In any case, reaction search must be done through a search on the reaction center, the atoms, and bonds involved in a reaction transformation.

One way to approach this problem is to use the InChI standard as a basis for describing reactions. The *Reaction InChI* (*RInChI*) is being developed as an open descriptor of reaction processes and may become a useful way of labeling reactions [16].

4.2.2.4.2 Using Data

Even if we can find something relevant in a database, is it true? Information will usually have been peer reviewed if it is in the academic literature, but peer reviewers in synthetic chemistry check for reasonable data and consistency, rather than repeating the experiments. Errors will be present. Reactions from patents are reviewed less carefully than those in the academic literature. Important information for a reaction to proceed might be missing.

Even if all of the reactions in the databases were recorded with absolute precision, there would still be a problem. Similar molecules do not always react in similar ways. There are databases of tens of millions of reactions. The number of molecules of interest is many orders of magnitude greater than this. As a result, it is unlikely that any search for a reaction will find a precise match for a reaction of interest, unless the specific molecules used in the reaction are already well known. A similar reaction may well be the best possibility, and it is not clear how similar is similar enough.

The details of reactions that have been recorded in the literature are critical to synthesis design, but are not, by themselves, enough to design effective new syntheses.

4.2.2.5 Computational and Experimental Analyses

4.2.2.5.1 Simulating Reactions

Since literature data give such an incomplete survey of chemical reactions, we need to have an alternative approach to predicting reactivity. Fortunately, such methods are available, and reaction mechanisms may be calculated by solving Schrödinger's equation for the molecular system. However, the calculations are so computationally demanding and the number of interesting systems so large, that a brute-force approach will never be successful, even if Moore's law speed increases continue for decades.

This can be addressed either by using less accurate calculations, which give approximate answers more quickly, or by selecting the systems for calculation with great care. In practice, a combination of the two simplifications is often most effective.

Calculations of reaction mechanisms usually focus on the key bond-forming and bond-breaking events, because these are the processes that lead to new molecules. However, the details of these changes depend on the interactions of other parts of the system. Nonbonded interactions, where a part of one molecule pushes against a part of another, can have a product-determining effect and are very hard to calculate precisely. The effects of solvents are also hard to quantify.

Computational simulations of reaction pathways are a key part of modern chemistry, and they give insights into chemical processes that could be obtained no other way. They do not, however, provide a complete and tractable solution to the problems of reaction prediction and synthesis design.

4.2.2.6 Challenges in Reaction Prediction

4.2.2.6.1 Chemoselectivity and Regioselectivity

If we consider only small molecules with just one reactive part on each, working out how they will interact is relatively straightforward. However, each starting material could, in principle, react in a number of different ways to form a variety of products. Ensuring that the desired reaction happens and competing reactions do not is a key part of reaction prediction and essential for synthesis planning.

If there are two different reactive groups in a molecule with different chemical properties, reacting one and not the other (chemoselectivity) may be possible, either by reacting the more active group first or by using its reactivity to protect it so that the less reactive group can do the desired reaction. Further, the reactivity of groups of atoms depends on their surroundings in a complex way. While the relative reactivity of two groups may be known when they are compared in isolation, the environment of the same groups in a new molecule may perturb the relative reactivity sufficiently to reduce or even reverse the selectivity.

If the same reactive group occurs twice in one molecule, it may be possible to get one of them to react in the presence of the other (regioselectivity), provided the environment of each is sufficiently different to differentiate their reactivity. The confident prediction of regioselectivity for complex novel molecules requires considerable experience and is not always possible.

In many cases, it may be possible to calculate the sense and degree of selectivity [17], even in very complex cases [18]. However, extensive experimental studies demonstrate that calculations still have some way to go before they can be trusted, unquestioningly, for all reactions [19]. Surprises occur, even in the best planned synthetic routes [20].

4.2.2.6.2 Stereoselectivity and Enantioselectivity

Achieving stereoselectivity has been a central challenge for synthetic chemistry over the last decades, and many excellent methods have been developed to ensure that the right molecule is made. However, the difference between the desired and the undesired product is small, and the molecular environment will have a big effect on the process.

4.2.2.6.3 Physical Properties

Even if it is possible to predict the reactivity and the selectivity of a new reaction, it may still not be a good synthetic procedure. All of the relevant molecules must be sufficiently soluble for them to interact in the reaction. The products must be separable from both the starting materials and the by-products of the reaction, requiring them to have sufficiently distinct physical properties that they can be purified using crystallization, distillation, or chromatography.

4.2.2.6.4 Yield

The yield of a reaction is a key property and one of the hardest to predict. The literature usually reports the purified yield, and the measurement includes the conversion of the starting materials to the product, an assessment of how hard it is to separate the products from the reaction mixture, information about the difficulty of handling the product, and some data about the level of purity that is acceptable to the experimenter. All of this will differ for reactions of similar compounds, and all parts of this are somewhat unpredictable. However, a reaction is only useful if it produces a reasonable yield, and so this is a key part of reaction prediction.

4.2.3 Synthesis Design

Given the extreme difficulty of predicting reactions, designing syntheses may seem to be a hopeless task. Most syntheses, in practice, begin with a promising route, which needs to be refined and redesigned as new information becomes available. Is there any hope that we could design a synthesis from scratch or write a computer program to do this for us? This challenge has been taken up by chemists and computer scientists in the late 1960s and early 1970s. These endeavors on developing computer-assisted synthesis design (CASD) systems can be considered as one of the beginnings of chemoinformatics. While humans are still better than computers and the challenges in synthesis are immense, useful computer approaches to synthesis design are available now.

4.2.3.1 Approaches and Problems

4.2.3.1.1 How Hard Is Synthesis?

Organic synthesis is sometimes compared to a game of chess. Both represent difficult challenges, but only chess has so far succumbed to computational analysis, with the best chess computers able to beat the best human players. The number of positions in chess is smaller than the number of molecules that satisfy Lipinski's rule. Each square on a chessboard can be empty or be occupied by 1 of 12 different types of pieces. This gives an upper bound of $13^{64} = 10^{71}$ on the number of possible positions. This number is far too big, since it includes positions with 64 kings and many other ridiculous possibilities. More sophisticated analyses suggest that 10^{50} is a more reasonable estimate. This is much smaller than the number of possible small molecules (perhaps 10^{60} *vide supra*) and much, much smaller than the number of interesting molecules.

Each synthetic transformation is like a move in chess, except that in chess we always know exactly what every move will do, and there are a finite, and fairly small, number of possibilities at each step. Every synthetic transformation is much more exciting. Will it work? Slightly different reaction conditions might change the outcome of a reaction.

If it was a reaction on a similar substrate, the uncertainty is much greater than if it were the same substrate. An apparently inconsequential difference may turn out to be critical. There may be a competing reaction, or a change in selectivity, or a change in physical properties.

Unlike chess, we do not know the exact starting point, our data is unreliable, our certainty that an attempt to make a particular move will lead to the expected result is low, the number of possible moves at each stage is huge, and our definition of success is complex and sometimes hard to define. Synthesis is, and will remain, a very hard problem.

4.2.3.1.2 How Do People Design Syntheses?

Retrosynthetic analysis, thinking about the synthesis backward from the product to the starting material, makes it possible to discover ways of simplifying structures and moving them toward available starting materials [21]. This approach also makes it possible to experiment with different approaches to the target molecule. Searching through the literature, now using computerized tools, is a key part of this process.

Papers that describe syntheses will usually report the syntheses of challenging structures, for which the success of the project was not assured, syntheses illustrating the use of a new reaction, or else processes for which there is a specific interesting step around which the synthesis is focused. The processes described in papers of this type advance the science of synthesis, but are not necessarily ideal for the production of general methods of constructing new molecules.

In general, an approach to a target molecule must take account of all the constraints that apply to the particular project:

- What is the balance between cost, scalability, speed, and purity?
- Do issues of intellectual property and patents need to be taken into account?
- How will the starting materials be sourced? A longer synthesis may be better if it starts from more reliable and accessible reagents.
- How will the hazards of the processes be managed?

A synthesis that is good for one group of constraints may be poor for another. A chemist working in discovery might want a small quantity of material as fast as possible. The small scale might substantially reduce the hazard, and it is acceptable to use expensive reagents and purification processes. A process chemist, working on a larger scale, with less flexible purification processes and much greater concern about sustainability, waste, and overall cost, is very unlikely to favor the same or even a similar route.

An excellent synthetic design may become less effective, because of a change in the availability of starting materials.

In choosing between competing synthetic routes, analysis of the workup may well influence the choice of the best pathway. Synthesis plans should take account of this nontrivial issue [22].

4.2.3.1.3 Finding Pathways

Beginning with the product, finding a pathway to starting materials, the problem can be addressed retrosynthetically, working out what substrates could undergo reactions to be turned into the product and then working backward toward available starting materials. This requires a way of assessing whether the precursor is really closer to available starting materials than the product, but this can probably be assessed reasonably reliable by considering what complexity is added to the system by the transformation.

Each retrosynthetic step opens up a huge number of possibilities, none of which can be transformed into the desired product with absolute certainty. Every step toward the starting materials opens up vast new swathes of chemical space. This combinatorial explosion makes exhaustive searching of the possible pathways impossible.

Fortunately, however, exhaustive searching through all possible pathways is not needed. Any reasonable route would be useful, even if not optimal. A variety of sensible suggestions would inspire synthetic chemists and illustrate the diversity of possible approaches. Finding a reasonable pathway is a much easier problem than finding an optimal, or even an excellent, pathway. It is still a major challenge.

4.2.3.1.4 Checking Pathways

Even the best-designed synthetic pathway needs to be tested. Reaction prediction is not so good that the outcome of a new reaction can be forecast with absolute certainty. A well-designed synthetic pathway is usually one that has possibilities for redesign with the new information that becomes available as the procedure is followed.

The world's best synthetic chemists are unlikely to be able to write down a synthesis of a complex molecule that proceeds exactly as expected, at every step, with high yields and good selectivities.

4.2.3.1.5 What Is a Good Synthesis?

The ideal synthesis may be defined in different ways, but a small number of steps are usually desirable. Hudlicky's analysis of syntheses of morphine and related compounds outlines more than 30 syntheses developed over more than half a century [23]. A five- or six-step synthesis, which Hudlicky describes as being "almost ideal," has yet to be developed. Continuing work in the area has led to short and innovative syntheses [24], but the perfect synthesis remains elusive.

A short synthesis is often a good synthesis, but the assessment of synthetic quality is not so simply defined. The environmental impact and the toxic effects of the overall process (life cycle assessment) both play a role in determining the excellence, or disadvantages, of particular pathways. Metrics [25] and algorithms [26] are becoming available to assess these issues.

4.2.3.2 State of the Art

4.2.3.2.1 How Can This Be Mechanized?

Is there any hope for computerization? The problem of synthetic design is challenging for people and requires good communication skills as well as a thorough understanding of chemistry. A variety of programs are available to plan syntheses. None of them has a dominant position, and they are used only by a minority of synthetic chemists, but they all have the potential to be helpful. CASD systems are tools that help chemists plan their syntheses in a more efficient way. They are not designed to replace chemists. Computers are good in tirelessly exploring a variety of reactions and pathway, while chemists are better at creatively combining disparate information and making imaginative leaps. The computer and the chemist, therefore, make a good team.

4.2.3.2.2 Literature Analysis: Consulting Reaction Databases

The analysis of the literature is the key starting point both for reaction prediction and for synthesis design. The key tools for this are currently the reaction databases Chemical Abstracts Service's SciFinder, and Elsevier's Reaxys (see *Methods Volume*, Chapter 6). Both of these give access to very large databases of reaction data, the former based on Chemical Abstracts and the latter built from the Beilstein Handbook. Both contain data on tens of millions of molecules and reactions, which represent a substantial proportion of all the reactions that have ever been carried out and published in the scientific literature. Translating from a publication to a database is a slightly subjective process, and so the two databases sometimes interpret results in a different way. These resources are essential for the efficient design of syntheses, and both contain programs that suggest retrosynthetic routes based on literature precedent. These are valuable tools to help synthetic chemists design and check reactions and syntheses, but their principal function is to support and not to replace the human chemist.

4.2.3.2.3 Reaction Prediction Systems

Programs have been developed that predict the outcome of reactions. A special issue of *Accounts of Chemical Research* recently focused on computational catalysis using quantum mechanical methods, and the editorial outlined the many successes in this area [27]. Studies of this type usually require a large amount of computer time to analyze a small group of related reactions and can generate both quantitative data and qualitative understanding. These approaches cannot, in general, be given a random set of reagents and conditions and make predictions with modest computer power. The programs listed later can do this. A lower level of precision in the predictions balances their speed and generality.

CAMEO CAMEO, written by Jorgensen group, predicts the products of reactions from the starting materials and the conditions using a mechanistic analysis [28, 29]. Reactions are divided into broad classes that have general applications to a wide range of substrates. This has the result that the program can, in principle, correctly predict the outcomes of reactions that have never been tried in a laboratory.

ROBIA The ROBIA program predicts organic reactivity for a limited range of processes, including aldols and retroaldols and acetal and hemiacetal formation [14]. The program's strength is its ability to compare large numbers of similar pathways and calculate how stereochemistry can affect the outcome. Its application lead to an *in silico* inspired synthesis of a natural product [30].

SOPHIA The SOPHIA (System for Organic reaction Prediction by Heuristic Approach system) [31] was developed by Funatsu *et al.* It is not constrained by accepted definitions and conventions of reaction groups, but tries to predict all possible reaction pathways from arbitrary reactants.

IGOR The IGOR program, developed by Ugi *et al.* [32], operates with formal reaction generators and so is not restricted to known chemical processes. This approach has been used to generate new chemical reactions as well as to analyze and predict known chemistry [33].

EROS The system EROS (Elaboration of Reactions for Organic Synthesis) was designed to perform both types of strategies, a forward and a retrosynthetic search, that is, to allow reaction prediction and synthesis design [34, 35]. It worked with formal reaction generators, breaking bonds, making bonds, and shifting electrons. The formal reaction generators allowed the simulation of both known and completely novel reactions. The selection of feasible reactions from among all the conceivable ones was largely based on physicochemical evaluations such as calculating heats of reactions [36], partial atomic charges [37], and values for the inductive [38], resonance, and polarizability effects [39]. The values calculated for these physicochemical effects were used with different emphasis depending on whether a problem of reaction prediction or of synthesis design was studied.

4.2.3.2.4 Synthesis Design

Many programs are available that create retrosynthetic pathways, and some of the most popular are outlined in the following text. There have been recent reviews of the field by Warr [1] and by Cook et al. [40]. The programs usually depend on rules that describe how molecules can react and differ in the methods by which these rules are developed and in the processes used to find good overall strategies. This section does not intend to give a comprehensive overview of all programs developed for synthesis design. Rather, the major milestones in the development of CASD system are presented. Details on the various systems can be obtained from the references given.

Synthesis design programs can be ordered chronologically, beginning with Corey's development of retrosynthesis and its encoding in a computer program, OCSS [41], which developed into LHASA [42]. The chronological approach then becomes rather complicated, as many programs were developed, some over several decades.

Ordering programs by strategy can also be helpful. Should the analysis go forward from the starting materials or backward from the product? Should transformations be limited to those with strong literature precedent of reliability, or might more speculative processes be considered? How should such transformations be selected and encoded? Should the program produce an answer unaided, or is interaction with a synthetic chemist desirable? The retrosynthetic approach LHASA was not designed to invent new chemistry, but to base its analysis on a database of known reactions selected by experts [42]. Any reaction, however well established, might create new chemistry when applied to a new substrate. The strategy of assuming that well-established reactions will do the expected thing is a simplification, one that makes it possible to generate useful and interesting synthetic strategies with a tractable amount of computer power.

Is it important for the user to know, in detail, how the program works? A complete account of the algorithm and the data used to generate synthetic routes would, in principle, enable chemists to assess the strengths and limitations of the results. The data and algorithms are likely to be very complicated, however, so this would not be easy in practice. Commercial synthesis design programs usually keep the details of their algorithms and databases secret, and so the precise efficacy of different approaches to synthesis design cannot be easily assessed.

LHASA The first such program was developed at Harvard [41, 42]. Corey's seminal paper lays down many of the ideas that subsequent programs developed and explored. A key feature of the program was the use of interactive computer graphics. Decades on, this now seems routine, but much of the original paper was focused on describing this key feature. The program has a series of manually derived heuristics for planning synthetic strategy and rules for synthetic transformations that can be used to implement the strategies. Finally, and most difficult, the evaluation of the strategies was considered. In the original program, OCSS, human intervention was crucial at this stage.

All of the ideas that were used in this earliest program have been tested and developed by subsequent programs. Computers have become more powerful, computer graphics have become ubiquitous, and human–computer interaction is no longer an activity worthy of comment. All of these changes have increased the power of machine-assisted synthetic analysis.

SynGen Hendrickson developed SynGen at Brandeis University, starting almost 50 years ago [43, 44]. Recognizing the intractably large number of possible synthetic routes that could lead to a target molecule, this program focuses on the skeletal disconnections that chop the target molecule up into available starting materials as rapidly as possible. It provided a practical and effective way of finding key strategic approaches to synthetic transformations, demonstrating that this was possible even with the available computer power of decades ago. The forward synthetic route could then be investigated with the aid of literature databases, looking for precedents in transformations similar to the ones suggested by SynGen.

WODCA Based on the experiences gained by the work on EROS, the WODCA (Workbench for the Organization of Data for Chemical Applications) system [45, 46] was developed by the Gasteiger group at Erlangen-Nuremberg, representing the second generation of synthesis planning programs, adding global strategy based on similarity analysis to the first-generation tools of synthesis planning and reaction prediction. WODCA offers a number of tools that aid the synthetic chemist, including a connection to large databases of starting materials and algorithms for the identification of strategic bonds in the synthesis (see Figure 4.5). This represented a major step forward in computer-aided organic synthesis and laid the groundwork for ideas that could become ever more powerful with larger databases of chemistry and faster computers. Like LHASA, the program encouraged interaction between a synthetic chemist and the computer tools, and this interface could now be much more sophisticated than before. Part of the design strategy was to separate the heuristics from the data used, so these two key processes could be developed independently.

ARChem Route Designer ARChem Route Designer is a tool to help chemists develop synthetic routes for target molecules [47]. The organic chemistry system is derived from databases including Wiley's ChemInform Reaction Library, and machine learning algorithms are used to develop the rules used to suggest syntheses.

ICSynth Like ARChem, ICSynth is based on the idea of rules for transformations [48] also building on the foundation laid by LHASA. The data used comes

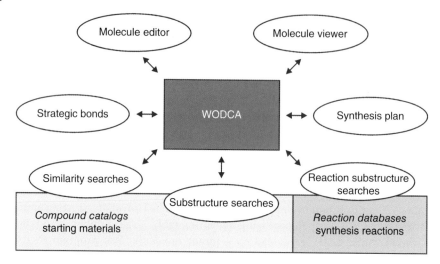

Figure 4.5 An overview of the methods in WODCA.

from InfoChem, and the program is used to help generate ideas for syntheses. Information technology does not replace the human synthetic chemists, but it can help their work by generating ideas that may not have occurred without the use of this tool. Like ARChem, ICSynth is a useful tool for synthetic chemists, but not one that is likely to replace them any time in the foreseeable future.

CHIRON The CHIRON computer program, developed by Hanessian et al., searches for common subunits to help design syntheses, with a particular emphasis on stereogenic centers [49]. While it does not construct a full retrosynthetic scheme, it suggests chiral pool starting materials that will be useful for the synthesis of enantiomerically pure compounds.

SynChem Like other programs, SynChem, which was developed by Gelernter et al. [50], uses a synthetic chemistry knowledge base of reactions that are known to work well and that are assumed to be predictable. In the 1990 paper, this database contained about a thousand reaction schema. The program is entirely self-guided, unlike the normal operation ARChem and ICSynth. This should be an ultimate aim of computer-aided synthesis design, but it means that the computer must do all of the hard work, including the final stage identified by Corey: evaluation of competing synthesis pathways. Which is the best route?

Chematica The most recent synthesis planning program, Grzybowski's *Chematica* [51], builds on existing programs and a very large database of chemical transformations. The program has access to dramatically more computer power and data than was possible for LHASA, decades ago. It is obvious that synthetic chemistry is very complicated. How complicated, exactly? Is it possible to fit all the knowledge necessary for complex syntheses within currently accessible computer power? Is it possible to search and apply such knowledge to generate new synthetic routes for molecules that are difficult for human synthetic chemists to do themselves? The *Chematica* program claims to have reached this stage of the development of computer-aided synthesis design.

4.2.3.2.5 Synthetic Accessibility

Predicting reaction outcomes and designing good syntheses are both very demanding questions. There is a much easier question, which is also very useful: How difficult will it be to synthesize a particular molecule? One way to approach this question would be to generate a synthesis and assess the challenges it presents. However, an estimate of synthetic accessibility, without the generation of a full synthesis, may be much easier to obtain and be useful in itself. If a large number of molecules are being considered as potential synthesis targets, then it would be helpful to screen these so that the most synthetically tractable are presented before the structures, which are likely to be more challenging to construct. The idea that calculating synthetic accessibility might be useful is a key insight in this area. Unlike synthesis design, a program that is good but imperfect could have a major impact on the field, because it would allow the synthetic chemist to prioritize their time toward the compounds that are most likely to be valuable. This is of particular interest in ranking large datasets of molecules obtained by virtual screening or *de novo* design.

One approach to this problem is to analyze the complexity of the target molecule and to use this as a measure of the synthesizability using knowledge of molecular features that are challenging to synthesize. Gasteiger *et al.* have taken measures of structural complexity and similarity to available starting materials and an assessment of strategic bonds for decomposing target compounds to define values for synthetic accessibility [52]. The approach led to calculated values for synthetic accessibility that agree as well with the values suggested by synthetic chemists as the synthetic chemists agreed with each other. The corresponding program SYLVIA is available free of charge for evaluation [53]. Ertl has reported a related approach based on both analysis of the fragments that comprise a molecule and an overall complexity score [54]. This program ranked compounds with a score from one (easy to make) to ten (hard to make), and these scores correlated well with manually estimated values for synthetic accessibility. Synthesizability has also been predicted based on comparison of the compound with databases of commercially available compounds [55]. Li and Eastgate pointed out that the assessment of synthetic accessibility is dependent both on molecular complexity and on the current synthesis technology and therefore defined a current complexity that improves as the methodology of chemical reactions progresses [56, 57].

An alternative approach to the problem is to construct molecules using a set of reactions, so that the molecules generated are all expected to be synthesizable. SYNOPSIS is a program that does this and has been used to generate synthesizable HIV inhibitors [58]. More recently, an evolutionary algorithm has been used to generate synthetically accessible analogs for a large set of small molecule drugs [59]. While the synthetic pathways were constrained, the pathways were rated well by synthetic chemists.

4.2.3.3 Future Directions

4.2.3.3.1 Open Access and Open Data

We need to get access to more records of synthetic experiments. These data exist, and much of it belongs to publishing companies. The technology to extract useful synthetic information from these data is well established. It seems not to be in the commercial interests of the publishers to provide licenses to use these data that are both flexible and affordable so that organic synthesis can develop as rapidly as possible.

4.2.3.3.2 Better Analytical Data

Analytical data on synthetic intermediates is usually only available in a summarized form in papers. Original data used to be discarded but is increasingly stored in electronic laboratory notebooks. These data could also make a huge difference to synthetic design. The original data may contain details of the structures, and hints of by-products, which are not mentioned in the conventional publication. Further, experiments that do not work are often omitted from papers, and yet this may be an important result. Most reactions work well for a series of substrates and may not work at all for others. Publications should include information about the groups that do not work. Inevitably, authors worry that others might get the reactions to work for these groups, but it is far better to stimulate research that leads to an effective transformation than to discount discouraging results that may provide new insights into reactivity.

4.2.3.3.3 Synthesis Machines

Ultimately, automated synthesis design should be used to program synthesis machines. Machines that make molecules exist now [60, 61], but do not yet have the generality and affordability required to make their use routine. These would then make the molecules while constantly generating analytical data that is more precisely defined and more reproducible than the data we are currently gathering. Once this happens, the virtuous circle between design and testing can become wholly automatic, and synthesis design programs should develop much more quickly and in synchronization with the machines that are actually creating molecules.

4.2.4 Conclusion

Maitotoxin may be unrealistically complicated example for the current generation of synthesis design programs. Computers are better at designing routes to simpler targets, but so are people. There is no immediate prospect – and even no intention – of replacing all skilled synthetic chemists with computers and robots.

The automated design of general, effective synthetic pathways is currently impossible. However, computational tools are becoming ever more effective at supporting synthetic chemists by providing methods to retrieve data, by analyzing mechanistic details of past and future reactions, and by suggesting possible pathways. The future of synthesis design will continue to merge human skills with computational skills, leading to ever improving synthetic methods.

References

[1] Warr, W. (2014) *Mol. Inf.*, **33**, 469–476.
[2] Judson, P. (2009) *Knowledge-Based Expert Systems in Chemistry: Not Counting on Computers*, Royal Society of Chemistry, Cambridge, 222pp.
[3] Nicolaou, K.C. and Sorensen, E.J. (1996) *Classics in Total Synthesis: Targets, Strategies, Methods*, John Wiley & Sons, Inc., New York, NY, 821pp, ISBN: 978-3-527-29231-8.
[4] Nicolaou, K.C. and Snyder, S.A. (2003) *Classics in Total Synthesis II: More Targets, Strategies, Methods*, John Wiley & Sons, Inc., New York, NY, 658pp, ISBN: 978-3-527-30684-8.
[5] Nicolaou, K.C. and Chen, J.S. (2011) *Classics in Total Synthesis III: Further Targets, Strategies, Methods*, John Wiley & Sons, Inc., New York, NY, 770pp, ISBN: 978-3-527-32957-1.
[6] Nicolaou, K.C., Heretsch, P., Nakamura, T., Rudo, A., Murata, M., and Konoki, K. (2014) *J. Am. Chem. Soc.*, **136**, 16444–16451.
[7] Towle, M.J., Salvato, K.A., Budrow, J., Wels, B.F., Kuznetsov, G., Aalfs, K.K., Welsh, S., Zheng, W., Seletsky, B.M., Palme, M.H., Habgood, G.J., Singer, L.A., DiPietro, L.V., Wang, Y., Chen, J.J., Quincy, D.A., Davis, A., Yoshimatsu, K., Kishi, Y., Yu, M.J., and Littlefield, B.A. (2001) *Cancer Res.*, **61**, 1013–1021.
[8] Bohacek, R.S., McMartin, C., and Guida, W.C. (1996) *Med. Res. Rev.*, **16**, 3–50.
[9] Heller, S.R., McNaught, A., Pletnev, I., Stein, S., and Tchekhovskoi, D. (2015) *J. Cheminf.*, **7**, 23.
[10] The website of the InChI trust: http://www.inchi-trust.org/ (accessed January 2018).
[11] Kolb, H.C., Finn, M.G., and Sharpless, K.B. (2001) *Angew. Chem. Int. Ed.*, **40**, 2004–2021.
[12] Massa, A. (2012) *Synlett*, **23**, 524–530.
[13] Rappoport, D., Galvin, C.J., Zubarev, D.Y., and Aspuru-Guzik, A. (2014) *J. Chem. Theory Comput.*, **10**, 897–907.
[14] Socorro, I.M. and Goodman, J.M. (2006) *J. Chem. Inf. Model.*, **46**, 606–614.
[15] Emami, F.S., Vahid, A., Wylie, E.K., Szymkuc, S., Dittwald, P., Molga, K., and Grzybowski, B.A. (2015) *Angew. Chem. Int. Ed.*, **54**, 10797–10801.
[16] Grethe, G., Goodman, J.M., and Allen, C.H.G. (2013) *J. Cheminf.*, **5**, 45.
[17] Kruszyk, M., Jessing, M., Kristensen, J.L., and Jørgensen, M. (2016) *J. Org. Chem.*, **81**, 5128–5134.
[18] Tantillo, D.J. (2016) *Org. Lett.*, **18**, 4482–4484.
[19] Mayr, H. and Ofial, A.R. (2015) *SAR QSAR Environ. Res.*, **26**, 619–646.
[20] Horn, E.J., Silverston, J.S., and Vanderwal, C.D. (2016) *J. Org. Chem.*, **81**, 1819–1838.

[21] Corey, E.J. and Cheng, X.-M. (1995) *The Logic of Chemical Synthesis*, John Wiley & Sons, Inc., New York, NY.
[22] Hill, G.B. and Sweeney, J.B. (2015) *J. Chem. Educ.*, **92**, 488–496.
[23] Reed, J.W. and Hudlicky, T. (2015) *Acc. Chem. Res.*, **48**, 674–687.
[24] Chu, S., Munster, N., Balan, T., and Smith, M.D. (2016) *Angew. Chem. Int. Ed.*, **55**, 14306–14309.
[25] Andraos, J. (2015) *J. Chem. Educ.*, **92**, 1820–1830.
[26] Eckelman, M.J. (2016) *Green Chem.*, **18**, 3257–3264.
[27] Tantillo, D.J. (2016) *Acc. Chem. Res.*, **49**, 1079.
[28] Salatin, T.D. and Jorgensen, W.L. (1980) *J. Org. Chem.*, **45**, 2043–2051.
[29] Jorgensen, W.L., Laird, E.R., Gushart, A.J., Fleischer, J.M., Gothe, S.A., Helson, H.E., Paderes, G.D., and Sinclair, S. (1990) *Pure Appl. Chem.*, **62**, 1921–1932.
[30] Currie, R.H. and Goodman, J.M. (2012) *Angew. Chem. Int. Ed.*, **51**, 4695–4697.
[31] Satoh, H. and Funatsu, K. (1996) *J. Chem. Inf. Comput. Sci.*, **36**, 173–184.
[32] Ugi, I., Bauer, J., Bley, K., Dengler, A., Dietz, A., Fontain, E., Gruber, B., Herges, R., Knauer, M., Reitsam, K., and Stein, N. (1993) *Angew. Chem. Int. Ed. Engl.*, **32**, 201–227.
[33] Bauer, J., Herges, R., Fontain, E., and Ugi, I. (1985) *Chimia*, **39**, 43–53.
[34] Gasteiger, J. and Jochum, C. (1978) *Top. Curr. Chem.*, **74**, 93–126.
[35] Gasteiger, J., Hutchings, M.G., Christoph, B., Gann, L., Hiller, C., Löw, P., Marsili, M., Saller, H., and Yuki, K. (1987) *Top. Curr. Chem.*, **137**, 19–73.
[36] Gasteiger, J. (1979) *Tetrahedron*, **35**, 1419–1426.
[37] Gasteiger, J. and Marsili, M. (1980) *Tetrahedron*, **36**, 3219–3228.
[38] Hutchings, M.G. and Gasteiger, J. (1983) *Tetrahedron Lett.*, **24**, 2541–2544.
[39] Gasteiger, J. and Hutchings, M.G. (1984) *J. Chem. Soc. Perkin*, **2**, 559–564.
[40] Cook, A., Johnson, A.P., Law, J., Mirzazadeh, M., Ravitz, O., and Simon, A. (2012) *WIREs Comput. Mol. Sci.*, **2**, 79–107.
[41] Corey, E.J. and Wipke, W.T. (1969) *Science*, **166**, 178–192.
[42] Corey, E.J., Long, A.K., and Rubenstein, S.D. (1985) *Science*, **228**, 408–418.
[43] Hendrickson, J.B. (1971) *J. Am. Chem. Soc.*, **93**, 6847–6854.
[44] Hendrickson, J.B., Grier, D.L., and Toczko, A.G. (1985) *J. Am. Chem. Soc.*, **107**, 5228–5238.
[45] Ihlenfeldt, W.-D. and Gasteiger, J. (1995) *Angew. Chem. Int. Ed. Engl.*, **34**, 2613–2633.
[46] Sitzmann, M. and Pförtner, M. (2003) Computer-assisted synthesis design, in *Chemoinformatics – A Textbook* (eds J. Gasteiger and T. Engel), Wiley-VCH Verlag GmbH & Co. KGaA, Weinheim, Section 10.3.2., 567–596.
[47] Law, J., Zsoldos, Z., Simon, A., Reid, D., Liu, Y., Khew, S.Y., Johnson, A.P., Major, S., Wade, R.A., and Ando, H.Y. (2009) *J. Chem. Inf. Model.*, **49**, 593–602.
[48] Bøgevig, A., Federsel, H.-J., Huerta, F., Hutchings, M.G., Kraut, H., Langer, T., Loew, P., Oppawsky, C., Rein, T., and Saller, H. (2015) *Org. Process Res. Dev.*, **19**, 357–368.
[49] Hanessian, S., Botta, M., Larouche, B., and Boyaroglu, A. (1992) *J. Chem. Inf. Comput. Sci.*, **32**, 718–722.

[50] Gelernter, H., Rose, J.R., and Chen, C.H. (1990) *J. Chem. Inf. Comput. Sci.*, **30**, 492–504.

[51] Szymkuć, S., Gajewska, E.P., Klucznik, T., Molga, K., Dittwald, P., Startek, M., Bajczyk, M., and Grzybowski, B.A. (2016) *Angew. Chem. Int. Ed.*, **55**, 5904–5937.

[52] Boda, K., Seidel, T., and Gasteiger, J. (2007) *J. Comput.-Aided Mol. Des.*, **21**, 311–325.

[53] Molecular Networks GmbH, https://www.mn-am.com/products/sylvia (accessed January 2018).

[54] Ertl, P. and Schuffenhauer, A. (2009) *J. Cheminf.*, **1**, 8.

[55] Fukunishi, Y., Kurosawa, T., Mikami, Y., and Nakamura, H. (2014) *J. Chem. Inf. Model.*, **54**, 3259–3267.

[56] Li, J. and Eastgate, M.D. (2015) *Org. Biomol. Chem.*, **13**, 7164–7176.

[57] Gasteiger, J. (2015) *Nat. Chem.*, **7**, 619–620.

[58] Vinkers, H.M., de Jonge, M.R., Daeyaert, F.F.D., Heeres, J., Koymans, L.M.H., van Lenthe, J.H., Lewi, P.J., Timmerman, H., Van Aken, K., and Janssen, P.A.J. (2003) *J. Med. Chem.*, **46**, 2765–2773.

[59] Masek, B.B., Baker, D.S., Dorfman, R.J., DuBrucq, K., Francis, V.C., Nagy, S., Richey, B.L., and Soltanshahi, F. (2016) *J. Chem. Inf. Model.*, **56**, 605–620.

[60] Li, J., Ballmer, S.G., Gillis, E.P., Fujii, S., Schmidt, M.J., Palazzolo, A.M.E., Lehmann, J.W., Morehouse, G.F., and Burke, M.D. (2015) *Science*, **347**, 1221–1226.

[61] Adamo, A., Beingessner, R.L., Behnam, M., Chen, J., Jamison, T.F., Jensen, K.F., Monbaliu, J.-C.M., Myerson, A.S., Revalor, E.M., Snead, D.R., Stelzer, T., Weeranoppanant, N., Wong, S.Y., and Zhang, P. (2016) *Science*, **352**, 61–67.

4.3 Explorations into Biochemical Pathways

Oliver Sacher[1] and Johann Gasteiger[2]

[1] Molecular Networks GmbH, Neumeyerstr. 28, 90411 Nürnberg, Germany
[2] Computer-Chemie-Centrum, Universität Erlangen-Nürnberg, Nägelsbachstr. 25, 91052 Erlangen, Germany

4.3.1 Introduction

Arguably, the most important chemical reactions are those that keep living species alive in a sometimes hostile environment. The central metabolism, also called endogenous metabolism, consists of biochemical reactions that break down the nutrients and then convert them into other small molecules that eventually may be assembled into macromolecules such as proteins, carbohydrates, or nucleic acid. One essential purpose of the endogenous metabolism is to generate energy to keep the temperature of a species at a constant value even if the environment is at a much lower temperature. As the product of one biochemical reaction may be the substrate of another reaction, these reactions are concatenated into series of reactions and into pathways.

With the deciphering of the human genome, the blueprint of the human organism, and its functions, interest has shifted to the role the products of genes, the proteins, play. Thus, genomics has given way to proteomics. Many of these proteins are enzymes that catalyze biochemical reactions, and therefore much research has centered on the metabolism of nutrients; metabolomics has entered the stage.

However, way before metabolomics, proteomics, and genomics appeared, much insight into biochemical pathways had been accumulated through more traditional biochemical research. Much of this work has been accumulated in a concise and beautiful manner on the wall chart "Biochemical Pathways" distributed by Roche in more than 700,000 hard copies (Figure 4.6).

This project had been started in 1965 by Dr. Gerhard Michal and his team, first at Boehringer Mannheim and then at Roche, and has been continuously updated. A Biochemical Pathways Atlas has been produced to provide more background information on all those reactions on the poster [1, 2]. Presently, the wall chart consists of two charts: Part I – Metabolic Pathways and Part II – Cellular and Molecular Processes. It is now also available on the Internet [3].

Figure 4.6 Biochemical Pathways wall chart (https://www.roche.com/pathways [3] – accessed January 2018).

4.3 Explorations into Biochemical Pathways

The poster "Biochemical Pathways" provides an overview of the biochemistry of unicellular organisms and fungi, higher plants, animals as well as humans, and general pathways. Through color coding the reactions in the different families of species are distinguished (Figure 4.7).

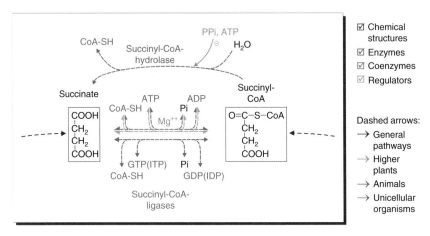

Figure 4.7 Details on a reaction on the Biochemical Pathways wall chart.

However valuable the information on biochemical reactions and metabolites on the Biochemical Pathways wall chart is, it is nevertheless quite often difficult to find specific information. Furthermore, a two-dimensional (2D) medium such as a wall chart has its inherent drawbacks for information that is in reality multidimensional, for, the metabolites in biochemical pathways are highly interconnected. As an example, L-glutamate, an important intermediate in the central metabolism, can be found at 32 positions of the wall chart (Figure 4.8). Clearly, it will be quite difficult to find all those positions (in fact, their detection was assisted by a structure search in the BioPath.Database (*vide infra*)). This example already indicates that more modern search possibilities are required to fully exploit the information content on the wall chart. In other words, the information on the wall chart has to be stored into a database to subject it to the full potential of chemoinformatics methods. Multidimensional information asks for multifaceted search possibilities.

The question is now, how can the research results embodied in the wall chart be combined with the results from genomics, proteomics, and metabolomics? Clearly, biochemical reactions are at the core of the information on the wall chart. Biologists, bioinformaticians, chemists, and chemoinformaticians have quite different ways of looking at chemical reactions. Figure 4.9 attempts to illustrate that the bioinformatics approach starts with a gene that expresses a protein, in this case an enzyme that is characterized by an EC number, a unique code assigned to an enzyme [4]. Such an enzyme converts compounds into other compounds that are characterized by their names; most of the times no deeper analysis of these compounds is made.

For a chemist, and accordingly also for a chemoinformatician, a reaction is an event catalyzed by an enzyme that converts compounds into other compounds

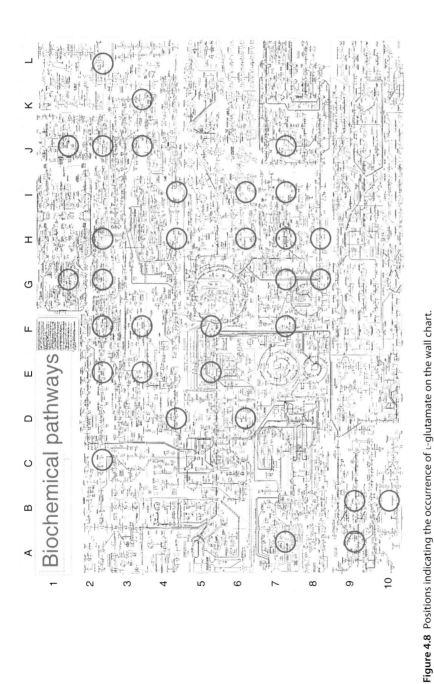

Figure 4.8 Positions indicating the occurrence of L-glutamate on the wall chart.

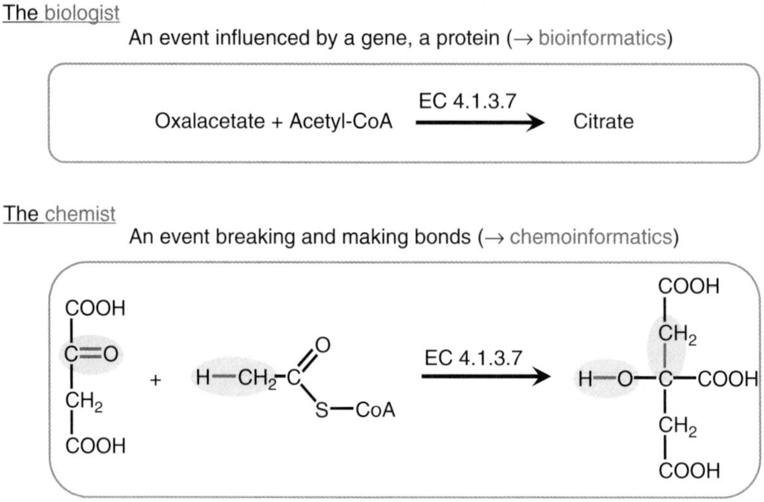

4.3 Explorations into Biochemical Pathways

The biologist
An event influenced by a gene, a protein (→ bioinformatics)

$$\text{Oxalacetate + Acetyl-CoA} \xrightarrow{\text{EC 4.1.3.7}} \text{Citrate}$$

The chemist
An event breaking and making bonds (→ chemoinformatics)

Figure 4.9 Different views on a biochemical reaction.

by breaking and making bonds. The hope is then whether new insights into biochemical reactions can be gained when bringing these two different viewpoints together. However, again, a chemoinformatics approach to biochemical reactions asks for storing them in a database and that is exactly what has been done by building the BioPath.Database.

4.3.2 The BioPath.Database

The project of storing the information contained on the Biochemical Pathway wall chart in a reaction database, the BioPath.Database, was initiated in 1998. At that time no reaction database solely devoted to biochemical reactions existed. From the very beginning our objective was to produce a high-quality reaction database. That required that all molecules were stored with atomic resolution, that is, as connection tables (see *Methods Volume*, Chapter 3) and all reactions should be completely stoichiometrically balanced. Furthermore, the reaction sites (see *Methods Volume*, Chapter 4) should be marked, and all atoms should be mapped from the substrates to the products. These requirements were strictly enforced during the building of the BioPath.Database [5].

Molecules, or metabolites, are given by names and synonyms and are stored with atomic resolution as connection tables providing access to each atom and bond of a molecule. Furthermore, stereochemical information has been stored. The 3D coordinates for all molecules calculated by the 3D structure generator CORINA (see *Methods Volume*, Section 3.5.2) [6, 7] were integrated into the database, and thus, a 3D model for each molecule is available. Other properties computed by chemoinformatics methods such as free energies are also available.

Reactions are stoichiometrically balanced, even to the point of storing protons when consumed or produced in a biochemical reaction. Thus, a complete mass balance of each reaction is provided. Reaction centers and the atoms and bonds directly involved in a reaction have been manually annotated. For all atoms in the substrates and the products, a complete atom-to-atom matching is provided. At a later stage this manual reaction center mapping was checked against a method that automatically determines the reaction center information and the atom-to-atom mapping numbers [8, 9]. For a dataset of 1542 reactions from the BioPath.Database, a 98.4% agreement between the manually annotated mappings and the computational predictions was found. The 24 reactions where discrepancies were observed could well be rationalized. Reaction center mapping is an essential prerequisite for correct reaction retrieval (see *Methods Volume*, Chapter 4).

Enzymes that are known to catalyze a reaction are given in the form of their names and the EC numbers [4].

Pathways are sequences of reactions that are given commonly agreed names and are specified accordingly.

It should be noted that the BioPath.Database stores more information than what is contained on the wall chart because additional information reported in the Biochemical Pathways Atlas [1, 2] and in the primary literature was also included in the database.

Presently the BioPath.Database comprises about 14,000 molecules, 4,000 reactions, enzymes, and pathways. It is made available at https://www.mn-am.com/databases/biopath/

Since the establishment of the BioPath.Database, other databases have been conceived that store biochemical reactions and pathways, the most notable ones being the KEGG REACTION database [10, 11], Reactome [12], and the databases at BioCyc [13]. The KEGG REACTION database is part of the Kyoto Encyclopedia of Genes and Genomes. It was initiated and is further extended and maintained by the group of Prof. Kanehisa at Kyoto University. It contains a much larger number of reactions and pathways than BioPath.Database, but the reactions are not as deeply annotated as in the BioPath.Database. For example, the reactions are not always mass-balanced, charges are not always equalized, the reaction site annotation does not directly consider bonds, and their atom-to-atom matching is only concentrating on essential atoms at the reaction site, not considering all atoms.

Reactome is maintained by a large group of specialists at the European Bioinformatics Institute and by scientists contributing to Reactome in the course of EU-funded projects. Small molecules are cross-referenced to the ChEBI database [14].

4.3.3 BioPath.Explore

To exploit the full potential of the BioPath.Database, a retrieval system BioPath.Explore has been developed and is made accessible at https://webapps.molecular-networks.com/biopath3/

BioPath.Explore provides a wide range of search methods.

Molecules can be searched by name and full structure search (a molecule editor for graphical input is also provided [15]).

Reactions can be searched by specifying the substrate or product or both or by the catalyzing enzyme name or EC number.

Pathways can be searched by their names or by looking for the shortest path between two molecules.

A variety of computed properties such as the 3D structure of molecules generated by CORINA [6, 7], their molecular weight, log P value, or Gibbs free energies [16] are given. Note that also molecules of that size are given in their full atomic resolution. All information about metabolic molecules, reactions, enzymes, and pathways are fully cross-linked that facilitates deep and broad exploration of biochemical pathways. Pathways are interactively linked and can be visualized.

4.3.4 Search Results

An overview of some search capabilities and results by a forerunner of BioPath.Explore has been published [17]. Examples of a structure search, a reaction search, and gross-formula searches were illustrated. A search for L-glutamate provided 32 hits that were used to indicate where L-glutamate can be found on the wall chart (see Figure 4.8).

4.3.4.1 Searching for Information on Molecules

Molecules can be searched by inputting a name in the field "Molecule query." If the name is contained as a substring in a series of names, the list of these molecule names can be obtained. Table 4.1 shows the first part (30 names) of a list of 65 molecule names that was obtained by inputting "glutamate" in the field "Molecule query."

In Figure 4.10 results for a search on chorismate will be given. This information has been obtained by inputting either the name "Chorismate" or sketching chorismate by the integrated molecule editor JSME [15].

Figure 4.10 shows details on the molecule chorismate, including some of its names, access to either the 2D or 3D structure, various calculated properties, and links to web-based simulation methods for the calculation of the ^{1}H NMR and the ^{13}C NMR spectrum.

Furthermore, alternative structure coding for cross-referencing is given such the SMILES, InChI, and InChIKey codes as well as links to the KEGG, ChEBI, and PubChem database. Furthermore, the three most similar molecules to chorismate, based on the Tanimoto index, are given. In addition, lists of reactions and pathways that chorismate participates in are presented and can be followed.

4.3.4.2 Searching for Information on Reactions

Figure 4.11 gives the list of the first 26 of a total of 120 reactions that are catalyzed by monooxygenase enzymes.

Either by clicking on the RXN00173 string in this hit list or by inputting "1.13.12.4" in the "EC, enzyme name or EC" field, the results shown in

Table 4.1 The first 30 instances (from 65) of molecule names that contain the substring "glutamate".

List of molecule names

Your query string "glutamate" matches the following 65 molecule names:
(0.02 seconds)
- (4S)-4-Hydroxy-4-methyl-L-glutamate
- (S)-Glutamate
- (S)-Glutamate-1-semialdehyde
- (S)-Glutamate-1-semialdhyde
- 10-Formyltetrahydrofolyl_L-glutamate
- 10-Formyltetrahydrofolylpolyglutamate
- 4-Hydroxy-4-methylglutamate
- 4-Hydroxy-L-glutamate
- 4-Methyl-L-glutamate
- 4-Methylene-L-glutamate
- 5,10-Methenyltetrahydrofolylpolyglutamate
- 5,10-Methenyltetrahydrofolypolyglutamate
- 5,10-Methylenetetrahydrofofolylpolyglutamate
- 5,10-Methylenetetrahydrofolypolyglutamate
- 5-Formiminotetrahydrofolylpolyglutamate
- 5-Formyltetrahydrofolylpolyglutamate
- 5-Methyltetrahydrofolate-L-glutamate
- 5-Methyltetrahydrofolylpolyglutamate
- 5-Methyltetrahydropteroyltri-L-glutamate
- D-Glutamate
- D-Homoglutamate
- D-Methylglutamate
- Formylisoglutamate
- Glutamate
- Glutamate-1-semialdehyde
- Indole-3-acetyl-glutamate
- Isoglutamate
- L-4-Hydroxyglutamate_semialdehyde
- L-Glutamate
- L-Glutamate_1-semialdehyde

Figure 4.12 are obtained. Observe that links are provided to the Rhea and EC-PDB databases. Furthermore, it can be followed how this reaction is embedded in some reaction pathways.

4.3.4.3 Searching for Information on Pathways

Activating the "Pathway" part and then clicking on the Submit button when all input fields are empty provides a list of all 640 pathways contained in the BioPath.Database. The user can then choose the biosynthesis or pathway he/she is interested in and obtain the sequence of reactions in the corresponding pathway. Alternatively, names of Start molecule and End molecule can be given, and the shortest sequence of reactions from this Start molecule to the End molecule

Molecule Detail Page

Molecule name

Chorismate

Additional molecule names

English
Chorisimate
CHOR
Chinese
分支酸

Molecular Structure:

2D 3D

Download

Properties calculated from structure

Property	Value
Formula	$C_{10}H_{10}O_6$
Weight	226.183 [g/mol]
Complexity	392.11
XLogP	0.13
TPSA	104.06
#HDonors	3
#HAcceptors	6
Gibbs Energy	-166.1 [kcal/mol]
^{1}H NMR spectrum	Predict NMR spectrum (requires installed Java plugin) Predict NMR spectrum (requires HTML5 compatible browser)
^{13}C NMR spectrum	Predict NMR spectrum (requires HTML5 compatible browser)
Isotopic distribution	Predict Isotopic distribution

?

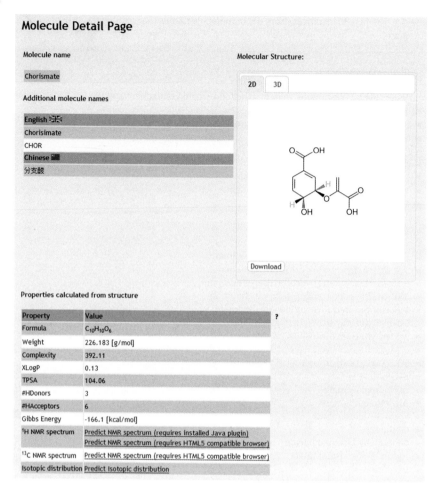

Figure 4.10 Details on the query molecule chorismate.

List of reactions

Your query string 'monooxygenase' matches the following 120 reaction conditions:

RxnID	EC	Enzyme name
RXN00010	1.14.16.1	phenylalanine 4-monooxygenase
RXN00118	1.14.99.10	steroid 21-monooxygenase
RXN00142	1.14.13.15	cholestanetriol 26-monooxygenase
RXN00152	1.14.15.6	cholesterol monooxygenase
RXN00173	1.13.12.4	lactate 2-monooxygenase
RXN00273	1.14.99.9	steroid 17alpha-monooxygenase
RXN00297	1.14.13.15	cholestanetriol 26-monooxygenase
RXN00335	1.14.16.2	tyrosine 3-monooxygenase
RXN00339	1.13.12.4	lactate 2-monooxygenase
RXN00369	1.13.12.1	arginine 2-monooxygenase
RXN00404	1.14.18.1	monophenol monooxygenase

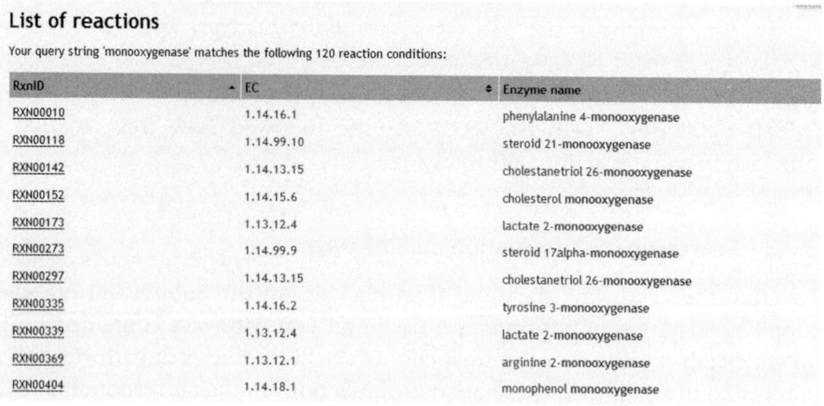

Figure 4.11 Results of a search for monooxygenases in the "Reaction" field.

4.3.4 Search Results

RXN00440		24-monooxygenase
RXN00466	2.7.11.6	[tyrosine 3-monooxygenase] kinase
RXN00543		24-monooxygenase
RXN00709	1.14.99.36	beta-carotene 15,15'-monooxygenase
RXN00737		(glutamat g-carboxylase) phylloquinone monooxygenase
RXN00762		24-monooxygenase
RXN00856	1.14.99.10	steroid 21-monooxygenase
RXN00897	1.14.15.5	corticosterone 18-monooxygenase
RXN00926	1.14.13.9	kynurenine 3-monooxygenase
RXN00962	1.14.15.6	cholesterol monooxygenase (side-chain-cleaving)
RXN01021	1.14.99.9	steroid 17alpha-monooxygenase
RXN01118	1.14.16.1	phenylalanine 4-monooxygenase
RXN01207	1.14.13.25	methane monooxygenase
RXN01216	1.14.99.9	steroid 17alpha-monooxygenase
RXN01218	1.14.16.4	tryptophan 5-monooxygenase

Figure 4.11 (Continued)

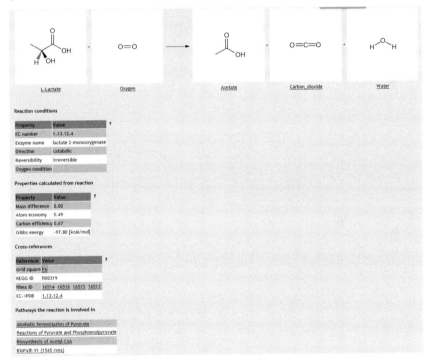

Figure 4.12 Results for searching for enzyme EC "1.13.12.4."

will be given. Figure 4.13 shows this for the pathway from farnesyl-diphosphate to artemisinin.

4.3.4.4 Some Statistics on Search Results

Some statistical results provide important insights into the essentials of metabolism and its organizational structure. Table 4.2 gives those molecules and ions that show a high occurrence in the database.

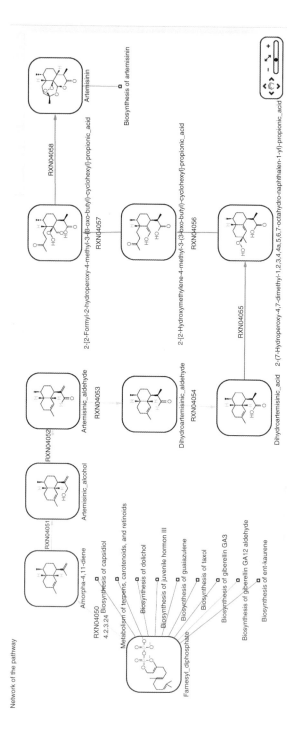

Figure 4.13 Shortest pathway from farnesyl-diphosphate to artemisinin.

Table 4.2 Number of occurrences in the BioPath.Database.

	Number of reactions
H^+	925
H_2O	724
ATP	254
NAD^+	208
ADP	206
$NADP^+$	202
NADPH	193
NADH	190
O_2	173
PO_4^{3-}	161
$P_2O_7^{4-}$	147
CO_2	144

It is interesting to note that all these molecules and ions are somehow involved in energy production. This emphasizes one of the most important tasks of the central metabolism: to produce energy.

Table 4.3 shows those metabolites that most often are substrates or products of reactions in the BioPath.Database.

The metabolites in Table 4.3 can be called hubs of biochemical pathways because they are very often accessed in these reactions much in the same way as certain highly frequented airports are called hubs. Hubs of biochemical pathways play a central role in the metabolism; they have a high turnover rate.

4.3.5 Exploitation of the Information in BioPath.Database

4.3.5.1 Classification of Enzymes

The Nomenclature Committee of the International Union of Biochemistry and Molecular Biology (NC-IUBMB) [4] assigns a unique EC number to each enzyme. Enzymes are classified into six main classes: oxidoreductases (i), transferases (ii), hydrolases (iii), lyases (iv), isomerases (v), and ligases (vi). Each main class for its side is further classified into several subclasses, which are further distinguished into subsubclasses, and finally a serial number is given. In this EC system each enzyme has its unique EC number, EC a.b.c.d, where "a" refers to one of the six classes, "b" indicates the subclass, "c" is the subsubclass, and "d" is the serial number in the subsubclass. This classification takes account of the overall chemical change produced by the complete enzyme reaction, but the details of the mechanism of the reaction and the formation of intermediate complexes of the reaction are not taken into account. In effect, varying criteria are considered in

4.3 Explorations into Biochemical Pathways

Table 4.3 Metabolites most frequently found as substrates or products in the database.

Metabolite	Structure	Number of reactions
L-Glutamate		71
2-Oxo-glutarate		45
Pyruvate		43
Acetate		23
Formate		22
Glycine		21
Succinate		21

the EC system such as reaction types, substrates, transferred groups, and acceptor groups.

However, it would certainly be quite useful to base the classification on more mechanistic criteria, or at least on factors that consider the reacting bonds. The BioPath.Database reports for each enzyme-catalyzed reaction the bonds broken and made during that reaction. Thus, this information could offer a good basis for enzyme classification. In order to investigate this, we have chosen a dataset of hydrolases (EC 3.b.c.d), for in the classification of hydrolases by the EC system, the notation is somehow based on the structural changes of the reaction. Thus, both the EC system and the classification based on mechanistic criteria attempted here should lead to comparable results.

The dataset retrieved from BioPath.Database in the initial study [18] comprised 135 reactions under catalysis by hydrolases from the subclasses EC 3.1.c.d to EC 3.8.c.d. In order to bring some mechanistic considerations into the classification, each reacting bond on the substrate side was characterized by six physicochemical descriptors: difference in partial atomic charges of the atoms of the bond, difference in σ-electronegativities, difference in π-electronegativities, effective bond polarizabilities, delocalization stabilization of negative charge, and delocalization stabilization of positive charge. In making these choices, we attempted to consider all major electronic effects influencing the reacting bonds, determining the reaction mechanism.

In order to have an unbiased approach to classification, an unsupervised learning method was chosen to base the classification only on the descriptors of the reacting bonds. The method of choice was a Kohonen neural network, a self-organizing map (SOM) (see *Methods Volume*, Chapter 11.2). An SOM produces a map where the reactions of the dataset are distributed in 2D.

In the map produced by the dataset of 135 reactions, the individual subclasses are quite well separated [18] (full map not shown). To go one level further down in the EC classification, the subsubclasses EC 3.1.1.d to EC 3.1.6.d were investigated [18]. Figure 4.14 shows only that section of the full SOM of the above study that contained these subsubclasses. One reaction (in neuron F1) is quite separated from the other reactions because it was the only one that broke two bonds. The reactions of the other subsubclasses are contained in coherent clusters as indicated by the coloring of these in Figure 4.14. The reason is that the bonds being hydrolyzed are contained in different substructures, a fact that is caught by the physicochemical descriptors used in this study. Thus, these physicochemical descriptors allow an automatic fine-tuned classification of enzyme-catalyzed reactions.

In a later study an extended dataset of 311 reactions catalyzed by hydrolases was investigated [19]. This larger dataset allowed the inclusion of reactions from the subclass 3.7.c.d. to 3.11.c.d in the study by an SOM. Furthermore, a support vector machine (SVM) and a hierarchical cluster analysis (HCA) (see *Methods Volume*, Section 11.1) were used to classify this dataset supporting the findings from the SOM study. By and large results similar to the previous study were obtained.

In the same publication, 651 reactions catalyzed by oxidoreductases, enzymes of the class EC 1.b.c.d, were investigated with all three methods: an SOM, an SVM, and an HCA. The classification results were similar to the EC system. However, the perception of similarity based on physicochemical descriptors of the reaction bonds showed finer details of the enzymatic reactions and thus could be used as a good basis for the comparison of enzymes. The advantage of the classification of enzyme-catalyzed reactions by chemoinformatics methods based on physicochemical descriptors of the reacting bonds is that this provides an unbiased automatic system that gives insights into electronic effects acting on biochemical reactions.

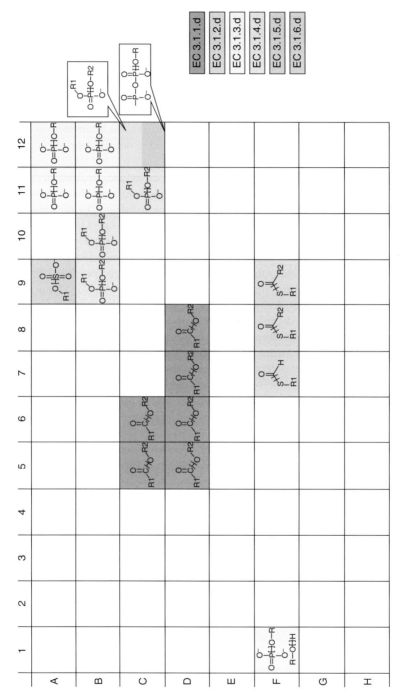

Figure 4.14 Section of the full SOM of the 135 reactions that shows the distribution of reactions of the subsubclasses of EC 3.1.c.d.

As a case in point, Figure 4.15 shows the SOM of a large dataset of reactions catalyzed by enzymes of the classes EC1.b.c.d (oxidoreductases) to EC 6.b.c.d (ligases). A closer inspection of the mapping of the oxidoreductases indicated that there are two clusters: one type of reactions that have mostly FAD (flavin adenine dinucleotide) as cofactor and another separated cluster that comprises reactions that have mostly NAD or NADP as cofactor. Apparently, the physicochemical descriptors catch an important electronic feature that is the basis for these enzymes to utilize different cofactors.

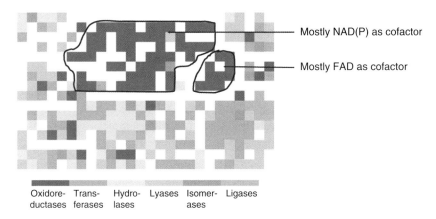

Figure 4.15 SOM of reactions from all classes of enzymes from EC 1.b.c.d. to EC 6.b.c.d.

4.3.5.2 Query Generation for Inhibitors of Enzymes

Clearly, an enzyme must bind the substrate of the reaction it catalyzes. However, even more important is that an enzyme tightly binds the transition state of the reaction because this leads to a substantial lowering of the activation energy. This is then responsible for the dramatic increases in reaction rates that are up to 10^9 faster than the reaction without the enzyme. Figure 4.16 indicates the effects of an enzyme on the energy landscape of a reaction.

Linus Pauling pointed this out already in 1948 that enzymes stabilize the transition states of a reaction [20]. He further postulated that then analogs of such transition states should act as potent inhibitors of enzymatic reactions. In order to investigate this postulate, information was extracted from the BioPath.Database. As an example, the conversion of adenosine monophosphate (AMP) to inosine monophosphate (IMP) catalyzed by the enzyme AMP deaminase (EC code 3.5.4.6) was investigated.

The investigation of the hypothesis that an inhibitor is a transition state analog clearly requires information on the 3D structure of the species involved. As the BioPath.Database contains the 3D structures of all its molecules generated by CORINA [6, 7], the 3D structures of AMP and IMP could be extracted. Access to the 3D structure of transition states can only be obtained by quantum mechanical methods. However, it can be postulated that the 3D structure of the transition state in the deamination (hydrolysis) of AMP may be structurally quite close to the intermediate that can be obtained by addition of a water molecule to AMP. For

4.3 Explorations into Biochemical Pathways

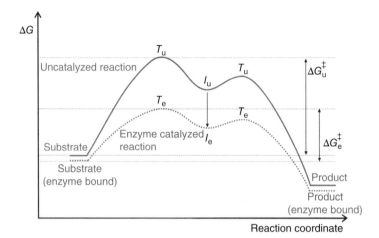

Figure 4.16 Effect of an enzyme on the energies of the substrate, the transition state, the intermediate, and the product of a reaction. The substantial lowering of the energies of the transition states and the intermediate is clearly distinguished.

this intermediate (Figure 4.17), also a 3D structure was generated by CORINA. Carbocyclic coformycin is an inhibitor of the deamination of AMP (Figure 4.17). A 3D structure for this inhibitor was also generated and then superimposed onto the substrate, the intermediate, and the product of the conversion of AMP to IMP.

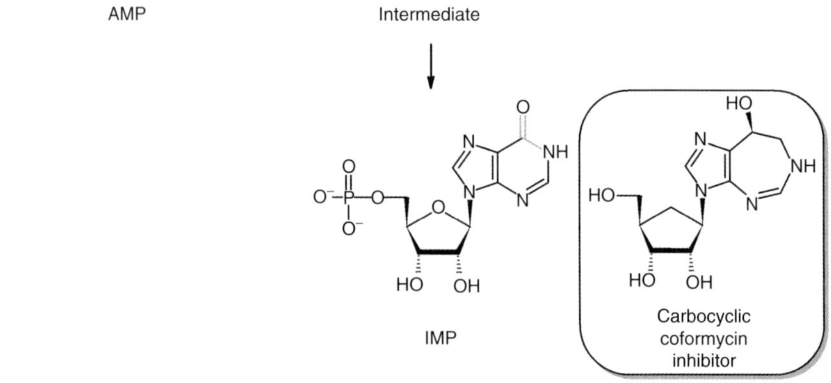

Figure 4.17 The intermediate of the conversion of AMP to IMP obtained by addition of water to AMP. The structure of the inhibitor of AMP deaminase, carbocyclic coformycin.

This superimposition of 3D structures was performed by the program GAMMA that is based on a genetic algorithm and also allows conformational flexibility in the superimposition process to provide a maximum fit [21]. Figure 4.18 shows the results of these three superimpositions. For all three superimpositions, the results in the root mean squares (RMS) deviations in the 3D coordinates of 16 atoms are given. The smallest deviation, that is, the best fit, is obtained in the superimposition of the inhibitor onto the reaction intermediate. The better fit of the intermediate (0.13 A vs 0.19 A or 0.20 A) might not seem to be very large. However, it is important that the fit is particularly good at the reaction site, especially that the OH group of the inhibitor and the OH group of the intermediate perfectly align (see circles in Figure 4.18). This may indicate that the enzyme enforces a very specific approach of the water molecule to AMP.

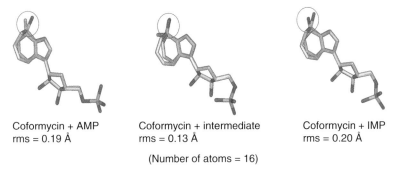

Coformycin + AMP
rms = 0.19 Å

Coformycin + intermediate
rms = 0.13 Å

Coformycin + IMP
rms = 0.20 Å

(Number of atoms = 16)

Figure 4.18 Superimpositions of the 3D structure of the inhibitor carbocyclic coformycin onto the substrate AMP, the intermediate, and the product IMP of the deamination of AMP.

Several other examples have been studied along these lines and provide further support for the transition state analog hypothesis [22]. Based on these studies the following procedure for finding new inhibitors of an enzyme can be derived:

- Generate an intermediate for the reaction that catalyzes the enzyme that should be inhibited.
- Generate a 3D structure for this intermediate.
- Use this 3D structure of the intermediate for a 3D structure search in a database of molecules.

4.3.5.3 Genome Clustering Based on Metabolic Capabilities

An organism's ability to survive in a specific environment is the result of the organism's regulatory and metabolic capabilities. In order to investigate this relationship, species were grouped together based on their similar metabolic capabilities. The habitats of these species were then mapped onto these groups [23].

In effect, this study was a combined application of bioinformatics and chemoinformatics data and methods. The bioinformatics information consisted of 214 annotated genomes of the corresponding species contained in the PEDANT database [24]; the chemoinformatic information consisted of 290 metabolic pathways from the BioPath.Database. In order to group genomes that show similar metabolic capabilities, the genomes were represented by their score-based pathway profiles and subjected to a hierarchical cluster algorithm. This clustering was then assigned to selected habitats, environmental conditions, and a few diseases. Several interesting, previously unknown insights into some diseases were obtained [23]. For example, analyzing the clustering of the genomes of microorganisms revealed unexpected, previously unreported interrelations between metabolism and disease.

In a similar study, the metabolic pathways relevant to phenotypic traits of microbial genomes were determined [25]. 266 microbial genomes from the PEDANT database were clustered based on the 290 pathway data from the BioPath.Database. A score-based metabolic reconstruction then allowed the generation of 290-dimensional pathway profiles for each genome (Figure 4.19). Various different phenotypes such as methanogenesis, causing periodontal disease, gram-positive, or obligate anaerobe were investigated. Several attribute subset selection methods were used to select pathways ranked by their relevance to a specific phenotype.

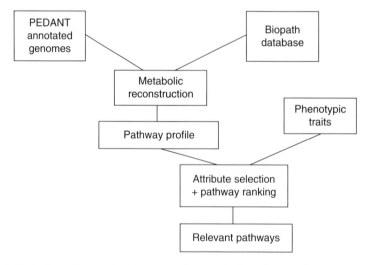

Figure 4.19 Outline of the method for uncovering metabolic pathways relevant to phenotypic traits of microbial genomes. (Kastenmüller et al. 2009 [25]. https://genomebiology.biomedcentral.com/articles/10.1186/gb-2009-10-3-r28. Licensed Under CC BY 2.0.)

The results for the phenotype "causing periodontal disease" are here discussed in more detail. The PEDANT database contained at the time of the study 15 fully sequenced oral genomes that included four of the six known periodontal pathogens.

Attribute selection and pathway ranking provided the following list of the nine most relevant pathways:

- Biosynthesis of coenzyme B12
- Glutamate fermentation
- Biosynthesis of 5-formino-THF

- Degradation of L-histidine to L-glutamate
- Glycolysis and glyconeogenesis (part)
- Biosynthesis of L-proline
- Urea cycle (part)
- Conversion of L-glutamate to L-proline
- Conversion of L-glutamate to L-ornithine

The first four obtained a higher pathway score than the last five ones. Several of these pathways, in particular the degradation of L-histidine, produce ammonia. Several clinical studies suggested ammonia as a mediator of periodontal disease. Thus, the findings here support what has been found in these clinical studies. The degradation of L-histidine is especially harmful as it generates three moles of ammonia per mole amino acid. This finding suggests that the histidine degradation process could be a possible new target for a more specific antibacterial treatment.

4.3.5.4 Prediction of Metabolic Pathways: Flavor-Forming Pathways

Chemical systems biology is a new discipline, linking chemical biology and systems biology. A classical systems biology approach for analyzing cellular processes starts with the reconstruction of biochemical networks based on annotated genomes. However, those metabolic models will contain numerous gaps because quite a few of the metabolites and the reactions leading to them will be unknown. Consequently, databases on biochemical or metabolic reactions are and always will be incomplete. In this bottleneck situation, methods for predicting potential metabolites and metabolic reactions are very much wanted. Many of the secondary metabolites such as flavor compounds are nonessential in metabolism, and many of their synthesis pathways are unknown.

A new approach, called reverse pathway engineering (RPE), has been developed, which combines chemoinformatics and bioinformatics analyses to predict the "missing links" between compounds of interest and their possible metabolic precursors by providing plausible chemical and/or enzymatic reactions [26]. The approach is based on a database of biochemical reaction types. These reaction types are obtained in a process of abstraction by classifying individual biochemical reactions into reaction types based on analyses of the reaction centers. Figure 4.20 indicates this by the conversion of the reaction database BioPath.Database into a database of reaction types that only contains the reaction center substructures together with their transformation rules.

Input of a query compound for which a metabolic pathway should be developed leads to an analysis of that compound for all reaction center substructures contained in this compound. The found reaction center substructures then initiate the corresponding transformation contained in the rules of the database of reaction types and thus suggest biochemical reactions leading to potential precursors of the query compound. For each suggested reaction, links are provided to the metabolic reactions that have been the basis for the derivation of the employed reaction type. These original reactions in the BioPath.Database also allow suggestions to be made which enzyme could achieve the suggested reaction. Repetition of this process with the precursors as query eventually generates an entire metabolic sequence or network.

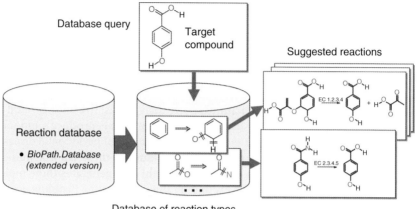

Figure 4.20 Outline of the chemoinformatics analysis of the reverse pathway engineering approach.

This is then followed by bioinformatics analyses that first retrieve information on the suggested candidate enzymes from databases such as BioPath.Database, BRENDA [27], or PubMed [28]. This information is then augmented by searches in genome and protein sequence databases. By comparative genomics analyses, putative enzymes are predicted and the retrieved protein sequences were used to search against the microbial genomes [29].

The approach was demonstrated by using flavor-forming pathways in cheese by lactic acid bacteria (LAB) as an example. Established routes leading to the formation of flavor compounds from leucine were successfully replicated. The main flavor products of leucine degradation are 3-methylbutanal, 3-methylbutanol, and 3-methylbutanoic acid, which provide cheesy, malty, and sweaty odors, respectively. When 3-methylbutanoic acid was input, three routes were predicted by the RPE approach (Figure 4.21). The two routes, as shown in Figure 4.21a and b, are the same as described in literature. The third one from Figure 4.21c is a novel route. The conversion of α-keto-isocaproate to α-hydroxy-isocaproate is a known reaction; the step from α-hydroxy-isocaproate to 3-methylbutanoic acid is novel.

In order to further analyze this proposed reaction, the output of the RPE method was inspected more closely. The most closely related reaction found in BioPath.Database was the oxidation of L-lactate by lactate-2-monooxygenase (EC 1.13.12.4) (see Figure 4.22).

As some enzymes have broader substrate specificities and may catalyze analogous reactions, it was hypothesized that lactate-2-monooxygenase can also catalyze the oxidation reaction of α-hydroxy-isocaproate. In order to further substantiate this hypothesis, additional bioinformatics studies were performed. Using the protein sequence of lactate-2-monooxygenase from *Mycobacterium smegmatis*, its homologs in LAB were identified by BLASTP (see *Methods Volume*, Chapter 13). By performing comparative genomics analysis, the orthologs in LAB that are currently annotated as L-lactate oxidase (LOX) were identified (Figure 4.23). Interestingly a mutant form of LOX from *Aerococcus*

4.3.5 Exploitation of the Information in BioPath.Database

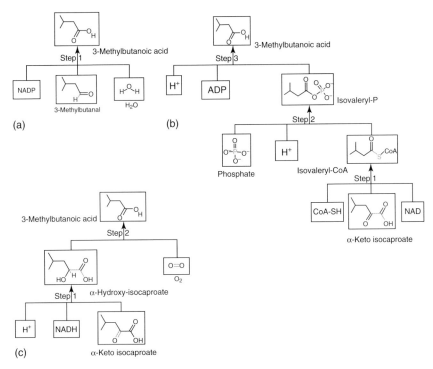

Figure 4.21 Three pathways to 3-methylbutanoic acid, generated by the RPE approach. The sequences (a) and (b) are known; sequence (c) is novel.

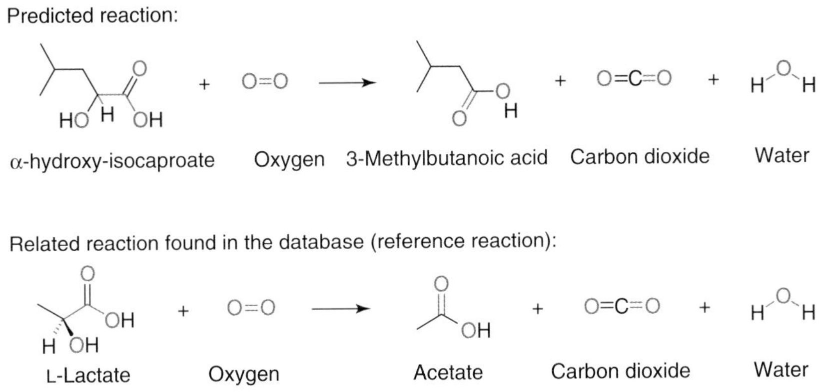

Predicted reaction:

α-hydroxy-isocaproate + Oxygen → 3-Methylbutanoic acid + Carbon dioxide + Water

Related reaction found in the database (reference reaction):

L-Lactate + Oxygen → Acetate + Carbon dioxide + Water

BioPath reaction ID	RXN00173, search BioPath.Explore
Enzyme name	Lactate 2-monooxygenase
EC number	1.13.12.4, search BioPath.Explore

Figure 4.22 Suggested reaction from the sequence in Figure 4.21c and the reference reaction from BioPath.Database.

viridans had been found by Yorita et al. in which alanine 95 was replaced by glycine that had the same oxidase activity on lactate as the wild-type LOX as well as an additional enhanced oxidase activity toward longer-chain hydroxy acids [30]. That suggests that broader substrate specificity toward longer-chain hydroxyl acids could be obtained if the conserved alanine 95 residue in LOX from LAB is mutated to glycine.

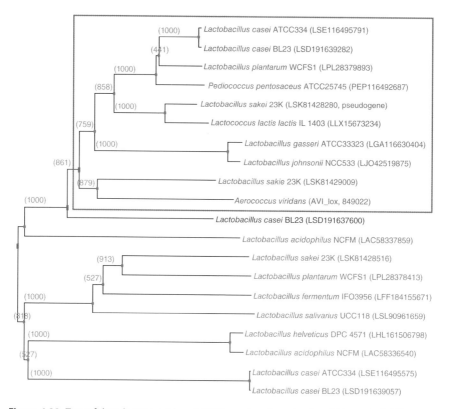

Figure 4.23 Tree of the L-lactate oxidase (LOX) homologs from lactic acid bacteria (LAB).

In a further study by the RPE approach, the synthesis of the flavor compound dimethylsulfide was investigated. This investigation successfully predicted the enzymes involved in the formation of dimethylsulfide but also provided new plausible reaction steps [26].

The RPE approach is unique in this field because it connects chemical data all the way back to genomic data. The merit of the RPE approach was here illustrated with the BioPath.Database. However, it can be extended to other application fields due to its flexibility of using other reaction databases and target compounds. Potential applications are in the field of biomarker discovery or in the interplay of host–microbial co-metabolism. Another important field could be found in synthetic biology, supporting the *in silico* design of biosynthesis pathways to construct microbial cell factories [31].

4.3.6 Summary

BioPath.Database is a rich database on biochemical and metabolic reactions that is deeply annotated with chemical information. This distinguishes it from other databases on metabolism that might contain more molecular structures and reactions and have more direct links to genomic information. However, such links can also be followed from the BioPath.Database as was shown in the RPE approach. The merits of the enhanced chemical information in the BioPath.Database were demonstrated by the applications in the examples given in the exploitation chapter such as classification of enzymes, query generation for inhibitors of enzymes, genome clustering based on metabolic capabilities (relevant pathways of diseases), and prediction of metabolic pathways (flavor-forming pathways).

Essentials

- Less than one part in 10^{50} of possible molecules have been made.
- Distinct potential molecular structures are not only highly numerous but also highly diverse.
- *Click chemistry* is a small group of reactions that usually work very well.
- Searching for reactions is much harder than searching for structures.
- Even if we can find something relevant in a database, is it true?
- A brute-force approach to simulating possible reactions will never be successful.
- Organic synthesis is much harder than chess and is not completely predictable.
- The best synthetic chemists are currently better at synthesis than the best computer programs.
- Synthesis prediction programs provide a valuable resource to aid and to inspire synthesis design.
- The methods for processing chemical reactions can also be applied to biochemical reactions.
- A database of biochemical reactions provides many new and interesting insights into the processes in living species.
- Enzymes have to tightly bind the transition state of a biochemical reaction.
- 3D models of the intermediates of biochemical reactions can be used for searching for inhibitors of enzymes.
- Information on the reaction site and the bonds broken and made in a reaction can be used for the classification of biochemical reactions.
- This classification of reaction can favorably be compared with the EC nomenclature.
- The combination of databases and methods from bioinformatics and chemoinformatics can provide insights into essential pathways of diseases.
- Building a database of biochemical reaction types allows one to search for reactions similar to a considered reaction.
- A reverse pathway engineering (RPE) approach can be utilized to construct flavor-forming pathways.
- The RPE approach can find new flavor-forming pathways.

Available Software and Web Services (accessed January 2018)

- http://cheminf.cmbi.ru.nl/cheminf/lhasa/
- Pathway Tools Software: http://brg.ai.sri.com/ptools/
- https://webapps.molecular-networks.com/biopath3/
- http://www.genome.jp/kegg/
- http://www.reactome.org/
- http://biocyc.org/
- http://www.hmdb.ca/
- https://en.wikipedia.org/wiki/List_of_biological_databases

Selected Reading

- Corey, E.J. and Cheng, X.-M. (1995) *The Logic of Chemical Synthesis*, John Wiley & Sons, Inc., New York. ISBN: 978-0-471-11594-6, 456 pp.
- Judson, P. (2009) *Knowledge-Based Expert Systems in Chemistry: Not Counting on Computers*, RSC, Cambridge, UK, 222 pp.
- Kastenmüller, G., Schenk, M.E., Gasteiger, J., and Mewes, H.-W. (2009) *Genome Biol.*, **10**, R28.
- Liu, M., Bienfait, B., Sacher, O., Gasteiger, J., Siezen, R.J., Nauta, A., and Geurts, J.M.W. (2014) *PLoS One*, **9**, e84769.
- Michal, G. and Schomburg, D. (eds) (2012) *Biochemical Pathways, An Atlas of Biochemistry and Molecular Biology*, 2nd edn, Hoboken, NJ, John Wiley & Sons, Inc., 416 pp.
- Reitz, M., Sacher, O., Tarkhov, A., Trümbach, D., and Gasteiger, J. (2004) *Org. Biomol. Chem.*, **2**, 3226–3237.
- Reitz, M., von Homeyer, A., and Gasteiger, J. (2006) *J. Chem. Inf. Model.*, **46**, 2333–2341.
- Warr, W.A. (2014) A short review of chemical reaction database systems, computer-aided synthesis design, reaction prediction and synthetic feasibility. *Mol. Inf.*, **33**, 469–476.

References

[1] Michal, G. (ed.) (1999) *Biochemical Pathways, Biochemistry Atlas*, Spektrum Akademischer Verlag, Heidelberg.
[2] Michal, G. and Schomburg, D. (eds) (2012) *Biochemical Pathways: An Atlas of Biochemistry and Molecular Biology*, 2nd edn, John Wiley & Sons, Inc., Hoboken, NJ, 416 pp.
[3] http://biochemical-pathways.com/#/map/1 (accessed January 2018).
[4] http://www.chem.qmul.ac.uk/iubmb/ (accessed January 2018).

[5] https://webapps.molecular-networks.com/biopath3/ (accessed January 2018).
[6] Sadowski, J. and Gasteiger, J. (1993) *Chem. Rev.*, **93**, 2567–2581.
[7] https://www.mn-am.com/products/corina (accessed January 2018).
[8] Körner, R. and Apostolakis, J. (2008) *J. Chem. Inf. Model.*, **48**, 1181–1189.
[9] Apostolakis, J., Sacher, O., Körner, R., and Gasteiger, J. (2008) *J. Chem. Inf. Model.*, **48**, 1190–1198.
[10] http://www.genome.jp/kegg/ (accessed January 2018).
[11] https://en.wikipedia.org/wiki/KEGG (accessed January 2018).
[12] http://www.reactome.org/ (accessed January 2018).
[13] http://biocyc.org/ (accessed January 2018).
[14] https://www.ebi.ac.uk/chebi/ (accessed January 2018).
[15] Bienfait, B. and Ertl, P. (2013) *J. Cheminf.*, **5**, 24.
[16] Rother, K., Hoffmann, S., Bulik, S., Hoppe, A., Gasteiger, J., and Holzhütter, H.-G. (2010) *Biophys. J.*, **98**, 2478–2486.
[17] Reitz, M., Sacher, O., Tarkhov, A., Trümbach, D., and Gasteiger, J. (2004) *Org. Biomol. Chem.*, **2**, 3226–3237.
[18] Sacher, O., Reitz, M., and Gasteiger, J. (2009) *J. Chem. Inf. Model.*, **49**, 1525–1534.
[19] Hu, X., Yan, A., Tan, T., Sacher, O., and Gasteiger, J. (2010) *J. Chem. Inf. Model.*, **50**, 1089–1100.
[20] Pauling, L. (1948) *Nature*, **161**, 707–709.
[21] Handschuh, S., Wagener, M., and Gasteiger, J. (1998) *J. Chem. Inf. Comput. Sci.*, **38**, 220–232.
[22] Reitz, M., von Homeyer, A., and Gasteiger, J. (2006) *J. Chem. Inf. Model.*, **46**, 2333–2341.
[23] Kastenmüller, G., Gasteiger, J., and Mewes, H.-W. (2008) *Bioinformatics*, **24**, i56–i62.
[24] Frishman, D. *et al.* (2003) *Nucleic Acid Res.*, **31**, 207–211.
[25] Kastenmüller, G., Schenk, M.E., Gasteiger, J., and Mewes, H.-W. (2009) *Genome Biol.*, **10**, R28.
[26] Liu, M., Bienfait, B., Sacher, O., Gasteiger, J., Siezen, R.J., Nauta, A., and Geurts, J.M.W. (2014) *PLoS One*, **9**, e84769.
[27] http://www.brenda-enzymes.info/ (accessed January 2018).
[28] http://www.ncbi.nlm.nih.gov/pubmed (accessed January 2018).
[29] Liu, M., Nauta, A., Francke, C., and Siezen, R.J. (2008) *Appl. Environ. Microbiol.*, **74**, 4590–4600.
[30] Yorita, K., Aki, K., Ohkuma-Soyejima, T., Kokubo, T. *et al.* (1996) *J. Biol. Chem.*, **271**, 28300–28305.
[31] Medema, M.H., van Raaphorst, R., Takano, E., and Breitling, R. (2012) *Nat. Rev. Microbiol.*, **10**, 396–403.

5 Structure–Spectrum Correlations and Computer-Assisted Structure Elucidation

Joao Aires de Sousa

Universidade Nova de Lisboa, Departamento de Quimica, Faculdade de Ciencias e Tecnologia, 2829-516 Caparica, Portugal

Learning Objectives

- To identify the main methods and tools available for the computer prediction of spectra from the molecular structure
- To briefly discuss computer-assisted structure elucidation (CASE) from spectral data
- To realize that a proper representation of the molecular structure is crucial for the prediction of spectra and to know the main approaches for structure representation in the context of structure–spectrum correlations
- To apply structure–spectrum correlations in automatic structure verification and dereplication, that is, the identification of compounds in complex mixtures

Outline

5.1 Introduction, 133
5.2 Molecular Descriptors, 135
5.3 Infrared Spectra, 137
5.4 NMR Spectra, 140
5.5 Mass Spectra, 150
5.6 Computer-Aided Structure Elucidation (CASE), 153

5.1 Introduction

Our current knowledge of molecular structures is largely derived from the interpretation of spectroscopic data. The investigation of molecular structures and of their properties is one of the most fascinating topics in chemistry with an immense impact in chemistry-related industries, environmental science, or analytical services.

In this chapter, both methods for spectra simulation from the molecular structure and for structure elucidation from spectroscopic data are explained. Several techniques and computer programs have been proposed under the generic name computer-assisted structure elucidation (CASE) for the proposal of a structure by the use of rule-based systems, which require a technique for assembling a complete structure from substructure fragments that have been predicted. Computational approaches have the potential to accelerate the tasks of structure elucidation, avoid human errors in the proposal of new structures, and attack challenging problems with less experimental requirements – a highly ambitious goal indeed! A general CASE strategy is to collect preliminary information from the available sources (e.g., molecular mass and fragments suggested by characteristic spectral features), which is then used to exhaustively generate structures in agreement with the molecular formula. The candidate structures are eventually filtered on the basis of constraints imposed by the user and by the comparison of experimental and predicted spectra [1].

Lindsay et al. [2] introduced the first program that was able to enumerate all acyclic structures from a molecular formula. The DENDRAL project was initiated in 1965 by Edward Feigenbaum, Joshua Lederberg, and Bruce Buchanan. It is generally considered as the first approach to using artificial intelligence to solving chemical problems. It began as an effort to make the concept of scientific reasoning and the formalization of scientific knowledge available with the computer. DENDRAL used a set of knowledge- or rule-based methods to deduce the molecular structure of organic chemical compounds from chemical analysis and mass spectrometry (MS) data. The DENDRAL system for structure elucidation from mass spectral data was the first expert system (ES) developed in chemistry. DENDRAL proved to be fundamentally important in demonstrating how rule-based reasoning could be developed into powerful knowledge engineering tools [3]. This is true in spite of the fact that the DENDRAL project was eventually abandoned. The enormous developments in this area of research, which has also incorporated data from newly available spectroscopic techniques, are explained in the last section of this chapter.

A different relevant task is *structure validation*. In synthetic organic chemistry, the product of a reaction is often anticipated, for example, because the reaction is known for similar substrates. In these cases, chemists have to confirm that the expected reaction actually occurred, that is, the product has the postulated molecular structure. This verification can be supported by spectroscopic data, namely, ^{1}H and ^{13}C NMR spectra, if the spectra of the expected structure are available or can be predicted. When many reactions are run simultaneously, as in multiple parallel synthesis and combinatorial chemistry, the whole process must be automated, which implies the availability of software for spectrum prediction and comparison [4, 5]. NMR spectroscopy has been proposed for this endeavor as it provides rich interpretable data and can be automated, and (as shown in this chapter) reliable predictions of spectra can be obtained with available software. More recent works also incorporate 2D NMR data [6, 7].

Chemistry has a language of its own for molecular structures, which has been developed from the first alchemy experiments to modern times. With the improvement of computational methods for chemical information processing, several descriptors for the handling of molecular information have been developed and used in a wide range of applications.

One of the most important tasks in the handling of molecular data is the evaluation of "hidden" information in large chemical datasets. Different from conventional database queries, data mining techniques generate new data that are used subsequently to characterize molecular features in a more general way. Generally, it is not possible to hold all the potentially important information in a dataset of chemical structures. Thus, the extraction of relevant information and the production of reliable secondary information are important topics.

Finding the adequate descriptors for the representation of chemical structures is one of the basic problems in chemical data analysis, for example, in the automatic establishment of relationships between structures and spectra. Several methods have been developed for the description of molecules including their chemical or physicochemical properties [8] (see also *Methods Volume*, Chapter 10).

Molecules are usually represented as 2D diagrams or three-dimensional (3D) molecular models. While the 3D coordinates of atoms in a molecule are sufficient to describe the spatial arrangement of atoms, they exhibit two major disadvantages as molecular descriptors: the number of descriptors depends on the size of a molecule, and they do not describe additional properties (e.g., atomic properties). The first feature is most important for the computational analysis of data. Even a simple statistical function, for example, a correlation, requires the information of all molecules of a dataset to be represented in equally sized vectors of a fixed dimension. A solution to this problem is a mathematical transformation of the Cartesian coordinates of a molecule into a vector of fixed length. The second point can be overcome by including the desired properties in the transformation algorithm. Examples of such transforms are the descriptors based on radial distribution functions (RDFs) (see *Methods Volume*, Section 10.3.4.4). RDF descriptors grew out of the research on structure–spectrum correlations [9].

5.2 Molecular Descriptors

5.2.1 Fragment-Based Descriptors

A widely used method in structure–spectrum correlations is fragment-based coding. With this approach, the molecule to be encoded is divided into several substructures that represent the typical information necessary for the task. Many authors have used this method for the automated interpretation [10, 11] and prediction [12] of spectra with artificial neural networks (NN)s, in expert systems (ES)s for structure elucidation [13], and with pattern recognition methods [14]. For example, a descriptor in the form of a binary vector is used simply to define the presence or absence of functional groups that exhibit important spectral features in the corresponding infrared (IR) spectrum. The main disadvantage of this method is that it imposes a restriction on the number of substructures represented. As an example, for the correlation of structures with IR spectra, a reasonable number of substructures varies from 40 to 720, depending on the user's more or less subjective view of the problem. Affolter *et al.* [15] showed that a simple assignment of IR-relevant substructures and corresponding IR bands does not describe spectrum–structure correlation to an adequate accuracy. This is mainly due to the effect of the chemical environment on the shape and position of absorption signals.

5.2.2 Topological Structure Codes

An enhancement of the simple substructure approach is the "fragment reduced to an environment that is limited" (FREL) method introduced by Dubois *et al.* [16] With the FREL method several centers of the molecule are described, including their chemical environment. By taking the elements H, C, N, O, and halogens into account and combining all bond types (single, double, triple, aromatic), the authors found descriptors for 43 different FREL centers that can be used to characterize a molecule.

One of the most widely used – and successful – representations of the constitution, the topology, of a molecule for the simulation of spectra is the hierarchical ordered description of the substructure environment (HOSE) code [17]. It is an atom-centered code taking into account the spheres of neighbors of each atom – it describes the environment of an atom in several virtual spheres – and has proven to be particularly successful for the interpretation and prediction of ^{13}C NMR spectra (see Section 5.4.2). The first layer in a HOSE code is defined by all the atoms that are one bond away from the central atom; the second layer includes the atoms within the two-bond distance; and so on. The rationale for this substructure code is that the structural environment of a certain atom in a molecule, for example, of a ^{13}C atom, determines the chemical shift of this atom. The length (and resolution) of the code depends on the number of spheres of neighbor atoms used for the structure description around the focus atom. Figure 5.1 illustrates the derivation of the HOSE code of the carbon atom in the carboxylic group of the ethyl ester of phenylalanine, for the first, second, and third spheres of neighbors.

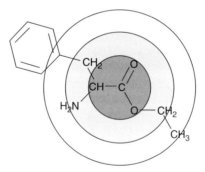

Figure 5.1 The HOSE code for a selected carbon atom describes its structural environment in hierarchical order by walking through the molecule in spheres. Only the non-hydrogen atoms are explicitly considered.

For ^{13}C NMR spectra, a HOSE code for each carbon atom in a molecule has to be determined. The spectral signals are assigned to the HOSE codes that represent the corresponding carbon atom. This approach has been used in ^{13}C NMR prediction and in algorithms that allow the automatic creation of "substructure–subspectrum" databases and are used in systems for proposing chemical structures directly from ^{13}C NMR [1].

The basic HOSE code ignores stereochemical information such as *cis–trans* isomer interaction that can contribute significantly to the chemical shift values. Robien adapted the HOSE code method [18] by extending it with descriptors for three-, four-, and five-bond interactions and with information about axial/equatorial substitution patterns.

In another way to encode the topology of a molecule and to characterize the complete arrangement of its atoms, the entire molecule is regarded as a connectivity graph where the edges represent the bonds and the nodes represent the atoms. An application of this concept can be found in some approaches to automatic interpretation of mass spectra that search the molecular graph for possible substructures (subgraphs) corresponding to the *m/z* values of a peak list [19].

5.2.3 Three-Dimensional Molecular Descriptors

Descriptors for the 3D arrangement of atoms in a molecule can be derived. The Cartesian coordinates of the atoms in a molecule can be calculated by semiempirical quantum mechanical or molecular mechanics (force field) methods. For larger datasets, fast 3D structure generators are available that combine data- and rule-driven methods to calculate Cartesian coordinates from the connection table of a molecule (e.g., CORINA [20, 21], see *Methods Volume*, Chapter 3).

There are some prerequisites for a 3D structure descriptor. It should be as follows:

- Independent of the number of atoms, that is, the size of a molecule
- Unambiguous regarding the 3D arrangement of the atoms
- Invariant against translation and rotation of the entire molecule

Further prerequisites depend on the chemical problem to be solved. Some chemical effects have an undesired influence on the structure descriptor if the experimental data to be processed do not account for them. A typical example is the conformational flexibility of a molecule, which has a profound influence on a 3D descriptor based on Cartesian coordinates. For the application of structure descriptors for structure–spectrum correlation problems such as in vibrational spectroscopy, two other points are desirable: the descriptor should contain physicochemical information related to vibrational states, and it should be possible to gain structural information or the complete 3D structure from the descriptor.

5.3 Infrared Spectra

5.3.1 Overview

Since IR spectroscopy monitors the vibrations of atoms in a molecule in 3D space, information on the 3D arrangement of the atoms should somehow be contained in an IR spectrum.

A series of monographs and correlation tables exist for the interpretation of vibrational spectra [22–25]. However, the relationship of frequency characteristics and structural features is rather complicated, and the number of known correlations between IR spectra and structures is very large. In many cases, it is almost impossible to analyze a molecular structure from an IR spectrum without the aid of computational techniques. Existing approaches are mainly based on the interpretation of vibrational spectra by mathematical models, rule sets, and decision trees or fuzzy logic approaches. IR intensities and Raman activities can also be obtained with quantum chemistry calculations (Ref. [26] is suggested for the interested reader). This chapter, however, is devoted to data-driven chemoinformatics approaches.

ESs were designed to assist the chemist in structural problem solving, based on the approach of characteristic frequencies. Gribov and Elyashberg [27] suggested different mathematical techniques in which rules and decisions are expressed in an explicit form. Elyashberg [28] pointed out that system symbolic logic [29] is a valuable tool in studying complicated objects of a discrete nature as in the discrete modeling of the structure–spectrum. In agreement with that, Zupan showed [30] that the relationship between the molecular structure and the corresponding IR spectrum could be represented conditionally by a finite discrete model. These relationships can be formulated as if–then rules in the knowledge base of an ES. Systems based on those logical rules can be found in several reviews [31, 32] and publications [33–37].

Karpushkin et al. implemented a library of correlations between substructures and spectral features to be applied in computer-aided IR and Raman spectrum interpretation [38]. This approach uses the concept of fragment nucleus (representing an ensemble of environmental situations of the same vibrating group having common spectral characteristics); the environment typically includes so-called queries designating a set of possible neighborhoods. Some of the correlations were formulated with the assistance of pattern recognition methods, and automatic validation of structure–spectrum relationships was performed.

Woodruff and Smith introduced the ES PAIRS [39], a program that is able to analyze IR spectra in the same manner as a spectroscopist would. Andreev et al. developed the ES EXPIRS [40] for the interpretation of IR spectra. EXPIRS provides a hierarchical organization of the characteristic groups that are recognized by peak detection in discrete frames. Penchev et al. [41] introduced a computer system that performs searches in spectral libraries and a systematic analysis of mixture spectra. It is able to classify IR spectra with the aid of linear discriminant analysis, artificial NNs, and the method of k-nearest neighbors.

5.3.2 Infrared Spectra Simulation

Gasteiger et al. [9] implemented the following approach for full spectra simulation. The RDF code descriptor (see *Methods Volume*, Section 10.3.4.4) allows the representation of the 3D structure of a molecule by a fixed (constant) number of variables (Figure 5.2). By using the fast 3D structure generator CORINA [20, 21], they were able to study the correlation between any 3D structure and IR spectra using artificial NNs. Steinhauer et al. [42] used the RDF codes as structure descriptors, together with the IR spectrum, to train a counter propagation (CPG) NN and establish the complex relationship between an RDF descriptor (input) and an IR spectrum (output). After training, the simulation of an IR spectrum is performed using the RDF descriptor of the query compound as the information vector of the Kohonen layer, which determines the central neuron. On

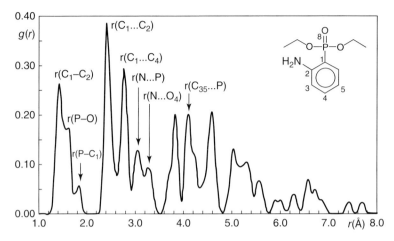

Figure 5.2 RDF code for a phosphonic ester using atomic numbers as atom property. (Figure from Ref. [45], permission obtained from Elsevier.)

input of this query RDF descriptor, the central neuron is selected and the corresponding IR spectrum in the output layer is presented as the simulated spectrum (Figure 5.3). Selzer et al. [43] described an application of this spectrum simulation method that provides rapid access to arbitrary reference spectra. Kostka et al. described a combined application of spectrum prediction and reaction prediction ESs [44]. The combination of the reaction prediction system EROS (Elaboration of Reactions for Organic Synthesis) and IR spectrum simulation proved to be a powerful tool for computer-assisted substance identification.

5.3.2.1 Structure Prediction

The CPG NN could also be applied in reverse mode, that is, to predict the RDF descriptor given an IR spectrum (Figure 5.4) [45]. The 3D structure for the predicted RDF can be searched in a database of RDF codes. An empirical modeling process is used to refine this 3D structure to obtain a 3D model of the molecular structure that fits the predicted RDF code and corresponds to the IR spectrum. Thus, we have here a successful approach for the prediction of the 3D structure from the IR spectrum.

5.3.2.2 QSPR Predictions of Specific Frequencies

For specific applications, the accurate prediction of the vibration frequency corresponding to a certain functional group is of importance. Quantitative structure–spectrum relationships that employ structural descriptors and regression techniques to predict the frequency were established. Examples are the prediction of C=O and P=O frequencies in Mannich bases with multilinear regressions using atomic electronegativity distance vector (VAED) as descriptors [46] and the prediction of the C=O frequency in carbonyl compounds of different types with both linear methods and NNs [47].

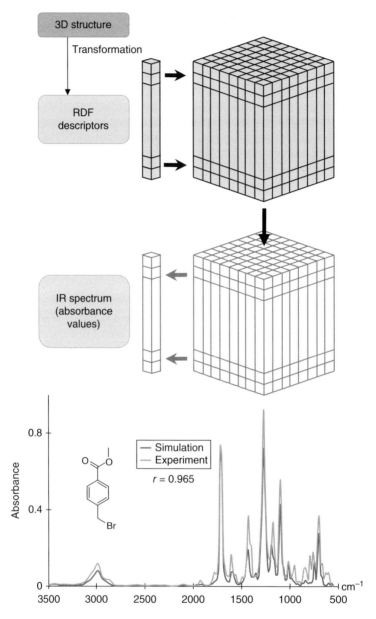

Figure 5.3 Training of a CPG NN to learn relationships between structures and IR spectra and example of a simulated spectrum.

5.4 NMR Spectra

NMR spectroscopy is probably the single most powerful technique for the confirmation of structural identity and for structure elucidation of unknown

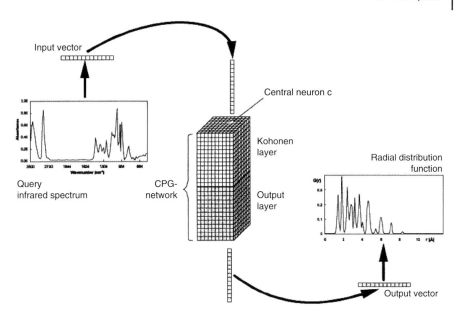

Figure 5.4 Scheme of a counterpropagation network for the derivation of 3D structures. (Figure from Ref. [45], permission obtained from Elsevier.)

compounds [48]. Additionally, the relatively low measurement times and the facility for automation contribute to its usefulness and industrial interest.

Thus, in several areas of chemistry, for example, combinatorial chemistry, many compounds are produced in short time ranges, and their structures have to be confirmed by analytical methods. A high degree of automation is required, which has fueled the development of software that can predict NMR spectra starting from the chemical structure and that calculates measures of similarity between simulated and experimental spectra [1]. These tools are obviously also of great importance to chemists working with just a few compounds at a time using NMR spectroscopy for structure confirmation.

Furthermore, the prediction of ^{1}H and ^{13}C NMR spectra is needed in systems for automatic structure elucidation [1]. In many such systems, structure candidates are automatically produced by a structure generator and are then filtered/ranked according to several criteria including the similarity between the spectrum predicted for each candidate and the experimental spectrum [49].

^{1}H NMR spectra are basically characterized by the chemical shifts and coupling constants of signals. The chemical shift for a particular atom is influenced by the 3D arrangement and bond types of the chemical environment of the atom and by its hybridization. The multiplicity of a signal depends on the coupling partners and on the bond types between atom and coupling partner.

The situation is less complex for ^{13}C NMR spectra. In fact, there is usually a good correlation between the 2D arrangement of atoms and their ^{13}C chemical shifts. In addition, ^{13}C NMR spectra are usually measured with the protons decoupled and then show few coupling effects except for fluorine and phosphorus atoms in the vicinity of the carbon atom. These are the reasons ^{13}C NMR spectra can be represented quite well as pairs of chemical shift values and intensities.

5.4.1 Quantum Chemistry Prediction of NMR Properties

When a molecule is submitted to a static magnetic field, the nuclear spin energy levels split. An oscillating magnetic field can then induce transitions between these energy levels and produce the NMR spectrum. In a molecule, the magnetic field at a nucleus results from the applied magnetic field and the shielding by the surrounding electrons which depends on neighbor atoms. The NMR chemical shift of a nucleus results from the difference in energy between the nuclear spin states, which can be calculated by quantum mechanical methods [50–53]. The effects of the external magnetic field on the nucleus of interest are added into the equations as a "perturbation." It is then possible to calculate the chemical shift, which is related to the total molecular energy, the applied magnetic field, and the nuclear magnetic moment.

Quantum mechanical calculations are particularly useful for the prediction of chemical shifts in chemical entities that are not frequently found in the available large databases of chemical shifts, for example, charged intermediates of reactions, radicals, and structures containing elements other than H, C, O, N, S, and P, halogens, and a few common metals.

The Gaussian program [54] is one of the most popular tools for quantum chemical calculation of ^{1}H and ^{13}C NMR chemical shifts. NMR shielding tensors may be computed with Gaussian using the gauge-invariant atomic orbital (GIAO) method and the Hartree–Fock, DFT, CCSD, or MP2 models [55]. NMR calculations may include effective core potentials (ECPs). At this high level of theory, the required computation times are considerable, and for many organic compounds the accuracy obtained is only comparable with faster empirical methods [56] described in the next sections.

NMR calculations are based on a given molecular geometry, which means that an experimental 3D structure must be available, or it has to be previously calculated. Because a rigid structure is used for the calculations, different chemical shifts are generally obtained for NMR-equivalent nuclei (e.g., the three protons of a methyl group), and, therefore, equivalent nuclei have to be specified if their chemical shifts are to be averaged at the end of the calculation.

Using semiempirical methods, which are based on approximate solutions of the Schrödinger equation but use parameterized equations, the computation times can be reduced by two orders of magnitude. HyperChem from Hypercube, Inc. [57], is an example of a software package that can calculate 3D geometries, chemical shifts, and coupling constants using semiempirical approaches.

5.4.2 NMR Spectra Prediction by Database Searching

A useful empirical method for the prediction of chemical shifts and coupling constants relies on the information contained in databases of structures with the corresponding NMR data. Large databases with millions of chemical shifts are commercially available and are linked to predictive systems, which basically rely on database searching [1, 58]. This has been done for different nuclei such as ^{1}H, ^{13}C, ^{31}P, ^{19}F, or ^{15}N [1]. Atoms (e.g., ^{1}H or ^{13}C) are internally represented by their structural environments, usually their HOSE codes [17].

When a query structure is submitted, a search is performed to find the atoms belonging to similar (overlapping) substructures. These are the atoms with the same HOSE codes as the atom in the query molecule. The prediction of the chemical shift is calculated from the chemical shift of the retrieved atoms (e.g., the average).

The similarity of the retrieved atoms to those of the query structure, and the distribution of chemical shifts among atoms with the same HOSE codes, can be used as measures of prediction reliability. When common substructures cannot be found for a given atom (within a predefined number of bond spheres), interpolations are applied to obtain a prediction; proprietary methods are often used in commercial programs.

The database approaches are heavily dependent on the size and quality of the database, particularly on the availability of entries that are related to the query structure. Such an approach is relatively fast, particularly in comparison with quantum chemical methods. The predicted values can be explained on the basis of the structures that were used for the predictions. Additionally, users can augment the database with their own structures and experimental data, allowing improved predictions for compounds bearing similarities to those added. In an advanced approach, parameters such as solvent information can be used to refine the accuracy of the prediction.

Examples of commercial implementations of this general approach are the Modgraph's NMRPredict [59] and ACD software [60] (in 2016, ACD calculations were based on 2.2 million ^{1}H chemical shifts and 3.0 million ^{13}C chemical shifts and can be made solvent specific).

The same concept has been implemented for specific families of compounds. For example, the program GlyNest estimates ^{1}H and ^{13}C NMR chemical shifts of glycans based on a database and a spherical environment encoding scheme for each atom [61]. Similarly, ^{1}H and ^{13}C NMR chemical shifts of proteins were predicted on the basis of sequence homology [62].

5.4.3 NMR Spectra Prediction by Increment-Based Methods

Most chemistry students have learned in spectroscopy textbooks how to estimate ^{13}C and ^{1}H NMR chemical shifts from tabulated values for classes of structures, which are corrected with additive contributions from neighboring functional groups or substructures. In such tables, initial chemical shifts are assigned to standard structure fragments (e.g., protons in substituted ethylenes). Substituents in specific positions (e.g., *gem*, *cis*, or *trans*) are assumed to make independent additive contributions to the chemical shift. The additive contributions are listed in a second series of tables.

This procedure has been implemented in computer programs [63, 64]. Commercial examples include ChemOffice [65] and NMRPredict. [59] Substructures can be encoded by using the additive model extensively developed by Pretsch *et al.* [66] and Clerc and Sommerauer [67]. The authors represented skeleton structures and substituents by individual codes and calculation rules. A more general additive model was introduced later by Small and Jurs [68] and was enhanced by Schweitzer and Small [69]. Several tables have been compiled for different types of atoms. The ChemOffice 2010 predictor was built based on 700 base values and about 2000 increments for ^{1}H and 4000 parameters for ^{13}C. It is claimed that about 90% of all CH_n groups can be estimated with a standard deviation of 0.2–0.3 ppm (^{1}H NMR) and 95% of the ^{13}C NMR shifts can be estimated with a standard deviation of 5.5 ppm [65]. Although mostly used for small organic compounds, increment-based methods were also designed for glycans and implemented in the program CASPER [70] for ^{13}C and ^{1}H chemical shifts. Once the tables and equations are defined, this family of methods is easy to implement, does not require database searching, and is extremely fast. On the other hand, the parent structure and the substituents must be tabulated, and interactions between substituents are neglected. Usually, slightly less accurate predictions are obtained than with the slower methods based on large databases.

A related approach was developed by Abraham and implemented in the CHARGE program [71]. It consists of the derivation of empirical equations for specific types of substructures, which calculate ^{1}H chemical shifts from physicochemical parameters of neighbor atoms/groups (such as electronegativity, polarizability, or anisotropy) and geometrical features such as angles and distances. While CHARGE was developed mostly for small organic molecules, specific equations and parameters were derived to predict chemical shifts of proteins [72] and nucleic acids [73].

5.4.4 NMR Spectra Prediction by Machine Learning Methods

Machine learning (ML) algorithms can build models automatically from a dataset of examples. Once trained, these models can be highly accurate and extremely fast in comparison with HOSE code predictions. An atom (e.g., ^{1}H, ^{13}C, ^{31}P) in a specific molecule can be encoded by a fixed-length representation of its intramolecular environment. An ML method, such as a neural network (NN), a support vector machine (SVM), or a random forest, can then be trained to predict the chemical shift from the atomic code. Several examples of this approach are available in the literature and have been incorporated into software packages. Some of these systems are illustrated here.

In the case of ^{1}H NMR chemical shifts, NNs have been trained to predict the chemical shift of protons on submission of a chemical structure. Two main issues play decisive roles: how a proton is represented and which examples are in the dataset.

A proton can be (numerically) represented by a series of topological and physicochemical descriptors, which account for the influence of the neighborhood on its chemical shift (e.g., partial charges, electronegativity, number of certain types of atoms in the neighborhood). Fast empirical procedures for the calculation of physicochemical descriptors are now easily accessible [74]. Geometric descriptors were added in the case of some rigid substructures, as well as for π-systems, to account for stereochemistry and 3D effects. Local RDF descriptors were used for that purpose (Figure 5.5).

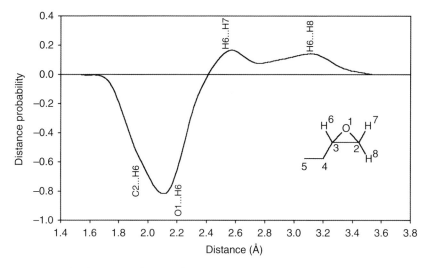

Figure 5.5 Radial distribution function for proton H-6 using partial atomic charge as the atomic property and indications of the distances contributing to each peak. (Figure from Ref. [75], permission obtained from ACS.)

With a relatively small training set of 744 ^{1}H NMR chemical shifts for protons from 120 molecular structures, CPG NN were used to establish relationships between protons and their ^{1}H NMR chemical shifts [75]. The protons were divided into four groups: protons attached to atoms in aromatic rings, non-aromatic π systems, rigid aliphatic systems, and nonrigid aliphatic systems. The selection of descriptors was performed with genetic algorithms. Later, the same group reported improved predictions with associative neural networks (ASNNs) trained with a much larger database (>18,000 chemical shifts) [76]. ASNN are ensembles of feed-forward NNs linked to an additional memory [77]. The ensemble prediction of the chemical shift is corrected based on the observed errors for the k-nearest neighbors in an additional memory of experimental chemical shifts. A mean absolute error of 0.2–0.3 ppm was usually observed for independent test sets. Finally, a procedure was envisaged to estimate coupling constants between ^{1}H atoms with the ASNNs trained for chemical shifts, in which a second memory was linked consisting of coupled protons and their experimental coupling constants. An ASNN finds the pairs of coupled protons most similar to a query, and these are used to estimate coupling constants. Using a diverse general dataset of 618 coupling constants, mean absolute errors of 0.6–0.8 Hz could be achieved in different experiments. These models were incorporated into the SPINUS software for full-spectrum prediction (Figure 5.6). The method and a web interface are freely available at http://joao.airesdesousa.com/spinus

Having in view the CASE of metabolites, Kuhn et al. [78] investigated several ML techniques to predict ^{1}H NMR chemical shifts based on experimental data for 18,692 protons from the NMRShiftDB database. Hydrogen atoms were represented by a combination of the physicochemical, topological, and geometrical descriptors from the works of Aires-de-Sousa et al. [75] and descriptors used to

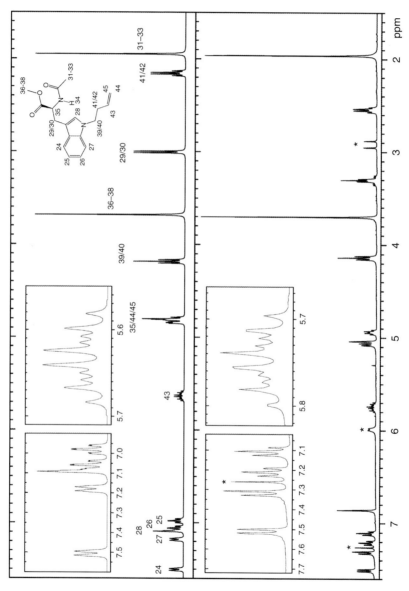

Figure 5.6 Experimental ^{1}H NMR spectrum (below) of the structure in the upper right corner compared to the full spectrum predicted by SPINUS (above) for the same structure. (*) In the experimental spectrum, the signal at 7.26 ppm is from the solvent (CDCl$_3$), the signal at 5.98 ppm is from the exchangeable NH proton, and the peaks at 2.85–2.95 ppm are from residues of DMF. (Figure from Ref. [76], permission obtained from ACS.)

predict chemical shifts of proteins from a study by Meiler [79]. The latter included descriptors for the atoms the hydrogen atom is bound to, for the 16 atoms closest by bond (topological) and for the 16 atoms closest in space (geometric). Comparing different ML and statistical techniques – multilinear regressions, k-nearest neighbors, decision trees, SVMs and random forests – the best predictions were obtained by random forests, SVM, and decision trees with an overall similar accuracy. The best classifier achieved a mean absolute error of 0.18 ppm in the 10-fold cross-validation procedure. The authors additionally applied HOSE codes and observed predictions slightly superior to those of the best ML technique.

For the prediction of ^{13}C NMR chemical shifts, approaches using NNs or linear regression modeling have been tried which used topological, physicochemical, or geometric parameters as independent variables [80–82]. While early studies focused on specific classes of compounds, the following systems are currently available having a more general scope. Meiler *et al.* [82] trained feed-forward NNs with a database of >500,000 ^{13}C chemical shifts. The environment of a carbon atom (the central atom) was represented by the frequency (count) of 28 atom types at specific bond distances (spheres) from the central atom. The total number of hydrogen atoms and the number of ring closures in each sphere were also used. Five spheres were defined, as well as a "sum sphere" for atoms at distances higher than five bonds. In order to take into consideration the importance of conjugated π-electronic systems, counts were additionally defined for the same spheres but including only atoms in "π-contact" with the central atom. NNs were independently trained for the nine types of carbon atoms. The authors reported a mean absolute error of 1.79 ppm for an independent test set with >15,000 chemical shifts and an acceleration of 1000 times in comparison with HOSE code database predictions.

Williams *et al.* further developed this scheme for the encoding of atomic environments and applied ML techniques to predict carbon and proton chemical shifts [83]. The training databases were composed of approximately 2 million ^{13}C and 1.2 million ^{1}H chemical shift values, and predictions for test sets were achieved with mean deviations of 1.5 and 0.18 ppm for the of ^{13}C and ^{1}H chemical shift, respectively. Partial least squares (PLS) methods were found to provide approximately the same accuracy as NNs and were 2–3 times faster. The results demonstrate that ML methods can be used to provide faster predictions with essentially the same accuracy of database searching methods. Several software packages (such as Mestrelab Mnova, Figure 5.7) combine different methods (database, ML, and additive schemes) to maximize performance and accuracy.

ML algorithms were also applied to predict chemical shifts of nuclei other than ^{1}H and ^{13}C. Examples include the prediction of ^{19}F chemical shifts with a fluorine fingerprint descriptor and a distance-weighted k-nearest neighbors algorithm [84], as well as the application of NNs to predict ^{195}Pt [85] and ^{31}P chemical shifts [86].

NMR chemical shifts are important sources of structural information on proteins. Their prediction from the molecular structure can assist in the selection and refinement of 3D models for new proteins. However, the complicated conformational dependencies of the chemical shifts make this endeavor rather challenging. The SPARTA+ [87] and PROSHIFT [79] programs rely on NNs

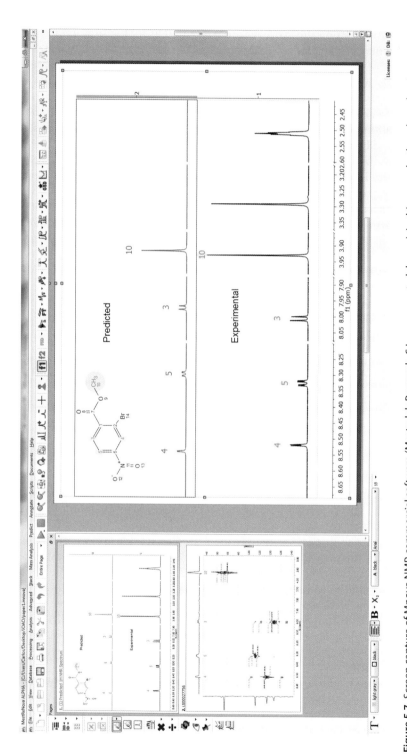

Figure 5.7 Screen capture of Mnova NMR commercial software (Mestrelab Research, S.L, www.mestrelab.com). In this example, the selected active page contains one experimental ^{1}H NMR spectrum, fully processed and analyzed, stacked together with its ^{1}H NMR predicted counterpart. Automatic assignment of the signals (possible with ^{1}H, ^{13}C, HSQC, and COSY experiments) is also depicted; atom labels are color-coded depending upon the quality of the assignment as derived from a fuzzy logic expert system.

to provide fast predictions for H, C, and N atoms of proteins. In SPARTA+, the 113 input signals to the network consist of tripeptide structural parameter sets including the backbone and side-chain torsion angles of the residue to be predicted and its two immediate neighbors, 20 amino acid-type similarity scores for each residue, information on interactions such as H-bonding, ring-current effects and electric field effects, and predicted backbone flexibility. The NN trained with a 580-protein structural database yielded rapid chemical shift prediction, with standard deviations of 2.45, 1.09, 0.94, 1.14, 0.25, and 0.49 ppm for N, C′, C_a, C_b, H_a, and H_N, respectively, between the predicted and experimental shifts for a set of 11 validation proteins.

In PROSHIFT, an NN was designed to receive the code of a single atom environment and predict its chemical shift. It was trained with ~65,000 ^{13}C, ~16,000 ^{15}N, and ~88,000 ^{1}H chemical shifts. The input consists of four different groups of descriptors representing

- The atom in the focus (which is always non-hydrogen)
- All atoms (up to 16) that are less than three bonds away from the focus in the covalent structure
- The 16 atoms that are closest in space
- Protein and sample-dependent parameters

The output layer has three neurons – for H, N, and C atoms – in order to take advantage of the fact that chemical shift values of a hydrogen atom and its covalently linked carbon or nitrogen atom are highly correlated (because both nuclei have a similar chemical environment). Thus the network is built to always predict the shift of a linked hydrogen atom in parallel to the shift of the heavy atom if it is a carbon or nitrogen atom. Testing with an independent dataset achieved root mean square deviations of 0.3 ppm for hydrogen, 1.3 ppm for carbon, and 2.6 ppm for nitrogen chemical shifts.

ML predictions of RNA chemical shifts have been reported, which are particularly relevant to enable restrained molecular dynamics simulations. Frank et al. trained random forests to predict ^{13}C and ^{1}H chemical shifts from atomic descriptors including the type of base and nucleus, physicochemical effects (ring-current, local magnetic anisotropy and polarization), close contacts, number of hydrogen bonds between specific parts of the molecule, total stacking interactions, and dihedral angles [88]. Later, fast linear functions were proposed based on interatomic distances [89].

In conclusion, chemists have now at hand software to assist in real tasks for which the simulation, or the assignment, of NMR spectra is required, particularly for organic molecules and proteins. ML and database methods have reached a prediction accuracy essentially at the level of experimental uncertainty and are fast enough to process millions of structures in a few hours. Recommended web services and software are listed under "Available Software and Web Services." Concurrently, quantum chemistry methods shall be considered for compounds/substructures outside the diversity space of the available databases, when a few small/medium size compounds are required and the computational cost can be afforded.

5.5 Mass Spectra

In contrast to IR and NMR spectroscopies, the principle of MS is based on decomposition and reactions of organic molecules on their way from the ion source to the detector. Consequently, structure–MS correlation is basically a matter of relating reactions to the signals in a mass spectrum. The chemical structure information contained in mass spectra is difficult to extract because of the complicated relationships between MS data and chemical structures. In effect, a single MS contains information on many fragmented species; it is a 1 to n relationship: 1 MS, n fragments. The aim of spectrum evaluation can be either the identification of a compound or the interpretation of spectral data in order to elucidate the chemical structure [90–92].

5.5.1 Identification of Structures and Interpretation of MS

Identification of mass spectra is typically performed by searching for similarities of the measured spectrum to spectra stored in a library [93, 94] – a common procedure with electron impact (EI) mass spectra for which databases and corresponding software products are available.

A more challenging problem is the interpretation of mass spectra by means of computational chemistry, and different strategies have been explored for substructure recognition. The use of correlation tables containing characteristic spectral data together with corresponding substructures has been successfully applied to other spectroscopic methods. However, because of the complexity of chemical reactions that may occur in mass spectrometers and because of the lack of generally applicable rules, this approach has been less successful with MS.

Algorithms based on multivariate classification techniques were constructed for automatic recognition of structural properties from spectral data. These are based on the characterization of spectra by a set of spectral features. A spectrum can be considered as a point in a multidimensional space with the coordinates defined by spectral features. Exploratory data analysis and cluster analysis are used to investigate the multidimensional space and to evaluate rules to distinguish structure classes.

Multivariate data analysis usually starts with generating a set of spectra and the corresponding chemical structures as a result of a spectrum similarity search in a spectrum database. The peak data are transformed into a set of spectral features, and the chemical structures are encoded into molecular descriptors [92]. A spectral feature is a property that can be automatically computed from a mass spectrum. Typical spectral features are the peak intensity at a particular mass/charge value or logarithmic intensity ratios. The goal of transformation of peak data into spectral features is to obtain descriptors of spectral properties that are more suitable than the original peak list data. Spectral features and their corresponding molecular descriptors are then applied to mathematical techniques of multivariate data analysis, such as principal component analysis (PCA) for exploratory data analysis or multivariate classification for the development of spectral classifiers [95–98]. PCA results in a scatterplot that exhibits spectrum–structure relationships by clustering similarities in spectral traits and/or structural features [99, 100].

5.5.2 Prediction of MS

Gasteiger *et al.* followed an approach based on models of the reactions taking place in the spectrometer [101, 102]. Automatic knowledge extraction is performed from a database of spectra, and the corresponding structures and rules are saved. The rules concern possible elementary reactions and models relating the probability of these reactions with physicochemical parameters calculated for the structures. The knowledge can then be applied to chemical structures in order to predict

1. The reactions that occur in the spectrometer
2. The resulting products
3. The corresponding peaks in the mass spectrum

For a dataset of noncyclic alkanes and alkenes, Jalali-Heravi and Fatemi described a feed-forward NN for the simulation of MS spectra [103]. The NN received 37 topological molecular descriptors as the input and comprised 44 output neurons (to predict the values of 44 m/z positions).

Commercial software is available that implemented fragmentation rules to simulate mass spectra in different conditions (such as EI or chemical ionization with different reagents). Examples include ACD/MS Fragmenter [60] and Mass Frontier [104] – Figure 5.8.

5.5.3 Metabolomics and Natural Products

MS has attained prominence in metabolomics research due to the high sensitivity (orders of magnitude higher than NMR) and ability to analyze complex mixtures using hyphenated setups (e.g., liquid chromatography (LC)-MS) or tandem MS. LC-MS is mostly used (instead of gas chromatography (GC)-MS for which large reference databases exist) as several classes of biological compounds lack the required thermal stability. The mass spectrum is usually obtained with electrospray soft ionization (ESI) or with chemical ionization. In the latter case, an adduct ion of the studied compound (e.g., $[M+H]^+$) is generated that fragments by collision-induced dissociation (CID). The interpretation of these types of spectra is more complex and their reproducibility on different instruments is more limited than with EI MS [105]. For the interpretation of metabolomics data, the main bottleneck is the identification of unknown compounds, and computational MS plays a central role here [106]. Structure–spectrum relationships support the identification of compounds in complex mixtures – a crucial task in metabolomics and in the screening of natural product extracts. To identify a metabolite, its experimental mass spectrum can be compared with the spectra *simulated* for the structures of metabolites in large databases to find the best match [107]. This is possible in principle with the commercial software mentioned before. However, rule-based approaches often fail to suggest rationalizations for a significant proportion of observed product ions. ML has been proposed to model the MS/MS fragmentation process from databases of spectra and

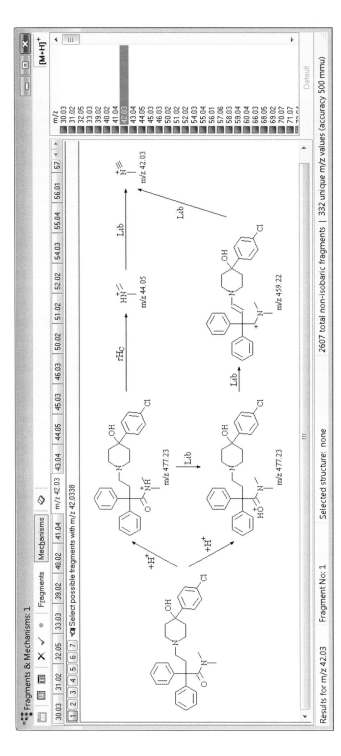

Figure 5.8 Screenshot of Mass Frontier software (HighChem, Ltd., www.highchem.com) showing predicted fragmentation mechanisms for a user provided compound.

structures, namely, to estimate the likelihood of any given fragmentation event occurring [108].

Instead of simulating the spectrum, a candidate compound can be evaluated by generating all its possible topological fragments and matching the fragment mass with the measured peaks. Candidates are then scored from the number of fragments that explain peaks in the measured spectrum and some estimation of the likelihood of the fragmentations to occur (e.g., bond dissociation energies) [19, 109–111]. This last approach does not necessarily create a mechanistically correct prediction of the fragmentation processes but performs a search in compound libraries using the measured fragments as additional structural hints. Another strategy for the identification of molecular structures (viz., metabolites) from MS/MS spectra is to predict a chemical fingerprint directly from an MS/MS spectrum (with an ML model) and then search a database of fingerprints for the most similar metabolite – CSI:FingerID method [112].

5.6 Computer-Aided Structure Elucidation (CASE)

DENDRAL [2, 113, 114], the first program for CASE, used an isomer generator software capable of generating exhaustive sets of isomers from a given molecular formula and implemented spectroscopic knowledge about spectrum–structure relationships. Substructures that have to be present in the query structure are collected in the so-called good list, while forbidden substructures are put into the "bad list." Structural information can be derived from low-resolution MS data, whereas high-resolution data can be used to determine the molecular formula. This program was the precursor to the first ESs for structure elucidation ever published: CONGEN [115] and GENOA [116]. These programs could handle any structure and enumerate the isomers of a molecular formula and were able to generate structures with more restrictive constraints, for example, isomers with specified molecular fragments. However, both GENOA and CONGEN use more heuristic than systematic algorithms. Later CASE programs based on a more systematic structure generation technique were the structure generators CHEMICS [117], ASSEMBLE [118], and COMBINE [119].

The problem of overlapping fragments was studied by Dubois *et al.* They developed the program DARC-EPIOS [120], which could retrieve structural formulas from overlapping ^{13}C NMR data. The software compares all sub-spectra of library fragments with the experimental ^{13}C NMR spectrum of the analyzed compound. Besides the chemical shifts and multiplicity of the signals, the signal area (i.e., the number of carbon atoms) is needed for automatic assignment. Substructures that have corresponding library sub-spectra coinciding with the query sub-spectra within specified deviation limits are selected and ranked according to their size (i.e., to the number of carbon and skeleton atoms). The chemical structure is generated by superimposing the atoms common to different fragments. Similar techniques have also been applied with the COMBINE program, while GENOA used a more general technique based on the determination of all possible combinations of nonoverlapping molecular fragments.

Artificial NNs have found an application in this strategy as a tool for spectral interpretation [121–123]. For example, Munk *et al.* trained a back-propagation NN to identify with high accuracy the presence of a broad range of substructural features from IR spectra, ^{13}C NMR spectra, and molecular formulas. These substructures were used as constraints on the structure generation process of the CASE program SESAMI [121].

All the CASE programs described above generate chemical structures by assembling atoms and/or molecular fragments, whereas COCOA [124] and GEN [125] operate by removing bonds from a hyperstructure that initially contains all the possible bonds between all the required atoms and molecular fragments (structure reduction).

Meiler and Will [126] introduced GENIUS, which implements genetic algorithms to approach the problem of finding structures consistent with an experimental ^{13}C NMR spectrum and a molecular formula. A population of structures with the given formula is randomly generated. Then an NN predicts a spectrum for each structure. Each structure is scored according to the similarity between the predicted spectrum and the experimental data. The population with the fittest structures survives, mutates, and mate, while that with the less fit structures are discarded (die). The procedure is repeated in an iterative way to optimize the constitution of the molecules until it produces the experimental spectra with a deviation as low as possible. This method avoids the exhaustive generation of all possible isomers (making it much faster), and at the same time it was able to find the correct structures for a set of simple examples.

The CASE approaches described so far heavily rely on the use of 1D NMR spectra and exhaustive generation of structures, which becomes impractical for large molecules. They were successful with modest-sized molecules, typically with less than 25 non-hydrogen atoms. The molecule size limitations were overcome when 2D NMR data became available to second-generation ESs. [127, 128] A variety of 2D NMR techniques reveal spin–spin couplings between magnetic nuclei in the molecule. HSQC and HMBC spectra exhibit couplings between the carbon nucleus and the protons bonded to this atom, thus providing a link between signals in the ^{1}H and ^{13}C spectra. COSY spectra reveal the interaction between protons separated by two or three bonds (although longer-range couplings are possible). Long-range proton–proton interactions transmitted through a chain of neighboring protons can be obtained with TOCSY spectra, while NOESY and ROESY techniques provide information regarding through-space interactions and the stereochemistry of the molecule.

2D NMR spectra enable CASE systems to deduce a molecular structure much more easily guided by the information about connectivity – and to tackle much more complex structures.

However, since uncertainty arises from the quality of analytical data, from the signal overlapping in ^{1}H and ^{13}C spectra, and from the fact that 2D NMR spectra can also exhibit signals corresponding to couplings between nuclei at "uncommon" distances, ESs require algorithms to process fuzzy data. Examples of such systems are LSD [129] and ACD structure elucidator (Figure 5.9). Successful applications include the revision of published structures for

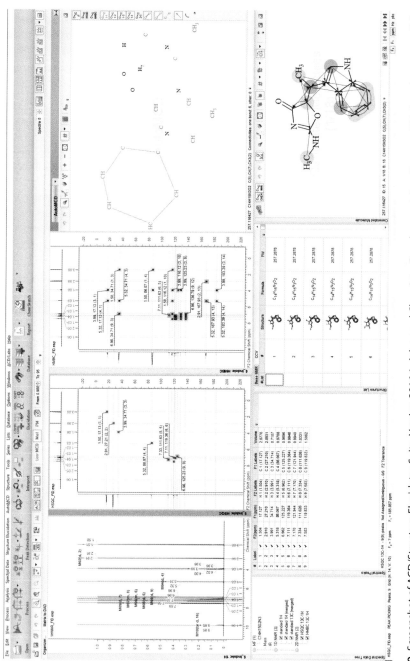

Figure 5.9 Screenshot of ACD/Structure Elucidator Suite, Version 2016.1, Advanced Chemistry Development, Inc., Toronto, ON, Canada, www.acdlabs.com.

highly complex natural products [130, 131] and the elucidation of structures "undecipherable" by classical NMR methods [132].

To conclude, the practical usefulness of CASE systems depends not only on their developments but also on their interaction with state-of-the-art spectroscopic techniques. They are highly recommended for the elucidation of new complex structures, combining information from MS, 1D ,and 2D NMR experiments – very challenging problems that are prone to mistakes in a purely human intellectual procedure.

Essentials

- A proper representation of the molecular structure is crucial for the prediction of spectra. Fragment-based methods, topological descriptors, physicochemical descriptors, and 3D descriptors have been used for this endeavor.
- NMR spectra have been predicted using quantum chemistry calculations, database searches, additive methods, regressions, and ML techniques.
- Several methods have been developed for establishing correlations between IR vibrational bands and substructure fragments. CPG NNs were used to make predictions of the full spectra from RDF codes of the molecules.
- Correlations between structure and mass spectra were established on the basis of multivariate analysis of the spectra, database searching, and the development of knowledge-based systems, some including explicit management of chemical reactions or molecular graphs.
- Robust implementations can currently propose correct structures from spectroscopic data, especially when the molecular formula and ^{13}C NMR spectrum are available for small molecules or from 2D NMR spectra in more complex situations.
- The development of CASE systems stood as one of the roots of chemoinformatics. Developments in recent years have revived this highly challenging field.

Available Software and Web Services (accessed January 2018)

- ACD software for spectra processing and prediction (NMR and MS), structure verification and automatic structure elucidation, Advanced Chemistry Development, Inc., www.acdlabs.com
- LSD (software to find all possible molecular structures of an organic compound that are compatible with its spectroscopic data); Jean-Marc Nuzillard, Reims Institute of Molecular Chemistry, France, http://eos.univ-reims.fr/LSD
- NMRPredict plug-in of Mnova (Prediction of NMR spectra); Mestrelab Research, S.L, www.mestrelab.com
- Mass Frontier (MS database management, spectra classification and prediction); HighChem, Ltd., www.highchem.com
- MetFrag (A combinatorial fragmenter for identifying product ions in mass spectra and searching chemical structure databases for matching molecules);

Leibniz Institute of Plant Biochemistry, Halle, Germany,
  http://c-ruttkies.github.io/MetFrag/
- CFM-ID (competitive fragmentation modeling for metabolite identification predicts MS spectra, assigns peaks and identifies metabolites in spectra generated by electrospray tandem mass spectrometry); Wishart research group, University of Alberta, Canada, https://sourceforge.net/projects/cfm-id
- A web database of NMR spectra, allowing for searching and prediction using HOSE codes and machine learning methods, http://www.nmrshiftdb.org
- Luc Patiny web service for NMR spectra processing, simulation and prediction, http://www.nmrdb.org
- Gasteiger and Aires-de-Sousa web service for ^{1}H NMR prediction, http://joao.airesdesousa.com/spinus
- ProShift Server for protein chemical shift prediction, http://www.meilerlab.org/index.php/servers/
- Leibniz Institute of Plant Biochemistry web service for the automatic assignment of MS peaks and matching against databases of molecules, http://msbi.ipb-halle.de/MetFrag
- Wishart CFM-ID web service for MS spectra prediction from the molecular structure, peak assignment and compound identification, http://cfmid.wishartlab.com
- Advanced Chemistry Development web services for spectra prediction, http://www.acdlabs.com/resources/ilab
- Bio-Rad movies on spectra processing and prediction, http://www.training.knowitall.com

Selected Reading

- Clerc, T.L. (1987) Automated spectra interpretation and library search systems, in *Computer-Enhanced Analytical Spectroscopy* (eds H.L.C. Meuzelaar and T.L. Isenhour), Plenum Press, New York, pp. 145–162.
- Elyashberg, M., Blinov, K., Molodtsov, S., Smurnyy, Y., Williams, A.J., and Churanova, T. (2009) Computer-assisted methods for molecular structure elucidation: realizing a spectroscopist's dream. *J. Cheminf.*, **1**, 3.
- Lindsay, R.K., Buchanan, B.G., Feigenbaum, E.A., and Lederberg, J. (1980) *Applications of Artificial Intelligence for Organic Chemistry – The DENDRAL Project*, McGraw-Hill, New York, 194pp.

Acknowledgement

This chapter is an updated and extended version of Hemmer, M., and Aires-de-Sousa, J. (2003) Structure-Spectra Correlations, in Chemoinformatics – A Textbook (eds. J. Gasteiger and T. Engel), Wiley-VCH, pp. 515–541

References

[1] Elyashberg, M.E., Williams, A.J., and Martin, G.E. (2008) *Prog. Nucl. Magn. Reson. Spectrosc.*, **53**, 1–104.

[2] Lindsay, R.K., Buchanan, B.G., Feigenbaum, E.A., and Lederberg, J. (1980) *Applications of Artificial Intelligence for Organic Chemistry – The DENDRAL Project*, McGraw-Hill, New York, 194pp.

[3] Luger, G.F. and Stubblefield, W.A. (1989) *Artificial Intelligence and the Design of Expert Systems*, Benjamin/Cummings Publishing, Redwood City, CA, 685pp.

[4] Griffiths, L. and Bright, J.D. (2002) *Magn. Reson. Chem.*, **40**, 623–634.

[5] Golotvin, S.S., Vodopianov, E., Lefebvre, B.A., Williams, A.J., and Spitzer, T.D. (2006) *Magn. Reson. Chem.*, **44**, 524–538.

[6] Keyes, P., Hernandez, G., Cianchetta, G., Robinson, J., and Lefebvre, B. (2009) *Magn. Reson. Chem.*, **47**, 38–52.

[7] Golotvin, S.S., Pol, R., Sasaki, R.R., Nikitina, A., and Keyes, P. (2012) *Magn. Reson. Chem.*, **50**, 429–435.

[8] Todeschini, R. and Consonni, V. (2002) *Handbook of Molecular Descriptors*, Wiley-VCH, Weinheim, 688pp.

[9] Gasteiger, J., Sadowski, J., Schuur, J., Selzer, P., Steinhauer, L., and Steinhauer, V. (1996) *J. Chem. Inf. Comput. Sci.*, **36**, 1030–1037.

[10] Munk, M.E., Madison, M.S., and Robb, E.W. (1991) *Mikrochim. Acta [Wien]*, **104**, 505–514.

[11] Cleva, C., Cachet, C., Cabrol-Bass, D., and Forrest, T.P. (1997) *Anal. Chim. Acta*, **348**, 255–265.

[12] Weigel, U.-M. and Herges, R. (1996) *Anal. Chim. Acta*, **331**, 63–74.

[13] Huixiao, H., Yinling, H., Xinquan, X., and Yufeng, S. (1995) *J. Chem. Inf. Comput. Sci.*, **35**, 979–1000.

[14] Weigel, U.-M. and Herges, R. (1992) *J. Chem. Inf. Comput. Sci.*, **32**, 723–731.

[15] Affolter, C., Baumann, K., Clerc, J.T., Schriber, H., and Pretsch, E. (1997) *Microchim. Acta*, **14**, 143–147.

[16] Dubois, J.E., Mathieu, G., Peguet, P., Panaye, A., and Doucet, J.P. (1990) *J. Chem. Inf. Comput. Sci.*, **30**, 290–302.

[17] Bremser, W. (1978) *Anal. Chim. Acta*, **103**, 355–365.

[18] Schütz, V., Purtuc, V., Felsinger, S., and Robien, W. (1997) *Fresenius' J. Anal. Chem.*, **359**, 33–41.

[19] Heinonen, M., Rantanen, A., Mielikainen, T., Kokkonen, J., Kiuru, J., Ketola, R.A., and Rousu, J. (2008) *Rapid Commun. Mass Spectrom.*, **22**, 3043–3052.

[20] Sadowski, J. and Gasteiger, J. (1993) *Chem. Rev.*, **93**, 2567–2573.

[21] Molecular Networks GmbH: CORINA, https://www.mn-am.com/online_demos/corina_demo (accessed January 2018).

[22] Bellamy, L.J. (1975) *The Infrared Spectra of Complex Molecules*, John Wiley & Sons, Ltd, Chichester, 433pp.

[23] Dolphin, D. and Wick, A. (1977) *Tabulation of Infrared Spectral Data*, John Wiley & Sons, Inc., New York, 568pp.
[24] Pretsch, E., Bühlmann, P., and Badertscher, M. (2009) *Structure Determination of Organic Compounds: Tables of Spectral Data*, Springer, Berlin, 433pp.
[25] Lin-Vien, D., Colthup, N.B., Fately, W.G., and Grasselli, J.G. (1991) *The Handbook of Infrared and Raman Characteristic Frequencies of Organic Molecules*, Academic Press, New York, 503pp.
[26] Zvereva, E.E., Shagidullin, A.R., and Katsyuba, S.A. (2011) *J. Phys. Chem. A*, **115** (1), 63–69.
[27] Gribov, L.A. and Elyashberg, M.E. (1979) *Crit. Rev. Anal. Chem.*, **8**, 111–220.
[28] Elyashberg, M.E. (1998) *Infrared spectra interpretation by the characteristic frequency approach*, in *The Encyclopedia of Computational Chemistry* (eds P. von Rague Schleyer, N.L. Allinger, T. Clark, J. Gasteiger, P.A. Kollman, H.F. Schaefer III,, and P.R. Schreiner), John Wiley & Sons, Ltd, Chichester, pp. 1299–1306.
[29] Grund, R., Kerber, A., and Laue, R. (1992) *MATCH*, **27**, 87–131.
[30] Zupan, J. (1989) *Algorithms for Chemists*, John Wiley & Sons, Inc., New York, 306pp.
[31] Luinge, H.J. (1990) *Vib. Spectrosc.*, **1**, 3–18.
[32] Warr, W. (1993) *Anal. Chem.*, **65**, 1087A–1095A.
[33] Elyashberg, M.E., Gribov, L.A., and Serov, V.V. (1980) *Molecular Spectral Analysis and Computers*, Nauka, Moskow (in Russian).
[34] Funatsu, K., Susuta, Y., and Sasaki, S. (1989) *Anal. Chim. Acta*, **220**, 155–169.
[35] Wythoff, B., Xiao, H.Q., Levine, S.P., and Tomellini, S.A. (1991) *J. Chem. Inf. Comput. Sci.*, **31**, 392–399.
[36] Andreev, G.N. and Argirov, O.K. (1995) *J. Mol. Struct.*, **347**, 439–448.
[37] Debska, B., Guzowska-Swider, B., and Cabrol-Bass, D. (2000) *J. Chem. Inf. Comput. Sci.*, **40**, 330–338.
[38] Karpushkin, E., Bogomolov, A., Zhukov, Y., and Borut, M. (2007) *Chemom. Intell. Lab. Syst.*, **88**, 107–117.
[39] Woodruff, H.B. and Smith, G.M. (1980) *Anal. Chem.*, **52**, 2321–2327.
[40] Andreev, G.N., Argirov, O.K., and Penchev, P.N. (1993) *Anal. Chim. Acta*, **284**, 131–136.
[41] Penchev, P.N., Kotchev, N.T., and Andreev, G.N. (2000) *Traveaux Scientifiques d'Université de Plovdiv*, **29**, 21–26.
[42] Steinhauer, L., Steinhauer, V., and Gasteiger, J. (1996) O*btaining the 3D structure from infrared spectra of organic compounds using neural networks*, in *Software-Entwicklung in der Chemie 10* (ed. J. Gasteiger), Gesellschaft Deutscher Chemiker, Frankfurt/Main, pp. 315–322.
[43] Selzer, P., Salzer, R., Thomas, H., and Gasteiger, J. (2000) *Chem. Eur. J.*, **6**, 920–927.
[44] Kostka, T., Selzer, P., and Gasteiger, J. (1997) *Computer-assisted prediction of the degradation products and infrared spectra of s-triazine herbicides*,

in *Software-Entwicklung in der Chemie 11* (eds G. Fels and V. Schubert), Gesellschaft Deutscher Chemiker, Frankfurt/Main, pp. 227–233.

[45] Hemmer, M.C., Steinhauer, V., and Gasteiger, J. (1999) *Vib. Spectrosc.*, **19**, 151–164.

[46] Liao, C., Chen, Z., Yin, Z., and Li, S.Z. (2003) *Comput. Biol. Chem.*, **27**, 229–239.

[47] Mu, G., Liu, H., Wen, Y., and Luan, F. (2011) *Vib. Spectrosc.*, **55**, 49–57.

[48] Elyashberg, M. (2015) *TrAC Trends Anal. Chem.*, **69**, 88–97.

[49] Elyashberg, M., Blinov, K., and Williams, A. (2009) *Magn. Reson. Chem.*, **47**, 371–389.

[50] Lodewyk, M.W., Siebert, M.R., and Tantillo, D.J. (2012) *Chem. Rev.*, **112**, 1839–1862.

[51] Ditchfield, R. (1974) *Mol. Phys.*, **27**, 789–811.

[52] Wolinski, K., Hinton, J.F., and Pulay, P. (1990) *J. Am. Chem. Soc.*, **112**, 8251–8260.

[53] For examples see: (a) Rychnovsky, S.D. (2006) *Org. Lett.*, **8**, 2895–2898; (b) Meng, Z. and Carper, W.R. (2002) *J. Mol. Struct.*, **588**, 45–53; (c) Czernek, J. and Sklenár, V. (1999) *J. Phys. Chem. A*, **103**, 4089–4093; (d) Barfield, M. and Fagerness, P. (1997) *J. Am. Chem. Soc.*, **119**, 8699–8711.

[54] Gaussian, Inc., http://www.gaussian.com (accessed January 2018).

[55] Toomsalu, E. and Burk, P. (2015) *J. Mol. Model.*, **21**, 24.

[56] Meiler, J., Lefebvre, B., Williams, A., and Hachey, M. (2002) *J. Magn. Reson.*, **157**, 242–252.

[57] Hypercube, Inc., http://www.hyper.com (accessed January 2018).

[58] For a description of the method in the context of 13C NMR seeRobien, W. (1998) *NMR data correlation with chemical structure*, in *The Encyclopedia of Computational Chemistry* (eds P. von Rague Schleyer, N.L. Allinger, T. Clark, J. Gasteiger, P.A. Kollman, H.F. Schaefer III,, and P.R. Schreiner), John Wiley & Sons, Ltd, Chichester, pp. 1845–1857.

[59] Modgraph Consultants Ltd, http://www.modgraph.co.uk (accessed January 2018).

[60] Advanced Chemistry Development, Inc., http://www.acdlabs.com (accessed January 2018).

[61] Loß, A., Stenutz, R., Schwarzer, E., and von der Lieth, C.-W. (2006) *Nucleic Acids Res.*, **34** (Web Server issue), W733–W737.

[62] Wishart, D.S., Watson, M.S., Boyko, R.F., and Sykes, B.D. (1997) *J. Biomol. NMR*, **10**, 329–336.

[63] Tusar, M., Tusar, L., Bohanec, S., and Zupan, J. (1992) *J. Chem. Inf. Comput. Sci.*, **32**, 299–303.

[64] (a) Schaller, R.B. and Pretsch, E. (1994) *Anal. Chim. Acta*, **290**, 295–302; (b) Fürst, A. and Pretsch, E. (1995) *Anal. Chim. Acta*, **312**, 95–105 and references cited therein.

[65] PerkinElmer Inc., http://www.cambridgesoft.com (accessed January 2018).

[66] Pretsch, E., Bühlmann, P., and Badertscher, M. (2009) *Structure Determination of Organic Compounds: Tables of Spectral Data*, 4th edn., Springer-Verlag, Berlin, 433pp.

[67] Clerc, J.T. and Sommerauer, H. (1977) *Anal. Chim. Acta*, **95**, 33–40.

[68] Small, G.W. and Jurs, P.C. (1984) *Anal. Chem.*, **56**, 1314–1323.

[69] Schweitzer, R.C. and Small, G.W. (1996) *J. Chem. Inf. Comput. Sci.*, **36**, 310–322.

[70] (a) Jansson, P.E., Kenne, L., and Widmalm, G. (1991) *J. Chem. Inf. Comput. Sci.*, **31**, 508–516; (b) Lundborg, M. and Widmalm, G. (2011) *Anal. Chem.*, **83**, 1514–1517.

[71] Abraham, R.J. (1999) *Prog. Nucl. Magn. Reson. Spectrosc.*, **35**, 85–152.

[72] Neal, S., Nip, A.M., Zhang, H., and Wishart, D.S. (2003) *J. Biomol. NMR*, **26**, 215–240.

[73] Wijmenga, S.S., Kruithof, M., and Hilbers, C.W. (1997) *J. Biomol. NMR*, **10**, 337–350.

[74] (a) Gasteiger, J. (1988) *Empirical methods for the calculation of physicochemical data of organic compounds*, in *Physical Property Prediction in Organic Chemistry* (eds C. Jochum, M.G. Hicks, and J. Sunkel), Springer-Verlag, Heidelberg, pp. 119–138; (b) PETRA software, Molecular Networks GmbH, https://www.mn-am.com/ (accessed January 2018).

[75] Aires-de-Sousa, J., Hemmer, M., and Gasteiger, J. (2002) *Anal. Chem.*, **74**, 80–90.

[76] Binev, Y., Marques, M.M., and Aires-de-Sousa, J. (2007) *J. Chem. Inf. Model.*, **47**, 2089–2097.

[77] Tetko, I.V. (2002) *J. Chem. Inf. Comput. Sci.*, **42**, 717–728.

[78] Kuhn, S., Egert, B., Neumann, S., and Steinbeck, C. (2008) *BMC Bioinformatics*, **9**, 400.

[79] Meiler, J. (2003) *J. Biomol. NMR*, **26**, 25–37.

[80] Clouser, D.L. and Jurs, P.C. (1996) *J. Chem. Inf. Comput. Sci.*, **36**, 168–172.

[81] Ivanciuc, O., Rabine, J.P., Cabrol-Bass, D., Panaye, A., and Doucet, J.P. (1997) *J. Chem. Inf. Comput. Sci.*, **37**, 587–598.

[82] Meiler, J., Meusinger, R., and Will, M. (2000) *J. Chem. Inf. Comput. Sci.*, **40**, 1169–1176 and references cited therein.

[83] Smurnyy, Y.D., Blinov, K.A., Churanova, T.S., Elyashberg, M.E., and Williams, A.J. (2008) *J. Chem. Inf. Model.*, **48**, 128–134.

[84] Vulpetti, A., Landrum, G., Rudisser, S., Erbel, P., and Dalvit, C. (2010) *J. Fluor. Chem.*, **131**, 570–577.

[85] Gabano, E., Marengo, E., Bobba, M., Robotti, E., Cassino, C., Botta, M., and Osella, D. (2006) *Coord. Chem. Rev.*, **250**, 2158–2174.

[86] West, G.M.J. (1993) *J. Chem. Inf. Comput. Sci.*, **33**, 577–589.

[87] Shen, Y. and Bax, A. (2010) *J. Biomol. NMR*, **48**, 13–22.

[88] Frank, A.T., Bae, S.H., and Stelzer, A.C. (2013) *J. Phys. Chem. B*, **117**, 13497–13506.

[89] Frank, A.T., Law, S.M., and Brooks, C.L. (2014) *J. Phys. Chem. B*, **118**, 12168–12175.

[90] Clerc, T.L. (1987) *Automated Spectra Interpretation and Library Search Systems in Computer-Enhanced Analytical Spectroscopy*, Plenum Press, New York, pp. 145–162.

[91] McLafferty, F.W., Loh, S.Y., and Stauffer, D.B. (1990) *Computer identification of mass spectra*, in *Computer-Enhanced Analytical Spectroscopy* (ed. H.L.C. Meuzelaar), Plenum Press, New York, pp. 163–181.

[92] Varmuza, K. (2000) *Chemical structure information from mass spectrometry*, in *Encyclopedia of Spectroscopy and Spectrometry* (eds J.C. Lindon, G.E. Tranter, and J.L. Holmes), Academic Press, London, pp. 232–243.

[93] McLafferty, F.W. and Hertel, R.H. (1994) *Org. Mass Spectrom.*, **8**, 690–702.

[94] Stein, S.E. and Scott, D.R. (1994) *J. Am. Soc. Mass Spectrom.*, **5**, 856–866.

[95] Adams, M.J. and Barnett, N.W. (2004) *Chemometrics in Analytical Spectroscopy*, 2nd edn, The Royal Society of Chemistry, Cambridge, 238pp.

[96] Beebe, K.R., Pell, R.J., and Seasholtz, M.B. (1998) *Chemometrics: A Practical Guide*, John Wiley & Sons, Inc., New York, 360pp.

[97] Massart, D.L., Vandeginste, B.G.M., Buydens, L.C.M., De Jong, S., and Smeyers-Verbeke, J. (1997) *Handbook of Chemometrics and Qualimetrics: Part A*, Elsevier, Amsterdam.

[98] Vandeginste, B.G.M., Massart, D.L., Buydens, L.C.M., De Jong, S., and Smeyers-Verbeke, J. (1998) *Handbook of Chemometrics and Qualimetrics: Part B*, Elsevier, Amsterdam, 876pp.

[99] Varmuza, K. (1998) *Chemometrics: multivariate view on chemical problems*, in *The Encyclopedia of Computational Chemistry* (eds P. von Rague Schleyer, N.L. Allinger, T. Clark, J. Gasteiger, P.A. Kollman, I.H.F. Schaefer, and P.R. Schreiner), John Wiley & Sons, Ltd, Chichester, pp. 346–366.

[100] Zupan, J. and Gasteiger, J. (1999) *Neural Networks in Chemistry and Drug Design*, 2nd edn, Wiley-VCH, Weinheim, 380pp.

[101] Gasteiger, J., Hanebeck, W., and Schultz, K.-P. (1992) *J. Chem. Inf. Comput. Sci.*, **32**, 264–271.

[102] Gasteiger, J., Hanebeck, W., Schultz, K.-P., Bauerschmidt, S., and Höllering, R. (1993) *Automatic analysis and simulation of mass spectra*, in *Computer-Enhanced Analytical Spectroscopy*, vol. 4 (ed. C.L. Wilkins), Plenum Press, New York, pp. 97–133.

[103] Jalali-Heravi, M. and Fatemi, M.H. (2000) *Anal. Chim. Acta*, **415**, 95–103.

[104] HighChem, Ltd., http://www.highchem.com (accessed January 2018).

[105] Rasche, F., Svatos, A., Maddula, R.K., Böttcher, C., and Böcker, S. (2011) *Anal. Chem.*, **83**, 1243–1251.

[106] Neumann, S. and Böcker, S. (2010) *Anal. Bioanal. Chem.*, **398**, 2779–2788.

[107] Hill, D.W., Kertesz, T.M., Fontaine, D., Friedman, R., and Grant, D.F. (2008) *Anal. Chem.*, **80**, 5574–5582.

[108] Allen, F., Greiner, R., and Wishart, D. (2015) *Metabolomics*, **11**, 98.

[109] Wolf, S., Schmidt, S., Müller-Hannemann, M., and Neumann, S. (2010) *BMC Bioinformatics*, **11**, 148.

[110] Ruttkies, C., Schymanski, E.L., Wolf, S., Hollender, J., and Neumann, S. (2016) *J. Cheminform.*, **8**, 3.

[111] Hill, A.W. and Mortishire-Smith, R.J. (2005) *Rapid Commun. Mass Spectrom.*, **19**, 3111–3118.

[112] Dührkop, K., Shen, H., Meusel, M., Rousu, J., and Böcker, S. (2015) *Proc. Natl. Acad. Sci. U. S. A.*, **112**, 12580–12585.

[113] Buchanan, B.G. and Feigenbaum, E.A. (1978) *Artif. Intell.*, **11**, 5–24.
[114] Lederberg, J. (1987) How DENDRAL was Conceived and Born. ACM Symposium on the History of Medical Informatics, National Library of Medicine. Later published in Blum, B.I. and Duncan, K. (eds) (1990) A History of Medical Informatics, Association for Computing, Machinery Press, New York, pp. 14–44.
[115] Carhart, R.E., Smith, D.H., Brown, H., and Djerassi, C. (1975) *J. Am. Chem. Soc.*, **97**, 5755–5763.
[116] Carhart, R.E., Smith, D.H., Gray, N.A., Nourse, J.B., and Djerassi, C. (1981) *J. Org. Chem.*, **46**, 1708–1718.
[117] Funatsu, K., Miyabayaski, N., and Sasaki, S. (1988) *J. Chem. Inf. Comput. Sci.*, **28**, 18–23.
[118] Shelley, C.A., Hays, T.R., Munk, M.E., and Roman, R.V. (1978) *Anal. Chim. Acta*, **103**, 121–132.
[119] Kalchhauser, H. and Robien, W. (1985) *J. Chem. Inf. Comput. Sci.*, **25**, 103–108.
[120] Carabedian, M., Dagane, I., and Dubois, J.E. (1988) *Anal. Chem.*, **60**, 2186–2192.
[121] Munk, M.E., Madison, M.S., and Robb, E.W. (1996) *J. Chem. Inf. Comput. Sci.*, **35**, 231–238.
[122] Luinge, H.J., van der Maas, J.H., and Visser, T. (1995) *Chemom. Intell. Lab. Syst.*, **28**, 129–138.
[123] Klawun, C. and Wilkins, C.L. (1996) *J. Chem. Inf. Comput. Sci.*, **36**, 69–81.
[124] Bangov, I.P. (1994) *J. Chem. Inf. Comput. Sci.*, **34**, 277–284.
[125] Contreras, M.L., Rozas, R., and Valdivias, R. (1994) *J. Chem. Inf. Comput. Sci.*, **34**, 610–619.
[126] Meiler, J. and Will, M. (2002) *J. Am. Chem. Soc.*, **124**, 1868–1870.
[127] Christie, B.D. and Munk, M.E. (1991) *J. Am. Chem. Soc.*, **113**, 3750–3757.
[128] Elyashberg, M., Blinov, K., Molodtsov, S., Smurnyy, Y., Williams, A.J., and Churanova, T. (2009) *J. Chem. Theory Comput.*, **1**, 3.
[129] Plainchont, B., Emerenciano, V.P., and Nuzillard, J.-M. (2013) *Magn. Reson. Chem.*, **51**, 447–453.
[130] Elyashberg, M., Williams, A.J., and Blinov, K. (2010) *Nat. Prod. Rep.*, **27**, 1296–1328.
[131] Elyashberg, M., Blinov, K., Molodtsov, S., and Williams, A.J. (2013) *J. Nat. Prod.*, **76**, 113–116.
[132] Elyashberg, M.E., Blinov, K.A., Molodtsov, S.G., and Williams, A.J. (2012) *Magn. Reson. Chem.*, **50**, 22–27.

6.1 Drug Discovery: An Overview

Lothar Terfloth[1], Simon Spycher[2], and Johann Gasteiger[3]

[1] Insilico Biotechnology AG, Meitnerstrasse 9, 70563 Stuttgart, Germany
[2] Eawag, Environmental Chemistry, Überlandstrasse 133, 8600 Dübendorf, Switzerland
[3] Computer-Chemie-Centrum, University of Erlangen-Nürnberg, Nägelsbachstr. 25, 91052 Erlangen, Germany

Learning Objectives

- To describe the drug discovery process
- To obtain a framework for Sections 6.2–6.11
- To identify a lead structure
- To discuss the impact of chemoinformatics on the drug discovery process
- To describe the "similar structure–similar property" principle
- To distinguish the difference between ligand-based and structure-based drug discovery
- To review Lipinski's "rule of five"
- To describe what ADMET properties are and what role they play in drug design

Outline

6.1.1 Introduction, 165
6.1.2 Definitions of Some Terms Used in Drug Design, 167
6.1.3 The Drug Discovery Process, 167
6.1.4 Bio- and Chemoinformatics Tools for Drug Design, 168
6.1.5 Structure-based and Ligand-Based Drug Design, 168
6.1.6 Target Identification and Validation, 169
6.1.7 Lead Finding, 171
6.1.8 Lead Optimization, 182
6.1.9 Preclinical and Clinical Trials, 188
6.1.10 Outlook: Future Perspectives, 189

6.1.1 Introduction

In this section we give an overview on the applications of chemoinformatics in the drug discovery process. After a brief introduction, definitions of some important terms used in drug design are given. Then the drug discovery process will be described. A detailed presentation of all aspects of the use of chemoinformatics methods during the drug design process is beyond the scope of this chapter. However, we were lucky to win experts for presenting some important tasks and problems in the drug design and development process that are tackled by chemoinformatics. These are presented in Sections 6.2–6.13. Other aspects are only cursory mentioned in Section 6.1. Readers interested in medicinal chemistry can obtain an excellent overview from the book *The Practice of Medicinal Chemistry* edited by Wermuth [1]. The development of new drugs is presented in two books, one in German by Böhm *et al.* [2] and a more recent one in English by Klebe [3]. An overview on different classes of drugs and their mechanism of action is given by Mutschler [4].

Many drugs such as the sulfonamides introduced by Domagk or penicillin by Fleming were discovered by serendipity and not as a result of rational drug design [5]. Up to the 1970s, hypothetical activity models dominated the syntheses of new compounds in drug research. The biological activity of these compounds was verified by experiments with isolated organs or animals. Accordingly, the throughput was limited by the speed of the biological tests. From about 1980 on, the development of *in vitro* models for enzyme inhibition and receptor binding studies attained a growing impact on drug research. In these years, the synthesis of compounds became the time-limiting factor. Based on the progress achieved in experimental techniques such as gene technology, combinatorial chemistry, and high-throughput testing, it became feasible to produce the proteins of interest and to rapidly obtain biological data. Today, genomics, proteomics, bioinformatics, combinatorial chemistry, and ultra-high-throughput screening (uHTS) provide an enormous amount of targets and data.

Concomitant with these achievements, the development in molecular modeling and other chemoinformatics methods has completely reshaped the drug development process. The application of data mining methods and virtual screening obtained a growing impact on the validation of "druggable" targets, lead finding, and the prediction of suitable absorption, distribution, metabolism, excretion, and toxicity (ADMET) profiles.

The development of a new drug is both a time-consuming and a cost-intensive process. It takes 12–15 years and costs up to €2.6 billion to bring a new drug to the market [6]. An "ideal" drug must be effective and safe and fulfill various additional criteria. The maximum daily dose should not exceed an amount of 200–500 mg. Drugs should be orally well absorbed and bioavailable. Metabolic stability should ensure a reasonable long half-life. Furthermore, a drug should be nontoxic and cause no or only minimal adverse effects. Finally, an ideal drug will distribute selectively to the target tissues. Despite the introduction of combinatorial chemistry and the establishment of high-throughput screening (HTS), the number of new chemical entities (NCEs) introduced into the world market did not change very much in the last decades. It was about 37 NCEs per year in the decade 1991–1999 [7, 8] and 34 per year in the period of 2010–2015 [9]. This standstill certainly has to be attributed to the fact that the requirements imposed on a new drug have become much more stringent.

Furthermore, it turned out that many compounds coming out of the drug design pipeline showed inappropriate ADMET properties and consequently caused their attrition in the preclinical or clinical phase. The estimated attrition rate in drug development is up to 90% [10]. The later in the drug discovery process the development of a new compound is discontinued, the higher is the financial loss for a pharmaceutical company. Therefore, the major reasons for the attrition of a new drug have to be addressed in drug design as early as possible. As a consequence thereof, the prediction of pharmacological properties is a central task of chemoinformatics in drug development in addition to lead finding and is integrated into the drug design process at an early stage.

6.1.2 Definitions of Some Terms Used in Drug Design

Before we continue with the description of the drug discovery process, we introduce some common terms used in drug design and give their definitions in the following list:

- *Lead structure*: According to Valler's and Green's definition, a lead structure is "a representative of a compound series with sufficient potential (as measured by potency, selectivity, pharmacokinetics, physicochemical properties, absence of toxicity, and novelty) to progress to a full drug development program" [11].
- *Ligand*: A ligand is a molecule binding to a biological macromolecule.
- *Enzyme*: Enzymes are endogenous catalysts converting one or several substrates into one or several products.
- *Substrate*: A substrate is the starting material of an enzymatic reaction.
- *Inhibitor*: A ligand preventing the binding of a substrate to its enzyme is called inhibitor.
- *Receptor*: Receptors are membrane-bound or soluble proteins or protein complexes exerting a physiological effect after binding of an agonist.
- *Agonist*: An agonist is a receptor ligand mediating a receptor response (intrinsic effect).
- *Antagonist*: An antagonist is a receptor ligand preventing the action of an agonist, in a direct (competitive) or indirect (allosteric) manner.
- *Ion channel*: A pore formed by proteins allowing the diffusion of certain ions through the cell membrane along a concentration gradient is called an ion channel. The channel opening is either ligand or voltage controlled.
- *Transporter*: A transporter is a protein that is transporting a molecule or ion through a cell membrane against a concentration gradient.
- *New chemical entity (NCE) (or new molecular entity)*: A compound that emerges from the process of drug discovery and has a promising activity against a particular biological target that is important in a disease.

6.1.3 The Drug Discovery Process

The drug discovery process comprises the following steps:

1. Target identification and validation
2. Hit and lead finding
3. Lead optimization
4. Preclinical testing
5. Clinical development
6. Drug approval (e.g., FDA approval)

Figure 6.1.1 indicates the steps involved in the design and development of a drug. This figure should be understood as outlining the tasks that have to be solved not necessarily as a clear-cut sequential step-by-step approach. In reality, quite often during one step, a reinvestigation of an earlier step will be performed to obtain deeper insight with knowledge acquired at a later stage, thus providing a feedback loop in the entire process. In essence, drug design is not a linear sequence of the steps indicated in Figure 6.1.1 but a highly interconnected process.

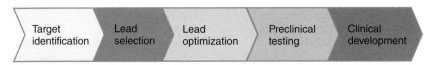

Figure 6.1.1 The drug discovery process.

6.1.4 Bio- and Chemoinformatics Tools for Drug Design

Drug design and development is presently the most important area for the application of bioinformatics and chemoinformatics methods. We will primarily focus on chemoinformatics methods but will point out at certain important steps also bioinformatics methods. Each major pharmaceutical company has a chemoinformatics department; however it is called: chemoinformatics, molecular modeling, chemical information, and so on. Even smaller companies have realized that they cannot do without the use of chemoinformatics methods. Pharmaceutical industry is the most important employer of chemoinformatics experts. Most new drugs are now developed with some assistance from bioinformatics and chemoinformatics methods.

Many chemoinformatics methods have been developed and improved in the course of over four decades. Table 6.1.1 lists the major methods and indicates at which step in the drug design process they are used. Note that some methods are used at several steps in this process, emphasizing that these steps are highly interconnected in the entire process.

Quite a few of these methods are detailed by competent authors in Sections 6.2–6.11. Some other methods are only briefly outlined in Section 6.1. This in no way says that they are less important. However, we had to limit ourselves in outlining the bioinformatics and chemoinformatics methods involved in the drug design process in order to not excessively expand the treatment of this albeit important field in this book. A few examples predominantly from our own research serve to illustrate the application of some of the methods not treated in separate sections.

6.1.5 Structure-based and Ligand-Based Drug Design

If the 3D structure of the target of interest is known from X-ray crystallography, NMR spectroscopy, or homology modeling, a suite of methods in structure-based drug design (SBD) can be applied. If an X-ray structure of the protein with a ligand is available, the binding mode of the ligand can be analyzed. Docking of new ligands with the same binding mode and ranking of these ligands by scoring functions guide the further drug development. Otherwise one has to apply

Table 6.1.1 Chemoinformatics and bioinformatics methods used at various stages of the drug discovery process.

Method	Target identification	Lead finding		Lead optimization		Preclinical testing
		LBD	SBD	LBD	SBD	
Bioinformatics	X					
Biochemical pathways	X					
Similarity search		X	X	X	X	
Lead hopping		X		X		
Natural products		X	X	X	X	
Combichem		X		X		
Synthetic accessibility		X	X	X	X	
High-throughput screening		X		X		
Pharmacophore		X	X	X	X	
Virtual screening		X	X	X	X	
Active site protection			X		X	
De novo design			X			
Docking/scoring			X		X	
QSAR				X	X	
ADME properties				X		X
Metabolism				X		X
Toxicity prediction				X		X

LBD, ligand-based design; SBD, structure-based design.

de novo design or perform a 3D search in a ligand database for a compound with a complementary shape and surface properties to the binding site of the receptor.

In an early stage of drug design without structural information about the target and without any knowledge about ligands binding to the target protein, one is obliged to use ligand-based design (LBD) methods such as to screen combinatorial and proprietary libraries by HTS. As soon as some ligands binding to the target are known, they serve as a starting point for similarity search in a ligand database and for the perception of a pharmacophore by superposition of the ligands. The various LBD and SBD methods are described in more detail in the following sections.

6.1.6 Target Identification and Validation

The process of target identification analyzes a complex disease process by dissecting it into its fundamental components. This allows one to identify the most decisive element for the manifestation of the disease. Target identification aims at understanding the biological processes related to a disease and to identify its mechanism and the structure of individual elements of the disease. Commonly, these individual elements are receptors, enzymes, and so on that become the target of new drugs.

After target selection, the properties of the target gene are then analyzed in disease models that may be cell models in test tubes and/or animal models. This process is called target validation. A target is validated when a specific action on the target has shown favorable effects in the disease models. Several factors have to be considered during target validation, for example, for the development of new antibiotics, acceptable targets will be either those that are essential for the life of the pathogen or are virulence factors. In addition, the target must be divergent between the pathogen and the host, so that modification and/or disruption of the target will attenuate or kill the microbe and/or inhibit its virulence without a detrimental impact on the host.

The steps target identification and target validation are assisted by genomics and functional as well as structural proteomics. Essential bioinformatics methods used in the drug design process are presented in more detail in Section 6.2. The number of drug targets for the existing drugs is currently about 500. An overview on the most important targets and the mechanism of drug action is given in Table 6.1.2. The distribution of different biochemical classes of drug targets is depicted in Figure 6.1.2 [12].

According to the knowledge on the sequence of the human genome, there are over 30,000 potential targets. So far, there is only little knowledge about these

Table 6.1.2 Drug targets and mechanisms of drug action.

Drug target	Mechanism of drug action
Receptors	Agonists and antagonists
Enzymes	Reversible and irreversible inhibitors
Ion channels	Blocker and opener
DNA	Alkylating agents, intercalating agents, wrong substrates (Trojan horses, e.g., 5-fluorouracil)

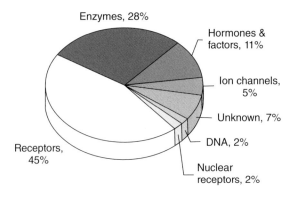

Figure 6.1.2 Distribution of drug targets.

new targets. The question arises how many of these targets can be modulated by potent, small "drug-like" molecules, that is, how many targets are "druggable"? Furthermore, what makes a target druggable? Currently, the estimated number of druggable targets is about 3,000 [13, 14].

An understanding of how drug candidates influence biochemical reactions of the endogenous metabolism is essential for the further development of drugs particularly so if the drug target is an enzyme or some metabolic disease. In this context databases of biochemical reactions such as those presented in Section 4.3 play an important role.

Pharmaceutical research in Western countries has primarily focused on the model

$$one\ drug \rightarrow one\ target$$

In contrast, indigenous knowledge on the merits of plants for curing diseases, particularly in Asian countries, has relied on using entire plants, that is, the full range of chemicals contained in a natural product. Only recently scientists in Western countries started to accept the notion that the secret in medicinal knowledge inherent in approaches such as Traditional Chinese Medicine (TCM) might lay in a disease being attacked at a series of target. Thus, the model

$$one\ drug \rightarrow many\ targets$$

has emerged in an approach called polypharmacology [15–17]. Polypharmacology suggests that more effective drugs can be developed by specifically modulating multiple targets. It is believed that drugs that hit multiple sensitive modes belonging to a network of interacting targets offer the potential for higher efficacy and may limit drawbacks arising from the use of a single target drug or a combination of drugs. Polypharmacology also aims at discovering unknown off-targets for existing drugs, an effect called drug repurposing [18].

In order to eliminate the bottleneck in identifying biological targets, the chemogenomics approach has been developed in recent years [19–21]. This new strategy screens classes of compounds against entire families of functionally related proteins. Chemogenomics has been used to identify the mode of action (MOA) of Traditional Chinese Medicines and of Ayurveda [22]. It has further shown success in identifying totally new therapeutic targets [23] and in identifying genes in biological pathways [24].

6.1.7 Lead Finding

6.1.7.1 Ligand-Based Drug Design (LBD)

6.1.7.1.1 Subset Selection and Similarity/Diversity Search

A fundamental problem encountered in drug design is the search in the huge chemical space. Whereas the estimated number of small, possibly drug-like molecules is in the magnitude of 10^{80}, there are only about 10^8 known compounds. Corporate compound libraries contain 10^7 compounds. Drug databases comprise 10^4 compounds. 10^3 compounds are on the market as commercial drugs. Among these commercial drugs are about 10^2 profitable drugs. The probability of selecting a commercial drug from a corporate compound library assuming that all known drugs are included in this library is $(10^3/10^7)100\% = 0.01\%$ Therefore, powerful techniques for subset selection are necessary for both HTS and virtual screening. There has been a shift of paradigm in the pharmaceutical industry from screening huge (combinatorial) libraries toward more focused libraries allowing hit rates in HTS of up to 10%.

Following the "similar structure–similar property principle," structurally similar molecules are expected to exhibit similar properties or biological activities. In order to select compounds for focused libraries, 2D and 3D similarity searches are performed. A decisive step forward could be made with the advent of automatic 3D structure generators such as CORINA [25–27] that could generate a 3D structure for any conceivable ligand. The seed structure for similarity perception will either be a drug having the desired activity that should be replaced by a compound that could obtain a new patent or a ligand binding to the target protein of interest. The distance of all ligands in a database to the ligands known to bind to the target of interest is calculated as described in the *Methods Volume*, Section 10.2.3.1. Then, the ligands are ranked in reverse order of their distance. The ligands with a high rank are selected for the focused library.

Methods to analyze the diversity of the selected subset ensure that an appropriate chemical space is covered. Molecular descriptors that can be used for the analysis of the similarity *resp.* diversity of compounds have been introduced in the *Methods Volume*, Chapter 10. Frequently applied distance measures are the Hamming distance or the Tanimoto coefficient (see *Methods Volume*, Section 10.2.1.3). One way to quantify the diversity of a library is to determine the centroid of all compounds and to calculate the mean and standard deviation of the pairwise distance of all compounds within the library to the centroid. Quite often interest is in finding lead structures with new scaffolds. Thus, methods that allow lead hopping are of interest.

For thousands of years, humans have used natural products for curing diseases. Nature has produced a cornucopia of complex molecular structures with a variety of interesting scaffolds. Therefore, it is not surprising that the rich sources of chemical structures have been explored and exploited for the development of new drugs. It has been determined that about 40% of present drugs have been developed by modifications of natural products [28]. Section 6.3 presents chemoinformatics methods applied to natural products. Section 6.4 presents methods how the experience of thousands of years in Traditional Chinese Medicines can be exploited with modern methods.

Example: Distinguishing Molecules of Different Biological Activity and Finding of New Lead Structures
It is essential to choose such molecular descriptors for similarity perception that reflect the biological activity of interest. As each receptor has a specific shape of the binding pocket and exhibits specific interaction fields, the first step in the definition of molecular similarity is the search for problem-specific molecular descriptors. This will be illustrated with a dataset of 172 molecules showing activity in the central nervous system (CNS), being

either benzodiazepine agonists (BDA) (60 compounds) or dopamine agonists (DAA)(112 compounds) [29].

The structures were represented by topological autocorrelation (see *Methods Volume*, Section 10.3.3.2) with the topological distance between the two atoms *i* and *j* running from 2 to 8 (seven distances altogether). As no specific information on the receptor was available, a rather broad structure representation was used covering seven physicochemical effects in the autocorrelation vector (σ-atomic charges, (σ+π) atomic charges, σ-electronegativity, π-electronegativity, lone pair electronegativity, atomic polarizability, and an atomic property of one (to just represent the molecular graph)). Thus, each molecule was represented by a 49 (7×7)-dimensional vector. Similarity perception was delegated to a self-organizing neural network (self-organizing map (SOM), Kohonen network) (see *Methods Volume*, Section 11.2) of size 10×7 neurons with the software SONNIA [30]. The entire dataset provided a map (SOM) that was then marked by assigning colors to the neurons of the network depending on whether a neuron contains a dopamine agonist (DDA) or a benzodiazepine agonist (BDA) (Figure 6.1.3) [29].

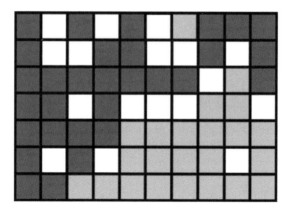

Figure 6.1.3 Kohonen map (10x7) obtained from a dataset of 112 dopamine agonists and 60 benzodiazepine agonists.

As can be seen, the two sets of molecules separate quite well. This is even more remarkable as the class membership was not used in training the network but only in visualizing the results of training (unsupervised learning!). This attests to the relevance of the chosen structure representation for reproducing effects that are responsible for the different binding of the two CNS active compounds, DAA and BDA.

Having now a structure representation that separates BDA and DAA quite well, these molecular descriptors can be used for searching for similar molecules that might serve as new lead structures. For illustration purposes, an entire catalog of 8,223 compounds available from a chemical supplier (Janssen Chimica) was added to this dataset of 172 molecules. (In real-life applications much larger datasets would be used.) With this larger dataset, one has to also increase the size of the SOM; a network of 40x30 neurons was chosen. Training this network with the same 49-dimensional structure representation as previously described, but now for all 8,395 structures provided the map (SOM) shown in Figure 6.1.4.

The DAA and BDA still separate quite well. What is, however, even more important is that we now know in which chemical space one would have to search for new lead structures for DAA or for BDA.

To further illustrate the point that this approach can also be used for lead hopping, the contents of molecules mapped into the neuron at position 5,8 of the SOM presented in Figure 6.1.4 are shown in Figure 6.1.5. This neuron obtained five BDA with two different scaffolds; thus if one would have started with one molecular skeleton, also compounds with a different skeleton would have been found.

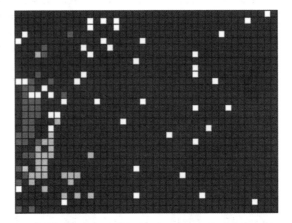

Figure 6.1.4 Kohonen map (40x30) of a dataset consisting of the dopamine and benzodiazepine agonists of Figure 6.1.3 and 8,223 compounds of a chemical supplier catalog.

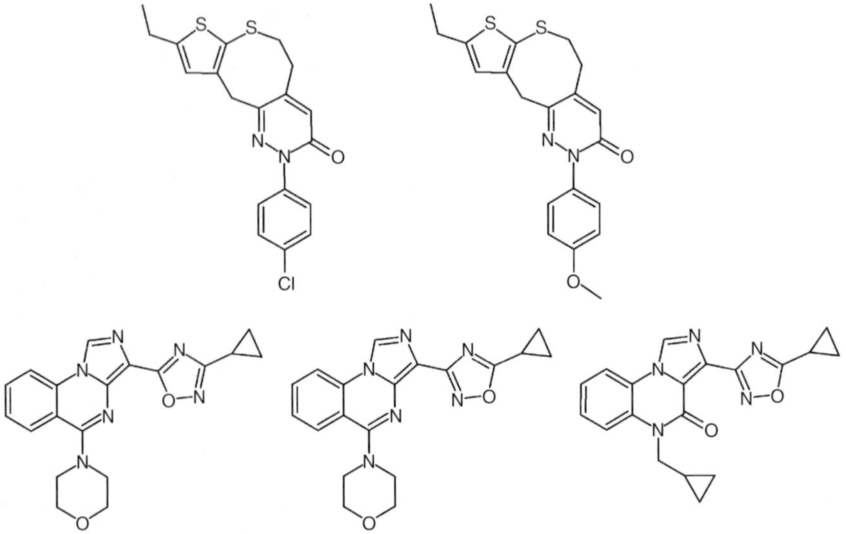

Figure 6.1.5 Structures that were mapped into the neuron at position 5,8 of the Kohonen map of Figure 6.1.4.

The results presented here imply that a similar approach can also be used for comparing two different libraries and for determining the degree of overlap between the compounds in these two libraries.

6.1.7.1.2 Design of Combinatorial Libraries

HTS data as well as virtual screening can guide and direct the design of combinatorial libraries. The number of compounds accessible by combinatorial synthesis quite often exceeds the number of compounds that can be synthesized by traditional methods. To reduce the number of products, a subset of fragments has to be chosen. Sheridan and Kearsley demonstrate the selection of a subset of amines for the construction of a tripeptoid library with a genetic algorithm using a similarity measure to a specific tripeptoid target as scoring function [31].

In order to synthesize libraries rich in biological motifs, databases of biological active compounds were fragmented by a system, RECAP, containing rules corresponding to eleven commonly used chemical reactions from combinatorial chemistry. This provided a library of building blocks that are highly useful for constructing biologically active compounds [32].

6.1.7.1.3 Analysis of HTS Data

A chemist synthesizes about 50 compounds per year by traditional organic synthesis. In combinatorial chemistry a series of homologs are synthesized. A reaction of the type $A^i + B^j \rightarrow A^i - B^j$ with $i \in \{1; 2; 3; \dots n\}$ and $j \in \{1; 2; 3; \dots m\}$ performed in parallel by robots gives access to $n \times m$ products in a single experiment. Thus, thousands of compounds are accessible in a short period of time.

With these massive amounts of data produced in HTS for combinatorial libraries, tools become necessary that allow one to navigate through these data and to extract the necessary information. A variety of methods has been developed by mathematicians and computer scientists addressing this task that has become known as data mining. Fayyad defined and described the term "data mining" as the "nontrivial extraction of implicit, previously unknown and potentially useful information from data, or the search for relationships and global patterns that exist in databases" [33]. Methods applicable for data analysis are presented in the *Methods Volume*, Chapter 11.

Data from HTS have dramatically increased in importance since the US National Institutes of Health under the National Center for Biotechnology Information released the PubChem database to the scientific community (see Section 6.5). This database stores data on assay screening results on chemical compounds. Presently it contains about 85 million unique chemical structures and about 1 million assay result tables. This massive amount of information is freely available and has become a key resource for drug design. Section 6.5 presents this database, its structure representation, and its search possibilities.

Example: Analysis of a Hydantoin Library

A combinatorial library was synthesized by first reacting 18 different amino acids with 24 aldehydes and then reacting all these products with 24 isocyanates providing $18 \times 24 \times 24 = 10,368$ hydantoins (Figure 6.1.6).

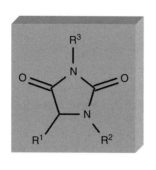

Building blocks:
R¹: 18 amino acids
R²: 24 aldehydes
R³: 24 isocyanates

10,368 compounds
(18 × 24 × 24)

HTS data of the selected assay:
Number of compounds: 5,513 of 10,368
Number of hits: 185
(%control < 50%)
Hit fraction: 3.4%

Figure 6.1.6 Synthesis and high-throughput screening results of a library of hydantoins.

HTS data from a specific assay were obtained for 5,513 of these 10,368 compounds. These data contained 185 positive hits (3.4%) in this assay. The task was then to develop a filter that would be able to extract the positive hits from such a library [34]. To this effect, six different structure representations were explored: autocorrelation of molecular surface properties (electrostatic potential (ESP), hydrogen-bonding potential (HBP), and hydrophobicity potential (HYP)) as well as three Daylight fingerprint representations of different sizes (256, 512, and 1024 dimensions) (see *Methods Volume*, Chapter 10). The performance of these six structure representations was analyzed with an SOM of size 60x45 neurons with the software SONNIA [30]. In Figure 6.1.7 the SOMs of these six

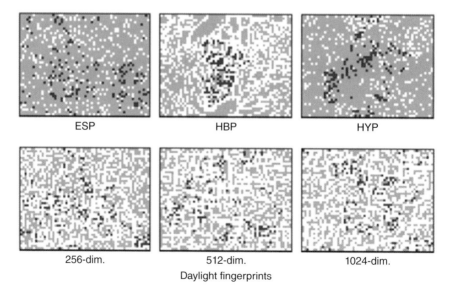

Figure 6.1.7 SOMs of a library of 5,513 hydantoins obtained through six different structure representations. ESP, electrostatic potential; HBP, hydrogen-bonding potential; HYP, hydrophobicity potential. Neurons that obtained a hit in dark gray; neurons with only non-hits in light gray.

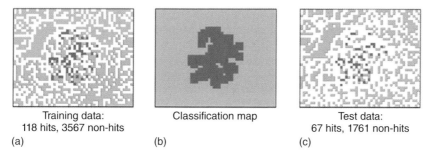

Figure 6.1.8 Development of a filter for hits in the hydantoin library. (a) SOM of the training set, (b) classification map obtained from the neurons with hits and their first-sphere neighbor neurons, and (c) SOM of the test set.

representations are shown, indicating in dark gray those neurons that contain one or more hits and those with non-hits in light gray color.

None of these maps show a complete separation of the hits from the non-hits. Only the map obtained from the autocorrelation of the HBP on the molecular surface shows a promising clustering of the hits. As the task was to make sure that all hits are found, an extension of the SOM was produced by assuming that not only the neurons having obtained a hit but also the neurons directly adjacent to these neurons might obtain a hit when scanning a new library. To investigate that point, the dataset of 5,513 compounds was split into two thirds for training an SOM and one third for testing the performance of the SOM as filter. In this separation of the dataset, it was made sure that hits and non-hits are split proportionally. Figure 6.1.8 shows the results of this study. The map (a) gives the trained SOM obtained from 118 hits and 3,567 non-hits, map (b) shows the classification map by coloring all neurons having obtained a hit as well as their first-sphere neighbor neurons in dark gray, and map (c) shows the map obtained from the test data with 67 hits and 1,761 non-hits. The classification map indicated in Figure 6.1.8b was able to extract 66 (96%) of the hits and 1,619 (92%) of the non-hits from the test set [34]. Thus, this work produced a highly successful approach for digging out the hits in a special assay from a library of hydantoins. Furthermore, it showed that HBP is very important in binding hydantoins in this assay, a result a chemist would immediately agree with by looking at the many sites in hydantoins available for hydrogen bonds.

6.1.7.1.4 Virtual Screening

The term "virtual screening" resp. "*in silico*" screening" is defined as a computational technique for the selection of compounds by evaluating very large libraries of compounds [35]. The computational models may aim at high potency, selectivity, appropriate pharmacokinetic properties, or favorable toxicology. Virtual screening may initially investigate large libraries of compounds, irrespective of whether they are already known or not yet synthesized and the focus on screening in-house libraries and compounds of external suppliers. Two different strategies can be applied:

- Diverse libraries can be used for lead finding by screening against several different targets. The selected compounds should cover the biological activity space well.
- Targeted or focused libraries are suited for both lead finding and optimization. If knowledge about a lead compound is available, compounds with a similar structure are selected for the targeted library. Targeted libraries are focused on a single target.

Virtual screening allows one to extend the scope of screening to external databases. By doing so, more and more diverse hits can be identified. The application of virtual screening techniques prior to or in parallel with HTS helps to reduce the assay-to-lead attrition rate observed from HTS. In addition, virtual screening is faster and less expensive than the experimental synthesis and biological testing. Both ligand-based and structure-based methods can be applied in virtual screening.

In general, the first step in virtual screening is the filtering by the application of Lipinski's "rule of five" [36]. Lipinski's work was based on the results of profiling the calculated physical property data on a set of 2,245 compounds chosen from the World Drug Index and studying their propensity for oral absorption. Statistical analysis of this dataset showed that approximately 90% of the remaining compounds had:

- A molecular weight less than 500 g/mol
- A calculated lipophilicity (log P) of less than five
- A number of H-bond donors less than five
- A number of H-bond acceptors (sum of all nitrogen and oxygen atoms) less than ten

An orally active drug has no more than one violation of one of these criteria. The cutoff values of this "rule of five" (the term arises from the multiple of the results of five) differ slightly within the pharmaceutical industry. Sometimes the "rule of five" is extended by a fifth condition:

- The number of rotatable bonds is less than ten.

In a more recent study, the physical properties of drugs in different development phases were compared [37]. It turned out that the molecular weight and lipophilicity are the properties showing the clearest influence on the successful passage of a candidate drug through the development process. The mean molecular weight of orally administered drugs decreases in the course of the development. Furthermore, the most lipophilic compounds are discontinued from development.

Other filters used for prefiltering account for lead- [38, 39] or drug-likeness [40–42], an appropriate ADMET profile [43–46], or favorable properties concerning receptor binding [47, 48].

6.1.7.1.5 Development of a Pharmacophore Model by 3D Structure Alignment

Pharmacophore perception for receptors with an unknown 3D structure can be carried out by comparing the spatial and electronic requirements of a set of ligands that are known to bind to the receptor of interest. The comparison of a set of ligands is performed by the structural alignment of these ligands. Section 6.6 presents the definition and applications of the pharmacophore concept.

Example: Superimposition of 3D Structures
A first idea about the 3D requirements of a series of ligands can be obtained by the superimposition of 3D structures to obtain the maximum common 3D substructure of several molecules. To this effect a program, GAMMA (Genetic Algorithm for Multiple Molecule Alignment), was developed that combined a genetic algorithm with a Newton optimizer [49, 50]. The genetic algorithm had as data structure two genes, one reflecting the number of atoms of the molecules that are superimposed and the other the torsional angles of the molecules to allow for conformational flexibility. Genetic operators were mutation and crossover as well as two problem-oriented operators, creep and crunch. Figure 6.1.9 shows the flexible superimposition of three muscle relaxants, namely, chlorpromazine, tolperisone, and tizanidine.

RMS = 0.81 Å
Size = 11 atoms

Figure 6.1.9 Flexible superimposition of the 3D structure of three muscle relaxants: chlorpromazine, tolperisone, and tizanidine.

6.1.7.2 Structure-Based Drug Design (SBD)

6.1.7.2.1 Exploration of the Binding Pocket

When the 3D structure of a target protein is known, the next challenge is to identify the binding pocket. Section 6.7 presents the methods for identifying the active site and analyzing their important features. Once the active site has been found and analyzed, several methods are available for using this information for lead finding. Fitting a ligand from a 3D structure database to the binding site of a target protein is called *docking*. The iterative building of new molecules in the binding site of a receptor is illustrated in the center and on the right-hand side of Figure 6.1.10. These procedures to find new leads are called *de novo design*. The building approach beginning with a single fragment and proceeding through the stepwise addition of further moieties is shown in the center. Alternatively, several small molecules already showing some affinity to the target protein are placed in the binding site of the protein and subsequently linked together (*linking*). To end up with high affinity ligands from SBD, a high degree of steric and electronic complementarity of the ligand to the target protein is required. Further on, a proper amount of the ligand's hydrophobic surface should be buried in the complex. A certain degree of conformational rigidity is important to ensure that the loss of entropy upon ligand binding is counterbalanced.

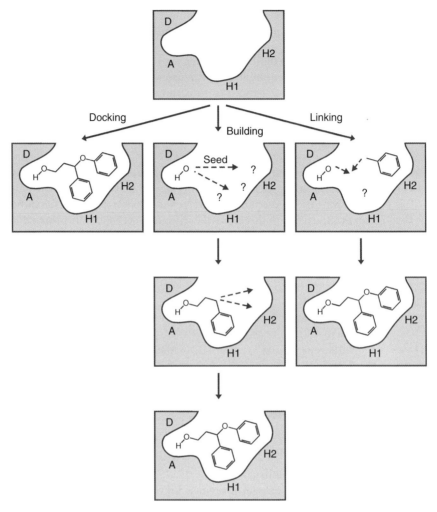

Figure 6.1.10 Different strategies to design a ligand in target-based drug discovery: docking (left), building (center), and linking (right). D = H-bond donor, A = H-bond acceptor, H1, H2 = hydrophobic regions of the protein.

The information on the binding site can also be used for the construction of a pharmacophore as presented in Section 6.6.

Example: Decomposition of Binding Interactions with Glutamate Racemase
Glutamate racemase (MurI), converting L-glutamate into D-glutamate, is essential for the biosynthesis of *Helicobacter pylori* cell walls. Therefore, MurI is a promising target for the design of antibacterial drugs. A dataset of 69 pyrazolopyrimidinediones' uncompetitive inhibitors of MurI (all active) was analyzed [51]. For one of these compounds, a co-crystallized structure with the MurI protein (PDB code: 2JFZ) was available. The entire dataset was then superimposed onto the structure of this ligand from the X-ray structure. Analysis of this dataset

with the molecular mechanics/generalized Born surface areas (MM/GBSA) approach [52] allowed the decomposition of the binding interactions into van der Waals, electrostatic, and polar solvation surfaces. These decomposed binding energies were correlated in a 3D quantitative structure–activity relationship (QSAR) approach with MurI inhibitory activity with partial least squares regression (PLSR). This combination, MM/GBSA-PLSR, is a novel method for structure-based 3D QSAR [51]. In the case of MurI inhibitors, a leave-one-out-cross-validation correlation coefficient (Q^2) of 0.822 was obtained, and an external test provided a predicted correlation coefficient (R^2_{pred}) of 0.817 with eight components. This study allowed the quantitative decomposition of the binding energies into van der Waals interactions (29.5%), electrostatic interactions (38.2%), and polar solvation interactions (32.3%). In order to better understand this model, ligand-based 3D QSAR studies were made with the CoMFA and CoMSIA methods. The CoMFA study provided a Q^2 of 0.684 and an R^2_{pred} of 0.561 with six components and the CoMSIA investigation a Q^2 of 0.687 and an R^2_{pred} of 0.748 with 12 components. The CoMFA model contained steric and electrostatic factors, whereas the CoMSIA model comprised electrostatic and H-bond acceptor components. Only the MM/GBSA-PLSR method was able to also predict the contribution of polar solvation [51]. Thus, with the MM/GBSA-PLSR approach, a method has become available that allows the decomposition of binding energies of ligands to their proteins into van der Waals, electrostatic, and polar solvation energies in a quantitative manner. This approach could also produce maps of these different contributions to ligand binding and thus lay the foundation for the definition of pharmacophore surfaces.

6.1.7.2.2 Screening

The suitability of these methods for large-scale virtual screening depends on the way how ligand flexibility is addressed. Methods such as a genetic algorithm, an exhaustive search, a Monte Carlo simulation, or a pseudo-Brownian sampling have been developed but require considerable computer power for large-scale virtual screening. An algorithm, Iterative Stochastic Elimination, initially developed for the exploration of the conformational space of the side chains of proteins [53] has also been applied to picking the best molecules for hitting a specific target [54]. Section 6.8 shows how the information on the 3D structure of the binding pocket and its interaction points can be used for the screening of datasets of molecules.

The prediction of the binding affinity of a ligand to a receptor is a challenging task. This becomes evident if the thermodynamics of ligand binding is analyzed in more detail (Figure 6.1.11).

Various enthalpic and entropic terms (ΔH, ΔS) of the protein, the ligand, and the bound and free water molecules have to be considered to determine the free energy (ΔG) of ligand binding.

The calculation of the binding affinity considering all these effects for virtual screening is highly challenging and not yet a solved problem. In order to circumvent the direct calculation of the entropy and enthalpy effects, scoring functions have been developed for use in docking programs, as detailed in Section 6.8.

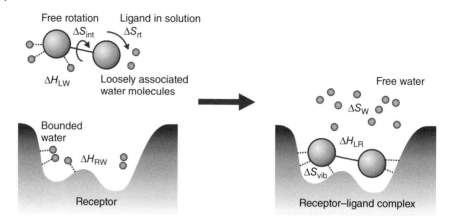

Figure 6.1.11 Thermodynamics of ligand binding.

During the step of lead finding, compound libraries are designed and synthesized. These compound libraries can also be screened in assays using the validated functional protein target. Compounds showing activity in a primary assay are labeled as "hits." The hits are validated in another screening experiment. The information generated in the screening assays is used to find compounds that are suitable for further development.

6.1.7.2.3 Synthetic Accessibility

Some of the methods mentioned previously, in particular *de novo* design, quite often generate molecular structures of rather high complexity. A medicinal chemist might consider them to be quite laborious or costly to synthesize requiring many synthetic steps of high specificity. Therefore, the selection of appropriate lead structures should also consider how difficult it might be to synthesize the corresponding structure. Computational methods have been developed to estimate the ease of organic synthesis, the synthetic accessibility, of a compound by considering structural complexity and similarity to available starting materials and the assessment of strategic bonds [55]. The values obtained by the program SYLVIA compare favorably with the evaluations of expert chemists but allow the automatic processing of large datasets of molecules [56]. Combination of this methodology with a *de novo* design method provided a comprehensive workflow to *de novo* design driven by the needs of computational and medicinal chemists [57].

A recent approach to estimate the complexity of organic molecules has been able to incorporate the progress in organic synthesis methodology into the evaluation of molecular complexity and thus has been termed current complexity [58, 59].

6.1.8 Lead Optimization

Whether a compound is suitable for serving for further development depends on several features. Some criteria affecting the search and optimization of a new lead structure are illustrated in Figure 6.1.12.

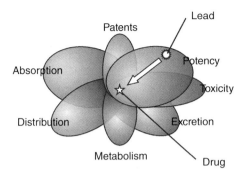

Figure 6.1.12 Factors affecting lead identification and optimization.

The lead optimization step aims at developing compounds with high potency and high selectivity, as well as acceptable pharmacokinetic profile, low toxicity, and absence of mutagenicity. Lead optimization is an iterative process. The initial lead compound is systematically modified. The analogs thus obtained are then tested in a biological assay. This information is then fed back into the design and development process.

Quite a few of the chemoinformatics methods used in the lead finding process and mentioned in Section 6.1.7 can also be used for lead optimization such as similarity searching, lead hopping, design of combinatorial libraries, determination of synthetic accessibility, HTS, virtual screening, pharmacophore analysis, *de novo* design, and docking and scoring, as also indicated in Table 6.1.1. Furthermore, this is also elaborated in Sections 6.4–6.7. However, as now more information than in the lead finding phase will be available, the lead optimization phase can build on more focused knowledge.

In addition, with a dataset of compounds and their measured biological activities available, QSAR models can be established. Such QSAR models may allow the identification of important features for high activity and thus guide the synthesis of highly active compounds.

6.1.8.1 ADMET Properties

Unsatisfactory and adverse ADMET profiles are a reason for attrition of new drug candidates during the development process [10, 60]. The major reasons for attrition of new drugs are:

- Lack of clinical efficacy
- Inappropriate pharmacokinetics
- Animal toxicity
- Adverse reactions in humans
- Commercial reasons
- Formulation issues

Historically, ADMET studies were performed after having settled on a highly active compound as drug candidate. In order to avoid costs, now, pharmaceutical companies evaluate the ADMET profiles of potential leads at an earlier stage of the development process. For the consideration of ADMET properties in virtual screening, computational methods for their prediction are needed. Section 6.9

presents some of the computational models developed for the calculation of important physical, chemical, or biological properties relevant for a drug.

Section 6.10 outlines methods that allow one to predict the degradation of drugs and drug candidates in the human body. This is important information for the prediction of the lifetime of drugs, of their excretion behavior, for predicting the interaction of metabolites in biochemical pathways, and for evaluating potential toxicities of metabolites.

6.1.8.2 Toxicity

Toxicity may be one of the most difficult properties to model. The difficulties arise because the effects of toxicants are species specific, organ specific, and time dependent (i.e., acute effects differ from chronic effects). This has the consequence that the concentration at which adverse effects occur can vary over several orders of magnitude depending on the species and the type of test. Several overviews on modeling toxicity have been published [61–63], and another one addressing more specific issues of ecotoxicology has been written by Escher and Hermens [64].

A detailed presentation of toxicity modeling is presented in Chapter 8. Here we introduce some important effects and show an approach on how to deal with one of the topics faced in toxicity modeling.

Toxic effects are measured through a wide variety of tests. One can distinguish two types: *in vivo* and *in vitro* tests. *In vivo* tests are carried out with organisms like rodents and in the case of ecotoxicology with birds, fish, water fleas, earthworms, and algae. *In vitro* tests are mostly performed with single cells, organelles like mitochondria or even just with enzymes that are affected by a toxicant. The classical *in vivo* test value for acute toxicity of a chemical is the LD_{50} value for terrestrial organisms and the LC_{50} value for aquatic organisms, that is, the dose or the concentration, respectively, at which 50% of the test species are killed by the toxic effects of a compound in a given time period. This is also one of the most common values to predict with QSAR equations.

In a classic study Corwin Hansch formulated a linear free energy model relating toxicity with a hydrophobicity term [65]. The most widely used descriptor for hydrophobicity in toxicology is lipophilicity as measured by the distribution coefficient between octanol and water, log P.

Figure 6.1.13 shows that indeed there is a strong correlation between toxicity and lipophilicity. The chemicals in this figure represent a quite diverse selection of aliphatic and aromatic hydrocarbons, about half of them with chloro substituents, the rest with either hydroxy or carbonyl groups. However, they all do not have specific effects on vital functions of the organism. The compounds are a subset of a toxicity dataset taken from literature [66].

Such direct correlations with lipophilicity can successfully model the toxicity of many common industrial chemicals. Figure 6.1.14, however, shows that reality is quite more complicated. Now all compounds of the previously mentioned dataset are shown [53]. The dataset actually was compiled with the aim of having highly diverse compounds and also a high diversity of toxic effects. All effects on targets of major importance in acute toxicity are covered – plus receptor-mediated

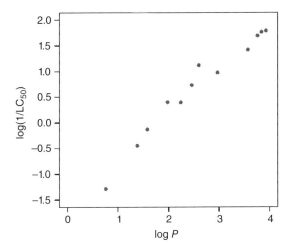

Figure 6.1.13 Compounds showing baseline toxicity (narcosis).

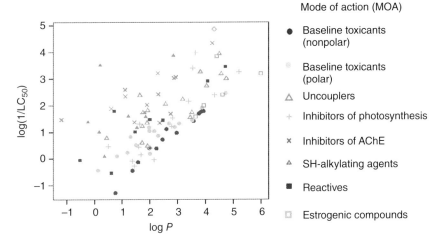

Figure 6.1.14 Compounds having a variety of toxic modes of action (MOA).

effects through environmental estrogens. These effects can be grouped into mode of actions (MOAs). An MOA is defined as a common set of physiological and behavioral signs characterizing a type of adverse biological response, while a toxic mechanism is defined as the biochemical process underlying a given MOA [67].

The toxicities represented in Figure 6.1.13 are caused solely by the tendency of the chemicals for moving into biological membranes and are often referred to as baseline toxicity or narcosis – a property that every compound has. However, in addition to that, many compounds can interact with a variety of more specific targets. Many of these chemicals in Figure 6.1.14 show toxicity values that are many orders of magnitude larger than baseline toxicity.

This wide distribution of toxicity values clearly emphasizes that it does not make sense to develop a universal QSAR equation for toxicity prediction valid for any chemical compound. Rather, first, the chemicals have to be classified into their MOA. QSAR equation can then be developed not for a set of chemicals from the same class but for chemicals having a common MOA [68].

Example: Prediction of the Toxic Mode of Action of Phenols

To illustrate the difference between a class of compounds and compounds having the same mode of action (MOA), a study is briefly reported. It will be shown how a counterpropagation (CPG) neural network (see *Methods Volume*, Section 11.2) can best be used to determine the MOA [69]. Emphasis in this example is placed on the choice of a network architecture. The dataset consisted of 220 phenols comprising four different MOA: polar narcotics (155 phenols), uncouplers of oxidative phosphorylation (19 phenols), soft electrophiles (22 phenols), and precursors to soft electrophiles (24 phenols) [70].

The phenols were initially represented by 6-dimensional topological autocorrelation vectors utilizing different physicochemical properties (see *Methods Volume*, Section 10.3.3.2), which were sent into a CPG network (see *Methods Volume*, Section 11.2) consisting of 11x11 neurons with SONNIA [30]. The autocorrelation vectors of the two best performing physicochemical properties (π-charge, q_π, and σ-electronegativity, χ_σ) were then concatenated to have an input vector of 12 variables giving 12 layers (one for each descriptor) in the input block and four layers in the output block corresponding to the four different MOAs. Figure 6.1.15 shows the architecture of the CPG network.

The distribution of the compounds in the four output layers of the CPG network was quite promising as shown in Figure 6.1.16, particularly so for the soft electrophiles (good clustering) and for the uncouplers of oxidative phosphorylation (pushed to the outskirts of the network).

The correct classification (fivefold cross-validation) with this structure representation amounted to 87.3%. It could be increased to 92.3% by adding nine descriptors from autocorrelating the HBP on the molecular surface (see *Methods Volume*, Section 10.3.6.2).

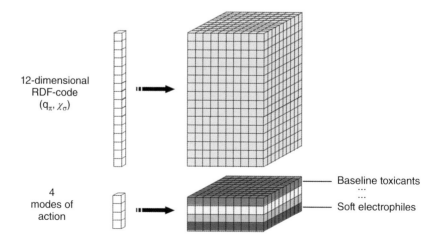

Figure 6.1.15 Architecture of a counterpropagation neural network for classifying phenols into four different MOAs.

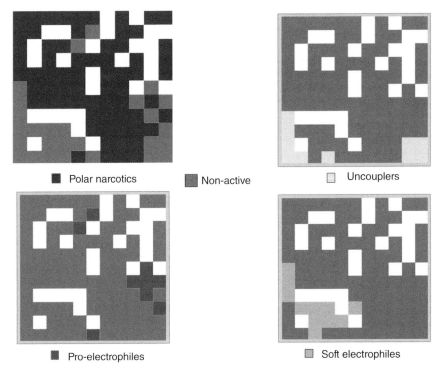

Figure 6.1.16 Distribution of phenols in the four output layers of the counterpropagation network.

In conclusion, this approach showed that a proper choice of the CPG architecture together with a structure representation that reflects the physicochemical effects active in these compounds led to an excellent classification of the compounds into their respective toxic MOA [69]. As an example, it could correctly be predicted that 2,6-dichlorophenol is a polar narcotic, whereas 2,3,4,5-tetrachlorophenol is an uncoupler of oxidative phosphorylation and thus much more toxic and problematic.

6.1.8.3 Toxicological Alerts

The previous example of phenols showed that the assignment of a compound to a certain class of substances is not sufficient for determining its toxic MOA. Nevertheless, as a first step to toxicity prediction, the analysis of the structural features of a compound is quite helpful. Since a long time, toxicological alerts have played an important role in the drug registration process [71]. A collection of toxicological alert, called ToxPrint, has been made available [72]. These structural features are encoded in a special notation, called chemotypes [73], that can be processed by a publicly available software, ChemoTyper [74]. The chemotypes have the additional potential that they can be enriched with physicochemical features assigned to the atoms and bonds of the substructures and thus fine-tune the toxicological alerts. Figure 6.1.17 shows an example for such a fine-tuning of a substructure by assigning partial charges to an atom of the toxicological alert.

In this way, truly toxic compounds can be separated from the nontoxic compounds within the same toxicological alert.

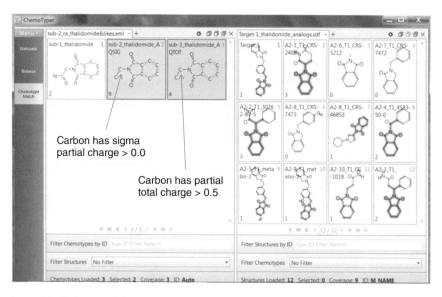

Figure 6.1.17 Screen of ChemoTyper: on the left are three chemotypes of the thalidomide skeleton with chemotypes differentiated by sigma charge (light gray) and total charge (dark gray); on the right-hand side is part of a dataset that indicates hits for the two different chemotypes.

6.1.9 Preclinical and Clinical Trials

The approval of a new drug requires preclinical and clinical studies and takes an average of 10 years to complete. The preclinical and clinical trials have to prove the safety and efficacy of every new medicine. The preclinical trials are performed in *in vitro* and in animal studies to assess the biological activity of the new compound. In phase I of the clinical trials, the safety of a new drug is examined, and the dosage is determined by administering the compound to about 20–100 healthy volunteers. The focus in phase II is directed onto the issues safety, evaluation of efficacy, and investigation of side effects in 100–300 patient volunteers. More than 1,000 patient volunteers are treated with the new drug in phase III to prove the efficacy and safety over long-term use.

Clearly, an enormous amount of data is acquired during preclinical and clinical studies. Their exploitation by bioinformatics and chemoinformatics methods can certainly bring a lot of insights. However, as these data are all collected and exploited in and for pharmaceutical companies, no details can be given here. In the future, the analyses of the data from preclinical and clinical studies by chemoinformatics and bioinformatics methods will certainly play a very important role and will have a major impact on drug development.

6.1.10 Outlook: Future Perspectives

A broad range of methods is now available in the field of chemoinformatics that will have a growing impact on drug design. Both LBD and SBD can use whatever information is available. The huge amount of data produced by HTS and combinatorial chemistry enforces the use of database and data mining techniques. Section 6.11 shows the methods that were collected and further developed at the computer-aided drug design group for projects internal to this public institution but also those that are made publicly available to assist academic researchers in their work on drug discovery.

Integrated systems suitable for processing the sometimes quite complex workflows more easily or automatically and to optimize new compounds in parallel for their potency, selectivity, and ADMET profile have to be developed. SBD will profit from faster algorithms allowing one to perform high-throughput docking of virtual libraries with high accuracy. Improved information management techniques may facilitate the connection and analysis of different sources of data.

Section 6.12 presents recent developments of simultaneously using a variety of data sources such as literature data, electronic databases, and experimental testing results in a data cycle for model development in drug discovery.

The treatment of conformational flexibility of both the ligand and the protein, the prediction of the binding affinity of a substrate to an enzyme, water desolvation of the ligand and of the binding pocket, the modeling of protein–protein interactions, the determination of the geometry as well as the calculation of the strength of hydrogen bonds, and the prediction of the 3D structure of proteins from the sequence are still challenging tasks that have to be addressed and improved in computer-assisted drug discovery.

The reliability of the *in silico* models will be improved and their scope for predictions will be broader as soon as more reliable experimental data are available. However, there is the paradox of predictivity versus diversity. The greater the chemical diversity in a dataset, the more difficult is the establishment of a predictive structure–activity relationship. On the other hand, a model developed based on compounds representing only a small subspace of the chemical space has no predictivity for compounds beyond its boundaries. Dramatic progress could be made when pharmaceutical companies open their archives and make their data available. With more and better data available from drug companies, academic researchers could enormously improve their models and methods. First steps in this direction have been made with the eTOX project [75] funded by the European Union within the Innovative Medicines Initiative.

In the future, the computer-assisted design of drug candidates and the computer-assisted design of syntheses for making make these compounds will become tightly integrated. This will also lead to a closer collaboration between chemoinformatics teams and medicinal chemists working in the laboratory. Section 6.13 presents the ideas of an academic researcher on the trends in drug discovery and emphasizes the integration of design, synthesis, and testing. It is also demonstrated how this process can be miniaturized and integrated into a production cycle.

As in any field of our daily live, also in drug discovery and development computer methods will continue to play an increasingly important role.

Essentials

- The drug discovery process comprises the following steps: (i) target identification, (ii) target validation, (iii) lead finding, (iv) lead optimization, (v) preclinical studies, (vi) clinical studies, and (vii) drug approval (FDA approval).
- A lead structure is "a representative of a compound series with sufficient potential (as measured by potency, selectivity, pharmacokinetics, physicochemical properties, absence of toxicity and novelty) to progress to a full drug development program."
- Chemoinformatics is primarily used in the steps lead finding and lead optimization within the drug discovery process. In particular the following tasks are involved:
  - Analysis of HTS data
  - Similarity search
  - Design of combinatorial libraries
  - Design of focused libraries
  - Comparison of the similarity/diversity of libraries
  - Virtual screening
  - Docking
  - *De novo* design
  - Pharmacophore perception
  - Prediction of binding affinities, physicochemical properties (such as solubility, log P, and pK_a), and pharmacokinetic properties (ADMET profile)
  - Establishment of QSAR models that can be interpreted and can guide the further development of a new drug
- Following the "similar property principle," structurally similar molecules are expected to exhibit similar properties or biological activities.
- The term "virtual screening" resp. "*in silico* screening" is defined as the selection of compounds by evaluating their desirability in a computational model. The desirability comprises high potency, selectivity, appropriate pharmacokinetic properties, and favorable toxicology.
- Lipinski's "rule of five":
  - Molecular weight < 500
  - Lipophilicity (log P) < 5
  - Number of H-bond donors < 5
  - Number of H-bond acceptors (# N + O) < 10

  has been propagated as determining compounds with a good oral bioavailability

Interesting Websites (accessed January 2018)

http://www.kubinyi.de/lectures.html
http://www.netsci.org (under The Science Center -> CompChem Articles)

Selected Reading

- Böhm, H.-J., Klebe, G., and Kubinyi, H. (1996) *Wirkstoffdesign*, Spektrum Akademischer Verlag GmbH, Heidelberg, 599pp.
- Böhm, H.-J. and Schneider, G. (2000) *Virtual screening for bioactive molecules*, in *Methods and Principles in Medicinal Chemistry*, vol. 10 (eds R. Mannhold, H. Kubinyi, and H. Timmerman), Wiley-VCH Verlag GmbH, Weinheim, 325pp.
- Klebe, G. (2013) *Drug Design*, Springer Verlag, Berlin, 901pp.
- Mutschler, E. (2001) *Arzneimittelwirkungen*, Wissenschaftliche Verlagsgesellschaft, Stuttgart, 1211pp.
- Wermuth, C.G., Aldous, D., Raboisson, P., and Rognan, D. (eds.) (2015) *The Practice of Medicinal Chemistry*, Academic Press, London, 902pp.

References

[1] Wermuth, C.G., Aldous, D., Raboisson, P., and Rognan, D. (eds.) (2015) *The Practice of Medicinal Chemistry*, Academic Press, London, UK, 902pp.
[2] Böhm, H.-J., Klebe, G., and Kubinyi, H. (1996) *Wirkstoffdesign*, Spektrum Akademischer Verlag GmbH, Heidelberg, 599pp.
[3] Klebe, G. (2013) *Drug Design*, Springer Verlag, Berlin, 901pp.
[4] Mutschler, E. (2001) *Arzneimittelwirkungen*, Wissenschaftliche Verlagsgesellschaft, Stuttgart, 1211pp.
[5] Kubinyi, H. (1999) *J. Recept. Sig. Transd.*, **19**, 15–39.
[6] Tufts Center for the Study of Drug Development News, http://csdd.tufts.edu/news/complete_story/tufts_csdd_rd_cost_study_now_published (accessed January 2018).
[7] Olsson, T. and Oprea, T.I. (2001) *Curr. Opin. Drug Discovery Dev.*, **4**, 308–313.
[8] Gaudillière, B. and Berna, P. (2000) *Annu. Rep. Med. Chem.*, **35**, 331–356.
[9] FDA Drugs, http://www.fda.gov/Drugs/DevelopmentApprovalProcess/DrugInnovation/ucm430302.htm (accessed January 2018).
[10] Prentis, R.A., Lis, Y., and Walker, S.R. (1988) *Br. J. Clin. Pharmacol.*, **25**, 387–396.
[11] Valler, M.J. and Green, D. (2000) *Drug Discovery Today*, **5**, 286–293.
[12] Brunton, L., Knollman, B., and Hilal-Dandan, R. (eds) (2017) *Goodman and Gilman's The Pharmacological Basis of Therapeutics*, 13th edn, McGraw-Hill, New York, 1440pp.

[13] Hopkins, A.L. and Groom, C.R. (2002) *Nat. Rev. Drug Discovery*, **1**, 727–730.
[14] Drews, J. (2000) *Science*, **287**, 1960–1964.
[15] Hopkins, A. (2008) *Nat. Chem. Biol.*, **4**, 682–690.
[16] Reddy, A.S. and Zhang, S. (2013) *Exp. Rev. Clin. Pharmacol.*, **6**, 41–47.
[17] Anighoro, A., Bajorath, J., and Rastelli, G. (2014) *J. Med. Chem.*, **57**, 7874–7887.
[18] Oprea, T.I. and Mestres, J. (2012) *AAPS J.*, **12**, 759–763.
[19] Bredel, M. and Jacoby, E. (2004) *Nat. Rev. Genet.*, **5**, 262–275.
[20] Kubinyi, H. (2006) *Ernst Schering Res. Found. Workshop*, **58**, 1–19.
[21] Gregori-Puigjane, E. and Mestres, J. (2008) *Comb. Chem. HTS*, **11**, 669–676.
[22] Mohd Fouzi, F., Koutsoukas, A., Lowe, R., Joshi, K., Fan, T.P., Glen, R.C., and Bender, A. (2013) *J. Chem. Inf. Model.*, **53**, 661–673.
[23] Bhattacharjee, B., Simon, R.M., Gangadharaia, C., and Karunakar, P. (2013) *J. Microbiol. Biotechnol.*, **23**, 779–784.
[24] Cheung-Ong, K., Song, K.T., Ma, Z., Shabtai, D., Lee, A.Y., Gallo, D., Heisler, L.E., Brown, G.W., Bierbach, U., Giaever, G., and Nislow, C. (2012) *ACS Chem. Biol.*, **7**, 1892–1901.
[25] Sadowski, J. and Gasteiger, J. (1993) *Chem. Rev.*, **93**, 2567–2581.
[26] Sadowski, J., Gasteiger, J., and Klebe, G. (1994) *J. Chem. Inf. Comput. Sci.*, **34**, 1000–1008.
[27] CORINA Classic – High-Quality 3D Molecular Models, is available at https://www.mn-am.com/products/corina and can be freely tested, (accessed January 2018).
[28] Newman, D.J. and Cragg, G.M. (2016) *J. Nat. Prod.*, **79**, 629–661.
[29] Bauknecht, H., Zell, A., Bayer, H., Levi, P., Wagener, M., Sadowski, J., and Gasteiger, J. (1996) *J. Chem. Inf. Comput. Sci.*, **36**, 1205–1213.
[30] SONNIA – Self-Organizing Neural Network Package, https://www.mn-am.com/products/sonnia (accessed January 2018).
[31] Sheridan, R.P. and Kearsley, S.K. (1995) *J. Chem. Inf. Comput. Sci.*, **35**, 310–320.
[32] Lewell, X.Q., Judd, D.B., Watson, S.P., and Hann, M.M. (1998) *J. Chem. Inf. Comput. Sci.*, **38**, 511–522.
[33] Fayyad, U.M., Piatetsky-Shapiro, G., and Smyth, P. (1996) *From data mining to knowledge discovery: an overview*, in *Advances in knowledge discovery and data mining* (eds U.M. Fayyad, G. Piatetsky-Shapiro, P. Smyth, and R. Uthurusamy), USA, AAAI Press, Menlo Park, CA, pp. 1–37.
[34] Teckentrup, A., Briem, H., and Gasteiger, J. (2004) *J. Chem. Inf. Comput. Sci.*, **44**, 626–634.
[35] Walters, W.P., Stahl, M.T., and Murcko, M.A. (1998) *Drug Discovery Today*, **3**, 160–178.
[36] Lipinski, C.A., Lombardo, F., Dominy, B.W., and Feeny, P.J. (1997) *Adv. Drug Delivery Rev.*, **23**, 3–25.
[37] Wenlock, M.C., Austin, R.P., Barton, P., Davis, A.M., and Leeson, P.D. (2003) *J. Med. Chem.*, **46**, 1250–1256.
[38] Teague, S.J., Davis, A.M., Leeson, P.D., and Oprea, T. (1999) *Angew. Chem. Int. Ed.*, **38**, 3743–3747.

too far from the active site are removed. Remaining poses are scored, and the top 100 scoring poses are locally optimized by applying small translations and rotations. TrixX uses a built-in conformer generator [41]. For each conformer as well as for the active site of the protein, a shape- and interaction-based descriptor [42] is calculated. The indexed descriptors for the conformers are stored in a database. The descriptor for the active site is used to query the database to identify valid poses. These poses are finally scored using the FlexX scoring function.

While these modern docking tools all address ligand flexibility, protein flexibility is often neglected since it dramatically increases the search space. Two different approaches are currently pursued to model protein flexibility [43]. The first and straightforward method is to generate an ensemble of protein conformers and dock the ligands into each conformer separately. This approach is referred to as *ensemble docking (ED)*. Every docking tool can in principle be used for *ED*. The generation of the right protein conformations and the selection of the best poses are however difficult. The second approach of modeling protein flexibility places the ligand into the cavity of a protein and then allows the protein to adapt its conformation to optimally fit to the ligand. This approach is called *induced fit docking (IFD)*. FITTED [44] is an example of a docking tool taking into account protein flexibility. It implements both *ED* and *IFD* and limits the search space by a GA. Despite these advances, there is still a need for more accurate and faster solutions to allow fully flexible protein–ligand docking routinely.

6.8.3 Scoring

Scoring functions are mathematical expressions for predicting the strength of non-covalent protein–ligand interactions. They are involved in three different application scenarios: pose prediction in docking, ranking of compounds in virtual screening, and prediction of binding affinity. The challenge in pose prediction is the identification of the ligand's natural binding mode within a protein active site from the numerous suggestions a docking algorithm offers. The major application is the ranking of compounds in virtual screening, aiming at discriminating biologically actives from inactives.

The most demanding application of a scoring function is binding affinity prediction since many different molecular interactions, forces, and effects have to be considered and balanced against each other. An evaluation of 16 different scoring functions according to their ability to manage these three tasks can be found in the review by Cheng *et al.* [45]. They concluded that today's scoring functions are quite good in pose prediction but that there is still a need of improvements in ranking and binding affinity prediction.

The physicochemical basis of all scoring functions is the prediction of the change in Gibbs free energy (ΔG) during the protein–ligand binding process. Upon binding, the energy of the formed complex is lower than the sum of the energies of the separate two molecules; it is a spontaneous process and therefore has a negative value. The Gibbs free energy can be calculated by the

Gibbs–Helmholtz equation (Eq. 6.8.1):

$$\Delta G = \Delta H - T\Delta S \tag{6.8.1}$$

ΔH and ΔS are the changes in enthalpy and entropy, respectively, and T is the absolute temperature in Kelvin. The enthalpic part (ΔH), which is made up mainly of the van der Waals, electrostatic, and hydrogen-bonding interactions between the ligand and the protein, is generally modeled by scoring functions. In comparison, the entropic part ($T\Delta S$) is often neglected due to the challenges in modeling the underlying effects. The release of unordered water molecules from the binding site and the freezing of torsional angles of the two molecules caused by ligand binding are two aspects of entropy. Only the latter one is explicitly modeled by some empirical scoring functions [46, 47].

The most common non-covalent interactions that are described in current scoring functions are hydrogen bonds, salt bridges, metal interactions, van der Waals interactions, aromatic interactions, and the hydrophobic effect (see Figure 6.8.1). Hydrogen bonds are directed interactions with distinct minima considering the interaction length and angles [48]. A prerequisite for a hydrogen bond is the desolvation of the hydrogen-bond donor and acceptor. Since desolvation of polar atoms is an energetically cost-intensive process, it is important that the buried polar atoms in the protein–ligand interface are saturated. Van der Waals interactions and the hydrophobic effect require a tight surface fit of the ligand to the protein. π–π interactions and cation–π interactions are another group of directed interactions, which were intensively studied by Diederich and coworkers [49]. Recently, a new group of interactions, the so-called weak polar interactions, was investigated according to their relevance in protein–ligand complexes. Among these are CH—O hydrogen bonds [50] and halogen interactions [51, 52]. Most of the interactions discussed previously play an important

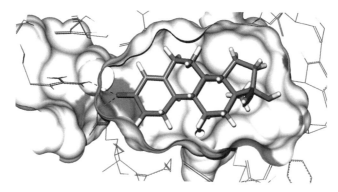

Figure 6.8.1 Interactions in a protein–ligand complex (PDB code: 1SQN). The major energetic contributions result from hydrogen bonds and hydrophobic effect. The protein surface is colored according to hydrophobicity (dark gray, hydrophilic atoms; white, hydrophobic atoms). The dominating interaction in the complex of norethindrone with the progesterone receptor is the hydrophobic effect, which is caused by the burial of the four aliphatic rings of the ligand in a deep hydrophobic pocket of the protein. The two hydrogen bonds (left side) contribute less to the overall binding affinity; they rather assist in orientating the ligand in the active site.

Interesting Websites (accessed January 2018)

http://www.kubinyi.de/lectures.html
http://www.netsci.org (under The Science Center -> CompChem Articles)

Selected Reading

- Böhm, H.-J., Klebe, G., and Kubinyi, H. (1996) *Wirkstoffdesign*, Spektrum Akademischer Verlag GmbH, Heidelberg, 599pp.
- Böhm, H.-J. and Schneider, G. (2000) *Virtual screening for bioactive molecules*, in *Methods and Principles in Medicinal Chemistry*, vol. 10 (eds R. Mannhold, H. Kubinyi, and H. Timmerman), Wiley-VCH Verlag GmbH, Weinheim, 325pp.
- Klebe, G. (2013) *Drug Design*, Springer Verlag, Berlin, 901pp.
- Mutschler, E. (2001) *Arzneimittelwirkungen*, Wissenschaftliche Verlagsgesellschaft, Stuttgart, 1211pp.
- Wermuth, C.G., Aldous, D., Raboisson, P., and Rognan, D. (eds.) (2015) *The Practice of Medicinal Chemistry*, Academic Press, London, 902pp.

References

[1] Wermuth, C.G., Aldous, D., Raboisson, P., and Rognan, D. (eds.) (2015) *The Practice of Medicinal Chemistry*, Academic Press, London, UK, 902pp.
[2] Böhm, H.-J., Klebe, G., and Kubinyi, H. (1996) *Wirkstoffdesign*, Spektrum Akademischer Verlag GmbH, Heidelberg, 599pp.
[3] Klebe, G. (2013) *Drug Design*, Springer Verlag, Berlin, 901pp.
[4] Mutschler, E. (2001) *Arzneimittelwirkungen*, Wissenschaftliche Verlagsgesellschaft, Stuttgart, 1211pp.
[5] Kubinyi, H. (1999) *J. Recept. Sig. Transd.*, **19**, 15–39.
[6] Tufts Center for the Study of Drug Development News, http://csdd.tufts.edu/news/complete_story/tufts_csdd_rd_cost_study_now_published (accessed January 2018).
[7] Olsson, T. and Oprea, T.I. (2001) *Curr. Opin. Drug Discovery Dev.*, **4**, 308–313.
[8] Gaudillière, B. and Berna, P. (2000) *Annu. Rep. Med. Chem.*, **35**, 331–356.
[9] FDA Drugs, http://www.fda.gov/Drugs/DevelopmentApprovalProcess/DrugInnovation/ucm430302.htm (accessed January 2018).
[10] Prentis, R.A., Lis, Y., and Walker, S.R. (1988) *Br. J. Clin. Pharmacol.*, **25**, 387–396.
[11] Valler, M.J. and Green, D. (2000) *Drug Discovery Today*, **5**, 286–293.
[12] Brunton, L., Knollman, B., and Hilal-Dandan, R. (eds) (2017) *Goodman and Gilman's The Pharmacological Basis of Therapeutics*, 13th edn, McGraw-Hill, New York, 1440pp.

[13] Hopkins, A.L. and Groom, C.R. (2002) *Nat. Rev. Drug Discovery*, **1**, 727–730.
[14] Drews, J. (2000) *Science*, **287**, 1960–1964.
[15] Hopkins, A. (2008) *Nat. Chem. Biol.*, **4**, 682–690.
[16] Reddy, A.S. and Zhang, S. (2013) *Exp. Rev. Clin. Pharmacol.*, **6**, 41–47.
[17] Anighoro, A., Bajorath, J., and Rastelli, G. (2014) *J. Med. Chem.*, **57**, 7874–7887.
[18] Oprea, T.I. and Mestres, J. (2012) *AAPS J.*, **12**, 759–763.
[19] Bredel, M. and Jacoby, E. (2004) *Nat. Rev. Genet.*, **5**, 262–275.
[20] Kubinyi, H. (2006) *Ernst Schering Res. Found. Workshop*, **58**, 1–19.
[21] Gregori-Puigjane, E. and Mestres, J. (2008) *Comb. Chem. HTS*, **11**, 669–676.
[22] Mohd Fouzi, F., Koutsoukas, A., Lowe, R., Joshi, K., Fan, T.P., Glen, R.C., and Bender, A. (2013) *J. Chem. Inf. Model.*, **53**, 661–673.
[23] Bhattacharjee, B., Simon, R.M., Gangadharaia, C., and Karunakar, P. (2013) *J. Microbiol. Biotechnol.*, **23**, 779–784.
[24] Cheung-Ong, K., Song, K.T., Ma, Z., Shabtai, D., Lee, A.Y., Gallo, D., Heisler, L.E., Brown, G.W., Bierbach, U., Giaever, G., and Nislow, C. (2012) *ACS Chem. Biol.*, **7**, 1892–1901.
[25] Sadowski, J. and Gasteiger, J. (1993) *Chem. Rev.*, **93**, 2567–2581.
[26] Sadowski, J., Gasteiger, J., and Klebe, G. (1994) *J. Chem. Inf. Comput. Sci.*, **34**, 1000–1008.
[27] CORINA Classic – High-Quality 3D Molecular Models, is available at https://www.mn-am.com/products/corina and can be freely tested, (accessed January 2018).
[28] Newman, D.J. and Cragg, G.M. (2016) *J. Nat. Prod.*, **79**, 629–661.
[29] Bauknecht, H., Zell, A., Bayer, H., Levi, P., Wagener, M., Sadowski, J., and Gasteiger, J. (1996) *J. Chem. Inf. Comput. Sci.*, **36**, 1205–1213.
[30] SONNIA – Self-Organizing Neural Network Package, https://www.mn-am.com/products/sonnia (accessed January 2018).
[31] Sheridan, R.P. and Kearsley, S.K. (1995) *J. Chem. Inf. Comput. Sci.*, **35**, 310–320.
[32] Lewell, X.Q., Judd, D.B., Watson, S.P., and Hann, M.M. (1998) *J. Chem. Inf. Comput. Sci.*, **38**, 511–522.
[33] Fayyad, U.M., Piatetsky-Shapiro, G., and Smyth, P. (1996) *From data mining to knowledge discovery: an overview*, in *Advances in knowledge discovery and data mining* (eds U.M. Fayyad, G. Piatetsky-Shapiro, P. Smyth, and R. Uthurusamy), USA, AAAI Press, Menlo Park, CA, pp. 1–37.
[34] Teckentrup, A., Briem, H., and Gasteiger, J. (2004) *J. Chem. Inf. Comput. Sci.*, **44**, 626–634.
[35] Walters, W.P., Stahl, M.T., and Murcko, M.A. (1998) *Drug Discovery Today*, **3**, 160–178.
[36] Lipinski, C.A., Lombardo, F., Dominy, B.W., and Feeny, P.J. (1997) *Adv. Drug Delivery Rev.*, **23**, 3–25.
[37] Wenlock, M.C., Austin, R.P., Barton, P., Davis, A.M., and Leeson, P.D. (2003) *J. Med. Chem.*, **46**, 1250–1256.
[38] Teague, S.J., Davis, A.M., Leeson, P.D., and Oprea, T. (1999) *Angew. Chem. Int. Ed.*, **38**, 3743–3747.

[39] Oprea, T.I., Davis, A.M., Teague, S.J., and Leeson, P.D. (2001) *J. Chem. Inf. Comput. Sci.*, **41**, 1308–1315.
[40] Clark, D.E. and Pickett, S.D. (2000) *Drug Discovery Today*, **5**, 49–58.
[41] Ajay, A., Walters, W.P., and Murcko, M.A. (1998) *J. Med. Chem.*, **41**, 3314–3324.
[42] Blake, J.F. (2000) *Curr. Opin. Biotechnol.*, **11**, 104–107.
[43] Li, A.P. and Segall, M. (2002) *Drug Discovery Today*, **7**, 25–27.
[44] Li, A.P. (2001) *Drug Discovery Today*, **6**, 357–366.
[45] Thompson, T.N. (2000) *Curr. Drug Metabol.*, **1**, 215–241.
[46] Keserü, G.M. and Molnár, L. (2002) *J. Chem. Inf. Comput. Sci.*, **42**, 437–444.
[47] Andrews, P.R., Craik, D.J., and Martin, J.L. (1984) *J. Med. Chem.*, **27**, 1648–1657.
[48] Bohacek, R.S. and McMartin, C. (1992) *J. Med. Chem.*, **35**, 1671–1684.
[49] Handschuh, S., Wagener, M., and Gasteiger, J. (1998) *J. Chem. Inf. Comput. Sci.*, **38**, 220–232.
[50] Handschuh, S. and Gasteiger, J. (2000) *J. Mol. Model.*, **6**, 358–378.
[51] Le, X., Gu, Q., and Xu, J. (2015) *RSC Adv.*, **5**, 40536–40545.
[52] Brown, R.A. and Case, D.A. (2006) *J. Comput. Chem.*, **27**, 1662–1675.
[53] Glick, M., Rayan, A., and Goldblum, A. (2002) *Proc. Natl. Acad. Sci. U.S.A.*, **99**, 703–708.
[54] Stern, N. and Goldblum, A. (2014) *Israel J. Chem.*, **54**, 1338–1357.
[55] Boda, K., Seidel, T., and Gasteiger, J. (2007) *J. Comput.-Aided Mol. Des.*, **21**, 311–325.
[56] SYLVIA – Estimation of the Synthetic Accessibility of Organic Compounds, https://www.mn-am.com/products/sylvia (accessed January 2018).
[57] Zaliani, A., Boda, K., Seidel, T., Herwig, A., Schwab, C.H., Gasteiger, J., Claussen, H., Lemmen, C., Degen, J., Pärn, J., and Rarey, M. (2009) *J. Comput.-Aided Mol. Des.*, **23**, 593–602.
[58] Li, J. and Eastgate, M.D. (2015) *Org. Biomol. Chem.*, **13**, 7164–7176.
[59] Gasteiger, J. (2015) *Nat. Chem.*, **7**, 619–620.
[60] Kennedy, T. (1997) *Drug Discovery Today*, **2**, 436–444.
[61] Schultz, T.W., Cronin, M.T.D., Walker, J.D., and Aptula, A.O. (2003) *J. Mol. Struct. (Theochem)*, **622**, 1–22.
[62] Richard, A.M., Yang, C., and Judson, R.S. (2008) *Toxicol. Mech. Meth.*, **18**, 103–118.
[63] http://www.warr.com/Lhasa_Symposium_2008_Report.pdf (accessed January 2018).
[64] Escher, B.I. and Hermens, J. (2002) *Environ. Sci. Technol.*, **36**, 4201–4217.
[65] Hansch, C. (1969) *Acc. Chem. Res.*, **2**, 232–239.
[66] Nendza, M. and Müller, M. (2000) *Quant. Struct.-Act. Relat.*, **19**, 581–598.
[67] Rand, G., Welss, P., and McCarthy, L.S. (1995) Introduction to aquatic toxicology, in *Fundamentals of Aquatic Toxicology* (ed. G. Rand), Taylor & Francis, Washington, DC, pp. 3–67.
[68] Bradbury, S.P. (1994) *SAR QSAR Environ. Res.*, **2**, 89–104.
[69] Spycher, S., Pellegrini, E., and Gasteiger, J. (2005) *J. Chem. Inf. Model.*, **45**, 200–208.

[70] Aptula, A.O., Netzeva, T.I., Valkova, I.V., Cronin, M.T.D., Schultz, T.W., Kuhne, R., and Schüürmann, G. (2002) *Quant. Struct.-Act. Relat.*, **21**, 12–22.

[71] Ashby, J. and Tenant, R.W. (1988) *Mutat. Res. Genet. Toxicol.*, **204**, 17–115.

[72] ToxPrint: https://toxprint.org (accessed January 2018).

[73] Yang, C., Tarkhov, A., Marusczyk, J., Bienfait, B., Gasteiger, J., Kleinoeder, T., Magdziarz, T., Sacher, O., Schwab, C.H., Schwoebel, J., Terfloth, L., Arvidson, K., Richard, A., Worth, A., and Rathman, J. (2015) *J. Chem. Inf. Model.*, **55**, 510–528.

[74] ChemoTyperThe ChemoTyper application, https://chemotyper.org (accessed January 2018).

[75] eTOX Project: http://www.etoxproject.eu (accessed January 2018).

6.2 Bridging Information on Drugs, Targets, and Diseases

Andreas Steffen and Bertram Weiss

Bayer Pharma Aktiengesellschaft, PH-DD-TRG-CIPL-Bioinformatics, Müllerstr. 178, Berlin 13342, Germany

Learning Objectives

- To discuss the importance of data curation and integration for enabling effective data analysis.
- To describe how integrative data analysis using proprietary and public data sources impacts pharma R&D along the value chain.

Outline

6.2.1 Introduction, 195
6.2.2 Existing Data Sources, 196
6.2.3 Drug Discovery Use Cases in Computational Life Sciences, 196
6.2.4 Discussion and Outlook, 201

6.2.1 Introduction

The pharmaceutical industry's core mission is to develop novel, innovative, efficacious, and at the same time cost-effective therapies [1]. In order to move beyond best-in-class drugs, pharmaceutical companies strive for first-in-class drugs and are in high need for new strategies that include external innovation in academia, precompetitive collaborations, biotech investments, in-licensing or acquisitions, and fostering innovation within the company's own R&D. One key strategy and competitive factor is the capability to maximally exploit the ever-growing life science data.

While traditionally new drug targets often emerged from textbook knowledge, biological reasoning, me-too approaches, or even serendipity, pharmaceutical companies now emphasize the importance of investing into innovative ways to identify novel and promising therapeutic concepts. The discipline of computational life sciences has matured and now impacts pharma R&D along the entire value chain. For bioinformatics target discovery, biomarker discovery, and indication expansion are three major fields, whereas chemoinformatics is focusing on library design, lead finding, and optimization. More data-driven and systematic approaches to drug and target discovery shall move pharma R&D beyond the often cited serendipity of the past [2].

Analysis of existing large experimental datasets coming from both internal and mainly external sources enable us today to quickly formulate actionable hypotheses that guide experimentalists to focus on the most promising

Applied Chemoinformatics: Achievements and Future Opportunities, First Edition.
Edited by Thomas Engel and Johann Gasteiger.
© 2018 Wiley-VCH Verlag GmbH & Co. KGaA. Published 2018 by Wiley-VCH Verlag GmbH & Co. KGaA.

experiments. However, the persistent rigid data silos and the often unstructured non-integrated nature of the data hamper data scientists to effectively leverage the large amount of existing data. Ideally this requires analysis-ready data so that computational scientists can derive hypothesis with minimal technical barriers. In this way most time is spent on identifying and algorithmically formulating the biological questions and their in-depth analysis and not on data cleansing and curation.

In the following sections we introduce key data sources for compounds, targets, and diseases (see Table 6.2.1), which are the three most relevant data entities for preclinical data science. Pharma R&D data is loosely structured around these three entities. Repositories do exist for all three; however, of particular strong interest are evidence data that relate these entities. Within the last decade new data repositories emerged that integrate the relations and provide evidences for them, for example, opentargets.org [3], ChEMBL [4], drug2gene [5], or PhenomicDB [6].

We refer our readers to the original publications for details and rather exemplify and discuss here the integrative uses of these data sources for the main bio- and chemoinformatics activities in target identification, patient stratification, indication expansion, and toxicity prediction. The discussion provides a perspective where we see future developments in the field.

6.2.2 Existing Data Sources

For key data sources of compounds, targets, and diseases please see Table 6.2.1.

6.2.3 Drug Discovery Use Cases in Computational Life Sciences

6.2.3.1 Target Mining

A healthy pipeline of a pharmaceutical company requires a constant flow of promising new targets. Bioinformaticians are asked to deliver new drug target proposals relevant either for so far unmet medical indications or for improving the therapeutic options of already but not optimally treated diseases. Coming up with new concepts to identify innovative new targets is of high importance for pharmaceutical companies as this can provide a competitive advantage in the search for new first-in-class drugs. In oncology, we are facing a comparably comfortable situation in terms of availability of genomic patient data that can be mined and exploited to identify novel druggable targets, whereas in most other indications this data avalanche is somewhat less pronounced [7].

New target identification campaigns might, for example, start with a question: what are frequently mutated genes in lung cancer that act as oncogenes and that could be targeted via small molecules? Mutation data of a large number of cancer patients can, for example, be obtained from the TCGA database [8]. The functional impact of a mutation has to be assessed in detail as gain-of-function

Table 6.2.1 Selected key data resources focusing on the three pharma R&D relevant entities: targets, compounds, and diseases.

Name	Short description	URL (January 2018)
Target-centric repositories		
NCBI Gene	All information about genes	ncbi.nlm.nih.gov/gene
Ensembl	All information about genes	www.ensembl.org
UniProt	All information about proteins	www.uniprot.org
Reactome	Pathway database	www.reactome.org
WikiPathways	Pathway database with public curation	www.wikipathways.org
Pathway Commons	Pathway meta-database	www.pathwaycommons.org
KEGG	Pathway database	www.genome.jp/kegg
Compound-centric repositories		
ChemSpider	Compound properties	www.chemspider.com
ZINC	Compound structures prepared for cheminformatics studies	zinc.docking.org
eMolecules	Compound ordering	www.emolecules.com
Disease-/phenotype-centric repositories		
OMIM	Catalogue of Mendelian diseases	www.ncbi.nlm.nih.gov/omim
Compound–target-centric repositories		
ChEMBL	Repository for biochemical activities	www.ebi.ac.uk/chembl
PubChem	Repository for biochemical activities	pubchem.ncbi.nlm.nih.gov
drug2gene	Meta-database for compound–target relations	
ChemBank	Repository for biochemical activities	chembank.broadinstitute.org
Compound–disease/phenotype-centric repositories		
DrugBank	Repository for drugs and their indications	
ClinicalTrials	Information about clinical trials	clinicaltrials.gov
Achilles	RNAi screening in cancer cell lines	portals.broadinstitute.org/achilles
CTD2	Compound informer set on cancer cell lines	ctd2.nci.nih.gov
Target–disease-centric repositories		
PhenomicDB	Meta-database for phenotype-disease relations	
opentargets	Repository around evidences for target–disease associations	www.opentargets.org
GWAS Catalog	Catalogue SNP to disease relations	www.ebi.ac.uk/gwas
HuGe Navigator	Gene-disease associations	www.cdc.gov/genomics/hugenet/hugenavigator.htm
CCLE	Cancer cell line encyclopedia	portals.broadinstitute.org/ccle
TCGA	Genomic data of cancer patients	tcga-data.nci.nih.gov

mutations can be directly targeted, whereas loss-of-function mutations that are more likely to occur in tumor suppressors are harder to address therapeutically. Many somatic genetic alterations are in fact due to increased genomic instability in tumors and contrary to so-called driver events do not contribute to tumor development [9]. These events are called passenger alterations and should usually be filtered out as they are not causal to tumor development. A range of bioinformatics algorithms such as MutSigCV, mutation assessor, or IntOGen are thus employed to assess the significance and the functional impact of a mutation [10–12]. One interesting and integrated approach is to look for spatial clusters of somatic mutations in order to identify functional hot spots that could point to a gain-of-function alteration [13–15].

Within the Achilles project, genome-wide RNAi experiments are conducted in a large number of cancer cell lines in order to uncover targetable vulnerabilities of tumors [16]. In contrast, the CTD2 project makes use of a so-called informer set of small molecules with known mode of action and provides experimentally determined sensitivity of a large number of cancer cell lines [17]. Such perturbator datasets can provide first insights into the phenotype of a gene's inhibition and thus are useful resources in the quest for new cancer targets.

Nowadays, text mining is a powerful tool to provide literature evidence for a gene's relevance to a disease of interest helping to quickly provide context of already existing knowledge about involvement of a gene in a disease and is especially useful for larger lists of candidate genes [18, 19].

Further, annotations such as the expression of the gene in the targeted tumor type and in the human body in general [20, 21], the availability of crystal structures in the PDB [22], disease–target association scores based on genetic linkage or genetic studies in model organisms [3, 23], the druggability of a target [24, 25], or competitor information from business intelligence databases can help to narrow down candidate lists into manageable sizes.

Researchers prioritize long lists of targets in a multiparametric fashion considering all these data and literature in order to prioritize drug target candidates systematically in a data-driven fashion and ultimately propose targets that will hopefully lead to new therapeutic options to the benefit of the patients. Integrative platforms to support this process exist and provide very effective ways to obtain target insights that support the target prioritization process [3, 26].

While the previous section provided a description of a target-centric approach to mine for new molecular entities to treating diseases, a comparably large part of new chemical starting points stems from so-called phenotypic screens, in which a desired phenotypic change of a cellular systems is screened for [27]. This is especially helpful to address pathways without suitable drug targets [28]. However, in many cases the phenotype-causing targets of the resulting screening hits are unknown and thus have to be elucidated with significant experimental efforts [29–31]. *In silico* target fishing approaches can be applied to focus experimental efforts by narrowing down the number of possible targets that could cause the phenotypic readout [30, 32, 33]. These approaches make use of large bioactivity data sources such as ChEMBL [34], ChemBank [35], or drug2gene [5] and the vast amounts of bioactivity data within pharmaceutical companies.

6.2.3.2 Biomarker Search for Patient Stratification

Once drug candidates are identified, preclinical research tries to develop hypotheses that patient populations could maximally benefit from a drug [36]. A biomarker identifying those patients has become a key value driver for successful drug development nowadays [37, 38]. In preclinical R&D often larger cell line sensitivity panels are performed. The Cancer Cell Line Encyclopedia (CCLE) provides genomic annotations of more than 1000 cell lines [40]. Further genomically characterized cell lines can be found in the Sanger cell line project and in the Genentech dataset [40, 41]. Differential sensitivities of drugs on these cell lines can then be related to genomic alterations of the cell lines via a range of algorithmic approaches, and predictive biomarker candidates for sensitivity can possibly be derived. The search for such biomarkers often starts with univariate approaches, to see whether simple and clinically exploitable markers for sensitivity exist, for example, the mutation or the copy number status of a known frequently altered gene. Multivariate approaches such as elastic net try to identify combinations of alterations that predict the sensitivity toward the compound [39, 42]. Often differential gene expression between the sensitive and the resistant cell lines are calculated [43–45]. Differentially expressed genes can then be further analyzed in terms of underlying biology, for example, the enrichment of genes from a signaling pathway or a known subtype of cancer. Tools such as gene set enrichment analysis [46] or the online platform DAVID [47] can be of great help to better understand the biology that causes certain patient populations to benefit from a drug and vice versa. Again, perturbator datasets such as Achilles or CTD2 can be useful resources for biomarker identification and might help to find genetic alterations that are predictive for sensitivity toward a drug-mediated inhibition of a target [16, 17].

While cellular disease models have proven useful to uncover genetic markers as predictors of drug sensitivity in cell lines, it is more and more understood that they can substantially deviate from tumors [48], and thus hypotheses derived from such models might not be translatable to the clinics [49]. New approaches try to improve the translatability of preclinical findings by establishing genomically characterized patient-derived primary cancer cells [50, 51] or patient-derived xenografts [52] and studying drug sensitivity on these models that are more likely to reflect the biology of real tumors.

6.2.3.3 Toxicity Prediction

Chemogenomic resources combined with sophisticated algorithmic approaches can be very useful to uncover reasons for compound-mediated toxicities or side effects. As described previously, based on the vast amounts of bioactivity data in publically available data sources and in-house repositories of pharmaceutical companies alike, target activity prediction models can be calculated. A range of toxicities can often be attributed to so-called off-target effects, and thus the set of the corresponding activity models can be applied to computationally predict target-mediated toxicities of compounds. In a recent study, Lounkine *et al.* presented a computational method termed similarity ensemble approach (SEA) to predict the activity of drugs on 73 unintended "side effect" targets

[53]. This method predicts whether a drug molecule could bind to an unwanted target based on the chemical similarity of the drug and the ensemble of all known binders of the unwanted target and a statistical model that controls for random similarity [54]. About half of the computational predictions could be experimentally verified, underlining the power of the combination of sophisticated algorithms and curated data to support drug discovery projects. Promising new algorithmic approaches that can digest large data resources might help to further positively impact drug discovery projects. In a recent FDA-initiated prospective competition termed Tox21, algorithmic groups were asked to predict toxic effects for 647 compounds using a training set comprising toxicity-compound associations for more than 11,000 chemical compounds [55]. Among all submissions, Mayr *et al.* ranked top by using a multitask deep learning approach [56].

While we believe that algorithmic innovations are key to intelligently mine existing data resources, well-curated data resources will always be much sought after and often remain the bottleneck. The consortium-based eTOX project aims to develop a semantically embedded database comprising industry legacy data and public toxicology data [57]. This precompetitive initiative allows for the mining of data from competitors using a broker model, so that ultimately all participating units benefit to bring safer compounds to the patient.

6.2.3.4 Indication Expansion

Once a candidate drug has shown to be efficacious and safe in animals and has thus been selected for clinical advancement, pharmaceutical companies try to identify further new upside indications besides the originally intended core indications [58]. In a first step, text and database mining and careful manual inspection is applied to reveal further target–disease associations mainly based on the concept of diseases sharing similar pathological mechanisms (e.g., inflammation in rheumatoid arthritis, endometriosis, or psoriasis; fibrosis in lung fibrosis, liver cirrhosis, or tissue scarring; angiogenesis in cancer, age-related macular disease, or endometriosis) that eventually might point to new patient populations. A new, well-integrated resource for target–disease relationships is opentargets.org [3].

Another promising approach to come to new indications is based on the Connectivity Map concept. In their landmark publication, Lamb *et al.* describe a systematic method to uncover functional connections of diseases, genes, and drugs [59]. This approach has been applied by pharmaceutical companies for indication expansion [60, 61]. Gene expression profile changes induced by a compound are compared with gene expression changes in diseased versus normal tissues (so-called gene signatures) in order to eventually identify those diseases whose gene expression change could possibly be counter-regulated by the compound. The hypothesis is that if a drug downregulates genes that are upregulated in a disease state (and vice versa), it is possible that patients with this disease could benefit from this drug as it *reverses* the transcriptomic changes of the disease mechanism. The antiepileptic drug topiramate was repositioned for inflammatory bowel disease (IBD). It was a significant hit when gene signatures of 164 drugs were compared with public gene signatures of IBD for anticorrelation [62].

In modern drug discovery it has become a value driver to exploit the potential of a compound in several diseases, and therefore it is a common task for computational groups to evaluate the upside potential of a target already during target identification.

6.2.4 Discussion and Outlook

Preclinical research is positively impacted by effective computational usage of relevant data sources. However, still much valuable internal data in pharmaceutical companies resides in rigid data silos lacking harmonization/standardization, common identifier spaces, or rigorously applied ontologies. It is often difficult to identify relevant data, to access it, and to overcome complex data structures. This severely hampers intelligent and efficacious data analysis. In recent years, many companies spent substantial efforts to enable researchers to creatively mine these data. They have understood that data need to be easily accessible, well integrated, and very importantly analysis ready so that most energy goes into the analyses and the interpretation as such. At the same time, publically available data sources described in Table 6.2.1 increase rapidly and provide substantial means for data-driven drug discovery in industry and academia alike. Easy and sustainable integration of public and internal data is a constant and costly challenge.

While the existence of data is key, it is the intelligent curation and integration that will make algorithmic mining possible. For example, if cellular models for diseases shall be selected, disease descriptions of both patient data and the corresponding cell line models need to be standardized so that they can be mapped onto each other. It is important that the corresponding entities in different data sources are described by standardized identifiers, for example, targets via NCBI Gene or Ensembl identifiers, so that integrative studies are not hampered by the necessity to investigate identifier mappings over and over again. For more complex entities like diseases, use of ontologies is a must. Currently, they are not consistently applied, and for diseases there are many ontologies and none is prevailing. Public funding organizations and precompetitive initiatives, for example, the Pistoia Ontologies Mapping project, start addressing some of these bottlenecks [63]. It is surprising that funding for generating data is readily available but often without considering dedicated budgets for proper dissemination to exploit it.

Whereas our chapter focused mainly on the organization and exploitation of the parts list of a cell and how molecules interact with each other, we have spared out the field of systems biology, where the approach is to understand all the parts and interactions as a whole. Although that is the ultimate goal of course, to our knowledge this Holy Grail has not yet shown to fully deliver sufficiently reliable results to significantly impact drug discovery projects. The foreseeable advent of wearable devices [64] and the digitalization of electronic health records [65] will produce very large amounts of phenotypic/clinical data and provide enormous computational opportunities and challenges. We foresee a future where we have common identifier spaces and ontologies for important entities (genes, drugs, diseases) and new datasets are semantically annotated right away easing integration. Analysis algorithms are registered and automatically applied to meaningful data. If statistically significant results are obtained, researchers who have registered for topics are alerted. By this, interesting new hypotheses are automatically generated, forwarded, and thus put under consideration to our skilled scientists.

Essentials

- Proper curation and integration of relevant data sources is key to effective data analysis in life sciences.
- Integrative data analysis requires common identifier spaces across datasets so that they can easily be joined and mined.
- Developing data standards, controlled vocabularies, and ontologies is important to cope with the rapidly growing scientific data across industry and academia.

Available Software and Web Services (accessed January 2018)

- http://www.opentargets.org
- http://www.cbioportal.org/
- http://www.ebi.ac.uk/chembl
- http://www.etoxproject.eu/

Selected Reading

- LaMattina, J.L. (2008) *Drug Truths: Dispelling the Myths about Pharma R&D*, John Wiley & Sons, Inc., Hoboken, NJ, 156 pp.
- Lengauer, T. (2008) *Bioinformatics – From Genomes to Therapies*, Wiley-VCH Verlag GmbH, Weinheim, Germany, 1814 pp.
- McKinney, W. (2012) *Python for Data Analysis*, O'Reilly Media, Sebastopol, CA, 463 pp.
- Wilkinson, M.D. *et al* (2016) The FAIR Guiding Principles for scientific data management and stewardship. *Sci. Data*, **3**. doi: 10.1038/sdata.2016.18

References

[1] Wang, L., Plump, A., and Ringel, M. (2015) *Drug Discovery Today*, **20**, 361–370.
[2] Kubinyi, H. (2006) Ernst Schering Research Foundation Workshop, www.kubinyi.de/schering58-2006.pdf (accessed January 2018).
[3] Opentargets https://www.targetvalidation.org/ (accessed January 2018).

[4] Bento, A.P., Gaulton, A., Hersey, A., Bellis, L.J., Chambers, J., and Davies, M. Krüger, F.A., Light, Y., Mak, L., McGlinchey, S., Nowotka, M., Papadatos, G., Santos, R., Overington, J.P. (2014) *Nucleic Acids Res.*, **42**, D1083–D1090.
[5] Roider, H.G., Pavlova, N., Kirov, I., Slavov, S., Slavov, T., Uzunov, Z., and Weiss, B. (2014) *BMC Bioinf.*, **15**, 68.
[6] Groth, P., Pavlova, N., Kalev, I., Tonov, S., Georgiev, G., Pohlenz, H.D., and Weiss, B. (2007) *Nucleic Acids Res.*, **35**, D696–D699.
[7] Gnad, F., Doll, S., Manning, G., Arnott, D., and Zhang, Z. (2015) *BMC Genomics*, **16** (Suppl. 8), S5.
[8] NIH Genomic Data Commons Data Portal, https://gdc-portal.nci.nih.gov/ (accessed January 2018).
[9] Vogelstein, B., Papadopoulos, N., Velculescu, V.E., Zhou, S., Diaz, L.A. Jr.,, and Kinzler, K.W. (2013) *Science*, **339**, 1546–1558.
[10] Lawrence, M.S., Stojanov, P., Polak, P., Kryukov, G.V., Cibulskis, K., and Sivachenko, A. (2013) *Nature*, **499**, 214–218.
[11] Sander, C., Schultz, N., Reva, B., Antipin, Y., and Sander, C. (2011) *Nature Commun.*, **39**, e118.
[12] Gonzalez-Perez, A., Perez-Llamas, C., Deu-Pons, J., Tamborero, D., Schroeder, M.P., and Jene-Sanz, A. (2013) *Nat. Methods*, **10**, 1081–1082.
[13] Kamburov, A., Lawrence, M.S., Polak, P., Leshchiner, I., Lage, K., and Golub, T.R. (2015) *Proc. Natl. Acad. Sci. U.S.A.*, **112**, E5486–E5495.
[14] Porta-Pardo, E., Garcia-Alonso, L., Hrabe, T., Dopazo, J., and Godzik, A. (2015) *PLoS Comput. Biol.*, **11**, e1004518.
[15] Engin, H.B., Hofree, M., and Carter, H. (2015) *Pac. Symp. Biocomput.*, 84–95.
[16] Cowley, G.S., Weir, B.A., Vazquez, F., Tamayo, P., Scott, J.A., and Rusin, S. (2014) *Sci. Data*, **1**, 140035.
[17] The Cancer Target Discovery and Development Network (2016) *Mol. Cancer Res.*, **14**, 675–682.
[18] Liu, Y., Liang, Y., and Wishart, D. (2015) *Nucleic Acids Res.*, **43**, W535–W542.
[19] Huang, C.C. and Lu, Z. (2016) *Briefings Bioinf.*, **17**, 132–144.
[20] Carithers, L.J., Ardlie, K., Barcus, M., Branton, P.A., Britton, A., and Buia, S.A. (2015) *Biopreserv. Biobanking*, **13**, 311–319.
[21] Stokoe, D., Modrusan, Z., Neve, R.M., de Sauvage, F.J., Settleman, J., and Seshagiri, S. (2015) *Nat. Biotechnol.*, **43**, D1113–D1116.
[22] Berman, H.M., Battistuz, T., Bhat, T.N., Bluhm, W.F., Bourne, P.E., and Burkhardt, K. (2002) *Acta Crystallogr., Sect. D: Biol. Crystallogr.*, **58**, 899–907.
[23] Groth, P., Leser, U., and Weiss, B. (2011) *Methods Mol. Biol.*, **760**, 159–173.
[24] Hopkins, A.L. and Groom, C.R. (2002) *Nat. Rev. Drug Discovery*, **1**, 727–730.
[25] Griffith, M., Griffith, O.L., Coffman, A.C., Weible, J.V., McMichael, J.F., and Spies, N.C. (2013) *Nat. Methods*, **10**, 1209–1210.
[26] Campbell, S.J., Gaulton, A., Marshall, J., Bichko, D., Martin, S., Brouwer, C., and Harland, L. (2010) *Drug Discovery Today*, **15**, 3–15.

[27] Swinney, D.C. and Anthony, J. (2011) *Nat. Rev. Drug Discovery*, **10** (7), 507–519.
[28] McMillan, M. and Kahn, M. (2005) *Drug Discovery Today*, **10**, 1467–1474.
[29] Lee, J. and Bogyo, M. (2013) *Curr. Opin. Chem. Biol.*, **17**, 118–126.
[30] Schirle, M. and Jenkins, J.L. (2016) *Drug Discovery Today*, **21**, 82–89.
[31] Wagner, B.K. and Schreiber, S.L. (2016) *Cell Chem. Biol.*, **23**, 3–9.
[32] Nettles, J.H., Jenkins, J.L., Bender, A., Deng, Z., Davies, J.W., and Glick, M. (2006) *J. Med. Chem.*, **49**, 6802–6810.
[33] Nidhi, Glick, M., Davies, J.W., and Jenkins, J.L. (2006) *J. Chem. Inf. Model.*, **46**, 1124–1133.
[34] Papadatos, G., Gaulton, A., Hersey, A., and Overington, J.P. (2015) *J. Comput.-Aided Mol. Des.*, **29**, 885–896.
[35] Seiler, K.P., George, G.A., Happ, M.P., Bodycombe, N.E., Carrinski, H.A., and Norton, S. (2008) *Nucleic Acids Res.*, **36**, D351–D359.
[36] Iorio, F., Knijnenburg, T.A., Vis, D.J., Bignell, G.R., Menden, M.P., and Schubert, M. (2016) *Cell*, **166**, 740–754.
[37] Plenge, R.M. (2016) *Sci. Transl. Med.*, **8**, 349ps15.
[38] Nelson, M.R., Tipney, H., Painter, J.L., Shen, J., Nicoletti, P., and Shen, Y. (2015) *Nat. Genet.*, **47**, 856–860.
[39] Barretina, J., Caponigro, G., Stransky, N., Venkatesan, K., Margolin, A.A., and Kim, S. (2012) *Nature*, **483**, 603–607.
[40] COSMIC Sanger Cell Line Project, http://cancer.sanger.ac.uk/cell_lines (accessed January 2018).
[41] Klijn, C., Durinck, S., Stawiski, E.W., Haverty, P.M., Jiang, Z., and Liu, H. (2015) *Nat. Biotechnol.*, **33**, 306–312.
[42] Chen, B.J., Litvin, O., Ungar, L., and Pe'er, D. (2015) *PLoS One*, **10**, e0133850.
[43] Ritchie, M.E., Phipson, B., Wu, D., Hu, Y., Law, C.W., Shi, W., and Smyth, G.K. (2015) *Nucleic Acids Res.*, **43**, e47.
[44] Love, M.I., Huber, W., and Anders, S. (2014) *Genome Biol.*, **15** (12), 550.
[45] Robinson, M.D., McCarthy, D.J., and Smyth, G.K. (2010) *Bioinformatics*, **26**, 139–140.
[46] Subramanian, A., Tamayo, P., Mootha, V.K., Mukherjee, S., Ebert, B.L., and Gillette, M.A. (2005) *Proc. Natl. Acad. Sci. U.S.A.*, **102**, 15545–15550.
[47] da Huang, W., Sherman, B.T., and Lempicki, R.A. (2009) *Nat. Protoc.*, **4**, 44–57.
[48] Domcke, S., Sinha, R., and Levine, D.A. (2013) *Nat. Commun.*, **4**, 2126.
[49] Lieu, C.H., Tan, A.C., Leong, S., Diamond, J.R., and Eckhardt, S.G. (2013) *J. Nat. Cancer Inst.*, **105**, 1441–1456.
[50] Pemovska, T., Kontro, M., Yadav, B., Edgren, H., Eldfors, S., and Szwajda, A. (2013) *Cancer Discovery*, **3** (12), 1416–1429.
[51] Pemovska, T., Johnson, E., Kontro, M., Repasky, G.A., Chen, J., and Wells, P. (2015) *Nature*, **519**, 102–105.
[52] Gao, H., Korn, J.M., Ferretti, S., Monahan, J.E., Wang, Y., and Singh, M. (2015) *Nat. Med.*, **21**, 1318–1325.
[53] Lounkine, E., Keiser, M.J., Whitebread, S., Mikhailov, D., Hamon, J., and Jenkins, J.L. (2012) *Nature*, **486**, 361–367.

[54] Keiser, M.J., Setola, V., Irwin, J.J., Laggner, C., Abbas, A.I., and Hufeisen, S.J. (2009) *Nature*, **462**, 175–181.
[55] Tox21 Tox21 Data Browser, https://tripod.nih.gov/tox21/ (accessed January 2018).
[56] Mayr, A., Klambauer, G., Unterthiner, T., and Hochreiter, S. (2016) *Front. Environ. Sci.*, **3** (80).
[57] Sanz, F., Carrio, P., Lopez, O., Capoferri, L., Kooi, D.P., and Vermeulen, N.P. (2015) *Mol. Inf.*, **34**, 477–484.
[58] Nielsch, U., Schafer, S., Wild, H., and Busch, A. (2007) *Drug Discovery Today*, **12**, 1025–1031.
[59] Lamb, J., Crawford, E.D., Peck, D., Modell, J.W., Blat, I.C., and Wrobel, M.J. (2006) *Science*, **313**, 1929–1935.
[60] Cheng, J., Yang, L., Kumar, V., and Agarwal, P. (2014) *Genome Med.*, **6**, 540.
[61] Sirota, M., Dudley, J.T., Kim, J., Chiang, A.P., Morgan, A.A., and Sweet-Cordero, A. (2011) *Sci. Transl. Med.*, **3**, 96ra77.
[62] Dudley, J.T., Sirota, M., Shenoy, M., Pai, R.K., Roedder, S., and Chiang, A.P. (2011) *Sci. Transl. Med.*, **3** (96), 96ra76.
[63] Barnes, M.R., Harland, L., Foord, S.M., Hall, M.D., Dix, I., and Thomas, S. (2009) *Nat. Rev. Drug Discovery*, **8**, 701–708.
[64] Gay, V. and Leijdekkers, P. (2015) *J. Med. Internet Res.*, **17**, e260.
[65] Jensen, A.B., Moseley, P.L., Oprea, T.I., Ellesoe, S.G., Eriksson, R., and Schmock, H. (2014) *Nat. Commun.*, **5**, 4022.

6.3 Chemoinformatics in Natural Product Research

Teresa Kaserer[1], Daniela Schuster[1], and Judith M. Rollinger[2]

[1] University of Innsbruck, Institute of Pharmacy, Department of Pharmaceutical Chemistry, Computer-Aided Molecular Design Group, Center for Molecular Biosciences Innsbruck, Innrain 80-82, 6020 Innsbruck, Austria
[2] University of Vienna, Department of Pharmacognosy, Faculty of Life Sciences, Althanstraße 14, 1090 Vienna, Austria

Learning Objectives

- To identify caveats in natural product research.
- To locate software and relevant data.
- To apply computational methods in natural product research.
- To report typical success stories in applying chemoinformatics methods to natural products.
- To distinguish the limits of computational methods in natural products research.

Outline

6.3.1 Introduction, 207
6.3.2 Potential and Challenges, 208
6.3.3 Access to Software and Data, 211
6.3.4 *In Silico* Driven Pharmacognosy-Hyphenated Strategies, 219
6.3.5 Opportunities, 220
6.3.6 Miscellaneous Applications, 228
6.3.7 Limits, 228
6.3.8 Conclusion and Outlook, 229

6.3.1 Introduction

The application of the computational techniques described in detail in the *Methods Volume* is not limited to a specific class of chemical structures. Accordingly, chemoinformatics has increasingly been employed in the field of natural product research during the last years. This research area differs considerably from synthetic chemistry. Natural products are usually characterized by complex shapes, with more stereochemical centers, thus comprise non-flat structures that often challenge natural product chemists. These characteristics, however, also implicate highly relevant three-dimensional (3D) aspects prone to interact on macromolecular drug targets [1]. Furthermore, natural product researchers are often confronted with complex mixtures of diverse constituents, and different overall research questions may need to be addressed. Section 6.3.2 provides a detailed description of challenges scientists have to face when investigating natural product-related issues.

Applied Chemoinformatics: Achievements and Future Opportunities, First Edition.
Edited by Thomas Engel and Johann Gasteiger.
© 2018 Wiley-VCH Verlag GmbH & Co. KGaA. Published 2018 by Wiley-VCH Verlag GmbH & Co. KGaA.

The seemingly straightforward use of many computational tools prompted scientists, also originally not coming from a computational chemistry-related field, to include *in silico* analysis in their studies. However, computational methods have certain caveats and liabilities. These may not be visible at first glance, but nevertheless have to be considered. For example, they require rigorous validation of *in silico* workflows prior to their application, as any newly implemented scientific technique. On the other hand, many computational chemists also investigate natural products, but may not be aware of the obstacles related to this specific research area. This chapter therefore aims to provide relevant and practically useful information for researchers with various scientific backgrounds.

6.3.2 Potential and Challenges

Natural products undisputedly continue to play a dominant role in drug discovery [2]. They are labeled as compounds endowed with the property of "metabolite likeness." This implicates that they have been trimmed by evolution to not only act as substrates or ligands of target proteins but also likely be substrates for one or more of the plethora of transport systems and are accordingly designed to reach their intracellular site of action [3]. Nevertheless, in the last two decades, the industry focused much more on chemical entities from synthesis. This reluctance in working with products from natural resources has – despite all the success stories – different reasons:

Major concerns are the access to and supply of natural material: There are estimates of numbers of plant species evaluated for some kind of biological activity that range between 5000 and 50,000 [4]. Considering the estimated number of total plant species on earth with certainly more than 300,000 species of vascular plants, hypothetic calculations reflect that only a minority of plant species have ever been considered in any kind of screening campaign. In addition, natural product sources are not limited to plants, but can also come from other organisms such as from marine species, sponges and mushrooms, lichen, or animals.

Although these estimates indicate a wide field of natural sources to be explored for bioactive compounds, easily accessible materials are often hackneyed with respect to biologically investigated extracts and isolates. On the other hand, the access to underexplored or untapped natural resources is often restricted by regulations of the United Nations Convention on Biological Diversity (https://www.cbd.int/information/parties.shtml) and the Nagoya protocol (https://www.cbd.int/abs/about/) or simple practical reasons. These regulations are important, since they care about the conservation of biodiversity and sustainability, as well as fair and equitable sharing of benefits arising from genetic resources.

The lack of compatibility with high-throughput screening programs has been regarded as a drawback when seeking for hits from nature. Several plant constituents (such as tannins or saponins) may interfere with sophisticated assay systems. Further, the repeated isolation of already known and well-investigated compounds from bioactive starting materials (extracts) is not only a disappointing endeavor but also a time- and resource-consuming process.

In recent years, several new technologies and strategies have significantly fueled the reemergence of natural products for drug discovery:

- Advances in hyphenated, high-resolution techniques for targeted isolation and dereplication (the latter enables the fast identification of known metabolites to avoid further isolation work on already well-studied compounds).
- Miniaturized technologies to save sample amount for the first evaluation in terms of sample identification and isolation (for structure elucidation and preliminary pharmacological tests).
- Appreciation and application of functional assays (high content screening, phenotypic assays), which are often more robust and meaningful than target-based assays. In addition, their readout gives a better view on the samples' bioavailability.
- Use of chemical scaffolds and biology-oriented synthesis as well as biotechnological production and genetic engineering to overcome resupply issue.
- Application of metagenomics and metabolomics for identifying novel classes of metabolites.
- Generation of sophisticated (drug-like enhanced) multicomponent screening libraries (extracts, fractions) and drug-like and diverse natural compound libraries.
- Application of diverse chemoinformatics tools, in particular for *in silico* prediction of ligand–target interactions and rationalization of bioactivities on a molecular level.

For further reading on new technological advances and strategies in natural product research and resupply issues, the authors refer to the surveys of Harvey *et al.* [1] and Atanasov *et al.* [5].

In the last 10 years, the application of chemoinformatics tools has led to outstanding findings in natural product research. It has pushed natural product drug discovery in an economic and target-oriented way. However, the newly implemented virtual screening achievements would have been successful only partly without the technological and experimental advances mentioned before. All the virtual predictions are only as good as the input information. This particularly refers to the reliability and accuracy of structural and biological information from both macromolecular targets and their ligands (and non-ligands), from which computer scientists build their hypothetic models. This also refers to the molecular databases of natural compounds and proteins used for virtual screening (see Section 6.3.3). All of them are based on millions of studies available mainly in the public domain and generated within decades of research efforts from thousands of scientists. To take advantage of these data in a proper scientific way, several steps have to be considered when using *in silico* tools, starting from the aim of the study, to the proper selection of the chemoinformatics tool (Figure 6.3.1), to the model generation and its strict and unbiased validation (theoretical and experimental validation, Figure 6.3.2), to its application in natural product research.

Adequate model validation heavily depends on the required kind of prediction as well as the precision level. Roughly, one can distinguish binding pose

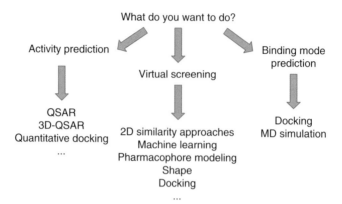

Figure 6.3.1 Selection of appropriate modeling tools depending on the aim of the study.

prediction (qualitative), activity classification (semi-quantitative, grouping, ranking), and quantitative calculations aiming at the actual estimation of the binding affinity. For an exhaustive description of validation metrics, the interested reader is referred to [6, 7]. An overview of benchmarking datasets for model and software validation is available in Ref. [8].

In the case of virtual screening studies, the screening method alone may still yield a large number of virtual hits. Independent from the screening method, additional computational filters can be applied to narrow down the hit lists to compounds with favorable physicochemical properties (Figure 6.3.3). Those filters are optional and can be modified according to the project. For example, the Lipinski's rule of five [9] may not be applicable in many natural product research scenarios. Some practical factors like a compound's accessibility also play important roles in the selection of test compounds. Finally, experimental validation of the selected hits is the last and very crucial step in model and workflow validation.

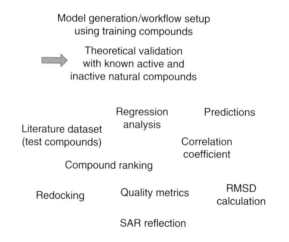

Figure 6.3.2 Depending on the selected methods, theoretical validation experiments are necessary to select the best performing models for making predictions.

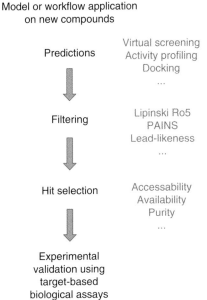

Figure 6.3.3 Virtual screening workflow including additional filtering and selection criteria.

6.3.3 Access to Software and Data

Various software solutions, both commercial and open source, are available for molecular modeling and virtual screening. A list of selected programs for commonly employed methods and a short description is provided in Table 6.3.1. In addition, exemplary *in silico* bioactivity profiling tools are included in the table. With these tools, a compound of interest can be screened against a whole set of targets, and the results suggest potential novel macromolecular interaction partners for the investigated molecule.

A comprehensive list of computational tools is provided at http://www.click2drug.org.

To apply these methods for virtual screening, chemical databases containing natural products are required. These can be created by drawing the structures of molecules, for example, for constituents that are physically available for testing. Special attention needs to be paid to the absolute configuration of the compounds after file conversion, as this step usually is error prone, and incorrect stereochemistry data lead to low-quality workflows [55].

In addition, a large number of ready-to-use compound collections are available. These databases can contain, for example, commercially available constituents, natural products reported in the literature, or compound collections assembled in research groups, and so on. A list of these databases, as well as a short description of the individual collections, is provided in Table 6.3.2. For applicability reasons, only databases were included in the table, where structural data can be downloaded for use on local machines. All collections of compounds, which can only be explored online, were excluded.

Table 6.3.1 Selected chemoinformatics tools.

Name	Description	Availability (accessed January 2018)	References
Pharmacophore modeling			
Discovery studio	Structure- and ligand-based modeling and virtual screening tool implemented in a large modeling suite	http://www.3ds.com/ (commercial)	[10]
LigandScout	Structure- and ligand-based modeling and virtual screening tool	http://www.inteligand.com (commercial)	[11]
Phase/Schrödinger	Ligand-based modeling and virtual screening tool implemented in a large modeling suite	http://www.schrodinger.com (commercial)	[12, 13]
MOE	Structure- and ligand-based modeling and virtual screening tool implemented in a large modeling suite	https://www.chemcomp.com (commercial)	[14]
Pharmer	Structure- and ligand-based modeling and virtual screening tool	http://sourceforge.net/projects/pharmer/ (opensource)	[15]
ZincPharmer	Web-based structure- and ligand-based modeling, which can directly be used to screen the ZINC database	http://zincpharmer.csb.pitt.edu/ (opensource)	[16]
PharmaGist	Ligand-based modeling tool; can be installed locally or used as web-based application; ligands can be uploaded to the online tool and automatically generated pharmacophore models are then sent *via* e-mail to the user; virtual screening applications are limited to the locally installed version	http://bioinfo3d.cs.tau.ac.il/PharmaGist/ (opensource)	[17–19]
Shape-based modeling			
ROCS	Locally installed shape-based modeling and virtual screening application; shape and color features can be combined	http://eyesopen.com/ (commercial)	[20, 21]

Name	Description	URL	Ref
Phase shape/Schrödinger	Locally installed shape-based modeling and virtual screening application; "pure" shape applications; atom types can be included; shape can also be composed of pharmacophore sites	http://www.schrodinger.com/ (commercial)	[22]
SHAFTS	Locally installed shape-based modeling and virtual screening application; combines shape and pharmacophore features	http://lilab.ecust.edu.cn/chemmapper/ (opensource)	[23, 24]
ShaEP	Locally installed shape-based modeling and virtual screening application; combines shape and electrostatic potential	http://users.abo.fi/mivainio/shaep/ (opensource)	[25]
Docking			
GOLD	Locally installed docking program; based on a genetic algorithm	www.ccdc.cam.ac.uk (commercial)	[26, 27]
GLIDE/Schrödinger	Locally installed docking program; grid based	http://www.schrodinger.com/ (commercial)	[28–30]
FlexX	Locally installed incremental docking program	http://www.biosolveit.de/ (commercial)	[31]
FRED	Locally installed docking program; based on shape complementarity and chemical feature alignment	http://eyesopen.com/ (commercial)	[32, 33]
AutoDock	Locally installed docking program; grid based; grids have to be calculated prior to docking	http://autodock.scripps.edu/ (opensource)	[34]
AutoDock Vina	Locally installed docking program; no grids have to be calculated before docking	http://vina.scripps.edu/ (opensource)	[35]
Panther	Locally installed docking program; based on shape-electrostatic models of ligands and binding site	http://www.jyu.fi/panther (opensource)	[36]
PLANTS	Locally installed docking program; based on the ant colony optimization algorithm	http://www.mnf.uni-tuebingen.de/ (opensource)	[37–40]
SwissDock	Web-based docking tool	http://www.swissdock.ch (opensource)	[41, 42]

(Continued)

Table 6.3.1 (Continued)

Name	Description	Availability (accessed January 2018)	References
iScreen	Web-based docking tool; can be employed to screen the TCM database in Taiwan	http://iscreen.cmu.edu.tw/intro.php (opensource)	[43]
Bioactivity profiling tools			
SEA	2D similarity-based machine learning tool; employs a statistical model which is corrected for random similarity	Both commercial and open source, the freeware is web-based and available @ http://sea.bkslab.org/	[44]
PASS	2D similarity-based machine learning tool; employs naive Bayes probabilities	Both commercial and open source, the freeware is web-based and available @ http://www.pharmaexpert.ru/passonline/	[45]
SPiDER	Self-organizing map-based prediction of drug equivalence relationships (SPiDER); combines the concepts of self-organizing maps, consensus scoring, and statistical analysis (web-based)	http://www.cadd.ethz.ch/software/spider.html (opensource)	[46]
ChemGPS-NP	Web-based tool tuned for navigation in biologically relevant chemical space using eight principal components describing physicochemical properties	http://chemgps.bmc.uu.se/batchelor/about.php (opensource)	[47]
TarFisDock	Web-based target identification tool that docks the query molecule into 698 protein targets from the Potential Drug Target Database (PDTD) [48]	https://bio.tools/tarfisdock (opensource)	[49]
Inte: Ligand HypoDB	A collection of >2700 manually generated ligand- and structure-based pharmacophore models for > 320 targets	http://www.inteligand.com/ (commercial)	[50–52]
PharmaDB	A collection of ~140,000 automatically generated structure-based pharmacophore models	http://www.3ds.com/ (commercial)	[53]
PharmMapper	Web-based tool, a collection of >7000 structure-based pharmacophore models for >1600 targets	http://59.78.96.61/pharmmapper/ (opensource)	[54]

Table 6.3.2 Natural products databases.

Name	Number of compounds	Description	Source (accessed January 2018)	References
3DMET	8,718 (2015-12-03)	A collection of natural metabolites extracted from the KEGG compound database	http://www.3dmet.dna.affrc.go.jp/	[56]
AfroCancer	390	African plant constituents that were associated with cancer were manually assembled from the original literature	Supporting information of the original reference (DOI: 10.1021/ci5003697)	[57]
AfroDB	954	A diverse collection of African plant constituents	http://journals.plos.org/plosone/article?id=10.1371/journal.pone.0078085#pone.0078085.s005	[58]
Analyticon Discovery Natural Products			http://www.ac-discovery.com/	
FRGx	216 in release 151015	Fragment library covering 50 natural product scaffolds		
MACROx	1,757 in release 151015	Semisynthetic macrocycles covering 8 chemical scaffolds		
MEGx	4,721 in release 151015	Compounds isolated from plants or microorganisms with a purity of >90%		
NATx	25,794 in release 151015	Semisynthetic natural products		
AntiBas2 2012	>40,000	Commercial collection of metabolites found in higher fungi and microorganisms	http://www.wiley-vch.de/stmdata/antibase.php	
CamMedNP	1,859	Pure constituents of Cameroonian medicinal plant, which have been manually assembled from original articles, PhD theses, and so on. For every compound also spectroscopic data, collection sites, plant sources, and known biological activities are provided	Supporting information of the original reference (http://bmccomplementalternmed.biomedcentral.com/articles/10.1186/1472-6882-13-88)	[59]

(Continued)

Table 6.3.2 (Continued)

Name	Number of compounds	Description	Source (accessed January 2018)	References
Cardiovascular disease herbal database (CVDHD)	35,230	Constituents from medicinal plants associated with cardiovascular diseases		[60]
Chem-TCM	12,070	Commercial database that connects plant constituents used in TCM with ethnobotanical uses, known targets, and botanical information	http://www.chemtcm.com/	
Chinese Ethnic Minority Traditional Drug Database (CEMTDD)	~4,060 constituents, ~621 herbs	Constituents from plants used in traditional medicine of Chinese minorities (e.g., Uyghurs or Kazakhs) was manually collected from books; data about diseases and confirmed and predicted targets are provided	http://www.cemtdd.com/index.html	[61]
Greenpharma DB	150,000	Commercial database	http://www.greenpharma.com/services/greepharma-core-database-gpdb/	
Herbal ingredients *in vivo* metabolism	1,270 (2013-08-12)	Herbal ingredients and their metabolites	The dataset can be downloaded from the ZINC homepage (http://zinc.docking.org/catalogs/himnp), whereas a detailed description of the molecules is provided at the HIM homepage (http://58.40.126.120:8080/him/index.html)	[62]
Human Metabolome Database (HMDB) – Plant	149 (2014-03-11)	A subset of plant metabolites extracted from the human metabolome database	This subset is available from http://zinc.docking.org/catalogs/hmdbplant. Detailed information concerning individual entries is provided at the Human Metabolome Database Homepage (http://www.hmdb.ca/)	[63, 64]

Name	Size	Description	Website	Ref
IBS natural compound library	60,813 screening compounds and 532 natural scaffolds (September 2015)	Contains natural products isolated from plants, microorganisms, marine species and other sources, derivatives of natural products, and natural product mimetics	https://www.ibscreen.com/natural-compounds	
MarinLit	Not reported	Commercial collection of marine natural products extracted from >28,000 original research articles	http://pubs.rsc.org/marinlit/	
Super Natural II	325,500	Information about 2D structures, physicochemical properties, predicted toxicity class and potential vendors	http://bioinf-applied.charite.de/supernatural_new/index.php	[65]
Naturally occurring Plant-based Anticancerous Compound-Activity-Target DataBase (NPACT)	1,574 (2013-09-17)	Natural products with anticancer activity manually curated from original literature	The dataset can be downloaded from the ZINC homepage (http://zinc.docking.org/catalogs/npactnp), whereas a detailed description of the molecules is provided at the NPACT homepage (http://crdd.osdd.net/raghava/npact/index.html)	[66]
Nuclei of Bioassays, Biosynthesis, and Ecophysiology of Natural Products (NuBBE) Database	643 (2013-01-28)	A collection of secondary metabolites from Brazil	The dataset can be downloaded from the ZINC homepage (http://zinc.docking.org/catalogs/nubbenp), whereas a detailed description of the molecules is provided at the NPACT homepage (http://nubbe.iq.unesp.br/portal/nubbedb.html)	[67]
Pan-African natural products library (p-ANAPL)	538	A library of isolated and physically available African natural products	Supporting information of the original reference (http://journals.plos.org/plosone/article?id=10.1371/journal.pone.0090655#pone.0090655.s001)	[68]
Specs natural products	859 compounds are available in an amount >1 mg in version Dec2015	Isolated or synthesized natural products and derivatives from plants, fungi, marine organism, bacteria, and so on	www.specs.net	

(Continued)

Table 6.3.2 (Continued)

Name	Number of compounds	Description	Source (accessed January 2018)	References
TCM Database @ Taiwan	>60,000	The whole database (without duplicates) can be downloaded, or subsets of constituents sorted by properties and/or origin	http://tcm.cmu.edu.tw/	[69]
TimTec				
Flavonoid derivatives collection	~500	Derivatives of 9 flavonoid scaffolds with good bioavailability properties	http://www.timtec.net/flavonoid-derivatives.html	
Gossypol and its derivatives	88	Gossypol derivatives from members of the *Gossypium* genus, Malvaceae	http://www.timtec.net/gossypol-and-derivatives.html	
Natural Derivatives Library	3,040	Natural derivatives, semi-synthetic compounds, mimetics, and analogs	http://www.timtec.net/ndl-3000-natural-derivatives-library.html	
Natural Products Library	800	Pure natural compounds	http://www.timtec.net/natural-compound-library.html	
Plant extracts	~130 extracts from stock, material of 2,600 plants is available for customized extracts, 9,000 plants can be collected on demand	Traditional use or therapeutic application, constituents, geography, and so on, is annotated for extracts on stock	http://www.timtec.net/plant-extracts.html	
TM-MC	~14,000	Data was extracted from Korean, Chinese, and Japanese Pharmacopeias and cross-linked to Pubmed entries, species origin, and application fields are annotated to compound names, no structures are provided!	Excel and OWL files can be downloaded from http://informatics.kiom.re.kr/compound/index.jsp	[70]
Universal Natural Products Database	229,358	A comprehensive collection of natural products for virtual screening		[71]

6.3.4 *In Silico* Driven Pharmacognosy-Hyphenated Strategies

There was a distinct delay in the application of *in silico* methods for the identification of bioactive compounds from natural sources in comparison to its application in medicinal chemistry. This circumstance has several reasons that mirror the challenges for the implementation of *in silico* tools in pharmacognostic workflows, for example, complexity of multicomponent mixtures and necessity of physical availability of broad natural compound libraries, which is hampered by limited access to rare metabolites and their costly and time-consuming isolation procedures.

This is particularly true when following a classical target-oriented drug discovery approach, where a validated, druggable target is virtually screened for thousands or millions of molecular structures able to comply with the mandatory features of the search query to enrich the pool of hit candidates. The idea is to narrow down the pool of compounds worth to be experimentally tested and thus to focus on compounds with a higher propensity to interact with the target under investigation. This workflow, however, requires an almost unlimited access to the predicted virtual hits for biological evaluation. Only in this way, real hits can be identified from virtual hits, and the underlying screening hypothesis can be critically validated. Depending on the predictive power of the search query, which again is based on the amount and quality of input information, the percentage of true hits within virtual hits may vary tremendously.

Although there is an increasing amount of providers that offer not only databases with natural product structures (Table 6.3.2) but also the compounds themselves, it has been shown that 83% of the core ring scaffolds that are present in natural products are not commercially available compounds [3]. For those virtual hits, which first have to be synthesized or isolated from natural sources before they can be subjected to experimental validation, it is uncertain if the efforts in the isolation procedure are worth to be done (plant selection, access to the starting material, collection, extraction, analytics, dereplication, chromatographic isolation, and structural identification).

At this point, it is highly recommended to take advantage of the *privileged structures* displaying multiple biological activities evolved in nature to bind to and influence the function of several protein targets [72]. They are prone to provide hits, even against more difficult screening targets, for example, protein–protein interactions [73]. It requires a sensible evaluation of information from various disciplines and their implementation in pharmacognostic workflows as reviewed before [74, 75].

In particular, the empirical knowledge about beneficial effects from herbal preparations may give valuable clues, for example, as derived from traditional medicine (see Section 6.4), clinical trials with multicomponent mixtures, or phenotypic extract screenings. Although their results reflect a response from a multicomponent mixture and/or a multi-target disease, this information is of high value as a criterion for the selection of starting material. They can give precious hints with respect to involved molecular targets as well as to secondary plant metabolites that may contribute to one of the virtually predicted ligand–target interactions. The poly-pharmacological (pleiotropic) character inherent to many natural products is best evaluated in virtual parallel screening filtering experiments, which provide insights into targets of physiological relevance in terms of possibly new applications and to disclose potential anti-targets [76, 77]. Several examples taking advantage from the abovementioned combinations from empirical and theoretical approaches are presented in more detail in the next section.

6.3.5 Opportunities

Computational tools can be employed in natural product research to address a multitude of aspects. This section provides case studies for common application fields (an overview is provided in Figure 6.3.4), where *in silico* investigations have successfully been applied to complement and support experimental investigations.

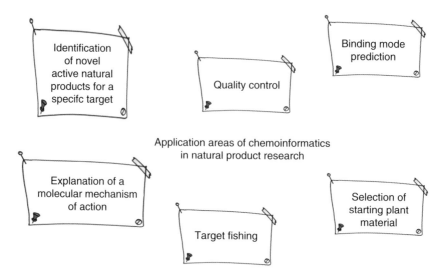

Figure 6.3.4 Implementation of chemoinformatics in natural product research.

6.3.5.1 Virtual Screening to Identify Novel Bioactive Natural Products for a Specific Target

Multi-conformational molecular databases consisting of natural products, for example, as listed in Table 6.3.2, can be explored to discover novel bioactive constituents for the target of interest (Figure 6.3.5). For this purpose, models representing a specific target are employed for screening the compound collection, and selected mapping virtual hits can be subjected to experimental investigations. For example, Su *et al.* [78] applied a docking-based virtual screening for the identification of novel rho kinase inhibitors from natural origin,

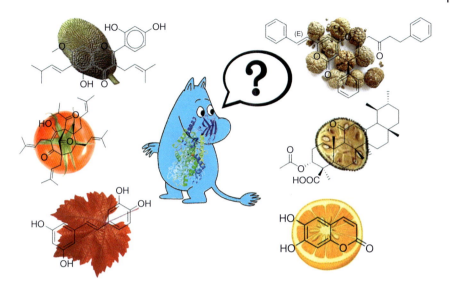

Figure 6.3.5 Virtual screening approach for the identification of novel active constituents for the target of interest.

which may be useful for the treatment of pulmonary hypertension. Starting from a collection of over 100,000 natural constituents, several preprocessing steps were undertaken to finally retrieve a dataset of about 25,000 chemically diverse and drug-like molecules. Before the prospective docking, the docking protocols generated in AutoDock and DOCK were theoretically validated *via* re-docking of the co-crystallized ligands. Whereas DOCK failed to reproduce the experimentally determined ligand pose, AutoDock successfully fitted the ligands in a similar manner into the binding pocket. The authors therefore used the faster DOCK algorithm to prefilter the dataset of 25,000 compounds, and the highest ranked 1000 molecules were subjected to the more accurate AutoDock screening protocol. The results were then reevaluated with a QSAR (quantitative structure-activity relationships) model. Out of the top-100 ranked compounds, six were finally purchased and investigated with an experimental kinase assay. For five of these compounds, the biological activity could be confirmed, and the two most active compounds baicalein and phloretin inhibited rho kinase with submicromolar IC_{50} values of 0.95 and 0.22 µM, respectively [78].

In 2012, Bauer et al. [79] applied pharmacophore models for microsomal prostaglandin E2 synthase1 (mPGES-1) inhibitors to identify novel natural products as potential anti-inflammatory compounds (Figure 6.3.6). The two pharmacophore models employed for this purpose had been extensively validated in retrospective and prospective experiments using synthetic compounds beforehand [80]. Virtual screening of the Chinese Herbal Medicines Database retrieved only a low number of virtual hits, of which several were depsides from different lichen species. Due to this finding, additional lichen metabolites isolated previously were screened as well. However, only two of them mapped an mPGES-1 model. Out of the ten compounds that were subjected to the

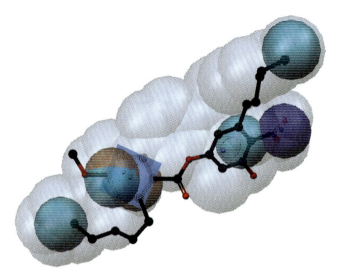

Figure 6.3.6 Perlatolic acid fitted into a pharmacophore model for mPGES-1 inhibitors. Chemical features are color-coded: hydrophobic, gray; negatively ionizable group, dark blue; aromatic ring (brown and blue plain). A steric restriction (cyan shape) is depicted as light gray cloud.

experimental testing, four were predicted as active by the pharmacophore-based screening (two initial hits and two more from the second run). Of these four compounds, three were active in a concentration-dependent manner in the cell free assay. In particular, IC_{50} values of 0.43, 0.4, and 1.15 µM were determined for physodic acid, perlatolic acid, and olivetoric acid, respectively. Importantly, none of the compounds that did not map a model was found to be active *in vitro* [79].

In the search for natural inhibitors of the enzyme 11β-hydroxysteroid dehydrogenase 1 (11β-HSD1), extracts of six well-established anti-diabetic medicinal plants were probed in an *in vitro* assay for their inhibitory potential. 11β-HSD1 catalyzes the conversion of inactive 11-ketoglucocorticoids to active 11β-hydroxyglucocorticoids and was found to be therapeutically useful to decrease blood glucose concentration and to ameliorate metabolic abnormalities in type 2 diabetes [81–84]. Among the tested samples, the leaf extracts of *Eriobotrya japonica* (Thunb.) Lindl. (i.e., loquat) showed a dose-dependent inhibition of 11β-HSD1 and a preferential inhibition of 11β-HSD1 versus 11β-HSD2 [84]. Virtual screening of the DIOS database [85] using a previously generated and experimentally validated pharmacophore model [86] clearly pointed toward the virtual hit corosolic acid to interact with the binding site of 11β-HSD1. Following this prediction toward one of the main ingredients of loquat, that is, corosolic acid, and the *in vitro* activity of the crude extract, this plant material was selected for in-depth phytochemical studies to subject its isolates to biological testing. Indeed, corosolic acid was confirmed as selective inhibitor of 11β-HSD1 with an IC_{50} value of 0.8 µM (measured in lysates of cells expressing recombinant human 11β-HSD1). Further bioactive pentacyclic triterpene acids of the ursane type were identified to synergistically act as

11β-HSD1-inhibitors in this herbal remedy and helped to further improve the predictive tool [87].

6.3.5.2 Target Fishing

Target fishing can be employed to identify potential macromolecules a natural product can interact with (Figure 6.3.7). To identify potential interaction partners of 16 constituents isolated from the aerial parts of *Ruta graveolens*, the compounds were screened against a large collection of over 2200 pharmacophore models representing more than 280 macromolecular targets [88]. The predicted biological activities on selected targets were further investigated with *in vitro* assays, thereby confirming several of the *in silico* results. Out of the five constituents matching human rhinovirus coat protein models, three were active, one was inactive, and one was cytotoxic. The computational workflow, however, failed to identify two further active compounds. In the case of acetylcholine esterase, all compounds that were active in the experimental testing also mapped a model. In return, a higher number of false positive hits were retrieved. Finally, rutamarin was the only secondary metabolite that mapped a cannabinoid receptor-2 model, and it was also the only compound that was active in the biological testing. The results of the prospective investigation in this study strongly supported the validity of the *in silico* workflow [88].

In a similar approach, leoligin, the main lignan from the alpine plant *Leontopodium alpinum*, was screened against the model collection [89]. Among the predicted targets for this secondary metabolite was the cholesterylester transfer protein (CETP). *In vitro* assays showed that leoligin activated human and rabbit CETP at picomolar and nanomolar levels, respectively, and blocked it at higher concentrations. In addition, an activating effect was observed also in a transgenic mouse model [89].

In a similar manner, target fishing can be employed to elucidate the molecular mechanism underlying the observed biological effect of a compound. In 2014,

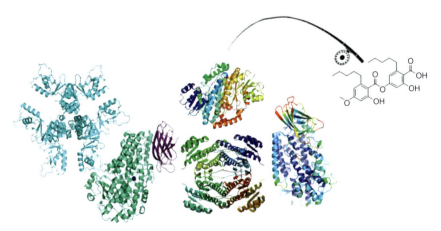

Figure 6.3.7 Target fishing approach for the identification of macromolecular targets for a specific compound.

Reker et al. [90] presented a novel method for target fishing, which is independent of the target structure. In this approach, the topological pharmacophore features of query compound fragments are compared to pre-calculated drug compound clusters. The constituent is then assigned to the cluster with the smallest Euclidian distance. Target information for the cluster was derived from confirmed interaction partners of reference drugs within the cluster. As prospective application example, the macrolide archazolide A (ArcA) was investigated. This compound exerts potent anticancer effects. ArcA inhibits the ion pump vacuolar-type H^+-ATPase at the nanomolar level; however, it was suggested that additional targets might be responsible for the pronounced antitumor effect. The analysis predicted several targets involved in arachidonic acid-associated signaling cascades as potential interaction partners, and subsequent biological testing confirmed a concentration-dependent effect of ArcA on half of these targets. In addition, weak effects on two further targets were observed. The experimental results validate the applicability of the natural product-derived fragment-based approach for the identification of novel macromolecular targets. Remarkably, all newly identified interaction partners of ArcA have also been linked to anticancer effects [90].

A variety of sesquiterpene lactones (STLs) possess considerable anti-inflammatory activity. It has been shown that they exert this effect by inhibiting the activation of the transcription factor NF-κB. In an investigation of a dataset of 103 structurally diverse STLs by a counterpropagation neural network various combinations of structure descriptors were explored [91]. It was found that a single model based on the 3D structure descriptors produced by a radial distribution function [92] based on π-charges had the best predictability for NF-κB activity of these STLs. This allowed the inference that the structural element of an α,β-unsaturated carbonyl group is responsible for this effect. This finding confirms the hypothesis that the attack of cystein-38 of the p65/NF-κB subunit is responsible for the activity of STLs. Thus, the combination of 3D structure descriptors that can be interpreted with the powerful data modeling method of a counterpropagation neural network supports what is supposed about the molecular mechanism of action.

In a similar study, STLs that inhibit serotonin release were investigated [93]. With a dataset of 54 STLs, the combination of 3D structure descriptors from radial distribution functions and molecular surface potentials [94] with a counterpropagation neural network allowed for the development of a structural model. Some descriptors model the structural requirements for both activities; other descriptors can be used to decide whether an STL is more active to NF-κB or to serotonin release.

Gong et al. [95] reported the isolation of two novel sterol metabolites from the marine sponge *Theonella swinhoei*. Both compounds showed cytotoxic effects on two human cancer cell lines. To unravel potential targets mediating the antitumor activity, a previously described reverse docking approach [96, 97] was employed. In detail, the compounds were docked into the structures of 211 cancer-related targets, and biological effects of the compounds were tested on the ten best-ranked targets. Thereby, the histone acetyltransferase p300 was identified as target of one of the novel sterols [95]. These examples highlight the support of chemoinformatics for the investigation of compounds with a known effect but elusive molecular mechanism.

In addition, computational methods are also useful when the effect of a natural product preparation is well documented, but the constituents actually responsible for the effect have not been identified so far. In the case of mastic gum, the oleoresin derived from *Pistacia lentiscus*, both the exact compounds and the exact targets involved in the therapeutic effect were unknown. Traditionally, mastic gum has been employed as remedy against diabetes. To investigate whether 11β-HSD1 may be involved in the therapeutic effects of mastic gum and that constituents are furthermore responsible for the activity, a natural compound database was screened against 11β-HSD1 models. Among the virtual hits were multiple *Pistacia* and several *P. lentiscus* triterpenoids. Therefore, the mastic gum sample, an acidic fraction of the oleoresin, and the two main triterpenes isolated from mastic gum were subjected to experimental testing. All four samples inhibited 11β-HSD1 in a concentration-dependent manner, and for the two pure compounds, IC_{50} values in the low micromolar range were determined [98].

6.3.5.3 Elucidation of the Binding Mode

Computational methods can provide a valuable support in predicting the potential interactions between a molecule and its macromolecular interaction partner (Figure 6.3.8). In an ethnobotanical screening approach, Atanasov *et al.* [99] identified polyacetylenes from *Notopterygium incisum* as novel PPARγ (peroxisome proliferator-activated receptors gamma) partial agonists. To get further insights into potential binding modes of the newly identified bioactive constituents, docking studies were conducted. The applied docking workflow was validated *via* re-docking of the co-crystallized ligand magnolol, and the program successfully placed the compound with a root-mean-square deviation of ~0.55 Å compared to the crystal structure into the PPARγ binding pocket.

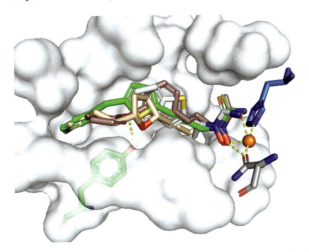

Figure 6.3.8 Computational methods as tools to provide insight into the molecular ligand binding interaction.

Docking of the polyacetylenes predicted a similar binding mode for these constituents, which included hydrophobic contacts with arms I and II of the Y-shaped binding pocket and the entrance region. In addition, the docking results suggested the formation of two hydrogen bonds with the residues Cys285 and Glu295 [99].

In a study by Temml *et al.* [100], the three structurally similar lignans arctigenin, trachelogenin, and matairesinol isolated from *Carthamus tinctorius* were investigated. The biological testing of the three constituents revealed a pronounced inhibitory effect of arctigenin on the enzyme indoleamine 2,3-dioxygenase, whereas trachelogenin was weakly active and matairesinol inactive. To establish a structure–activity relationship and rationalize the molecular mechanism underlying the distinct biological effects of these close derivatives, a docking protocol and subsequent pharmacophore-based analyses of the predicted interaction patters were employed. The results identified a crucial interaction of both active compounds with Ser235, which was not observed for the inactive matairesinol. Additionally, the pose of trachelogenin lacked the interaction with the iron of the heme group, which may cause the reduced potency compared to arctigenin [100].

6.3.5.4 Identification of Starting Plant Material

In silico studies can also aid in the identification of appropriate plant material for further phytochemical and biological investigations (Figure 6.3.9). In a recent study, Grienke *et al.* [101] used two previously identified plant-derived influenza neuraminidase (NA) inhibitors as templates for a 3D similarity screening performed with Rapid Overlay of Chemical Structures (ROCS) to disclose novel natural scaffolds able to interact with the NA binding site. Among the top-ranked molecules, a significant number of licorice ingredients could be retrieved. Accordingly, the roots of *Glycyrrhiza glabra* L. (Fabaceae) were identified as a plant source containing constituents that share structural commonalities with previously identified NA inhibitors from other natural sources. Guided by this prediction, licorice root extracts were phytochemically investigated. Among the 12 isolates, four constituents not only were able to block the NA of different influenza strains in the low micromolar range but also, more importantly, exhibited a distinct anti-influenza effect in the cytopathic effect assay. Three of the experimentally confirmed NA-inhibiting licorice constituents gave high scores in the shape-focused virtual screening.

Similarly, Ikram *et al.* [102] employed a docking-based strategy to identify novel plant material with bioactivity toward influenza NA. The applied docking workflow was extensively validated *prior* to its prospective use. First, the co-crystallized ligand was re-docked into the binding pocket to investigate whether the protocol can identify the experimentally determined ligand binding mode. In a second step, also a dataset containing confirmed active NA inhibitors and ~2000 decoys (compounds with unknown activity, but considered as inactive) were docked. An area under the ROC curve of 0.99 was retrieved, thereby suggesting that the docking workflow was highly suitable for the identification of novel bioactive molecules. Accordingly, the protocol was employed for the prospective docking of ~3000 natural products from Malaysian plants. The

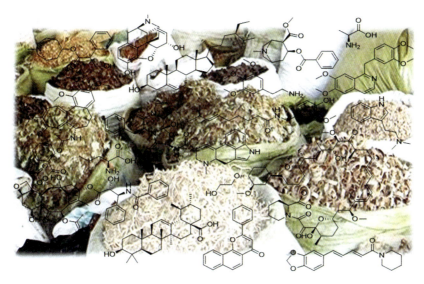

Figure 6.3.9 Computational methods for the selection of plant material as promising starting point for experimental investigations.

top-ranked compounds were mainly constituents from five different plants. Extracts and fractions of these plants were evaluated *in vitro*, and all had at least moderate inhibitory activity. In the following, 12 compounds were isolated and five of them showed NA inhibitory activity. Although the novel phytochemicals are only moderately active, the approach in general proved to be successful for the identification of plants with the desired biological effect [102].

Chagas-Paula *et al.* [103] combined machine learning methods and LC-MS (liquid chromatography-mass spectrometry) data to predict the anti-inflammatory effects of Asteraceae extracts. A decision tree was generated to identify biomarkers in the LC-MS spectra of 57 Asteraceae samples, which were responsible for dual 5-lipoxygenase and cyclooxygenase inhibition. Based on these biomarkers, a predictive artificial neuronal network (ANN) model was constructed and rigorously validated. This ANN model can be employed to identify potentially anti-inflammatory Asteraceae species exclusively by analyzing the LC-MS-derived metabolomic profiles of their extracts [103].

6.3.5.5 Quality Control

In a similar approach, Yu *et al.* [104] developed a fast analysis tool to identify the cultivation region of *Panax quinquefolius* L. and detect potential adulterations with other *Panax* species. The diversity of 14 saponins among the HPLC (high performance liquid chromatography) chromatograms of *P. quinquefolius* L., *P. ginseng*, and *P. notoginseng* samples were investigated with hierarchical cluster analysis and principal component analysis, resulting in a clear discrimination of the three species. In addition, a linear discrimination function was used to distinguish the cultivation region of *P. quinquefolius* L. using the saponin contents from 31 samples originating from three different countries, that is, the United

States, Canada, and China. The validity of the workflow was further evaluated with 12 *P. quinquefolius* samples collected from the three different countries. These samples could successfully be discriminated from the other two species by the developed tools, and in addition, the correct cultivation areas were annotated. Finally, six commercial samples labeled as *P. quinquefolius* were investigated, and the results revealed that two of the commercial products were adulterated with *P. ginseng* [104].

6.3.6 Miscellaneous Applications

Natural products are the results of secondary metabolisms. Different plant species and particularly different plant tribes show slight differences in their secondary metabolic pathways and thus produce different sets of metabolites and different natural products. It was therefore investigated whether STLs produced by various tribes of the plant family Asteraceae can be used to distinguish between different families of that tribe [105]. A dataset of 144 STLs was used to distinguish between the tribes of Eupatorieae, Heliantheae, and Veronineae of the Asteraceae family. 3D structure descriptors were chosen to characterize the STLs, and the dataset was analyzed by a self-organizing map (SOM). This allowed for the classification of the Asteraceae family into tribes and subtribes. Furthermore, predictions of the occurrence of STLs from a plant species according to the taxa they belong to were also performed by the neural network.

In further work, a much larger dataset of 921 STLs isolated from seven tribes was investigated [106]. As many STLs are produced in different tribes, an STL may have to be assigned to more than one tribe. Thus, the methodology had to be extended to a multilabeled classification. Such an approach was developed and its performance compared to a single-label classification and analyzed from a chemotaxonomic point of view. The multilabeled approach allowed for the modeling of reality as closely as possible and to significantly decrease the number of plant sources that have to be considered for finding a certain STL.

6.3.7 Limits

The main limitation of computational tools with respect to natural products research is their high dependency on the availability of structural data for phytochemicals. It is obvious that only structurally defined compounds can be leveraged for *in silico* prediction. For many natural products, however, this data is not or only partially available so far. Although efforts are ongoing to elucidate the structures of secondary metabolites, nature provides such a vast resource of diverse constituents in which still the majority of compounds need to be identified. In addition, many constituents may be reported exclusively in the original literature and are not yet implemented in molecular databases for virtual screening, which makes the large-scale investigation of many diverse constituents from various sources difficult. The increasing number of freely accessible natural products databases can provide a valuable support to overcome these challenges.

Especially when chemoinformatics tools are used to predict the biological activities or molecular targets for natural product extracts, two challenges emerge. First, the drawbacks described above concerning limited data about constituents also apply to extracts: Even for well-studied extracts, usually not all constituents have been reported in databases or the scientific literature. Second, additive biological effects of multicomponent mixtures are especially challenging to predict. Although there is currently much effort in developing predictive tools for compound mixtures [107], this is definitely one of the challenging fields for the future.

Despite the many success stories about the application of chemoinformatics in natural product research, computational tools still produce many false positive virtual hits or incorrect results. This is also true when applied to synthetic compounds; however, natural products pose a specific challenge for virtual screening programs and algorithms. The reason for this lies in the development and validation phase of the software. Screening algorithms are developed and tested mostly on synthetic compound data – mainly because there are much more biological data available on synthetic compounds. While the programs may work accurately within this "chemical training space," their performance in other parts of the chemical universe such as the natural products area is largely unknown. So in order to get reliable predictions, *in silico* workflows have to be rigorously validated before they are applied in prospective experiments. In addition, the *in silico* results should be confirmed with experimental methods. We [108] and others [109] have also reported evidence in literature to support our *in silico* results. However, for an ultimate proof of the results, biological testing is inevitable.

6.3.8 Conclusion and Outlook

As shown by numerous success stories like the ones outlined above, chemoinformatics tools are useful and efficient tools to enhance natural product research. Efforts to discover novel activities for natural constituents can be guided in a straightforward manner to promising biological tests. When the true positive success rates of high-throughput screening and virtual screening are compared, virtual screening followed by experimental testing of hits usually identifies a higher true positive hit rate: For high-throughput screening, hit rates of 0.021% for protein tyrosine phosphatase inhibitors [110], 0.0007% for cruzain inhibitors [111], 0.55% for glycogen synthase kinase-3β [112], and 0.1% for formyl peptide receptor ligands [113] were reported. In comparison, a summary of 20 prospective virtual screening campaigns followed by biological testing of hits [114] revealed hit rates between 2.5% and 100% (mean 20.6%, median 14.3%) – a considerable boost in efficiency. On the other hand, virtual screening may also miss highly active compounds [115]. Therefore, both high-throughput screening and chemoinformatics approaches, complementing each other, are needed in natural product research. Especially in the natural products field, where ethnobotanical information provides further hints on the properties of constituents of a remedy, the experimental efforts can further be guided by this historical information [116].

Properly used, chemoinformatics provides valuable tools for estimating the physicochemical properties of natural products and selecting substances and macromolecular targets for testing as well as predicting binding poses. Especially in the future, when more natural compounds are available not only as database entries but also as easily accessible (purchasable) compounds, when chemoinformatics tools have been properly trained and validated for natural products, and when activity predictions of multicomponent mixtures have evolved, those tools will be even more convenient to use in natural product research. Still, chemoinformatics approaches will always require a sensible implementation in pharmacognostic workflows that should ideally be based on the combined expertise from natural product researchers and computational chemists. But we are convinced – and hopefully could also demonstrate by the examples presented herein – that computation has a role to play in the prospering field of rationalized phytotherapy and natural product lead discovery.

Essentials

- *In silico* applications enhance traditional, experimental natural product research in the selection of plant material for research, elucidation of the mechanism of actions of a given natural product, quality control, binding rationalization, and identification of novel ligands for a given, pharmacologically interesting target.
- Although with powerful and technically easy-to-use tools, the scientists need to thoroughly validate each calculation and ensure high data quality to produce reliable models and results.

Available Software and Web Services (accessed January 2018)

- ZINC database, subset natural product catalogs: zinc.docking.org/browse/catalogs/natural-products
- List of commercial and freely available data processing, molecular modeling, and virtual screening tools: click2drug.org
- Protein Data Bank: http://www.wwpdb.org
- Typical docking programs: AutoDock, AutoDock Vina, DOCK, GOLD.
- Software Suits: Discovery Studio, LigandScout, Molecular Operating Environment – MOE, Schrödinger Software Suite.
- Visualization: Pymol.
- See Table 6.3.1 for an extended list.

Selected Reading

- Atanasov, A.G., Waltenberger, B., Pferschy-Wenzig, E.M., Linder, T., Wawrosch, C., Uhrin, P., Temml, V., Wang, L., Schwaiger, S., Heiss, E.H., Rollinger, J.M., Schuster, D., Breuss, J.M., Bochkov, V., Mihovilovic, M.D., Kopp, B., Bauer, R., Dirsch, V.M., and Stuppner, H. (2015) Discovery and resupply of pharmacologically active plant-derived natural products: a review. *Biotechnol. Adv.*, **33**, 1582–1614.
- Rollinger, J.M., Stuppner, H., and Langer, T. (2008) Virtual screening for the discovery of bioactive natural products. *Prog. Drug Res.*, **65**, 213–249.
- Schuster, D. and Wolber, G. (2010) Identification of bioactive natural products by pharmacophore-based virtual screening. *Curr. Pharm. Des.*, **16**, 1666–1681.

References

[1] Harvey, A.L., Edrada-Ebel, R., and Quinn, R.J. (2015) *Nat. Rev. Drug Discovery*, **14**, 111–129.
[2] Newman, D.J. and Cragg, G.M. (2012) *J. Nat. Prod.*, **75**, 311–335.
[3] Hert, J., Irwin, J.J., Laggner, C., Keiser, M.J., and Shoichet, B.K. (2009) *Nat. Chem. Biol.*, **5**, 479–483.
[4] Miller, J.S. (2013) *Planta Med.*, **79**, IL45.
[5] Atanasov, A.G., Waltenberger, B., Pferschy-Wenzig, E.-M., Linder, T., Wawrosch, C., Uhrin, P., Temml, V., Wang, L., Schwaiger, S., Heiss, E.H., Rollinger, J.M., Schuster, D., Breuss, J.M., Bochkov, V., Mihovilovic, M.D., Kopp, B., Bauer, R., Dirsch, V.M., and Stuppner, H. (2015) *Biotechnol. Adv.*, **33**, 1582–1614.
[6] Braga, R.C. and Andrade, C.H. (2013) *Curr. Top. Med. Chem.*, **13**, 1127–1138.
[7] Cherkasov, A., Muratov, E.N., Fourches, D., Varnek, A., Baskin, I.I., Cronin, M., Dearden, J., Gramatica, P., Martin, Y.C., Todeschini, R., Consonni, V., Kuz'min, V.E., Cramer, R., Benigni, R., Yang, C., Rathman, J., Terfloth, L., Gasteiger, J., Richard, A., and Tropsha, A. (2014) *J. Med. Chem.*, **57**, 4977–5010.
[8] Lagarde, N., Zagury, J.-F., and Montes, M. (2015) *J. Chem. Inf. Model.*, **55**, 1297–1307.
[9] Lipinski, C.A., Lombardo, F., Dominy, B.W., and Feeney, P.J. (1997) *Adv. Drug Delivery Rev.*, **23**, 3–25.
[10] Dassault Systèmes BIOVIA, *Discovery Studio Modeling Environment*; San Diego, CA: Dassault Systèmes.
[11] Wolber, G. and Langer, T. (2005) *J. Chem. Inf. Model.*, **45**, 160–169.
[12] Dixon, S., Smondyrev, A., Knoll, E., Rao, S., Shaw, D., and Friesner, R. (2006) *J. Comput.-Aided Mol. Des.*, **20**, 647–671.
[13] Dixon, S.L., Smondyrev, A.M., and Rao, S.N. (2006) *Chem. Biol. Drug Des.*, **67**, 370–372.

[14] Chemical Computing Group *Molecular Operating Environment (MOE)*, Chemical Computing Group Inc., 1010 Sherbooke St. West, Suite #910, Montreal, QC, Canada, H3A 2R7.

[15] Koes, D.R. and Camacho, C.J. (2011) *J. Chem. Inf. Model.*, **51**, 1307–1314.

[16] Koes, D.R. and Camacho, C.J. (2012) *Nucleic Acids Res.*, **40**, W409–W414.

[17] Dror, O., Schneidman-Duhovny, D., Inbar, Y., Nussinov, R., and Wolfson, H.J. (2009) *J. Chem. Inf. Model.*, **49**, 2333–2343.

[18] Inbar, Y., Schneidman-Duhovny, D., Dror, O., Nussinov, R., and Wolfson, H. (2007) Deterministic pharmacophore detection via multiple flexible alignment of drug-like molecules, in Proceeding of RECOMB 2007 – Lecture Notes in Computer Science (eds T. Speed and H. Huang), Springer Verlag, Berlin Heidelberg.

[19] Schneidman-Duhovny, D., Dror, O., Inbar, Y., Nussinov, R., and Wolfson, H.J. (2008) *Nucleic Acids Res.*, **36**, W223–W228.

[20] OpenEye Scientific Software, Santa FE, NM, vROCS, http://www.eyesopen.com (accessed January 2018).

[21] Hawkins, P.C., Skillman, A.G., and Nicholls, A. (2007) *J. Med. Chem.*, **50**, 74–82.

[22] Sastry, G.M., Dixon, S.L., and Sherman, W. (2011) *J. Chem. Inf. Model.*, **51**, 2455–2466.

[23] Liu, X., Jiang, H., and Li, H. (2011) *J. Chem. Inf. Model.*, **51**, 2372–2385.

[24] Lu, W., Liu, X., Cao, X., Xue, M., Liu, K., Zhao, Z., Shen, X., Jiang, H., Xu, Y., Huang, J., and Li, H. (2011) *J. Med. Chem.*, **54**, 3564–3574.

[25] Vainio, M.J., Puranen, J.S., and Johnson, M.S. (2009) *J. Chem. Inf. Model.*, **49**, 492–502.

[26] GOLD, CCDC, Cambridge, UK, www.ccdc.cam.ac.uk (accessed January 2018).

[27] Jones, G., Willett, P., Glen, R.C., Leach, A.R., and Taylor, R. (1997) *J. Mol. Biol.*, **267**, 727–748.

[28] Friesner, R.A., Banks, J.L., Murphy, R.B., Halgren, T.A., Klicic, J.J., Mainz, D.T., Repasky, M.P., Knoll, E.H., Shelley, M., Perry, J.K., Shaw, D.E., Francis, P., and Shenkin, P.S. (2004) *J. Med. Chem.*, **47**, 1739–1749.

[29] Halgren, T.A., Murphy, R.B., Friesner, R.A., Beard, H.S., Frye, L.L., Pollard, W.T., and Banks, J.L. (2004) *J. Med. Chem.*, **47**, 1750–1759.

[30] Friesner, R.A., Murphy, R.B., Repasky, M.P., Frye, L.L., Greenwood, J.R., Halgren, T.A., Sanschagrin, P.C., and Mainz, D.T. (2006) *J. Med. Chem.*, **49**, 6177–6196.

[31] Rarey, M., Kramer, B., Lengauer, T., and Klebe, G. (1996) *J. Mol. Biol.*, **261**, 470–489.

[32] OpenEye Scientific Software, Santa FE, NM, FRED, http://www.eyesopen.com (accessed January 2018).

[33] McGann, M. (2011) *J. Chem. Inf. Model.*, **51**, 578–596.

[34] Morris, G.M., Huey, R., Lindstrom, W., Sanner, M.F., Belew, R.K., Goodsell, D.S., and Olson, A.J. (2009) *J. Comput. Chem.*, **30**, 2785–2791.

[35] Trott, O. and Olson, A.J. (2010) *J. Comput. Chem.*, **31**, 455–461.

[36] Niinivehmas, S.P., Salokas, K., Lätti, S., Raunio, H., and Pentikäinen, O.T. (2015) *J. Comput.-Aided Mol. Des.*, **29**, 989–1006.

[37] Korb, O., Stützle, T., and Exner, T. (2006) Plants: application of ant colony optimization to structure-based drug design, in *Ant Colony Optimization and Swarm Intelligence* (eds M. Dorigo, L. Gambardella, M. Birattari, A. Martinoli, R. Poli, and T. Stützle), Springer, Berlin Heidelberg.

[38] Korb, O., Stützle, T., and Exner, T. (2007) *Swarm Intell.*, **1**, 115–134.

[39] Korb, O., Stützle, T., and Exner, T.E. (2009) *J. Chem. Inf. Model.*, **49**, 84–96.

[40] Korb, O., Möller, H.M., and Exner, T.E. (2010) *ChemMedChem*, **5**, 1001–1006.

[41] Grosdidier, A., Zoete, V., and Michielin, O. (2011) *J. Comput. Chem.*, **32**, 2149–2159.

[42] Grosdidier, A., Zoete, V., and Michielin, O. (2011) *Nucleic Acids Res.*, **39**, W270–W277.

[43] Tsai, T.-Y., Chang, K.-W., and Chen, C.-C. (2011) *J. Comput.-Aided Mol. Des.*, **25**, 525–531.

[44] Keiser, M.J., Roth, B.L., Armbruster, B.N., Ernsberger, P., Irwin, J.J., and Shoichet, B.K. (2007) *Nat. Biotechnol.*, **25**, 197–206.

[45] Filimonov, D.A., Lagunin, A.A., Gloriozova, T.A., Rudik, A.V., Druzhilovskii, D.S., Pogodin, P.V., and Poroikov, V.V. (2014) *Chem. Heterocycl. Compd.*, **50**, 444–457.

[46] Reker, D., Rodrigues, T., Schneider, P., and Schneider, G. (2014) *Proc. Natl. Acad. Sci. U.S.A.*, **111**, 4067–4072.

[47] Larsson, J., Gottfries, J., Muresan, S., and Backlund, A. (2007) *J. Nat. Prod.*, **70**, 789–794.

[48] Gao, Z., Li, H., Zhang, H., Liu, X., Kang, L., Luo, X., Zhu, W., Chen, K., Wang, X., and Jiang, H. (2008) *BMC Bioinf.*, **9**, 1–7.

[49] Li, H., Gao, Z., Kang, L., Zhang, H., Yang, K., Yu, K., Luo, X., Zhu, W., Chen, K., Shen, J., Wang, X., and Jiang, H. (2006) *Nucleic Acids Res.*, **34**, W219–W224.

[50] Steindl, T.M., Schuster, D., Laggner, C., and Langer, T. (2006) *J. Chem. Inf. Model.*, **46**, 2146–2157.

[51] Steindl, T., Schuster, D., Wolber, G., Laggner, C., and Langer, T. (2006) *J. Comput.-Aided Mol. Des.*, **20**, 703–715.

[52] Markt, P., Schuster, D., Kirchmair, J., Laggner, C., and Langer, T. (2007) *J. Comput.-Aided Mol. Des.*, **21**, 575–590.

[53] Meslamani, J., Li, J., Sutter, J., Stevens, A., Bertrand, H.O., and Rognan, D. (2012) *J. Chem. Inf. Model.*, **52**, 943–955.

[54] Liu, X., Ouyang, S., Yu, B., Liu, Y., Huang, K., Gong, J., Zheng, S., Li, Z., Li, H., and Jiang, H. (2010) *Nucleic Acids Res.*, **38**, W609–W614.

[55] Kirchmair, J., Markt, P., Distinto, S., Wolber, G., and Langer, T. (2008) *J. Comput.-Aided Mol. Des.*, **22**, 213–228.

[56] Maeda, M.H. and Kondo, K. (2013) *J. Chem. Inf. Model.*, **53**, 527–533.

[57] Ntie-Kang, F., Nwodo, J.N., Ibezim, A., Simoben, C.V., Karaman, B., Ngwa, V.F., Sippl, W., Adikwu, M.U., and Mbaze, L.M.a. (2014) *J. Chem. Inf. Model.*, **54**, 2433–2450.

[58] Ntie-Kang, F., Zofou, D., Babiaka, S.B., Meudom, R., Scharfe, M., Lifongo, L.L., Mbah, J.A., Mbaze, L.M.a., Sippl, W., and Efange, S.M.N. (2013) *PLoS One*, **8**, 1–15.

[59] Ntie-Kang, F., Mbah, J., Mbaze, L.M.a., Lifongo, L., Scharfe, M., Hanna, J.N., Cho-Ngwa, F., Onguene, P., Owono, L.C.O., Megnassan, E., Sippl, W., and Efange, S.M.N. (2013) *BMC Complement. Altern. Med.*, **13**, 2–10.

[60] Gu, J., Gui, Y., Chen, L., Yuan, G., and Xu, X. (2013) *J. Cheminf.*, **5**, 1–6.

[61] Huang, J., Zheng, Y., Wu, W., Xie, T., Yao, H., Pang, X., Sun, F., Ouyang, L., and Wang, J. (2015) *Oncotarget*, **6**, 17675–17684.

[62] Kang, H., Tang, K., Liu, Q., Sun, Y., Huang, Q., Zhu, R., Gao, J., Zhang, D., Huang, C., and Cao, Z. (2013) *J. Cheminf.*, **5**, 1–6.

[63] Wishart, D.S., Jewison, T., Guo, A.C., Wilson, M., Knox, C., Liu, Y., Djoumbou, Y., Mandal, R., Aziat, F., Dong, E., Bouatra, S., Sinelnikov, I., Arndt, D., Xia, J., Liu, P., Yallou, F., Bjorndahl, T., Perez-Pineiro, R., Eisner, R., Allen, F., Neveu, V., Greiner, R., and Scalbert, A. (2013) *Nucleic Acids Res.*, **41**, D801–D807.

[64] Irwin, J.J., Sterling, T., Mysinger, M.M., Bolstad, E.S., and Coleman, R.G. (2012) *J. Chem. Inf. Model.*, **52**, 1757–1768.

[65] Banerjee, P., Erehman, J., Gohlke, B.-O., Wilhelm, T., Preissner, R., and Dunkel, M. (2015) *Nucleic Acids Res.*, **43**, D935–D939.

[66] Mangal, M., Sagar, P., Singh, H., Raghava, G.P.S., and Agarwal, S.M. (2013) *Nucleic Acids Res.*, **41**, D1124–D1129.

[67] Valli, M., dos Santos, R.N., Figueira, L.D., Nakajima, C.H., Castro-Gamboa, I., Andricopulo, A.D., and Bolzani, V.S. (2013) *J. Nat. Prod.*, **76**, 439–444.

[68] Ntie-Kang, F., Amoa Onguéné, P., Fotso, G.W., Andrae-Marobela, K., Bezabih, M., Ndom, J.C., Ngadjui, B.T., Ogundaini, A.O., Abegaz, B.M., and Meva'a, L.M. (2014) *PLoS One*, **9**, 1–9.

[69] Chen, C.Y.-C. (2011) *PLoS One*, **6**, 1–5.

[70] Kim, S.-K., Nam, S., Jang, H., Kim, A., and Lee, J.-J. (2015) *BMC Complement. Altern. Med.*, **15**, 1–8.

[71] Gu, J., Gui, Y., Chen, L., Yuan, G., Lu, H.-Z., and Xu, X. (2013) *PLoS One*, **8**, 1–10.

[72] Lachance, H., Wetzel, S., Kumar, K., and Waldmann, H. (2012) *J. Med. Chem.*, **55**, 5989–6001.

[73] Drewry, D.H. and Macarron, R. (2010) *Curr. Opin. Chem. Biol.*, **14**, 289–298.

[74] Rollinger, J.M. and Wolber, G. (2011) Computational approaches for the discovery of natural lead structures, in *Bioactive Compounds from Natural Sources*, 2nd edn (ed. C. Tringali), CRC Press, London.

[75] Rollinger, J.M., Langer, T., and Stuppner, H. (2006) *Curr. Med. Chem.*, **13**, 1491–1507.

[76] Rollinger, J.M. (2009) *Phytochem. Lett.*, **2**, 53–58.

[77] Schuster, D. (2010) *Drug Discovery Today*, 7, e205–e211.

[78] Su, H., Yan, J., Xu, J., Fan, X.-Z., Sun, X.-L., and Chen, K.-Y. (2015) *Pharm. Biol.*, **53**, 1201–1206.

[79] Bauer, J., Waltenberger, B., Noha, S.M., Schuster, D., Rollinger, J.M., Boustie, J., Chollet, M., Stuppner, H., and Werz, O. (2012) *ChemMedChem*, **7**, 2077–2081.

[80] Waltenberger, B., Wiechmann, K., Bauer, J., Markt, P., Noha, S.M., Wolber, G., Rollinger, J.M., Werz, O., Schuster, D., and Stuppner, H. (2011) *J. Med. Chem.*, **54**, 3163–3174.

[81] Gathercole, L.L., Lavery, G.G., Morgan, S.A., Cooper, M.S., Sinclair, A.J., Tomlinson, J.W., and Stewart, P.M. (2013) *Endocrinol. Rev.*, **34**, 525–555.

[82] Grundy, S.M. (2008) *Arterioscler. Thromb. Vasc. Biol.*, **28**, 629–636.

[83] Masuzaki, H., Paterson, J., Shinyama, H., Morton, N.M., Mullins, J.J., Seckl, J.R., and Flier, J.S. (2001) *Science*, **294**, 2166–2170.

[84] Gumy, C., Thurnbichler, C., Aubry, E.M., Balazs, Z., Pfisterer, P., Baumgartner, L., Stuppner, H., Odermatt, A., and Rollinger, J.M. (2009) *Fitoterapia*, **80**, 200–205.

[85] Rollinger, J.M., Steindl, T.M., Schuster, D., Kirchmair, J., Anrain, K., Ellmerer, E.P., Langer, T., Stuppner, H., Wutzler, P., and Schmidtke, M. (2008) *J. Med. Chem.*, **51**, 842–851.

[86] Schuster, D., Maurer, E.M., Laggner, C., Nashev, L.G., Wilckens, T., Langer, T., and Odermatt, A. (2006) *J. Med. Chem.*, **49**, 3454–3466.

[87] Rollinger, J.M., Kratschmar, D.V., Schuster, D., Pfisterer, P.H., Gumy, C., Aubry, E.M., Brandstötter, S., Stuppner, H., Wolber, G., and Odermatt, A. (2010) *Bioorg. Med. Chem.*, **18**, 1507–1515.

[88] Rollinger, J.M., Schuster, D., Danzl, B., Schwaiger, S., Markt, P., Schmidtke, M., Gertsch, J., Raduner, S., Wolber, G., Langer, T., and Stuppner, H. (2009) *Planta Med.*, **75**, 195–204.

[89] Duwensee, K., Schwaiger, S., Tancevski, I., Eller, K., van Eck, M., Markt, P., Linder, T., Stanzl, U., Ritsch, A., Patsch, J.R., Schuster, D., Stuppner, H., Bernhard, D., and Eller, P. (2011) *Atherosclerosis*, **219**, 109–115.

[90] Reker, D., Perna, A.M., Rodrigues, T., Schneider, P., Reutlinger, M., Mönch, B., Koeberle, A., Lamers, C., Gabler, M., Steinmetz, H., Müller, R., Schubert-Zsilavecz, M., Werz, O., and Schneider, G. (2014) *Nat. Chem.*, **6**, 1072–1078.

[91] Wagner, S., Hofmann, A., Siedle, B., Terfloth, L., Merfort, I., and Gasteiger, J. (2006) *J. Med. Chem.*, **49**, 2241–2252.

[92] Hemmer, M.C., Steinhauer, V., and Gasteiger, J. (1999) *Vib. Spectrosc.*, **19**, 151–164.

[93] Wagner, S., Arce, R., Murillo, R., Terfloth, L., Gasteiger, J., and Merfort, I. (2008) *J. Med. Chem.*, **51**, 1324–1332.

[94] Wagener, M., Sadowski, J., and Gasteiger, J. (1995) *J. Am. Chem. Soc.*, **117**, 7769–7775.

[95] Gong, J., Sun, P., Jiang, N., Riccio, R., Lauro, G., Bifulco, G., Li, T.-J., Gerwick, W.H., and Zhang, W. (2014) *Org. Lett.*, **16**, 2224–2227.

[96] Lauro, G., Romano, A., Riccio, R., and Bifulco, G. (2011) *J. Nat. Prod.*, **74**, 1401–1407.

[97] Cheruku, P., Plaza, A., Lauro, G., Keffer, J., Lloyd, J.R., Bifulco, G., and Bewley, C.A. (2012) *J. Med. Chem.*, **55**, 735–742.

[98] Vuorinen, A., Seibert, J., Papageorgiou, V.P., Rollinger, J.M., Odermatt, A., Schuster, D., and Assimopoulou, A.N. (2015) *Planta Med.*, **81**, 525–532.

[99] Atanasov, A.G., Blunder, M., Fakhrudin, N., Liu, X., Noha, S.M., Malainer, C., Kramer, M.P., Cocic, A., Kunert, O., Schinkovitz, A., Heiss, E.H., Schuster, D., Dirsch, V.M., and Bauer, R. (2013) *PLoS One*, **8**, 1–9.

[100] Temml, V., Kuehnl, S., Schuster, D., Schwaiger, S., Stuppner, H., and Fuchs, D. (2013) *FEBS Open Bio.*, **3**, 450–452.

[101] Grienke, U., Braun, H., Seidel, N., Kirchmair, J., Richter, M., Krumbholz, A., von Grafenstein, S., Liedl, K.R., Schmidtke, M., and Rollinger, J.M. (2014) *J. Nat. Prod.*, **77**, 563–570.

[102] Ikram, N.K.K., Durrant, J.D., Muchtaridi, M., Zalaludin, A.S., Purwitasari, N., Mohamed, N., Rahim, A.S.A., Lam, C.K., Normi, Y.M., Rahman, N.A., Amaro, R.E., and Wahab, H.A. (2015) *J. Chem. Inf. Model.*, **55**, 308–316.

[103] Chagas-Paula, D.A., Oliveira, T.B., Zhang, T., Edrada-Ebel, R., and Da Costa, F.B. (2015) *Planta Med.*, **81**, 450–458.

[104] Yu, C., Wang, C.-Z., Zhou, C.-J., Wang, B., Han, L., Zhang, C.-F., Wu, X.-H., and Yuan, C.-S. (2014) *J. Pharm. Biomed. Anal.*, **99**, 8–15.

[105] Da Costa, F.B., Terfloth, L., and Gasteiger, J. (2005) *Phytochemistry*, **66**, 345–353.

[106] Hristozov, D., Gasteiger, J., and Da Costa, F.B. (2008) *J. Chem. Inf. Model.*, **48**, 56–67.

[107] Gu, S., Yin, N., Pei, J., and Lai, L. (2013) *Mol. Biosyst.*, **9**, 1931–1938.

[108] Grienke, U., Kaserer, T., Pfluger, F., Mair, C.E., Langer, T., Schuster, D., and Rollinger, J.M. (2015) *Phytochemistry*, **114**, 114–124.

[109] Mohd Fauzi, F., Koutsoukas, A., Lowe, R., Joshi, K., Fan, T.-P., Glen, R.C., and Bender, A. (2013) *J. Chem. Inf. Model.*, **53**, 661–673.

[110] Doman, T.N., McGovern, S.L., Witherbee, B.J., Kasten, T.P., Kurumbail, R., Stallings, W.C., Connolly, D.T., and Shoichet, B.K. (2002) *J. Med. Chem.*, **45**, 2213–2221.

[111] Ferreira, R.S., Simeonov, A., Jadhav, A., Eidam, O., Mott, B.T., Keiser, M.J., McKerrow, J.H., Maloney, D.J., Irwin, J.J., and Shoichet, B.K. (2010) *J. Med. Chem.*, **53**, 4891–4905.

[112] Polgár, T., Baki, A., Szendrei, G.I., and Keserűu, G.M. (2005) *J. Med. Chem.*, **48**, 7946–7959.

[113] Young, S.M., Bologa, C., Prossnitz, E.R., Oprea, T.I., Sklar, L.A., and Edwards, B.S. (2005) *J. Biomol. Screening*, **10**, 374–382.

[114] Hein, M., Zilian, D., and Sotriffer, C.A. (2010) *Drug Discovery Today*, 7, e229–e236.

[115] Schuster, D., Spetea, M., Music, M., Rief, S., Fink, M., Kirchmair, J., Schütz, J., Wolber, G., Langer, T., Stuppner, H., Schmidhammer, H., and Rollinger, J.M. (2010) *Bioorg. Med. Chem.*, **18**, 5071–5080.

[116] Rollinger, J.M., Haupt, S., Stuppner, H., and Langer, T. (2004) *J. Chem. Inf. Comput. Sci.*, **44**, 480–488.

6.4 Chemoinformatics of Chinese Herbal Medicines

Jun Xu

Sun Yat-Sen University, School of Pharmaceutical Sciences, 132 East Circle at University City, Guangzhou 510006, P. R. China

Learning Objectives

- To evaluate the conceptual differences between Chinese herbal medicine and Western medicine.
- To combine the two different concepts with chemoinformatics methods.
- To transfer the TCM knowledge into new anti-T2D agents with chemoinformatics.

Outline

6.4.1 Introduction, 237
6.4.2 Type 2 Diabetes: The Western Approach, 237
6.4.3 Type 2 Diabetes: The Chinese Herbal Medicines Approach, 238
6.4.4 Building a Bridge, 238
6.4.5 Screening Approach, 240

6.4.1 Introduction

In the past years, chemoinformatics has been applied in Chinese herbal medicine studies such as creating databases [1, 2], mining the active components [3], or classifying the Chinese herbal medicines into hot or cold according to the theory of Chinese medicine [4]. It is, however, quite clear that Chinese herbal medicine has conceptually major differences to Western medicine. In Western medicine, a drug is acting against at least one disease, while a Chinese herbal medicine is taken against at least one symptom that can be caused by many diseases.

It will be shown here how strategies and information from both schools of thought can be brought together to provide new approaches to drug discovery. The new process makes use of chemoinformatics methods providing a bridge between the two basic concepts. The platform will be illustrated with type 2 diabetes (T2D).

6.4.2 Type 2 Diabetes: The Western Approach

T2D is a systemic disease involving multiple mechanisms and molecular targets. Each target is associated with a number of small molecular regulators (inhibitors or activators). There are many mechanisms of actions (MOAs) for treating T2D. Each MOA can be associated with more than one anti-T2D drug target. Currently, there are more than seventy anti-T2D drug targets, and new targets are still emerging [5].

6.4.3 Type 2 Diabetes: The Chinese Herbal Medicines Approach

For Chinese herbal medicine, however, T2D is associated with "Xiaoke" ("emaciated thirst"), which refers to a group of syndromes, including thirst, dry mouth, sugary urine, polydipsia, polyorexia, polyphagia, emaciation, and fatigue [6]. Currently, there are more than eighty Chinese herbal medicines approved by the Chinese FDA (SFDA) for "Xiaoke" treatments. The top 10 herbal medicines are *Radix Astragali* (Huangqi), *Radix Rehmanniae* (Dihuang), *Radix Trichosanthis* (Gualou), *Radix Ophiopogonis* (Maidong), *Fructus Schisandrae* (Wuweizi), *Radix Puerariae* (Gegen), *Rhizoma Anemarrhenae* (Zhimu), *Rhizoma Dioscoreae* (Shanyao), *Radix Ginseng* (Renshen), and *Fructus Lycii* (Gouqizi). They have been studied for their chemistry, biology, and therapeutic uses [7].

6.4.4 Building a Bridge

Chemoinformatics approaches have been used to study the relationships between classical biomolecular disease targets and the Chinese herbal medicine and translate the knowledge of Chinese medicine into the therapies of Western pharmacology. To do this, we constructed a database (HDB) for Chinese herbal medicines against "Xiaoke" (Figure 6.4.1) and a database for chemical agents (ADB) against T2D (Figure 6.4.2). HDB consisted of data regarding herbal prescriptions against

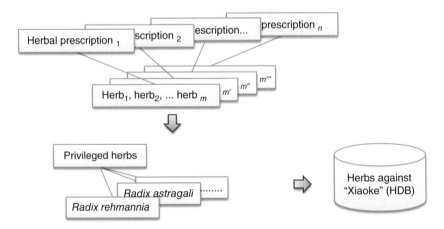

Figure 6.4.1 Creating a herbal prescription database, which contains data regarding herbs and their chemical structures of active constituents against "Xiaoke."

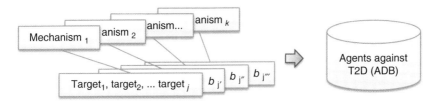

Figure 6.4.2 Creating anti-T2D compound database (ADB) from scientific literatures, which contains the chemical structures of anti-T2D agents, the mechanisms, and targets.

"Xiaoke," the active compound chemical structures, potential targets, and other experimental data. ADB consisted of anti-T2D agents, their chemical structures, and drug targets.

By comparing chemical substructures from the two databases, we were able to figure out an herb-chemome-MOA network (HCMN), which consisted of herbs, chemomes (or privileged chemotypes), and MOAs for T2D. The network is a bridge connecting Chinese medical concepts with molecular targets through different MOAs (Figure 6.4.3).

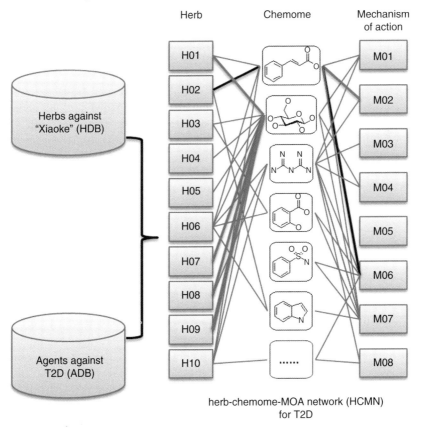

Figure 6.4.3 HCMN. The network demonstrates the relations among herbs, chemotypes, and mechanisms of actions.

In Figure 6.4.3, H01–H10 represent *Radix Astragali* (Huangqi), Radix *Rehmanniae* (Dihuang), *Radix Trichosanthis* (Gualou), *Radix Ophiopogonis* (Maidong), *Fructus Schisandrae* (Wuweizi), *Radix Puerariae* (Gegen), *Rhizoma Anemarrhenae* (Zhimu), *Rhizoma Dioscoreae* (Shanyao), *Radix Ginseng* (Renshen), and *Fructus Lycii* (Gouqizi). M01–M08 represent insulin secretion enhancement, insulin sensitivity enhancement, glucose uptake enhancement in adipose and muscle tissues, glucose absorption inhibition from the intestine, glucose production inhibition from hepatocytes, prevention of diabetic complication via inflammatory or glycation mechanisms, improvement of glycemic control or lipid profile or blood pressure, and prevention of β-cell degradation via inhibition of human islet amyloid polypeptides. Through mining these databases, six privileged chemotypes (guanidines, cinnamyl-acid derivatives, benzenesulfone derivatives, indoyl derivatives, salicyl derivatives, and glycoside derivatives) were revealed by applying an approach proposed by Xu [8]. These chemotypes are major chemical building blocks (found in herbal medicine) regulating multiple drug targets against T2D. Based upon the network, Chinese medical concepts and the herb remedies are rationalized at the biomolecular level.

For example, *Radix Rehmanniae* (Dihuang) is one of the top herbs recommended for "Xiaoke" treatments. The frequently used herbal compounds (herbal compounds are recipes of Chinese herbal medicine) are *Xiaoke Wan* (Bolus), *Jade Maid Brew*, and *Liu Wei Di Huang Wan*, all having Dihuang as a major ingredient. The significant bioactive constituents of *Rehmanniae* are catalpols, jioglutins, jionosides, and flavonoids. The most common structural features of these bioactive constituents are the cinnamyl-acid-like chemotypes. Dihuang inhibits glycation *in vitro* through chemical compounds with cinnamyl-acid-like chemotypes. The network demonstrates that glycation can be associated with M06 (prevention of diabetic complication via inflammatory or glycation mechanisms), which involves targets such as aldose reductase-2 (ALR-2), NF-κB-p65, inducible isoform of nitric oxide synthase (iNOS), and TNF-α. Inhibiting these protein targets can have anti-inflammatory effects so as to enhance diabetic wound healing.

6.4.5 Screening Approach

To translate the TCM knowledge in this example, into new anti-T2D agents, a focused compound library biased on cinnamyl-acid-like derivatives/mimics was virtually constructed and screened against ALR2 with a virtual screening approach employing molecular dynamics simulations [9] (see Figure 6.4.4).

The virtual screening protocol started with three ALR2–ligand complexes, which represent three types of static binding modes. The virtual screening campaign was executed by docking chemical structures into the refined three-dimensional structures. The virtual screening resulted in 71 virtual hits, which were subjected to *in vitro* recombinant human ALR2 inhibition assays (Figure 6.4.5). The bioassays confirmed that 26 of the virtual hits actively inhibited ALR2. Seven compounds from this set, (**4**, **7**, **14**, **22**, **23**, **24**, and **25**) (Figure 6.4.6) inhibited ALR2 with IC_{50}'s $\leq 10\,\mu M$; two compounds from

6.4.5 Screening Approach | 241

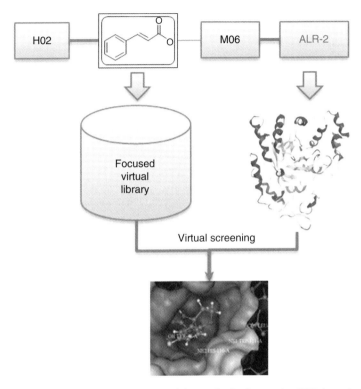

Figure 6.4.4 *Radix Rehmanniae* (H02) is one of the top herbs for treating T2D; its main active constituents have cinnamyl-acid-like common fragments/chemotypes, which are associated with mechanism group M06, particularly for ALR2 target. Library for virtual screening. The compounds selected for the virtual library are based on cinnamyl-acid-like scaffolds/mimics.

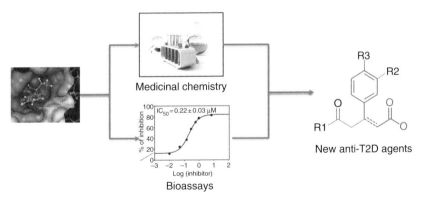

Figure 6.4.5 Discovering new anti-T2D agents. The virtual hits are confirmed by chemistry and bioassays.

them (**14** and **25**) inhibited ALR2 with sub-micromolar IC_{50} concentrations (0.22 and 0.89 μM). These compounds exhibit potencies comparable with the commercially available drug, epalrestat (Figure 6.4.7), while showing superior selectivity to epalrestat relative to inhibition of ALR-1.

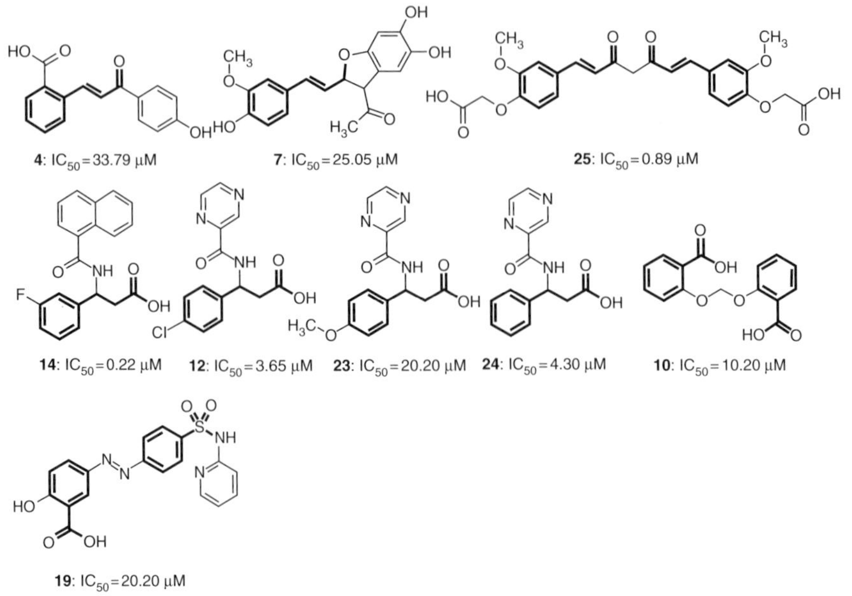

Figure 6.4.6 Cinnamyl-acid-like compounds as anti-T2D agents (ALR2 inhibitors).

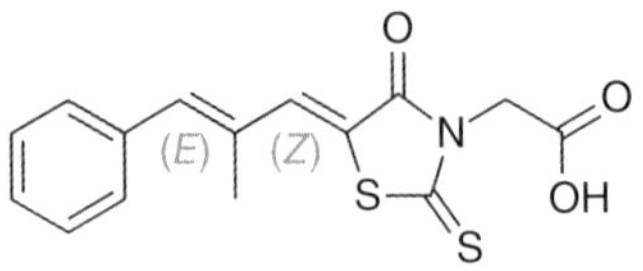

Figure 6.4.7 Epalrestat.

In silico experiments revealed that cinnamyl-acid-like chemotypes provided a productive scaffold to support the interactions between small molecular inhibitors and the "hot" residues in the ALR2 binding site. Compound **25** (Figure 6.4.7) can be viewed as the product of fusing two cinnamyl-acid-like fragments. On the other hand, compounds **14** and **25** can also be viewed as a β-amino-phenylpropanoic mimic (**14**) and curcumin mimic (**25**), which exist in Dihuang and other typical Chinese herbs against "Xiaoke." Usually, these fragments are connected with a glycoside (Figure 6.4.8). According to our database (Figure 6.4.2), many cinnamyl-acid-like compounds with or without glycosides are anti-T2D agents via several MOAs (Figure 6.4.3), including targeting PPAR-γ (M02), ALR2 (M06), and G6PT (M05). Biological experiments also revealed that several of the active compounds contained multiple anti-T2D privileged chemotypes. For examples, compound **10** ($IC_{50} = 10.2\,\mu M$ for ALR2) has two salicyl fragments linked by an ~O—CH_2—O~ group; compound **19** ($IC_{50} = 10.03\,\mu M$ for ALR2) can be viewed as a molecule with a salicyl fragment linked to a benzenesulfone fragment by an ~N=N~ group (Figure 6.4.6).

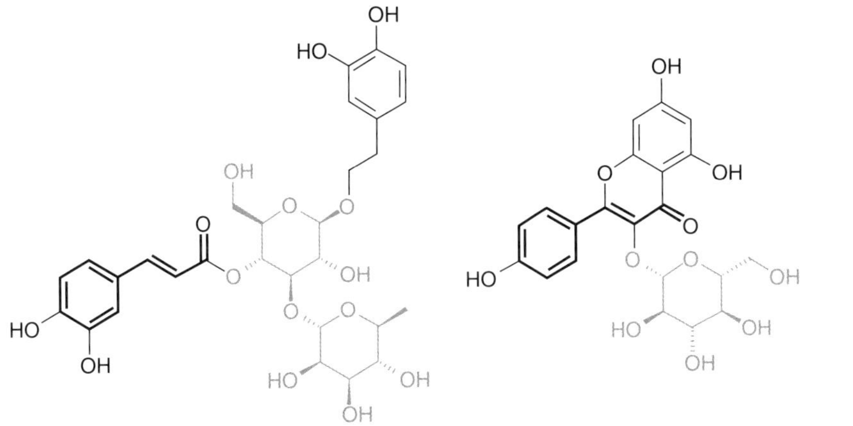

Acteoside (from *Radix Romania*)   Astragalin (from *Astragalus membranaceus*)

Figure 6.4.8 Anti-T2D agents derived from Dihuang and Huangqi. Cinnamyl-acid-like chemotypes are in bold. Glycosides are colored in light grey.

Thus, in summary, the relationship between the T2D targets and the Chinese medical concepts is articulated through HCMN (Figure 6.4.3). Based upon the privileged chemotypes in HCMN, the knowledge intrinsic to Chinese medicine can be translated into new anti-T2D agents through the steps in this protocol (Figures 6.4.1–6.4.5):

1. Elucidate herbal compatibilities.
2. Identify active constituents.
3. Derive privileged chemotypes.
4. Select targets by referencing HCMN.
5. Enumerate a virtual library by connecting the privileged chemotypes with chemical linkers.
6. Generate conformations for the compounds in the virtual library.
7. Virtually screen this library against the selected targets from step 4.
8. Confirm the hits by biological assays.

The networks have been incorporated into a website at http://www.rcdd.org.cn/PNDD. The aforementioned approach can offer a bridge between TCM herb treatments and the search for molecular targets, thus helping to discover new chemotherapies for other systemic diseases.

Essentials

- Elucidation of herbal compatibilities.
- Identification of active constituents.
- Deduction of privileged chemotypes.
- Selection of targets for active constituents.
- Enumeration of a virtual library by connecting the privileged chemotypes.

Available Software and Web Services (accessed January 2018)

- Traditional Chinese Medicine Database System: http://cowork.cintcm.com/engine/windex1.jsp
- ChemStable: Predicting chemical stability for compounds; www.rcdd.org.cn/chemstable
- PNDD: Ligand-target relationship network; www.rcdd.org.cn/PNDD
- LBVS: Ligand-based virtual screening for natural products; www.rcdd.org.cn/lbvs
- ASDB: Database for natural product scaffolds annotated with targets; www.rcdd.org.cn/asdb
- cBinderDB: Covalent binding inhibitor database; www.rcdd.org.cn/cbinderdb
- TCM Database@Taiwan: A comprehensive TCM database collecting small molecular structures and related information from TCM; tcm.cmu.edu.tw

Selected Reading

- Ge, H., Wang, Y., Xu, J., Gu, Q., Liu, H., Xiao, P., Zhou, J., Liu, Y., Yang, Z., and Su, H. (2010) Anti-flu agents from TCM. Natural Products Reports, Royal Society of Chemistry, pp. 1758–1780.
- Gu, P. and Chen, H. (2014) Modern bioinformatics meets traditional Chinese medicine. *Brief Bioinform.*, **15** (6), 984–1003.

References

[1] Chen, C.Y.-C. (2011) *PLoS One*, **6**, e15939.
[2] Ru, J., Li, P., Wang, J., Zhou, W., Li, B., Huang, C., Li, P., Guo, Z., Tao, W., Yang, Y., Xu, X., Li, Y., Wang, Y., and Yang, L. (2014) *J. Cheminform.*, **6**, 13.
[3] Cui, L., Wang, Y., Liu, Z., Chen, H., Wang, H., Zhou, X., and Xu, J. (2015) *J. Chem. Inf. Model.*, **55**, 2455–2463.
[4] Wang, M., Li, L., Yu, C., Yan, A., Zhao, Z., Zhang, G., Jiang, M., Lu, A., and Gasteiger, J. (2015) *Mol. Inf.*, **34**, 2–9.
[5] Moller, D.E. (2001) *Nature*, **414**, 821–827.
[6] Zhang, H.T., Wang, C., Xue, H., and Wang, S. (2010) *Eur. J. Integr. Med.*, **2**, 41–46.
[7] Gu, Q., Yan, X., and Xu, J. (2013) *J. Pharm. Pharma. Sci.*, **16**, 331–341.
[8] Xu, J., Gu, Q., Liu, H., Zhou, J., Bu, X., Huang, Z., Lu, G., Li, D., Wei, D., Wang, L., and Gu, L. (2013) *Sci. China Chem.*, **56**, 71–85.
[9] Wang, L., Gu, Q., Zheng, X.H., Ye, J.M., Liu, Z.H., Li, J.B., Hu, X.P., Hagler, A., and Xu, J. (2013) *J. Chem. Inf. Model.*, **53**, 2409–2422.

6.5 PubChem

Wolf-D. Ihlenfeldt

Xemistry GmbH, Hainholzweg 11, D-61462 Königstein, Germany

Learning Objectives

- To describe PubChem, a major chemical information hub on the Internet.
- To review the design of a database system for storing chemical structures and assay results.
- To identify the data items and query functionalities particular to chemical structure databases.
- To acquire techniques to make Internet-based chemistry information accessible both for human perusal and software-driven processing.

Outline

6.5.1 Introduction, 245
6.5.2 Objectives, 246
6.5.3 Architecture, 246
6.5.4 Data Sources, 247
6.5.5 Submission Processing and Structure Representation, 248
6.5.6 Data Augmentation, 249
6.5.7 Preparation for Database Storage, 249
6.5.8 Query Data Preparation and Structure Searching, 250
6.5.9 Structure Query Input, 253
6.5.10 Query Processing, 254
6.5.11 Getting Started with PubChem, 254
6.5.12 Web Services, 255
6.5.13 Conclusion, 255

6.5.1 Introduction

Since its original launch in 2004, the PubChem database (pubchem.ncbi.nlm.nih.gov) has quickly become a key source of freely accessible information on individual molecules and assay screening results pertaining to these [1, 2]. This reason is enough to look at this database and how its components mesh together in more detail in its own dedicated book section.

One word of caution before our exploration begins: Since its original launch, the appearance and features of PubChem have constantly been evolving and expanded. This text describes the database portal as it is presenting itself in March 2017. In all likelihood, there will already be minor tweaks and updates when the first printed issues of this book are shipped. However, the fundamental

Applied Chemoinformatics: Achievements and Future Opportunities, First Edition.
Edited by Thomas Engel and Johann Gasteiger.
© 2018 Wiley-VCH Verlag GmbH & Co. KGaA. Published 2018 by Wiley-VCH Verlag GmbH & Co. KGaA.

design principles of the database are essentially still the same as laid down in the development years before the original launch date and will remain so.

6.5.2 Objectives

The original intent behind the development of PubChem was to provide a repository for the assay screening results funded by the *Molecular Libraries Initiative* project of the National Institutes of Health (NIH). Soon, this initial limited scope was expanded to provide a general repository for any type of screening results outside the NIH initiative – and then to allow deposition of structures even without links to a specific assay.

PubChem is not a database that exists without context. Technically, it is just one of the many databases hosted by the US National Center for Biotechnology Information (NCBI), which include such diverse information sources as PubMed (medical and biological/biochemical literature references), GenBank (DNA sequences), or MeSH (a controlled dictionary used to describe and index biological and chemical topics). This context required to fit the new design into the existing Entrez query system infrastructure – the massive software system behind the NCBI database cluster. PubChem links back and forth between dozens of records in other databases within the NCBI ecosystem and supports cross-database queries within the cluster. Due to the nature of the information stored in PubChem, the standard Entrez query facilities are sometimes an awkward fit. Many standard chemoinformatics queries cannot be formulated at all in the general Entrez query language, so additional, separate interfaces for these had to be implemented.

6.5.3 Architecture

PubChem is not a single database but consists of three main databases.

6.5.3.1 The Compound Database

It contains (at the time of writing) about 85 million unique, structurally distinct chemical structures. Every structure in this database has a unique identifier, the compound identifier (CID). It is a simple integer, and these are assigned in numerically increasing order.

6.5.3.2 The Substance Database

Substances in this context are the depositor records for a chemical structure. Every new depositor record gets assigned a unique ID, the SID (structure ID), which is also a simple integer. Substances are not unique structures (i.e., there may be multiple identical structures in this database) because the same molecule may be entered by different submitters or even multiple times by the same submitter, for example, as part of different assays in which the same structure was tested.

After submission, substance records are tested for logical coherence and standardized. The latter means that the structure is recoded to a consistent representation, for example, regarding the use of ionic or pentavalent forms of nitro groups or the placement of wedge bonds at stereo centers. This is explained in more detail below (see also *Methods Volume*, Chapter 3). If the standardized structure corresponds to an existing compound record, a link between the SID and the CID is established. Otherwise, a new CID is generated. Many CIDs link to multiple SIDs, but every SID only links to a single (current) CID. The current size of the SID database is about 200 million entries.

The compound and substance databases are described in more detail in [3].

The third database is the *assay database*. It contains about one million assay result tables. Every row of test results (blank dummy or missing data rows excluded) links to a single SID. The unique identifier for an assay in this database is the AID (assay ID) – also an integer. The size of these assay tables is very variable. There are assays with less than 10 molecules tested and high-throughput assays approaching a million tested structures. Certain fields, which are a prerequisite for data processing across multiple assays, must be present for an assay deposition, but in addition to the core set, any number of additional columns may be stored. A very basic case of a required attribute is the summarized assay outcome of a test substance (active, inactive, or inconclusive) – but in addition to this very generic flag, assays typically provide more detailed activity records such as IC_{50} values, or even dose/response curves from multiple concentration points. The latter are valuable for the detailed data analysis within an assay or closely related assay sets, while the former is a basic tool for marking structures as generally biologically inactive, active, or overactive (e.g., promiscuous binders). The organization of the assay database has been documented in [4, 5].

It was a conscious design choice not to link assay rows directly to CIDs. Errors do happen, and there are instances where, for example, it was found out that an assay molecule was not what is was thought to be, or additional information, such as stereochemistry, becomes only available after initial submission. In that case, the assay record is updated, a new SID is assigned to the corrected structure record, and it is then resolved to an existing or a new CID. Another case is when the structure standardization software components are updated, which may lead to a different cleaned-up CID structure.

6.5.4 Data Sources

Besides submitting assay data where the structure information is a side item, it is also possible to submit pure structure data without assay links, but with collections of other properties. These properties include, for example, IDs and references to external databases outside the NCBI world. As a result, PubChem structure entries now include a comprehensive set of links to other databases. These include third-party chemical and biological or regulatory databases – from vendor catalogs to research databases sponsored by foreign governments to patent links. Submitter-provided links are displayed on the PubChem result pages, and by this mechanism PubChem has established itself as a central jump-off point to many different data sources.

At the time of writing, there were 450 different certified data sources from all over the world [6]. There are requirements for participant's professional qualifications and guarantees that the uploader possesses all required intellectual property rights for the provided data, but other than that, upload credentials are freely granted. Every data record, SID or AID, is linked to a specific submitter so that action may be taken in case of problems. There are also mechanisms to temporarily hide information items or delay their general visibility to support requirements such as publication, grant, or patent embargo dates. Submitted but not yet publicly visible data may be accessed by reviewers, by individual secret URLs, and so on. Submitters may also revoke their own data.

6.5.5 Submission Processing and Structure Representation

After submission, an important step is the initial processing of the structure and assay data before it is added to the database. This process is called standardization. In involves a comprehensive set of logical tests for compound sanity (such as charge distribution and overall charge), as well as recoding structural features (such as the already mentioned nitro groups) to a consistent style. PubChem internally stores molecules with a full hydrogen set. Historically, when memory and available disk space were orders of magnitude smaller than in today's computer systems, many structure databases did not encode hydrogens as explicit atoms. This is mirrored in legacy file formats such as the ubiquitous SD file, where it is customary to omit all hydrogens or hydrogens that are not typically displayed in structure drawings. This legacy encoding can, especially in case of elements that have multiple valid valency states, introduce uncertainty about the true nature of the structure. In PubChem, in the absence of explicit hydrogen encoding, the hydrogen state is determined once and for all during standardization, at a single point of failure, and at an early stage where submitters can verify, and, if necessary, fix it directly before the structure is committed to the database.

Stereochemistry is another common problem. The standardization process checks for spurious stereochemistry encoding (i.e., indicating stereochemistry at atoms or bonds, which are not stereogenic). One problem that cannot be automatically resolved is missing stereochemistry. From submission data alone it is often not possible to determine whether a tested compound is a mixture of enantiomers, or its stereochemistry is specified but either unknown at the time of the experiment or not encoded by the submitter. The internal PubChem structure encoding does support advanced stereochemistry such as those found in square–planar configurations or allenes, but it is often not simple or even impossible to transport this information robustly in legacy file formats.

The PubChem structure data manipulation engine can work with bonds which cannot be represented as classical electron-counted bond in the VB model. An attempt is made to recode metal complexes with *complex* bonds so that there are no unreasonable atomic charges and bond electron counts and to detect charge-pair electrostatic interactions, which should not be encoded as VB bonds. This extended bond information can be retrieved when downloading structure data – PubChem has defined a reasonably backward-compatible Molfile/SDF format extension for this purpose. The download of the native ASN.1-encoded structure data directly contains the extended bond information in its core data structures.

The structure standardization processor is publicly accessible as a separate web service [7].

6.5.6 Data Augmentation

After standardization, a standard set of structure attributes is computed. These are properties with MedChem significance (molecular weight, log P, rotatable bond count, hydrogen donor and acceptor counts, stereogenic or stereo-defined atom and bond counts, IUPAC name), standard structure encodings (such as canonical SMILES or InChI strings), as well as computed 2D and 3D atomic coordinates, including a conformer set for 3D searching. When switching between CIDs and their associated SID set, it is sometimes important to understand that 2D and 3D coordinates of an SID are displayed as submitted (if they were part of the submission, obviously a SMILES-based submission contains neither of these, and all coordinates are computed), while the same data shown for a CID is computed. For example, the computational layout enforces standard orientations for the orientation and configuration of well-known ring systems such as steroids (six-membered rings on the bottom, five-membered ring to the upper right) and can have an impact even on trivial compounds such as pyridine (nitrogen on the bottom) or aniline (nitrogen substituent on top), which submitters often do not deposit in standard 2D layout.

Besides handling plain submission structures, in many cases there are input structures that are not a single compound or a salt or other derivative of a base compound. In these cases, a couple of variants of the original structure are generated and processed as separate entities. These include, for example, desalted forms of the submission (e.g., normalizing a metal acetate to acetic acid), a canonic tautomer, or, in case of multicomponent submissions, splits into individual components. These normalizations and splits are tracked, and relationship links are stored in the database. In the user interface, this information is used to provide navigation links so that the processing path of a structure can be followed if desired.

6.5.7 Preparation for Database Storage

The final part of the standardization is the computation of a set of structure hash codes. The most comprehensive of these hash codes, which incorporates both stereochemistry and isotope labeling, is the actual primary key of the CID database. There is a 1 : 1 mapping between the comprehensive structure hash and a CID. Once established, the mapping is persistent. CIDs may become unused, for example, because all SIDs resolving to a specific CID have been withdrawn. But if at a later state another submission resolves to the dormant CID, it is reactivated. There are other hash codes besides the comprehensive one used for the primary key. These include hash codes that ignore stereochemistry or isotope labeling or detect tautomers. Their role is twofold: they are also used as navigation links (i.e., the jump to isotope-labeled variants of a structure uses the isotope-ignorant hash code, which only incorporates connectivity and stereochemistry) and they are part of the PubChem structural similarity computation algorithm (see below).

All information accumulated for a structure SID or CID or an assay AID is merged into a data blob, which conforms to the global NCBI data encoding standards. NCBI has historically decided to standardize all its electronic data in a universal schema-driven binary data format called ASN.1 [8]. Outside of NCBI it is mostly used in telecommunications applications. Similar to the better-known XML approach, the details of structure encoding are described in a specification file.

In the context of NCBI, those parts of the information on a structure or assay that have direct counterparts in other NCBI systems use shared encoding patterns. Over the decades, older core NCBI data encoding schemes have grown into elaborate setups to cover even most rare and exotic data representation needs. The encoding scheme for literature references, which are not only potentially a part of PubChem records but also, for example, are at the very heart of PubMed, is about ten times as complex as the newly added specification sections for encoding PubChem-specific assay and structure data.

PubChem structure and assay data can be downloaded in the original binary ASN.1 format, or an auto-generated ASCII variant thereof, and that is the only format that is guaranteed to not lose any information, which cannot be expressed in other, simpler translated export formats. Nevertheless, there is at least one third-party cheminformatics toolkit, which can process the PubChem structure and assay data in native ASN.1 format [9]. The ASN.1 encoding specifications can be freely downloaded [10].

All PubChem data, structures and assays alike, are also freely available for direct file download and may be used in in-house applications without involvement of the NCBI database systems [11].

6.5.8 Query Data Preparation and Structure Searching

For query processing purposes, the full data blob of an assay or structure is much too slow to access and decode. Therefore, queryable fields are extracted and stored in dedicated database columns. For normal string (textual) and numeric fields, such as CID, ID, compound names, external URL links, heavy atom count, or molecular weight, this is straightforward and not handled differently than in any other database in the NCBI Entrez cluster, and these fields can be queried with standard Entrez syntax [12]. For chemistry-specific queries, a couple of special-purpose fields are added as additional columns. These include the following:

The various *hash codes* computed from the structures, which were already mentioned as part of the standardization process. They are primarily used for fast full-structure searches but also play a role in the PubChem similarity search method.

6.5.8 Query Data Preparation and Structure Searching

Several *fingerprint bitvectors*, again computed from the structures. They encode the presence or absence of specific features, such as elements, atoms with environments, short atom sequences, or ring patterns in the database structures. Depending on the fingerprint type, either a fixed subset of features is tested and encoded as set or unset bits at specific positions in the vector, or a full set of features of certain classes are detected and encoded in a pseudo-random bitvector position. The size of the fingerprints is typically several hundred bits. These bitvectors serve two purposes:

The first one is the acceleration of substructure or superstructure searches. It is comparatively expensive to match a substructure graph (with atoms as vertices and bonds as connections) onto a structure graph, and it is also rather slow to set up the match data structures in the first place. Therefore, it is standard procedure in chemical substructure searches to avoid the atom and bond matching completely where possible.

A feature bitvector can help with this task. The (generally less populated) feature bitvector of the query structure is computed once at the beginning of a query. A comparison to the precomputed feature bitvectors of the database structures can often directly prove that a database structure cannot match and that therefore it does not need to be examined in detail. This is done by checking whether every set bit (which corresponds to a feature present) in the substructure is also set in the structure bitvector. If it is not set, the database structure does not contain the substructure feature and thus cannot match. Unset bits in the substructure bitvector are not tested, and neither are extra set bits (corresponding to additional features) in the database structures. The bits on the bitvector can be checked with simple processor instructions in parallel in batches of 64 bits or even larger words, so this is a very fast operation. For typical substructure queries, this process can often discard 99% of the database structures directly as nonmatching. The recognized name of this filtering method is *screening*.

For queries where the role of substructure and database structures is reversed (superstructure search), a little additional trickery is involved. Strictly speaking, a database structure with its complete hydrogen set cannot be a substructure to any query structure except by an identity match. If a database structure is considered a substructure, it is therefore meant to match without its hydrogen atoms. However, this means that you cannot simply take bitvector stored for screening because any feature bit which includes a hydrogen atom must not be tested. The solution is either to use a mask to hide those bits that involve hydrogens (which is straightforward only for fingerprints with a defined set of patterns) or to simply compute a second screening bitvector of all database structures, which excludes any hydrogen-containing patterns and features. PubChem uses the second approach.

For superstructure queries it is also helpful to compare the number of heavy atoms in the database structures (in their substructure role) against that of the query. If there are more heavy atoms in the database record, it cannot be a superstructure. For normal substructure queries, this could also be done, but is usually not very selective because typical query substructures are not large and most database structures exceed its size.

The second use of bitvectors is for similarity searching. A classical definition of similarity in chemoinformatics uses simple bitvector comparison methods to compute a similarity score between structures. Since set bits present in both bitvectors indicate common structural features, and bits present only in one of the bitvectors indicate structural features that are unique to one of the compared structures, this is an intuitive approach. There are many published methods to compare bitvectors (Euclidian distance, city block distance, cosine, Tversky, Tanimoto, Dice, Forbes, Hamman, Kulczynski, Pearson, Russel/Rao, Simpson, Yule, etc.). But it has been found that in standard chemoinformatics applications the actual algorithm is of minor importance for the MedChem significance or quality and ordering of the results. The Tanimoto algorithm has become more or less the standard method, and this is also what PubChem uses. Standard implementations compute a floating-point value between 0.0 (no similarity at all) and 1.0 (complete agreement of the bits in both vectors).

PubChem scales this value to an integer score between 0 and 100 and then adds a unique twist. In addition to computing the standard score, it also compares the hash codes of the query structure and the database structures. For full structural identity, the score is boosted to 104. In case either the isotope or stereochemistry does not match, but the structural connectivity is the same, the score is 103. For just an identical connectivity, the score is 102, and for a tautomer 101. If none of these additional structural identities can be established, the normal similarity score is reported.

In order to support molecular formula queries effectively, the element counts of the structures are also stored in a dedicated integer-vector class database column.

Another element in the database are precomputed structure renderings in various sizes, from tiny thumbnail-type previews that do not show any atom symbols but just colored squares for hetero atoms to fully expanded structure renderings in several styles.

The last step in preparing the database for structure searching is the encoding of the structure in a compact format, which is easy to decode into a data structure suitable for atom-by-atom matching. Structure line notations such as SMILES or SLN are certainly compact, but do not make a good choice for this purpose. Parsing these representations is comparatively expensive, and the decoded structure still lacks a lot of information, which is routinely part of substructure queries, such as the ring or aromaticity status of atoms or bonds, and computing these on the fly for every match test is prohibitively expensive. For PubChem, we developed a custom structure encoding called *Minimols*. In an average of about 150 bytes per structure record, just twice the length of a typical SMILES, not only the structural connectivity is encoded, but also about 20 often-used atom, bond, ring, and ring system query attributes are immediately available after decompression, which only requires a single memory allocation or static buffer.

For numerical and textual data fields, standard database column indices are set up. For link traversal functionality (e.g., from CID to SIDs or from AID to SIDs or to similar assays or structures), there are many dedicated and precomputed source ID/target ID mapping tables, which support the navigation within the database chemical structure space or assay collection without involving any processing beyond a simple indexed database table row lookup.

6.5.9 Structure Query Input

There are various ways to submit structure queries to PubChem. Experts may want to input SMARTS or another query pattern directly or via cut and paste from external sources. It is also possible to upload structure query files, for example, MDL Molfiles with query attributes, or ChemDraw/SymxyDraw files. The PubChem structure query processing engine is not limited to a single query specification format. Rather, a broad selection of commonly used query file formats can be read and their query content is translated into a universal internal format.

Another input method is to use the PubChem sketcher, a web application for the interactive drawing of query structures. Its development is an interesting story. NCBI has always had a strict policy to disallow the use of any Java applets due to limited help desk resources – in practical experience, these often create problems on client systems, which are not properly configured. Likewise, an input tool that requires the installation of any client software (such as a ChemDraw plug-in) was not a solution, and neither was the prescription of specific browsers or client computer platforms. Since NCBI has a lot of users from all corners of the world, with all kinds of outdated and unmaintained client computer setups, neither up-to-date browsers with advanced JavaScript support nor fast Internet connections could be expected or required. The baseline was the need to support browsers back to IE6 and to present a usable structure drawing tool over a dial-up connection, with bandwidth requirements not exceeding that of Internet radio.

This problem was solved in a custom development project. The PubChem sketcher is almost entirely server based, with minimal and very basic JavaScript components on the client and relying on continuously updated server-generated GIF and PNG images (HTML canvas objects and SVG are not supported on older browsers) to render the evolving structure drawing. There is no upload of massive JavaScript or Java bytecode libraries before start-up, and the structure images compress very well, because they have mostly white background, so its bandwidth requirements are surprisingly low. Twelve years after the initial launch, these technologies and restrictions are probably up for review, but until now the sketcher has served its purpose well and without almost no help desk incidents [13].

Besides supporting direct user-driven structure sketching, it also operates as a structure conversion hub. It can recognize and read about 50 chemistry file formats of uploaded data files and is linked to many web services, not just hosted by NCBI, for on-the-fly conversion of structure identifiers such as compound and trade names, CAS numbers, well-known database IDs, SMILES, InChI, or SLN strings into a structure that can then be either directly submitted or further edited. It is even possible to upload and decode images with structure drawings, which are submitted behind the scenes to a chemical OCR service.

While the sketcher shows a SMARTS query in a feedback field, this is not what is actually transmitted to the PubChem query processing system. The sketcher and the structure match engine share a code base and use a proprietary serialized object format for the lossless information transfer between these systems.

6.5.10 Query Processing

After a query structure has been received, as an upload file, a SMILES/SMARTS or other structure input, or as a data object from the sketcher, a query is constructed. The computational complexity of the actual queries is highly variable – from a simple ID lookup, which is nearly instantaneous because a database field index exists, to broad, complex, screening-insensitive structure queries that could potentially run for many minutes. More complex queries are not executed directly but buffered in a queuing system, and their execution may be delayed depending on the system load.

The PubChem databases are physically organized both as parallel database instances on many servers with an automatic load-balancing system and use segmentation to keep sections of the full database on different servers, which can be queried in parallel.

The generation of result pages is template based. Different templates are used depending on the context of the query.

6.5.11 Getting Started with PubChem

So after all these words, what are typical educational or research tasks one can address with PubChem, and how does one proceed to execute these tasks?

Instead of providing sample step-by-step instructions for study problems, which probably do not work any longer in exactly the described fashion shortly after the first printed book edition has shipped, it is preferable to point the reader to the official PubChem tutorials maintained and kept current by the PubChem staff.

A good starting point is the PubChem course pages [14]. A reader can either download the course material or even book a PubChem representative for an on-site one- or two-day course.

Educational material is available for both simple steps, such as starting a structure search [15], which are often used repeatedly as parts of a larger solution, as well as in the format of small, self-contained studies targeted at students and first-time PubChem users. Even on YouTube, there are plenty of tutorials to be found [16, 17].

An example of a simple biochemical exploration (the different binding modes of aspirin and Tylenol to the PLA2 protein), which uses not only PubChem but also other NCBI protein-related databases, can be found here [18]. Other instructive example studies are focusing on assay data related to the Gaucher disease [19], the toxicology of sarin and related compounds [20], and the characteristics of agonists of human serotonin receptors [21].

There are many peer-reviewed publications that describe the facilities of PubChem, which were used to obtain insights into real current MedChem problems. Papers [22–24] are especially instructive.

An example of using PubChem web services (see below) in the context of a larger project, including the building of a local database with retrieved PubChem data, can be found in [25].

6.5.12 Web Services

PubChem does not just provide an HTML-based user interface targeted at human users. It has also been designed as a resource which can be integrated into third-party applications. For use of PubChem as a data source by external software, web pages are a format that is often very difficult to use because extracting specific data from these is anything but straightforward. For this reason, PubChem implements various alternative interfaces [26] designed for use by software tools. It is also accessible by means of the general Entrez command line tools, which work across the NCBI database cluster.

For accessing general NCBI database functionality via software, the most basic gateway is the general-purpose Entrez e-utilities [27]. These include applications to query database status and configuration information, to submit standard numerical or text queries, to manage query results, or to retrieve specific result items. Query parameters are passed as URL arguments, and the result is XML data.

The most advanced gateway to PubChem-specific functionality is the Power User Gateway (PUG) [28]. Most of the queries that can be invoked via the web interface can also be submitted by this mechanism, and there are even advanced capabilities of PUG that are not mirrored by the user interface. The configuration of a PUG job is specified by an XML data blob, and the returned results are also (usually) XML. Besides query processing, PUG also provides access to structure processing steps in the PubChem pipeline, for example, structure standardization.

PUG is the most powerful programmatic access channel into PubChem, but setting up the XML job section is comparatively complex. For standard tasks, there is also a REST interface, where the parameters are simply passed, as with the Entrez e-utilities, as URL arguments [29].

6.5.13 Conclusion

PubChem has clearly reshaped the landscape of publically accessible chemical information. The availability of a comprehensive computer-readable collection of chemical structure data and assay screening results has been invigorating for scientific data exchange.

Fears that PubChem would significantly compete with or even eradicate commercial data providers (such as Chemical Abstracts Services (CAS) from the American Chemical Society) have not materialized, and intense political lobbying efforts to kill or restrict PubChem in its early years now appear quaint [30]. Rather, the importance of PubChem as a central hub for initial structure or assay lookup and portal for subsequent jumps into external sites (including paywalled systems) has been recognized by many commercial data providers, many of whom now routinely deposit link data directly into PubChem.

Funding appears to be secure for the foreseeable future, so at this time, it looks like PubChem is there to stay and continue to grow.

Essentials

- PubChem is a major chemical information hub on the Internet and a good model for the handling of chemical information in a database context.
- Handling chemical information in a database context requires special techniques, which are explained in the text.
- Making chemical information accessible and queryable by means of a web interface also requires domain-specific tools and software.
- Thanks to the high degree of interconnection between PubChem and other chemistry- and biologics-related Internet-accessible data sources, advanced questions regarding the functions and properties of chemical structures in many different scenarios, including research and education, can be addressed, which cannot be solved relying on a single information source.

Available Software and Web Services (accessed January 2018)

- PubChem: pubchem.ncbi.nlm.nih.gov. The source code of the PubChem (and generally NCBI) system, which was developed by US government employees, but not those parts licensed as components from third parties, is available on request. However, this is enormously complex and extremely site-specific software, which is not of much interest for normal educational or academic environments.
- NCBI multi-database portal: www.ncbi.nlm.nih.gov/
- ChemSpider, a structure search portal by the Royal Society of Chemistry with some overlapping functionality: www.chemspider.com
- The two major third-party providers of software modules to PubChem are OpenEye (www.eyesopen.com/) and Xemistry (www.xemistry.com). OpenEye does not offer free software access, but Xemistry does for academic and educational purposes (www.xemistry.com/academic). The Xemistry academic software package includes many functions used by PubChem and potentially interesting tools for working with PubChem data locally.

References

[1] Bolton, E., Wang, Y., Thiessen, P.A., and Bryant, S.H. (2008) Chapter 12, in *Annual Reports in Computational Chemistry*, vol. 4 (eds R.A. Wheeler and D.C. Spellmeyer), Elsevier, Oxford, UK, pp. 217–241.

[2] Wang, Y., Xiao, J., Suzek, T.O., Zhang, J., Wang, J., and Bryant, S.H. (2009) *Nucleic Acids Res.*, **37**, W623–W633.

[3] Kim, S., Thiessen, P.A., Bolton, E., Chen, J., Fu, G., Gindulyte, A., Han, L., He, J., He, S., Shoemaker, B.A., Wang, J., Yu, B., Zhang, J., and Bryant, S.H. (2016) *Nucleic Acids Res.*, **44**, 1202–1213.

[4] Wang, Y., Suzek, T., Zhang, J., Wang, J., He, S., Cheng, T., Shoemaker, B.A., Gindulyte, A., and Bryant, S.H. (2014) *Nucleic Acids Res.*, **42**, 1075–1082.

[5] Wang, Y., Bolton, E., Dracheva, S., Karapetyan, K., Shoemaker, B.A., Suzek, T.O., Wang, J., Xiao, J., Zhang, J., and Bryant, S.H. (2010) *Nucleic Acids Res.*, **38**, 255–266.

[6] NCBI Data Sources, https://pubchem.ncbi.nlm.nih.gov/sources/ (accessed January 2018).

[7] NCBI PubChem Standardization Service Help, https://pubchemdocs.ncbi.nlm.nih.gov/standardization-service (accessed January 2018).

[8] NCBI Structure, www.ncbi.nlm.nih.gov/Structure/asn1.html (accessed January 2018).

[9] Cactvs toolkit: www.xemistry.com (accessed January 2018).

[10] ftp://ftp.ncbi.nlm.nih.gov/pubchem/specifications/pubchem.asn (accessed January 2018).

[11] NCBI PubChem Download Facility Help, https://pubchemdocs.ncbi.nlm.nih.gov/downloads (accessed January 2018).

[12] Entrez Help [Internet]. Bethesda (MD): National Center for Biotechnology Information (US); 2005-. Entrez Help. 2006 Jan 20 [Updated 2016 May 31]. www.ncbi.nlm.nih.gov/books/NBK3837/(accessed January 2018).

[13] Ihlenfeldt, W.D., Bolton, E., and Bryant, S.H. (2009) *J. Cheminf.*, **1**, 20.

[14] NCBI Principles of PubChem, www.ncbi.nlm.nih.gov/Class/PubChem/course.html (accessed January 2018).

[15] NCBI Finding Compounds Using the Structure Sketcher, www.ncbi.nlm.nih.gov/Class/PubChem/powertools/sketcher.html (accessed January 2018).

[16] www.youtube.com/watch?v=piYf5QfJ8OM (accessed January 2018).

[17] DivCHED CCCE: Cheminformatics OLCC PubChem Advanced Search Tutorials, olcc.ccce.divched.org/PubChemAdvSearch (accessed January 2018).

[18] NCBI Comparing Binding Modes: Aspirin and Tylenol, https://www.ncbi.nlm.nih.gov/Class/PubChem/essentials/aspirin.html (accessed January 2018).

[19] NCBI Finding BioAssay Data: Gaucher Disease, www.ncbi.nlm.nih.gov/Class/PubChem/essentials/gaucher.html (accessed January 2018).

[20] NCBI Finding Toxicological Data: Sarin, www.ncbi.nlm.nih.gov/Class/PubChem/essentials/sarin_tox.html (accessed January 2018).

[21] NCBI Finding Active Compounds: Agonists of Human Serotonin Receptors, www.ncbi.nlm.nih.gov/Class/PubChem/essentials/serotonin.html (accessed January 2018).

[22] Zhou, Z., Wang, Y., and Bryant, S.H. (2010) *J. Mol. Graphics Modell.*, **8**, 714–727.

[23] Cheng, T., Li, Q., Wang, Y., and Bryant, S.H. (2011) *J. Chem. Inf. Model.*, **51**, 2440–2448.

[24] Fu, G., Ding, Y., Seal, A., Chen, B., Sun, Y., and Bolton, E. (2016) *BMC Bioinf.*, **17**, 160.

[25] http://master.bioconductor.org/packages/devel/bioc/html/bioassayR.html (accessed January 2018).
[26] Kim, S., Thiessen, P.A., Bolton, E., and Bryant, S.H. (2015) *Nucleic Acids Res.*, **43**, 605–611.
[27] NCBI Entrez Programming Utilities Help [Internet], www.ncbi.nlm.nih.gov/books/NBK25499/ (accessed January 2018).
[28] NCBI PubChem PUG Help, https://pubchemdocs.ncbi.nlm.nih.gov/power-user-gateway (accessed January 2018).
[29] PUG REST https://pubchemdocs.ncbi.nlm.nih.gov/pug-rest (accessed January 2018).
[30] Kaiser, J. (2005) *Science*, **308**, 774.

6.6 Pharmacophore Perception and Applications

Thomas Seidel[1], Gerhard Wolber[2], and Manuela S. Murgueitio[2]

[1] University of Vienna, Faculty of Life Sciences, Department of Pharmaceutical Chemistry, Althanstraße 14, 1090 Vienna, Austria
[2] Freie Universität Berlin, Institute of Pharmacy, Computer-Aided Drug Design, Pharmaceutical and Medicinal Chemistry, Königin-Luisestr. 2+4, 14195 Berlin, Germany

Learning Objectives

- To explain what pharmacophores are, how they are represented, and what they can (and cannot) tell us
- To describe the main strategies for the pharmacophore modeling of drug–target interactions depending on the nature of the input information
- To review the practical applications of the pharmacophore concept in the drug design process and differences to other techniques
- To assess the currently available software packages for computer-aided pharmacophore modeling and searching

Outline

6.6.1 Introduction, 259
6.6.2 Historical Development of the Modern Pharmacophore Concept, 260
6.6.3 Representation of Pharmacophores, 262
6.6.4 Pharmacophore Modeling, 268
6.6.5 Application of Pharmacophores in Drug Design, 272
6.6.6 Software for Computer-Aided Pharmacophore Modeling and Screening, 278
6.6.7 Summary, 278

6.6.1 Introduction

In the field of medicinal chemistry, the concept of *pharmacophores* has become increasingly popular in recent years and matured into a valuable and efficient basis for computer-aided drug design techniques. Due to their intuitive nature, pharmacophores are easy to understand and illustrative. This also renders them rather useful as a tool to describe and explain ligand–target binding mechanisms. However, some confusion around the term pharmacophore still exists, depending on the background and context; it is often attributed with different meanings. Historically, medicinal chemists used (and still use) the term pharmacophore to denote common structural or functional elements that are essential for the activity of a set of compounds toward a particular biological target. However, this

Applied Chemoinformatics: Achievements and Future Opportunities, First Edition.
Edited by Thomas Engel and Johann Gasteiger.
© 2018 Wiley-VCH Verlag GmbH & Co. KGaA. Published 2018 by Wiley-VCH Verlag GmbH & Co. KGaA.

view does not quite match the modern official IUPAC definition [1] of pharmacophores from 1998 that says:

> "A pharmacophore is the ensemble of steric and electronic features that is necessary to ensure the optimal supra-molecular interactions with a specific biological target structure and to trigger (or to block) its biological response."

According to this definition, pharmacophores do not represent specific structural motifs of molecules (β-lactames, dihydropyridine) or associations of functional groups (e.g., primary amines, sulfonamides), but are an abstract description of essential steric and electronic properties that are required for an energetically favorable interaction of a ligand with the receptor of the macromolecular target. Pharmacophores can thus be considered as the largest common denominator of molecules that show a similar biological profile and are recognized by the same binding site of the target [2].

6.6.2 Historical Development of the Modern Pharmacophore Concept

The concept of pharmacophores is not new in the medicinal chemistry and has already been successfully applied before computers were used in chemistry. Simple pharmacophoric patterns were described in the literature and used as a tool for the development of new drug molecules [2]. Considerations on structure–activity relationship were already developed in the 1940s based on the knowledge of bond lengths and van der Waals radii allowing for a construction of simple two-dimensional models. Notable with this respect is the recognition of the ability of p-aminobenzoic acid (PABA, a biological precursor of dihydrofolic acid) to reverse the bacteriostatic effect of p-aminobenzenesulfonamides, which led to the formulation of the fundamentals of the theory of metabolite antagonism by Woods and Fildes [3, 4]. As shown in Figure 6.6.1, PABA and the sulfonamides are isosteres and either the metabolite or its antagonist can

Figure 6.6.1 p-Aminobenzoic acid (PABA) and p-aminobenzenesulfonamide are isosteres and show similarities regarding interatomic distances that are critical for binding to the dihydrofolate reductase enzyme surface [2]. Binding of the sulfonamide instead of PABA thus inhibits the biosynthesis of tetrahydrofolic acid.

6.6.2 Historical Development of the Modern Pharmacophore Concept

attach to the critical area on the dihydrofolate reductase enzyme surface. If the latter happens, the metabolic process is interrupted and, in the case of bacteria, multiplication is inhibited.

Another early example was the development and pharmacological evaluation of *trans*-diethylstilbestrol. Diethylstilbestrol acts as an estrogenic agent due to its similarity with estradiol [5] (see Figure 6.6.2). Again the proposed model was two-dimensional even though the non-planarity of the estradiol conformations was already known at that time.

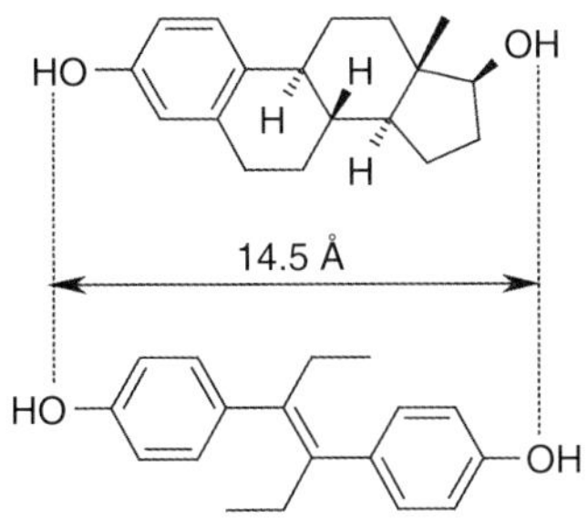

Figure 6.6.2 Analogy between estradiol and *trans*-diethylstilbestrol [2].

As molecular structure concepts became more widespread in the early twentieth century, including conformational and chiral effects on structure and reactivity, there was increasing success in explaining the interdependencies between activity and ligand structure. Similarly, it became obvious that the presence and connectivity of pharmacophoric groups alone is insufficient for the physiological activity of a molecule and that they also need to be presented to the receptor in an appropriate geometry for recognition. A step into this direction was the introduction of the three-point contact model by Easson and Stedman [6]. If an asymmetric center is present in a molecule, it is assumed that the substituents of the chiral center make a three-point contact with the target receptor. The three inequivalent contact sites on the receptor can form a complementary match with only one enantiomer of the drug molecule but not with the triangle of substituents in its optical antipode. Considering adrenaline as an example, the more active natural (R)-(−)-adrenaline forms a contact to the receptor via the three interactions shown in Figure 6.6.3a. The less active stereoisomer (S)-(+)-adrenaline is only capable of establishing a two-point contact (Figure 6.6.3b). This leads to the loss of the energetically favorable hydrogen-bonding interaction and thus results in an approximately 100-fold lower activity of (S)-(+)-adrenaline compared with the natural form (R)-(−)-adrenaline.

Confidence in drug–receptor theory grew even much stronger when the first crystal structures of ribonuclease and other enzymes became available [7]. The crystal structure of the dihydrofolate reductase enzyme, having the anticancer drug methotrexate bound to it, was particularly revealing [8]. With growing numbers of elucidated protein structures, many more structural models could thus be generated based on their homology to known crystal structures.

Another conceptual breakthrough was the answer to the question of whether chemical forces alone are sufficient to explain drug–receptor interactions and

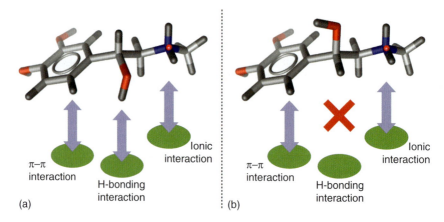

Figure 6.6.3 Interaction capabilities of natural (R)-(−)-adrenaline (a) and its stereoisomer (S)-(+)-adrenaline (b) [2].

other pharmacological aspects, or whether additional, yet unknown driving forces have to be discovered. A thorough investigation of intermolecular forces by Wolfenden [9] provided a rationale for explaining rates of drug–receptor interactions by chemical forces alone. This later evolved into present-day calculations of ligand–receptor binding energies, dynamic motion of ligand and receptor, and linear free energy perturbation estimation.

The examples and milestones outlined above, together with influential work done by Gund [10], Humblet and Marshall [11], and many others, paved the way for the modern pharmacophore concept and all its derived applications. The pharmacophore concept summarizes the insight into the effects of chemical structure on bioactivity and thus allows medicinal chemists to postulate pharmacophore models as the "essence" of the structure–activity knowledge they have gained in an extensive structural study of a series of active and inactive molecules for a given drug target.

6.6.3 Representation of Pharmacophores

To be a useful tool for drug design, pharmacophores have to represent the nature and location of functional groups involved in ligand–target interactions, as well as the different types of non-covalent bonding and their characteristics in a way that is uniform and easy to comprehend for humans. Furthermore, pharmacophores also have to show predictive power that, at its best, enables the design of novel chemical structures that are not evidently derived by the translation of structural features from one active series to the other, or even allows effective scaffold hopping [12].

Usually, pharmacophores are represented by a spatial arrangement of so-called chemical (or pharmacophoric) features that specify essential structural elements and/or observed ligand–receptor interactions in the form of geometric entities. Although the representation of pharmacophores in such a way is quite simple, it sufficiently fulfills the above requirements and has found general acceptance among medicinal chemist. However, special attention must be paid regarding the right level of abstraction of the types of chemical features used in the construction of pharmacophores. In early pharmacophore modeling techniques, such as the active analog approach described by Marshall *et al.* [13], features constituting a pharmacophore could represent any fragment or atom type. Recent techniques [14] use a more general way for building pharmacophore models, for example, a single geometric entity for all negative ionizable groups. Such general definitions result in models that are universal, at the cost of selectivity. Selectivity, however, is also an important issue for the quality of pharmacophore models, and features that are too general may need to be constrained to additionally include characteristics of the underlying functional groups instead of just reflecting universal chemical functionality. Being too restrictive, on the other hand, will increase the number of different feature types at the cost of comparability and the ability to identify novel, structurally unrelated chemical compounds. Current software packages for pharmacophore modeling like *Catalyst* [15], *LigandScout* [16], *MOE* [17], and *Phase* [18], always have to face a trade-off in the design of a generally applicable feature set that is universal and, at the same time, still selective enough to reflect all relevant types of observed ligand–receptor interactions.

In order to describe the different levels of universality and specificity of chemical features, a simple layer model according to Table 6.6.1 can be used [19–21].

Table 6.6.1 Classification of the abstraction levels of chemical features.

Layer	Classification	Universality	Specificity
1	Molecular graph descriptor (atom, bond) with geometric constraint	−−	+++
2	Molecular graph descriptor (atom, bond) without geometric constraint	−	++
3	Chemical functionality (hydrogen bond donor, acceptor) with geometric constraint	++	+
4	Chemical functionality (lipophilic area, positive ionizable group) without geometric constraint	+++	−

In this model, a lower layer number corresponds to higher specificity and, therefore, lower universality. Some concrete examples of chemical features together with the corresponding abstraction level are given below:

- *Layer 1*: A phenyl group facing another aromatic system within a distance of 2–4 Å
- *Layer 2*: A methyl group, a primary, secondary, or tertiary amine or a hydroxyl moiety
- *Layer 3*: A hydrogen bond acceptor vector including the acceptor location as well as the projected donor point; an aromatic ring system with location and orientation (ring plane)

- *Layer 4*: Hydrogen bond acceptor without a projected donor point; a lipophilic group

The most frequent reason for defining feature types on the low universality levels 1 and 2 is that a definition on a higher level is not sufficient to describe essential features occurring in the training set (see [22] for an example). Even if customization results in a layer 1 or layer 2 feature, there should be a possibility of including layer 3 or 4 information in order to categorize and thus increase comparability (e.g., a carboxylic acid as a layer 2 feature is a subcategory of "negative ionizable," which is a layer 4 feature).

The following sections give a brief overview of the most important types of ligand–receptor interactions and their corresponding geometric representation in pharmacophore models.

6.6.3.1 Hydrogen Bonding Interactions

Hydrogen bonding is an attractive interaction of electropositive hydrogen atoms with an electronegative atom (H-bond acceptor) like oxygen, fluorine, or nitrogen. The participating hydrogen must be covalently bound to another electronegative atom (H-bond donor) to create the hydrogen bond. Hydrogen bonds are the most important specific interactions in the formation of specific ligand–receptor complexes [22]. To capture the characteristics of hydrogen bonding interactions (see Figure 6.6.4), they are usually modeled as a position with a certain tolerance for the acceptor (donor) atom and a projected point (also with a certain tolerance) for the position of the corresponding donor (acceptor) atom. Together these two positions form a vector that constrains the direction of the H-bonding axis and also the location of the interacting atom in the target receptor. Donor and acceptor features with a direction constraint thus belong to layer 3 of the layer model. When the direction constraint is omitted, they become less specific layer 4 features that will match any acceptor/donor atom irrespective of whether essential geometric preconditions for the formation of a hydrogen bond are actually fulfilled.

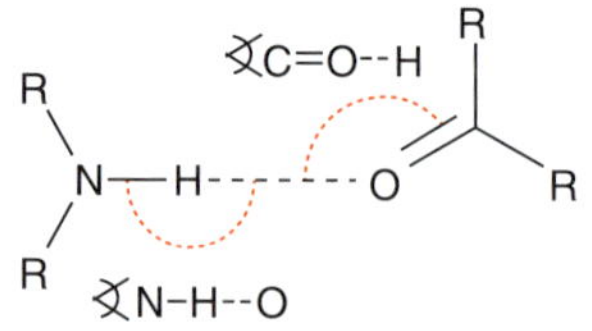

Figure 6.6.4 Hydrogen bonding geometry: the involved N, H, and O atoms are nearly linearly aligned. The N–O distance is typically between 2.8 and 3.2 Å. The N–H–O angle is >150° and the C=O–H angle between 100° and 180°.

6.6.3.2 Hydrophobic Contacts

Hydrophobic (lipophilic) contacts occur when nonpolar amino acid side chains in the protein come into close contact with lipophilic groups of the ligand. Such lipophilic groups are, for example, aromatic and aliphatic hydrocarbons, as well as halogen substituents (e.g., chlorine, fluorine) and many heterocycles (e.g., thiophene, furane). Since lipophilic areas on the protein and ligand surface are not capable of participating in any polar interactions, attractive forces are negligible for the effect of hydrophobic interactions. They are rather driven by the displacement of water molecules from nonpolar areas in the binding pocket to the outside of the protein (see Figure 6.6.5). This leads to a higher entropy of the system due to a gain in mobility and allows the now unconstrained water molecules to form energetically favorable hydrogen bonds. According to the formula $\Delta G = \Delta H - T\Delta S$, both contributions will lower the free energy change ΔG for the interaction and thus increase the ligand's overall binding affinity.

Since hydrophobic interactions are undirected, they can be represented as unconstrained layer 4 features in the form of tolerance spheres that are located in the center of hydrophobic atom chains, branches, or groups of the ligand.

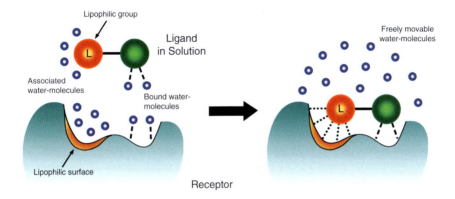

Figure 6.6.5 On the formation of lipophilic contacts (hydrophobic interactions), water molecules covering lipophilic areas of the binding pocket are forced to move to the outside of the ligand–receptor complex. This increases the entropy of the system due to a gain in mobility of the water molecules. The resulting contribution to the binding affinity is typically between −100 and −200 J/mol per Å² of the lipophilic contact surface. (Adapted from Böhm et al. 1996 [23].)

6.6.3.3 Aromatic and Cation–π Interactions

Electron-rich π-systems like aromatic rings are capable of forming strong attractive interactions with other π-systems (π stacking) and with adjacent cationic groups (e.g., metal ions, ammonium cations in protein side chains) [24]. The interaction energies can be of the same order of magnitude as hydrogen bonds and thus play an important role in various aspects of molecular biology (e.g., stabilization of DNA and protein structures, enzymatic catalysis, and molecular recognition). Since cation–π and π–π interactions require a certain relative geometric configuration of the interacting counterparts (see Figure 6.6.6), they belong to the class of directed interactions. In pharmacophore models, aromatic features are therefore at least represented by tolerance spheres located in the center of the aromatic ring system (→ layer 4 feature). To account for the directional aspects

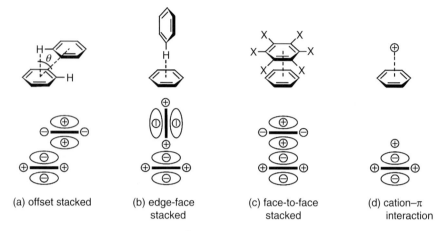

Figure 6.6.6 Steric configurations of π–π and cation–π interactions [25].

of aromatic interactions, they are often attributed with additional information about the spatial orientation of the aromatic ring system in the form of a ring plane normal or two points that define this vector ($\rightarrow$ layer 3 feature).

6.6.3.4 Ionic Interactions

Ionic interactions are rather strong (>400 kJ/mol) attractive interactions that occur between oppositely charged groups of the ligand and the protein environment. Positive or negative ionizable areas can be single atoms as, for example, metal cations. Groups of atoms that are likely to be protonated or deprotonated at physiological pH (e.g., carboxylic acids, guanidines, aromatic heterocycles) can also be represented by positively or negatively ionizable areas. Ionic interactions are of electrostatic nature and thus undirected that allows the corresponding pharmacophoric features to be represented by simple tolerance spheres (layer 4 features).

6.6.3.5 Metal Complexation

Some proteins contain metal ions as cofactors. A prominent example are metalloproteases [26] that contain Zn^{2+} ions that are coordinated to the protein via three amino acids as shown in Figure 6.6.7. In such proteins, a complexation of the metal ion with suitable electron-donating atoms or functional groups of the ligand is often the most important contribution to the overall binding affinity and essential for the ligand's mode of action. Functional groups and structural elements that exert a strong affinity for metal ions are, for example, thiols R–SH, hydroxamates R–CONHOH, or sulfur and nitrogen containing heterocycles. In pharmacophore models, metal binding interactions are usually represented by tolerance spheres located on single atoms or in the center of groups that are capable of interacting with the metal ions. To additionally constrain the location of the coordinated metal ion or to accommodate for a particular coordination geometry, a vector representation similarly to hydrogen bonding interactions is also often used.

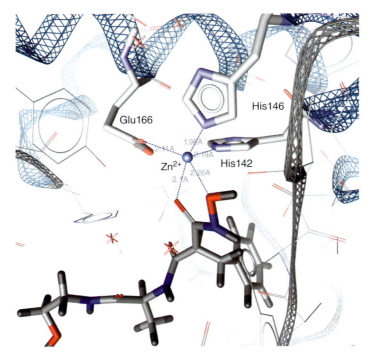

Figure 6.6.7 Thermolysin in complex with the hydroxamic acid inhibitor
N-[(2S)-2-benzyl-3-(hydroxyamino)-3-oxopropanoyl]-L-alanyl-N-(4-nitrophenyl)glycinamide
(BAN, PDB-code: 5TLN). The Zn^{2+} ion is penta-coordinated with the characteristic amino acids
Glu166, Hist142, and His146 of thermolysin, and the hydroxyl- and carbonyl-oxygen of the
hydroxamic acid moiety [27].

6.6.3.6 Ligand Shape Constraints

Features in a pharmacophore model represent necessary, but not sufficient chemical characteristics active molecules must possess to allow for a specific, high affinity binding to a given target receptor. A molecule may thus be capable of presenting a set of features that is entirely consistent with the pharmacophore model but still fails to bind. A possible reason is that some part of the molecule would experience a steric clash with the receptor if it were to bind in the mode described by the pharmacophore model. A common way to emulate this situation is exclusion volumes. They are usually represented by spheres with varying size that indicate regions of "forbidden" space, a structure may not occupy when it is aligned to the pharmacophore. The most reliable source of information for a proper placement of exclusion volumes is the crystallographic structure of the receptor. Such receptor-based exclusion volumes are centered on appropriate atoms of the binding-site surface with sizes dictated by the van der Waals radii of the corresponding atoms (see Figure 6.6.8). A clash of the aligned molecule with one of the excluded volume spheres directly corresponds to a steric overlap with an atom of the receptor surface and thus indicates a presumably poor fit of the molecule. When the three-dimensional (3D) receptor structure is not available (which is often the case), then the placement of exclusion volumes is less

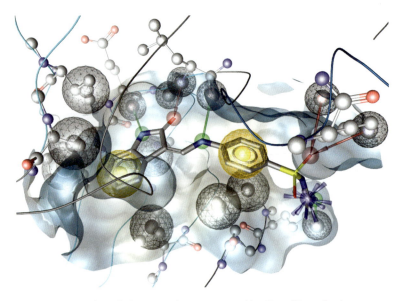

Figure 6.6.8 Receptor-based pharmacophore generated by *LigandScout* for the CDK2/inhibitor complex 1KE9. Gray spheres represent exclusion volumes that model the shape of the receptor surface. Yellow spheres represent hydrophobic, green arrows hydrogen bond donor, and red arrows hydrogen bond acceptor features. The blue spherical star represents a positive ionizable group in an ionic interaction.

straightforward. In this case, the location and size of the exclusion volumes must be assigned manually or one has to use computer-aided methods that distribute volume spheres based on the union of the molecular shapes of a set of aligned known actives.

6.6.4 Pharmacophore Modeling

Pharmacophore models can be created by a variety of methods such as manual construction, automated perception from the structure of one or more ligands, or receptor-based deduction from a crystallographic structure. The particular method or workflow that is best suited for modeling a given problem depends on a number of factors like the nature and quality of available data, the computational resources, and the aim and further use of the created pharmacophore model. The following sections will give an overview of the methods and show the applicability of the most commonly used approaches for pharmacophore modeling.

6.6.4.1 Manual Pharmacophore Construction

The simplest way (in terms of algorithmic complexity) to create a pharmacophore model is its manual construction based on information about the structure and/or special characteristics of a series of known active ligands. A manually constructed pharmacophore can be quite advantageous, particularly if it is derived from the X-ray structure of a ligand in its binding conformation or from a ligand with a low conformational flexibility. In either case, the locations of the pharmacophoric features are essentially pinned down, so one of the biggest uncertainties – conformational flexibility – is eliminated. However, there is still the question of the particular features to incorporate into the model, which is not always easy to infer without additional information such as the structure of a ligand–receptor complex. With the advent of powerful computer-aided methods for pharmacophore modeling, the importance of a manual pharmacophore construction from scratch has largely diminished. Manual changes to a pharmacophore model are nowadays often only limited to the refinement or fine-tuning of pharmacophores that were generated in an automated manner by a computer program.

6.6.4.2 Receptor-Based Pharmacophores

The availability of information about the 3D structure (e.g., from NMR experiments or crystallographic structure elucidation) of a ligand/receptor complex is a tremendous advantage when it comes to the development of high-quality pharmacophore models. The structure of the bound ligand directly provides its bioactive conformation that is indispensable for the correct placement of pharmacophoric features. Knowledge of the binding-site structure enables the pharmacophore to incorporate more detailed information about regions that are not accessible to the ligand. A fundamental step in the development of a receptor-based (often also called structure-based) pharmacophore model is the analysis of the binding site and its associated ligand to identify potential interaction points. A number of methods can be used to identify such regions [28]. These methods include programs such as *GRID* [29] that probe the binding site with small molecules or functional groups and calculate the interaction energies between the probe molecule and the atoms of the protein at points on a grid lattice to generate a molecular interaction field (MIF). These fields can then be contoured by energy-level iso-surfaces to find the most favorable regions for interactions with the receptor by an H-bond acceptor or donor (or any other type of feature). *LUDI* [30] and *SuperStar* [31] use a knowledge-based approach in which rules are used to generate a set of interaction sites for each atom or functional group of the binding site that is capable of participating in a non-bonded contact. The rules are largely based on statistical analysis of experimental structures and take into account the chemical nature of the atoms as well as preferences on the orientation of features such as hydrogen-bond donors and acceptors. With the positions of the interaction sites in hand, all possible 3D pharmacophores with three, four, or more features can then be generated. An important step in such a procedure is the subsequent validation of the generated pharmacophores (e.g., by enrichment-based methods) to filter out all models with no significance.

Commercially available programs that are able to perform the entire modeling process from structure to pharmacophore include *Structure-Based Focusing* [32–34] and *LigandScout* [16, 19, 20]. In *Structure-Based Focusing* a sphere with user-adjustable location and size is used to mark key residues in the binding site, and a *LUDI* interaction map is generated to describe favorable interactions in which a ligand is expected to engage. The interaction map is transferred to an interaction model that consists of a set of complementary points in the binding pocket, representing possible locations of pharmacophore features on the ligand. A user-defined density controls the number of points created, but it is usually quite large. Therefore, hierarchical clustering is performed to select a smaller number of representative features. After the addition of exclusion volumes, a database of known actives is searched to determine which pharmacophore models are most frequently matched. *LigandScout* takes a more direct approach and derives a pharmacophore model from a single ligand–receptor complex. After perceiving hybridization states, unsaturated bonds, and aromatic rings, the ligand and binding pocket structure is analyzed for the presence of atoms and groups that can take part in hydrogen bonding, hydrophobic, aromatic, ionic, and metal binding interactions. Pharmacophoric feature detection can be customized with respect to interaction-specific geometric characteristics like allowed distances and angle ranges. Whether a feature is incorporated into the final pharmacophore model depends on its location relative to a complementary feature in the binding site. For example, a hydrogen bond acceptor feature of the ligand is only included if there is an opposing hydrogen donor feature on the receptor side within a certain distance and angle range. After all complementary feature pairs of the complex have been detected and the corresponding ligand-side features have been put into the derived pharmacophore model, exclusion volume spheres are finally added to resemble the shape of the binding pocket. Figure 6.6.8 shows a typical receptor-based pharmacophore that was generated by *LigandScout* for the CDK2 complex 1KE9.

6.6.4.3 Ligand-Based Pharmacophores

When information about the 3D structure of the receptor is limited or not available but a sufficient number of actives has already been identified, then ligand-based methods provide an alternative way to leverage the available information and develop pharmacophore models that can help to search for new actives. An important condition for ligand-based methods to work and to deliver good models is that the ligands used for model generation must bind to the same receptor at the same location in the same way. Otherwise, the resulting pharmacophore models will not represent the correct mode of action and are essentially useless.

For the derivation of common pharmacophores from a set of ligands, many different algorithms were devised [28, 35], but on the whole they all follow the general workflow depicted in Figure 6.6.9. After the import and preparation of the input structures, the first crucial step is usually the generation of sufficiently large and diverse sets of ligand conformers. This has to be done because the bioactive conformation of the input ligands is normally not known. However, it can be assumed that one conformation in each set of generated conformers is at least a good approximation of the bioactive conformation. The next step is at the heart of the overall procedure and aims at finding a chemical feature pattern [21] that is common to all training set ligands and can be superimposed with at least one

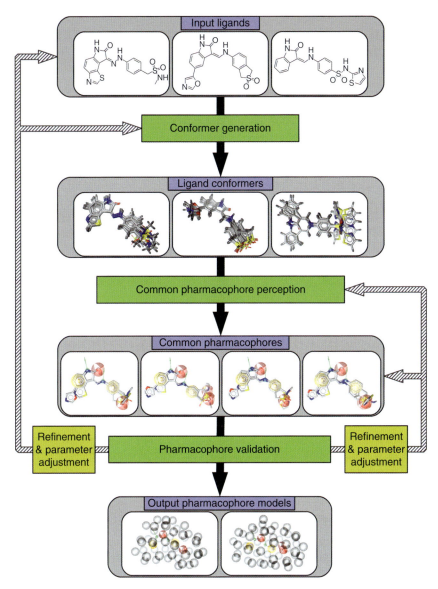

Figure 6.6.9 Ligand-based pharmacophore modeling workflow starting from a set of known actives.

conformation of each ligand. Since often more than one such pharmacophoric pattern can be found, the result is usually not only a single pharmacophore model but also a list of multiple possible solutions that get ranked according to some fitness function. From this list, the user typically has to select the best model(s) according to a carefully performed validation procedure [36]. Methods for pharmacophore validation can be divided into three categories [35]:

1. Statistical significance analysis and randomization tests.

2. Enrichment-based methods: Measuring the ability to recover active molecules from a test database in which a small number of known actives have been hidden among randomly selected compounds. Database mining and the utilization of receiver operating characteristic (ROC) curves [36] fall into this category.
3. Biological testing of matching molecules.

If pharmacophore validation results indicate a generally unsatisfying quality of the generated models, they may be refined manually (e.g., deletion/addition of features; applicable if only small changes are required), or the whole modeling procedure must be repeated with a different setup, for example, changes to the composition of the training and/or test set and tuning of ligand conformer and pharmacophore model generation parameters until acceptable results are obtained. The high number of variables that influence the procedure and the disregard of the receptor structure makes pure ligand-based modeling relatively error prone and leaves much room for interpretation. The algorithmic power of the employed software, a high expertise of the user, and a thorough validation of the obtained results are therefore critical for the successful application of this modeling approach.

6.6.5 Application of Pharmacophores in Drug Design

3D pharmacophores are applied in a variety of different areas of the drug design process. The probably most established use of 3D pharmacophores is virtual screening to identify new lead structures for drug development. In the next subsection, the application of pharmacophores for virtual screening in different use cases will be exemplified. Next, further common applications of 3D pharmacophores will be summarized. Finally, the currently emerging field of dynamic pharmacophores that combine the knowledge gained by molecular dynamics simulations with the concept of 3D pharmacophores will be introduced.

6.6.5.1 3D Pharmacophore-Based Virtual Screening

Due to their simplicity and abstract nature, 3D pharmacophores are an ideal tool for virtual screening of large compound libraries [37] and have successfully been employed to identify antiviral agents [38], G-protein-coupled receptor modulators [39] and nuclear receptor modulators [40], among others. The collections of small molecules that are commonly screened today often contain millions of compounds. The sparse pharmacophoric representation of the ligand–target interactions leads to a reduced computational complexity in the hit identification process and thus poses a big advantage for this method. In addition, 3D pharmacophore-based virtual screening allows for the discovery of bioisosteric small molecules that have the same biological effect but distinct chemical scaffolds and functional groups different from the original compounds [12]. This so-called scaffold hopping is of special interest for pharmaceutical companies as known active scaffolds may be covered by patents of competing companies or show undesired effects in terms of ADME–Tox properties.

6.6.5 Application of Pharmacophores in Drug Design

A 3D pharmacophore for virtual screening is usually created as a binding hypothesis based on the data available on the given target protein and its small molecule binders. Depending on the accessible information either a structure or a ligand based, or a combination of both approaches will be selected. If a crystal structure of the receptor with a bound ligand is available, a structure-based model will be generated. If only active molecules are known, the pharmacophore will be generated in a ligand-based approach. For cases in which both the 3D structure of the target protein and structures of small molecule binders are available, a combined approach might be the best choice.

In order to be useful for virtual screening, a pharmacophore model has to be capable of predicting bioactive molecules. The assessment of its predictive power is ideally performed retrospectively by screening databases of compounds previously shown to be actives and inactives in experiments. Like this it can be tested whether the model can distinguish between actives and inactives. If no knowledge on active ligands is available, the quality of the models has to be assessed in a more qualitative manner. A problem that is often faced when selecting a dataset for the statistical validation of a model is the fact that the amount of known inactives is often not sufficient, as negative results are only seldom published. Alternatively, a set of decoys can be generated and used. Different metrics are used to assess the predictive power of the model. The most commonly used metrics are introduced next. For an extensive review on this topic, we refer to the reviews by Truchon and Bayly [41] and Kirchmair *et al.* [42].

If a virtual screening protocol selects n molecules from a database with N entries, the hit list is composed of actives (true positives, TP) and inactives (false positives, FP). The active compounds that were not selected are called false negatives (FN), and the inactives that were correctly not retrieved are referred to as true negatives (TN) (Figure 6.6.10). The *selectivity* (Se, true positive rate, Eq. (6.6.1)) describes the ratio of retrieved actives (TP, *true positives*) toward the total amount of molecules screened that is the sum of *true positives* and *false*

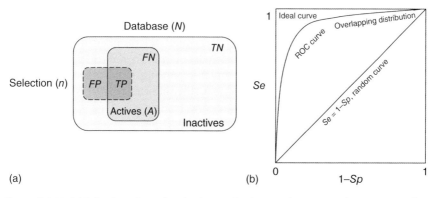

Figure 6.6.10 (a) Selection of n molecules from a database with N entries. (b) ROC curves for an ideal, an overlapping and a random distribution of actives and decoys.

negatives (FN):

$$Se = \frac{TP}{TP + FN} \tag{6.6.1}$$

The *specificity* (*Sp*, false positive rate, Eq. (6.6.2)) represents the amount of *true negative TN* compounds divided by the total number of negative compounds in the database that results from the sum of *TN* and *false positives* (*FP*):

$$Sp = \frac{TN}{TN + FP} \tag{6.6.2}$$

The *yield of actives* (*Ya*, Eq. 6.6.3) describes the ratio between the *TP* and the size of the hit list (*N*):

$$Ya = \frac{TP}{N} \tag{6.6.3}$$

The *enrichment factor* (*EF*, Eq. 6.6.4) is a descriptor for the improvement of the hit rate by virtual screening in comparison with random selection and is given by setting the *Ya* into relation with the coefficient between *actives* (*A*) and *inactives* (*N*):

$$EF = \frac{TP/n}{A/N} \tag{6.6.4}$$

A convenient and frequently used tool to assess the screening performance of a virtual screening protocol is ROC curves [42]. These curves display the *sensitivity* on the *x*-axis and 1−*specificity* on the *y*-axis. The course of the curve gives a good impression on the quality of the prediction. An ideal distribution vertically rises to the maximum and then horizontally continues to the right. A random distribution is represented by a diagonal line between the upper right and the lower left corners. An overlapping distribution of a model that selects more active than inactive molecules will run between these two curves (Figure 6.6.10).

Once the appropriate 3D pharmacophore for a given target has been generated and validated, it can be used for virtual screening. This is performed against databases of small molecules that can comprise commercially available compounds, natural products, or small molecules that potentially can be synthesized. Prior to virtual screening the compounds have to be transferred into multiconformer screening databases in order to take into account their conformational flexibility. The virtual screening will result in a list of virtual screening hits that match the 3D pharmacophore. To reduce the number of virtual screening hits and select the most promising molecules for biological validation, the hit list is often subjected to further filtering steps through different methods like molecular docking, shape- and feature-based screening, filtering for bioavailability, drug-likeness, and ease of synthesis.

One successful application of 3D pharmacophores for virtual screening involves the prediction of novel small organic molecules that prevent the dimerization of toll-like receptors subtype 2 (TLR2) [43]. TLRs recognize pathogens and play an important role by triggering the innate immune response. However, their uncontrolled signaling triggering massive inflammation has been shown to be involved in the development of life-threatening diseases such as sepsis. Thus, new TLR2 antagonists can serve as potential therapeutics to control this uncontrolled inflammation response. At the beginning of the study, neither small molecule TLR2 antagonists nor a small molecule binding site was known. Thus, the crystal structure of the receptor was the only possible starting point for the generation of a 3D pharmacophore. After the identification of a potential binding site for small molecules, a structure-based pharmacophore was derived from previously calculated MIFs using *LigandScout* [16, 20]. As no small molecule modulators of TLR2 were known at this point of the study, the models could not be retrospectively statistically validated. Instead the model was optimized by screening different small molecule libraries with the model and qualitatively assessing how well the screening hits fitted the TLR2 binding site after minimization. The final models were then used to perform virtual screening based on the 3D pharmacophore against a database of around 3 million commercially available compounds. In order to prove whether the identified compounds actually fulfilled the predicted interaction pattern, the list of virtual screening hits was subjected to molecular docking. The poses were inspected and five compounds selected for testing. One of these substances could be confirmed as a TLR2 antagonist (Figure 6.6.11).

Another screening campaign focuses on the identification of small molecule inhibitors of α-amylase [44]. This enzyme is responsible for the catalysis of starch hydrolysis and thus represents an interesting target for the treatment of obesity. In this study a structure-based 3D pharmacophore was derived from the enzyme bound to its substrate. For this purpose, first a central area of the catalytic site of the enzyme was identified to be crucial for ligand binding. As a next step the potential of the model to discriminate between actives and inactives was proved by screening a set of 19 actives and 55 inactives retrieved

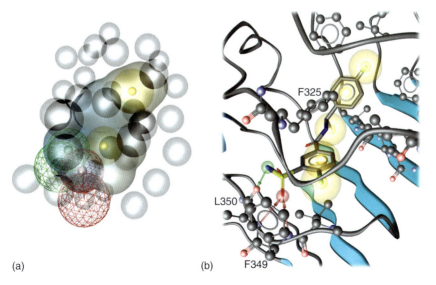

Figure 6.6.11 (a) 3D pharmacophore derived from the TLR2 binding site. (b) Binding mode of the TLR2 antagonist discovered by pharmacophore-based virtual screening.

from literature and calculating the achieved enrichment factor, specificity, and selectivity of the model. Once the pharmacophore had been validated, it was used to screen a library of approximately 1.8 million commercially available compounds. The pharmacophore-based virtual screening was combined with docking and filtering by physicochemical properties and finally resulted in the discovery of several α-amylase inhibitors.

6.6.5.2 Other Applications: *De Novo* Design and Evaluation of Binding Conformations

Apart from virtual screening, 3D pharmacophores can be applied for a variety of different purposes. They can be used to support the process of fragment-based *de novo* design [45]. Here, the models can be used to describe and constrain specific areas of a binding site and add information to the fragment growing process and to perform virtual screening for suitable molecular fragments. Additionally, pharmacophores can be used to support the pose selection after molecular docking. For example, this can be performed by rescoring docking poses by their accordance with a 3D pharmacophore in LigandScout [16, 19, 20] that represents important chemical features necessary for binding. Some docking applications even directly employ scoring functions based on this technique. A scoring function based on pharmacophore similarity in which the pharmacophore of a reference ligand is used to guide the placement of other ligands [46] was implemented for the docking program DOCK [47]. A similar workflow in which pose sampling and ranking are performed based on a structure-based pharmacophore is implemented in the program PharmDock [48]. Last but not least, biological profiling of small molecules to predict their activity is also be performed using collections of 3D pharmacophore models [49]. The advantage of this approach is the fact that the results stay easily interpretable.

6.6.5.3 Recent Developments: Dynamic Pharmacophores

The crystal structures of protein–ligand complexes from which structure-based 3D pharmacophores are derived only represent a static snapshot of a flexible protein. The fact that the models do not contain information on the flexibility of the system can become a disadvantage. Molecular dynamics simulation makes it possible to gain information on the conformational changes of a protein over time [50]. Thus, several attempts have recently been undertaken to take advantage of the information on protein flexibility gained by molecular dynamic simulations by integrating it into the generation and application of 3D pharmacophores [51–53]. In order to allow for a direct and in-depth use of the conformational and dynamic information retrieved by molecular dynamics in the generation of 3D pharmacophores, the novel concept of *dynophores* (dynamic pharmacophores) has been developed and applied [51, 53]. Dynophores are derived from molecular dynamics trajectories and include information on the full protein–ligand complex flexibility. This is achieved by extracting and collecting the information on pharmacophoric features from every single step of the simulation and joining it into a dynophore. The features from the different frames are grouped by their feature type, and the involved atoms are combined into so-called "superfeatures" that contain information on the type, the frequency, the geometry, and the sequence of interactions formed by a ligand. The "superfeatures" are graphically represented by a 3D feature density that represents the occurrence frequency of the interaction patterns with the target site. Dynamic pharmacophores have been applied to the prediction of phase II metabolism through sulfotransferase 1E1 [51]. Sulfotransferases are highly flexible enzymes, and integrating the information on protein flexibility into the generated predictive models allowed for an accurate prediction of the sulfonation of a set of virtual screening hits from DrugBank, a collection of 6494 experimental and approved drugs [54]. The dynophores were also applied to the analysis of the feature space, in order to understand how a specific feature of the sulfotransferase moves during the molecular dynamics simulation and how often it is present. This is then graphically represented in the form of feature clouds as depicted in Figure 6.6.12 [53]. The depicted example shows kaempferol bound to SULT1E1, its static pharmacophore (Figure 6.6.12a) and the dynophore resulting from a molecular dynamics simulation of the complex (Figure 6.6.12b). The comparison of both interaction patterns shows how the dynophore gives additional information on the presence of the features in the course of time. In the static pharmacophore two π stacking interactions are present. The dynophore contains only one of these features, showing that this specific interaction might be more pronounced in the dynamic system. The integration of such information can be very valuable when studying flexible proteins.

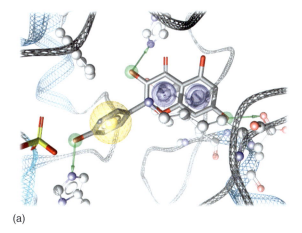

(a)

Figure 6.6.12 3D pharmacophore and dynophore of kaempferol bound to SULT1E1. (a) Static view of kaempferol bound to SULT1E1 with depicted 3D pharmacophore. (b) Kaempferol is represented with the resulting dynophore showed as spatial point clouds.

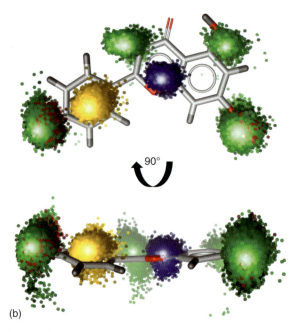

(b)

Figure 6.6.12 (Continued)

6.6.6 Software for Computer-Aided Pharmacophore Modeling and Screening

Today there are several software suites that can be used for molecular modeling and 3D pharmacophore-based virtual screening. The most popular ones are the software packages *Catalyst (now integrated in Discovery Studio)* [15], *LigandScout* [16], *MOE* [17], and *Phase* [18]. All of these software suites include implementations of the previously described feature types. However, their exact placement and definition differs [21]. This leads to a high variability of screening results when the pharmacophores are employed for virtual screening [55, 56]. The way in which the virtual screening itself is implemented also differs from software to software. While *LigandScout* and *Catalyst* require a pre-generation of multiconformer libraries of the compounds to be screened, *MOE* and *Phase* allow for on-the-fly calculation. However, this approach results in longer screening times [57]. The main difference between the different programs is the way the matching of 3D pharmacophores is implemented. *MOE* [58], *Phase* [59], and *Catalyst* [60] use cascading n-point pharmacophore fingerprints for pharmacophore molecule superpositioning. In *LigandScout* a pattern matching algorithm that allows for a higher geometric accuracy is implemented [19, 20].

6.6.7 Summary

3D pharmacophores represent a simple and efficient way to describe the essential steric and electronic properties needed by a small molecule to favorably interact with a given macromolecule. They describe the essential structural elements and ligand–receptor interactions through the abstract definition of the spatial arrangement of chemical or pharmacophoric features like H-bond donors, H-bond acceptors, positive ionizable areas, hydrophobic areas, and so on. A 3D pharmacophore can either be derived in a ligand- or structure-based manner, dependent on the information available. Due to their simplicity, 3D pharmacophores have proven to be useful for different steps of the drug design process. They have proven to be especially useful for identifying novel lead structures by virtual screening. Other application areas are *de novo* design, scaffold hopping, parallel screening, and so on. Nowadays, the concept of pharmacophores is evolving into integrating a new level of information. Through integrating the knowledge gained by molecular dynamics simulations into the concept of 3D pharmacophores so-called dynamic pharmacophores or dynophores are generated.

Essentials

- A 3D pharmacophore is an abstract description of essential steric and electronic properties necessary for an energetically favorable interaction of a ligand with a macromolecule.
- Pharmacophores are represented by the spatial arrangement of chemical or pharmacophoric features that describe essential structural elements and/or ligand–receptor interactions as H-bond acceptors, H-bond donors, hydrophobic areas, positive and negative ionizable areas, and so on.
- Pharmacophores are used in multiple areas of the drug design process. Their most established application is for virtual screening. Alternatively, they can also be used for *de novo* design, parallel screening, docking, and so on.

Available Software and Web Services (accessed January 2018)

- pharmacophore.org
- DS Catalyst, Biovia, Dassault systèmes, www.accelrys.com
- LigandScout, Inte:Ligand GmbH, www.inteligand.com
- Molecular Operating Environment (MOE), Chemical Computing Group Inc., www.chemcomp.com
- Phase, Schrödinger Inc., www.schrodinger.com

Selected Reading

- Böhm, H.-J., Klebe, G., and Kubinyi, H. (1996) *Wirkstoffdesign*, Spektrum Akademischer Verlag, Heidelberg, 599 pp.
- Güner, O.F. (ed.) (2000) *Pharmacophore Perception, Development, and Use in Drug Design*, International University Line, La Jolla, CA, 560 pp.
- Leach, A.R., Gillet, V.J., Lewis, R.A., and Taylor, R. (2010) Three-dimensional pharmacophore methods in drug discovery. *J. Med. Chem.*, **53**, 539–558.

- Klebe, G. (2013) *Drug Design*, Springer Verlag, Berlin, 901 pp.
- H. Kubinyi, *Success stories of computer-aided design* in *Computer Applications in Pharmaceutical Research and Development*, S. Ekins, Wiley Series in Drug Discovery and Development, Wiley-Interscience, New York, 2006, pp. 377–424
- Langer, T. and Hoffmann, R.D. (eds) (2006) *Pharmacophores and Pharmacophore Searches*, Wiley-VCH, Weinheim, Germany, 395 pp.
- Merz, K.M., Ringe, D., and Reynolds, C.H. (eds) (2010) *Drug Design: Structure- and Ligand-Based Approaches*, Cambridge University Press, New York, 286 pp.

References

[1] Wermuth, C.G., Ganellin, C.R., Lindberg, P., and Mitscher, L.A. (1998) *Pure Appl. Chem.*, **70**, 1129–1143.

[2] Wermuth, C.G. (2006) Pharmacophores: historical perspective and viewpoint from a medicinal chemist, in *Pharmacophores and Pharmacophore Searches*, vol. 32 (eds T. Langer and R.D. Hoffmann), Wiley-VCH Verlag GmbH & Co. KGaA, Weinheim, 1–13.

[3] Woods, D.D. (1940) *Br. J. Exp. Pathol.*, **21**, 74–90.

[4] Woods, D.D. and Fildes, P. (1940) *Chem. Ind.*, **59**, 133–134.

[5] Dodds, E.C. and Lawson, W. (1938) *Proc. R. Soc. London, Ser. B*, **125**, 122–132.

[6] Easson, L.H. and Stedman, E. (1933) *Biochem. J*, **27**, 1257–1266.

[7] Gund, P. (2000) Evolution of the pharmacophore concept in pharmaceutical research, in *Pharmacophore Perception, Development, and Use in Drug Design* (ed. O.F. Güner), International University Line, La Jolla, CA, 3–12.

[8] Mathews, D.A., Alden, R.A., Bolin, J.T., Filman, D.J., Freer, S.T., Hamlin, R., Hol, W.G., Kislink, R.L., Pastore, E.J., Plante, L.T., Xuong, N., and Kraut, J. (1978) *J. Biol. Chem.*, **253**, 6946–6954.

[9] Wolfenden, R. (1976) *Ann. Rev. Biophys. Bioeng.*, **5**, 271–306.

[10] Gund, P. (1979) *Annu. Rep. Med. Chem.*, **14**, 299–308.

[11] Humblet, C. and Marshall, G.R. (1980) *Annu. Rep. Med. Chem.*, **15**, 267–276.

[12] Hessler, G. and Baringhaus, K.-H. (2010) *Drug Discovery Today Technol.*, **7**, 263–269.

[13] Marshall, G.R., Barry, C.D., Boshard, H.E., Dammkoehler, R.A., and Dunn, D.A. (1979) The conformational parameter in drug design: the active analog approach, in *Computer-Assisted Drug Design*, vol. 112 (eds E.C. Olson and R.E. Christoffersen), American Chemical Society, Washington, DC, 205–226.

[14] Greene, J., Kahn, S., Savoj, H., Sprague, P., and Teig, S. (1994) *J. Chem. Inf. Comput. Sci.*, **34**, 1297–1308.

[15] DS Catalyst, *Discovery Studio Modeling Environment 4.5*, Dassault Systèmes BIOVIA, San Diego, CA, www.accelrys.com (accessed January 2018)

[16] G. Wolber, F. Bendix, T. Seidel, G. Ibis, M. Biely, R. Kosara, *LigandScout 4.0*, Inte:Ligand GmbH, Vienna, Austria, www.inteligand.com (accessed January 2018).

[17] Chemical Computing Group, Inc Molecular Operating Environment (MOE) 2015.10, Montreal, Canada, www.chemcomp.com (accessed January 2018).
[18] Phase, *4.5*, Schrödinger, LLC, New York, NY, 2015, www.schrodinger.com (accessed January 2018).
[19] Wolber, G. and Kosara, R. (2006) Pharmacophores from macromolecular complexes with LigandScout, in *Pharmacophores and Pharmacophore Searches*, vol. 32 (eds T. Langer and R.D. Hoffmann), Wiley-VCH Verlag GmbH & Co. KGaA, Weinheim, 131–150.
[20] Wolber, G. and Langer, T. (2005) *J. Chem. Inf. Model.*, **45**, 160–169.
[21] Wolber, G., Seidel, T., Bendix, F., and Langer, T. (2008) *Drug Discovery Today*, **13**, 23–29.
[22] Krovat, E.M. and Langer, T. (2003) *J. Med. Chem.*, **46**, 716–726.
[23] Böhm, H.-J., Klebe, G., and Kubinyi, H. (1996) Protein-ligand-Wechselwirkungen, in *Wirkstoffdesign*, Spektrum Akademischer Verlag, Heidelberg - Berlin - Oxford.
[24] Ma, J.C. and Dougherty, D.A. (1997) *Chem. Rev.*, **97**, 1303–1324.
[25] Waters, M.L. (2002) *Curr. Opin. Chem. Biol.*, **6**, 736–741.
[26] Böhm, H.-J., Klebe, G., and Kubinyi, H. (1996) Metalloprotease-hemmer, in *Wirkstoffdesign*, Spektrum Akademischer Verlag, Heidelberg - Berlin - Oxford.
[27] Matthews, B.W. (1988) *Acc. Chem. Res.*, **21**, 333–340.
[28] Leach, A.R., Gillet, V.J., Lewis, R.A., and Taylor, R. (2010) *J. Med. Chem.*, **53**, 539–558.
[29] Goodford, P.J. (1985) *J. Med. Chem.*, **28**, 849–857.
[30] Böhm, H.J. (1992) *J. Comput.-Aided Mol. Des.*, **6**, 61–78.
[31] Verdonk, M.L., Cole, J.C., and Taylor, R. (1999) *J. Mol. Biol.*, **289**, 1093–1108.
[32] Kirchhoff, P.D., Brown, R., Kahn, S., Waldman, M., and Venkatachalam, C.M. (2001) *J. Comput. Chem.*, **22**, 993–1003.
[33] Venkatachalam, C.M., Kirchhoff, P., and Waldman, M. (2000) Receptor-based pharmacophore perception and modeling, in *Pharmacophore Perception, Development, and Use in Drug Design* (ed. O.F. Güner), International University Line, La Jolla, CA, 339–350.
[34] Dixon, S.L. (2010) Pharmacophore methods, in *Drug Design: Structure- and Ligand-Based Approaches* (eds K.M. Merz, D. Ringe, and C.H. Reynolds), Cambridge University Press, New York, 137–150.
[35] Poptodorov, K., Luu, T., and Hoffmann, R.D. (2006) Pharmacophore model generation software tools, in *Pharmacophores and Pharmacophore Searches*, vol. 32 (eds T. Langer and R.D. Hoffmann), Wiley-VCH Verlag GmbH & Co. KGaA, Weinheim, 17–47.
[36] Triballeau, N., Bertrand, H.-O., and Achner, F. (2006) Are you sure You have a good model? in *Pharmacophores and Pharmacophore Searches*, vol. 32, 325–364 (eds T. Langer and R.D. Hoffmann), Wiley-VCH Verlag GmbH & Co. KGaA, Weinheim.
[37] Triballeau, N., Acher, F., Brabet, I., Pin, J.-P., and Bertrand, H.O. (2005) *J. Med. Chem.*, **48**, 2534–2547.

[38] Murgueitio, M.S., Bermudez, M., Mortier, J., and Wolber, G. (2012) *Drug Discovery Today Technol.*, **9**, 219–225.
[39] Bermudez, M. and Wolber, G. (2015) *Bioorg. Med. Chem.*, **23** (14), 3907–3912.
[40] El-Houri, R.B., Mortier, J., Murgueitio, M.S., Wolber, G., and Christensen, L.P. (2015) *Planta Med.*, **81** (6), 488–494.
[41] Truchon, J.F. and Bayly, C.I. (2007) *J. Chem. Inf. Model.*, **47** (2), 488–508.
[42] Kirchmair, J., Markt, P., Distinto, S., Wolber, G., and Langer, T. (2008) *J. Comput. Aid. Mol. Des.*, **22**, 213–228.
[43] Murgueitio, M.S., Henneke, P., Glossmann, H., Santos-Sierra, S., and Wolber, G. (2014) *ChemMedChem*, **9**, 813–822.
[44] Al-Asri, J., Fazekas, E., Lehoczki, G., Perdih, A., Gorick, C., Melzig, M.F., Gyemant, G., Wolber, G., and Mortier, J. (2015) *Bioorg. Med. Chem.*, **23**, 6725–6732.
[45] Mortier, J., Rakers, C., Frederick, R., and Wolber, G. (2012) *Curr. Top. Med. Chem.*, **12**, 1935–1943.
[46] Jiang, L. and Rizzo, R.C. (2015) *J. Phys. Chem. B*, **119** (3), 1083–1102.
[47] Moustakas, D.T., Lang, P.T., Pegg, S., Pettersen, E., Kuntz, I.D., Brooijmans, N., and Rizzo, R.C. (2006) *J Comput. Aid. Mol. Des.*, **20**, 601–619.
[48] Hu, B. and Lill, M.A. (2014) *J. Cheminform.*, **6**, 14.
[49] Vuorinen, A. and Schuster, D. (2015) *Methods*, **71**, 113–134.
[50] Mortier, J., Rakers, C., Bermudez, M., Murgueitio, M.S., Riniker, S., and Wolber, G. (2015) *Drug Discovery Today*, **20**, 686–702.
[51] Rakers, C., Schumacher, F., Meinl, W., Glatt, H., Kleuser, B., and Wolber, G. (2016) *J. Biol. Chem.*, **291**, 58–71.
[52] Wieder, M., Perricone, U., Boresch, S., Seidel, T., and Langer, T. (2016) *Biochem. Biophys. Res. Commun.*, **470**, 685–689.
[53] Sydow, D. (2015) Dynophores: novel dynamic pharmacophores – implementation of pharmacophore generation based on molecular dynamics trajectories and their graphical representation. Master thesis. Humboldt Universität zu Berlin.
[54] Wishart, D.S., Knox, C., Guo, A.C., Shrivastava, S., Hassanali, M., Stothard, P., Chang, Z., and Woolsey, J. (2006) *Nucleic Acids Res.*, **34**, 668–672.
[55] Spitzer, G.M., Heiss, M., Mangold, M., Markt, P., Kirchmair, J., Wolber, G., and Liedl, K.R. (2010) *J. Chem. Inf. Model.*, **50**, 1241–1247.
[56] Kirchmair, J., Ristic, S., Eder, K., Markt, P., Wolber, G., Laggner, C., and Langer, T. (2007) *J. Chem. Inf. Model.*, **47**, 2182–2196.
[57] Seidel, T., Ibis, G., Bendix, F., and Wolber, G. (2010) *Drug Discovery Today Technol.*, **7**, e221–e228.
[58] Labute, P., Williams, C., Feher, M., Sourial, E., and Schmidt, J.M. (2001) *J. Med. Chem.*, **44**, 1483–1490.
[59] Dixon, S.L., Smondyrev, A.M., and Rao, S.N. (2006) *Chem. Biol. Drug Des.*, **67**, 370–372.
[60] Guner, O., Clement, O., and Kurogi, Y. (2004) *Curr. Med. Chem.*, **11**, 2991–3005.

6.7 Prediction, Analysis, and Comparison of Active Sites

Andrea Volkamer[1], Mathias M. von Behren[2], Stefan Bietz[2], and Matthias Rarey[2]

[1] Charité – Universitätsmedizin Berlin, corporate member of Freie Universität Berlin, Humboldt-Universität zu Berlin, and Berlin Institute of Health, Institute of Physiology, Virchowweg 6, 10117 Berlin, Germany
[2] Universität Hamburg, ZBH - Center for Bioinformatics, Bundesstraße 43, 20146 Hamburg, Germany

Learning Objectives

This chapter shall help to understand the roles of protein active sites in the context of structure–function relationships and target assessment for computer-aided drug design:
- To describe the methods for active site prediction
- To distinguish and prioritize active sites
- To compare active sites based on their structure

Outline

6.7.1 Introduction, 283
6.7.2 Active Site Prediction Algorithms, 284
6.7.3 Target Prioritization: Druggability Prediction, 292
6.7.4 Search for Sequentially Homologous Pockets, 296
6.7.5 Target Comparison: Virtual Active Site Screening, 298
6.7.6 Summary and Outlook, 304

6.7.1 Introduction

Computational approaches have long entered the early drug development pipeline to accelerate and reduce costs in drug discovery [1]. While the classical computer-aided application is screening of large compound datasets for new lead structures (see Section 6.8), recent advances in structure elucidation and structural genomic projects enabled high-throughput approaches for target screening, for example, target prioritization, characterization, and comparison.

By nature, proteins are designed to interact with other molecules to carry out different functional roles in the cell. These interactions take place in cavities, so-called binding sites or active sites. Both terms describe the position on the protein surface where a small molecule can bind and will be used as such within this article. Literally, the term active site specifies more precisely an enzyme's catalytic site.

A precise annotation of cavities on the protein surface to which small molecules can bind is the prerequisite for subsequent calculation steps in most computer-assisted approaches for drug discovery such as target assessment and comparison (see Figure 6.7.1). If a bound ligand is known for a structure, it can

Applied Chemoinformatics: Achievements and Future Opportunities, First Edition.
Edited by Thomas Engel and Johann Gasteiger.
© 2018 Wiley-VCH Verlag GmbH & Co. KGaA. Published 2018 by Wiley-VCH Verlag GmbH & Co. KGaA.

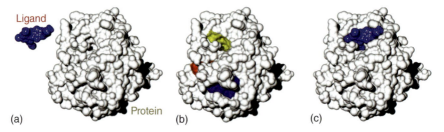

Figure 6.7.1 (a) Protein structure (gray) with unbound potential ligand (blue) in surface representation. (b) Prediction of potential binding sites (yellow, red, blue). (c) Protein–ligand complex structure, where ligand binds to the yellow binding site.

be used for binding site annotation. Nevertheless, if only apo structures of a protein are known or if additional allosteric sites are of interest, binding site detection methods are needed.

In the early drug design phase, target prioritization, that is, selecting those representatives from a set of disease-modifying targets that have the highest potential for modulation by low molecular weight compounds is of major interest. To predict the so-called target druggability [2], binding site descriptors are used in combination with machine learning techniques to extract patterns that separate (known) druggable from undruggable proteins [3]. Once a certain target or a group of targets is selected, it is essential for the success of a drug design campaign to gain as much knowledge about the target as possible. One important step of such target investigation attempts is the collection of alternative structures, which allows additional insights into a protein's intrinsic characteristics, for example, its conformational variability or its mutation sensitivity. Furthermore, binding site features can be used to compare proteins to shed light on functional relationships between them [4]. In the pharmaceutical context, binding site comparison can help in detecting potential polypharmacological or adverse effects and therefore can contribute to the design of inhibitors with distinct selectivity patterns [5]. Furthermore, the comparison of active sites of enzymes can assist in rational enzyme design. Generally, such high-throughput methods allow to think outside the box and to detect homologies between structures that would not have been found otherwise.

This chapter summarizes basic knowledge about active site detection techniques and target assessment approaches. Section 6.7.2 explains current active site prediction algorithms, followed by Section 6.7.3 introducing methods for target prioritization, that is, druggability annotation. Section 6.7.4 describes approaches for the enrichment of structural knowledge by active site ensemble generation. In Section 6.7.5, structure-based approaches for active site comparison are introduced. Finally, the topic is summarized.

6.7.2 Active Site Prediction Algorithms

The first computational approaches for active site detection date back to the early 1990s [6, 7]. Since then many methods have followed [8–11]. Although clear progress in algorithms and prediction accuracy has been achieved, the nature of proteins makes exact pocket and boundary definitions difficult. Proteins are flexible molecules resulting in a magnitude of potential cavity shapes. Cavities accommodate a wide range of binding partners, for example, ions, waters, small molecules, and peptides or even other proteins. Therefore, pockets exist from being small to large, shallow to deep, and homogeneous to highly branched [12].

Generally, pocket detection algorithms can be divided into sequence- and structure-based approaches. Due to the fact that, classically, sequences have been known before structures could be elucidated, sequence-based methods have been introduced first. Sequences are compared with respect to conserved residues following the hypothesis that protein function is encoded in evolutionary-driven conservation. Instruments to identify conserved residues are multiple sequence alignments, active site profiles, or homology-based knowledge transfer [13–15]. Although broadly used, one has to keep in mind that the success rates of sequence-based methods highly depend on the degree of sequence identity between the proteins of interest. Furthermore, most methods lack information about the spatial arrangement of active site residues, which is vital for ligand recognition.

Nowadays that more than 136,500 three-dimensional (3D) protein structures are freely available in the Protein Data Bank (PDB) [16]; these drawbacks can be overcome by investigating the structure of proteins. A common strategy within these approaches is the identification of points of interest on the protein surface that are subsequently clustered to pockets. The difference between the methods lies in the selection of these points encoding either geometric or energetic characteristics of the structure (Figure 6.7.2).

Geometry-based methods analyze the shape of the molecular surface to locate cavities and are solely based on the 3D coordinates of the protein atoms. They can further be divided into grid- and sphere-based methods. Grid-based methods embed the protein into a Cartesian grid with a usual spacing between $0.4\,\text{Å}$ [9] and $2.0\,\text{Å}$ [6]. Subsequently, grid points are labeled based on their solvent accessibility. A grid point is assigned as occupied if it lies within the van der Waals (vdW) radius of any surrounding protein atom, otherwise as free. POCKET [6] was the first published grid-based algorithm using buriedness information to find cavities. The environment of each grid point is scanned along the x-, y-, and z-axes to find those points that are deeply buried in the pocket. The hypothesis is that if a continuous volume of free solvent-exposed grid points is enclosed – on many sides – by protein atoms, there is a high chance that these points are located inside a cavity. Thus, for each grid point, the number of lines along the three-coordinate axis delimited by a protein atom in both directions (so-called protein–solvent–protein (PSP) events) is counted (see upper left corner of Figure 6.7.2). Grid points with more than the algorithm's specific threshold of PSP events are considered as buried and clustered to pockets. Due to discretization effects, the resulting pockets depend on the orientation of the protein in the grid and can vary with the used grid spacing. Besides the usage of a smaller grid spacing, these limitations have partly been overcome by considering more scan directions: LIGSITE [7] uses seven scan directions based on the faces and corners of a cube placed around each grid point and PocketPicker [8] scans

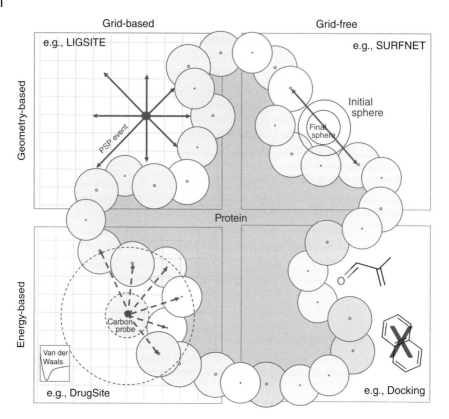

Figure 6.7.2 Exemplary illustration of pocket detection methods vertically grouped into grid-based and grid-free approaches and horizontally separated into geometry- and energy-based methods. Republished with permission of Future Science Group, from Future Med. Chem. (2014) 6(3), 319–31; permission conveyed through Copyright Clearance Center, Inc.

30 rays obtained from the triangulation of an octahedron. A novel geometric grid-based approach has been introduced with DoGSite [9] incorporating a difference of Gaussian (DoG) filter from image processing to identify cavities on the protein surface. Following the nature of this edge detection filter, rapid changes in DoG density allow locating cavity points on the grid that are clustered to subpockets. Subsequently, neighboring subpockets are assembled to pockets of typical size to accommodate a small-molecule ligand.

Grid-free geometric algorithms incorporate spherical objects to find protein cavities. SURFNET [17] and PASS [18], for example, place spheres on the protein surface. SURFNET starts with rather large so-called gap spheres between atom pairs. Subsequently, their radii are scaled down until they are free of clashes with any surrounding protein atom. The clusters of final gap spheres describe the shape and the size of the cavities (see upper right corner of Figure 6.7.2). In PASS, the protein is coated with layers of small virtual probes. Based on the number of protein atoms surrounding each probe, the buried probes are identified and unburied probes are discarded. Additional probe layers are attached until the cavities are filled and potential binding site centers are selected. In CAST [12], the concept of alpha shapes [19] has been introduced and has since then been used in several other approaches, for example, SiteFinder [20], APROPOS [21], and Fpocket [22]. The general idea is to partition the protein surface. In CAST [12], a Voronoi diagram forming a cell around each protein atom and its mathematical dual, the Delaunay triangulation [23], are calculated. Triangles with edges completely or partly outside the protein are labeled. Next, labeled triangles that are neighboring in space are combined via a discrete flow method, such that continuous clefts on the protein surface can be assigned. Fpocket [22] uses the same concept of alpha spheres, which describe spheres containing exactly four protein atoms on their boundary with no protein atom in its interior [24]. The radii of these alpha spheres describe the local curvature. Small spheres are located within the protein, while large spheres are at the exterior. Thus, intermediate alpha spheres fulfilling a specific minimum and maximum radius criterion are overrepresented in clefts and can, therefore, be used to identify pockets.

Energy-based methods compute the interaction of a probe or a molecular fragment with the protein and areas with favorable energetic responses are assigned to pockets [25–28]. In DrugSite [25], a grid-based approach, a carbon probe is placed on each grid point and the van der Waals energies between the probe and the surrounding protein atoms within a distance of 8 Å are calculated (see lower left part of Figure 6.7.2). Subsequently, grid points with unfavorable energy values are truncated and a moving average filter is applied to smoothen the energy potential on the grid. A contour level is calculated based on the mean energy over the grid points, as well as their standard deviation. Finally, grid points fulfilling this cutoff, that is, energetic hot spots, are merged to pockets. Similarly, Q-SiteFinder [26] positions a methyl probe on each grid point, and the interaction energy with the protein is calculated. Another grid-free energetically motivated approach is docking of fragments to a protein of interest [27]. Fragments are placed and a scoring function is incorporated to evaluate the fragments' position at the protein surface. Finally, pockets are assigned based on the quantity of fragments that bind to a specific area (see lower right part of Figure 6.7.2).

Since each individual method has some strengths and weaknesses, a reasonable strategy is to combine methods, therefore enhancing the prediction power. Most fused approaches combine sequence and structural information to overcome the individual drawbacks [29–31]. For example, LIGSITE^{CSC} [30] (CSC = Connolly surface and conservation) is an extension of LIGSITE. Firstly, instead of using PSP events as in the original method, surface–solvent–surface events capture buriedness. Secondly, a re-ranking of the three largest pockets by the degree of conservation of the closest surface residues follows. Residue conservation scores are obtained from the ConSurf-HSSP database [32]. A similar re-ranking procedure is applied in SURFNET-ConSurf [33], the successor of SURFNET [17]. Another approach, FINDSITE [34], starts with the sequence of a protein and uses a threading algorithm to identify similar ligand-bound template structures. Using structure alignment software, templates are superimposed onto the target of interest. Putative target binding sites are specified based on the location of the ligands bound to the threading templates. Eventually, these are ranked according to the number of templates that share the respective binding pocket. In contrast to combining sequence and structural information, SiteMap [35] is a structure-based approach that joins geometric and energetic information of the protein surface mapped onto a grid. Concerning geometry, the algorithm investigates the buriedness of grid points, similar to the previously described PocketPicker [8] approach using a higher number of scanned directions, that is, 110 rays. Additionally, energetic contributions are added by calculating the van der Waals energies of a carbon probe placed on the individual grid points, similar to the DrugSite [25] approach. Finally, potential binding sites are assembled by clustering grid points, fulfilling both buriedness and energetic criteria.

Finally, MetaPocket [36, 37] is a tool that simply combines previously published algorithms for pocket predictions enabling a consensus scoring. The first version, MetaPocket 1.0 [36], includes LIGSITE^{CS}, SURFNET [17], PASS [18], and Q-SiteFinder [26]. In the more recent MetaPocket 2.0 [37] version, Fpocket [22], ghecom [38], ConCavity, and POCASA [39] have been added. The calculation includes three steps. First, all eight methods are invoked and the three top-ranking pockets from each method are transferred to the second step, the generation of the meta-pocket. Then, the individual pockets are clustered by their center of mass and ranked by a reliability score. Finally, the meta-pocket is represented by the center of mass of the specific cluster, and the pocket residues are assigned accordingly.

To evaluate pocket detection methods, the ability to retrieve the true cavity of the co-crystallized ligand is tested in retrospective experiments. Therefore, for each structure, all pockets on the protein surface are predicted and sorted by size. A prediction is generally considered as correct, if the largest predicted pocket contains the co-crystallized ligand. Nevertheless, comparing the performance of existing methods for predicting the true binding site with high precision is difficult. Most algorithms are evaluated on diverse datasets and rely on individual criteria for what is considered a correct prediction; an overview is given in Table 6.7.1.

As shown in Table 6.7.1, dataset sizes range from 10 up to several thousands of protein structures, and success rates above 60% have been reported for all methods. Besides the difference in datasets used for evaluation, the intrinsic criteria for a correct prediction differ between methods. Success criteria range from requiring a certain overlap of predicted and experimentally defined active site residues, over finding the co-crystallized ligand in one of the largest pockets, to restricting the pocket center to be at most 4 Å apart from any atom of the bound ligand [35]. Weisel et al. [8] released the latter criterion together with a small dataset of 48 bound and unbound structures that has since then been used as a benchmark to compare the performance of various pocket detection algorithms (Table 6.7.2).

Nevertheless, this criterion does not fully cover the quality of a pocket in terms of volume and pocket boundary. Since binding site prediction is often a first step for further pocket analysis and comparison, a more precise definition is helpful. With the purpose of comparing computationally detected pockets of a protein for druggability and function annotation, it is important to assess which portion of the ligand is covered by the predicted pocket and *vice versa*. Therefore, criteria like mutual overlap [22], ligand occupied fractions of the pockets [46], or ligand and pocket coverage [9] have been introduced to rate the quality of a pocket.

Table 6.7.1 Nine pocket detection methods together with the category they belong to, the datasets they have been evaluated on, and the respective prediction accuracies. Besides the general ability to find a location that somehow overlaps with the co-crystallized ligand, accuracies with respect to more precise correctness criteria are added, if available in the respective publications.

Method	Category, type	Dataset	Prediction accuracy	Correctness criterion
LIGSITE [7]	Geometric, grid-based, buriedness	10 proteins	100%	All atoms in contact with the ligand found
SURFNET [17]	Geometric, grid-free, spheres	67 single-chained enzymes [40]	Top1 84%	Ligand atoms found in pocket
PASS [18]	Geometric, grid-free, spheres	30 protein–ligand complexes	Top1 63% Top3 86%	Distance of any pocket point ≤4 Å to any ligand atom
CAST [12]	Geometric, grid-free, alpha spheres/shapes	51 monomeric enzymes from SURFNET [40] subset	Top1 74% of 39 enzymes with one binding site	Criterion not further specified
Fpocket [22]	Geometric, grid-free, alpha spheres/shapes	20 protein structures from Cheng et al. [41]	Top1 75% (70%), Top3 95% (90%)	PPc [8]: In parentheses: MOc
		82 Astex structures [42]	Top1 67% (73%), Top3 82% (88%)	
DrugSite [25]	Energetic, grid-based, vdW energy	5616 protein–ligand complex pockets from 4711 PDB structures	RO > 0.0: 99% [of these 81% Top1, 96% Top3] RO > 0.8: 86%	RO between experimental and predicted pocket residues
SiteMap [35]	Geometric + energetic, grid-based, buriedness + vdW energy	538 pre-compiled monomeric PDBBind [43] subset	Top1$_{BS}$ 86%, Top3$_{BS}$ 91%	PPc [8]: Pockets are ranked by score (Top$_{BS}$)

Table 6.7.1 (Continued)

Method	Category, type	Dataset	Prediction accuracy	Correctness criterion
PocketPicker [8]	Geometric, grid-based, buriedness	48 complexes and respective apo structures	see Table 6.7.3	PPc [8]: Distance of geometric pocket center ≤ 4 Å to any ligand atom
		20 protein structures from Cheng et al. [41]	Top1 70%, Top3 80%	
		82 Astex structures [42]	Top1 59%, Top1 67%	
DoGSite [9]	Geometric, grid-based, Difference of Gaussian	828 PDBBind [43] structures	Top1 76%, Top3 93% (LC_{50} 90%, $LC_{50} + PC_{25}$ 70%)	At least one ligand atom within the pocket In parentheses: Top3 with LC_{50} (50% ligand coverage) and PC_{25} (25% pocket coverage) criteria
		6754 scPDB [44] structures	Top1 77%, Top3 92% (LC_{50} 88%, $LC_{50} + PC_{25}$ 66%)	

Abbreviations used: vdW, van der Waals; Top1, ligand was found in the largest pocket; Top3, ligand was found in one of the three largest pockets; PPc, PocketPicker criterion [8], pocket center at most 4 Å apart from any ligand atom; MOc, mutual overlap criterion, mutual overlap between ligand atoms and alpha spheres [22]; RO, relative overlap between experimental and predicted pocket residues [25]; LC, ligand coverage; PC, pocket coverage [9].

Table 6.7.2 Pocket prediction success rates for 48 bound and 48 unbound proteins, sorted by publication year. Success rates present the percentage of cases in which the correct active site was calculated by the respective algorithms.

Method	Year	Top1		Top3	
		Unbound	Bound	Unbound	bound
LISE [45]	2013	81	92	92	96
MetaPocket2.0 [37]	2011	80	85	94	96
VICE [46]	2010	83	85	90	94
DoGSite [9]	2010	71	83	92	92
MSPocket [47]	2010	75	77	88	94
POCASA [39]	2010	75	77	92	90
Fpocket [22]	2009	69	83	94	92
PocketPicker [8]	2007	69	72	85	85
LIGSITE^{CS} [30]	2006	60	69	77	87
Q-SiteFinder [26]	2005	52	75	75	90
PASS [18]	2000	60	63	71	81
CAST [12]	1998	58	67	75	83
LIGSITE [7]	1997	58	69	75	87
SURFNET [17]	1995	52	54	75	78

Top1, ligand was found in the largest pocket; Top3, ligand was found in one of the three largest pockets. Numbers are collected from recent publications [8–10].

As the trend in Table 6.7.2 shows, the success rates of the algorithms improved over the last years. Nevertheless, all of the presented pocket prediction methods have their drawbacks. Sequence-based methods depend on the homology between the structures and the quality of the multiple sequence alignments. Grid-based methods are sensitive to grid spacing, protein position, and orientation. Geometry-based methods encounter problems with precise cavity ceiling definitions and sphere-based methods with the detection of wide cavities. Finally, energy-based approaches, for example, highly depend on the selected probes and fragments as well as on the underlying scoring function. Although different strategies to parameterize the cutoff values within the methods have been investigated, they commonly suffer from the variability and different shapes of pockets, as well as the different protein architectures from mono- to multimers. One limitation in structure-based pocket prediction approaches results from the investigation of rigid crystal structures of, in reality, flexible molecules. Since the size and shape of pockets change with the molecule's conformation, first approaches modeling protein flexibility incorporate molecular dynamics (MD) calculations to sample several snapshots of the protein, which can then be used to analyze pocket ensembles [48], transient pockets [49, 50], or cryptic pockets [51].

In conclusion, the success rates of recent pocket prediction approaches clearly show the advances in accuracy and precision. Furthermore, due to advances in algorithms and computing power, calculation times per protein structure are nowadays in the range of seconds, allowing the application of such methods for high-throughput analyses.

6.7.3 Target Prioritization: Druggability Prediction

As expressed by the high cost and time expenses for drug discovery [1], it is of high practical importance to prioritize targets in early drug development phases. Failure rates of 60% in drug discovery projects, attributed to undruggability of the underlying targets, were reported in 2003 [52]. Due to the large amount of available structural data, computer methods are a vital instrument to accelerate *a priori* prediction of target accessibility and to assist in the selection of the most promising targets [2]. In this context, the term druggability has been coined [53]. Druggability is defined as the general ability of a disease-modifying target to be modulated by low molecular weight compounds. Ligandability, bindability, targetability, or chemical tractability [54–57] are other terms encompassing similar questions. All these concepts describe the ability of a target to bind a molecule; however, they vary in the inherent properties of the bound molecule. For ligandability or bindability, only binding a small molecule is required [54], while for druggability, depending on the definition, the molecule has to bind with high affinity, obey drug-likeness criteria [53], or be orally bioavailable [41].

Target prioritization has actively been researched over the last two decades including experimental and computational approaches. In pharmaceutical industry, high-throughput screening (HTS) has widely and successfully been applied for target assessment [56, 58, 59]. Biochemical or NMR screens are performed against large compound datasets. The number and chemical characteristics of compounds binding to a specific target are measured, and their quantities are used to assess the target's druggability [58].

Nevertheless, the reliability of these estimations highly depends on how well the compound sets sample the chemical space. A better space coverage has, for example, been provided by introducing NMR-based fragment screening [60], for which a correlation between hit rates and success rates in hit-to-lead programs has been ascertained. Although these screens are mostly automated, they require higher resources than computational approaches. The direct virtual counterpart of NMR-based methods is *in silico* screening. Binding site assessment by *in silico* compound or fragment screening was likewise reported to correlate with NMR-based hit rates [55]. Thus, success rates of drug-like ligands virtually docked into the active site of a target protein can be used for prioritization. For example, the FTMAP algorithm searches the entire protein surface for regions that bind a number of small organic probe molecules to identify druggable hot spots [61]. Note that ligand- or fragment-based NMR and *in silico* screens rather address ligandability than druggability [27].

The largest group of computational methods relies on the extraction of specific target features that potentially imply druggability. As already discussed in the last section, classically the protein sequence is known before structure. Thus, first computational methods derived druggability information from sequence properties [62, 63]. Nevertheless, prediction accuracies for separating druggable from undruggable targets with respect to sequence properties were reported not to exceed 70% [2], implying the need for alternative models and descriptors. Concomitant with advances in structure elucidation, the attention has been shifted to methods incorporating structural active site information. Descriptor-based approaches generally share three requirements: a description of the discriminating features of the active site, a classification algorithm, and a labeled dataset for model training. Initially, the active site has to be specified. Active site annotation can be done using a known ligand or via an active site detection algorithm (see Section 6.7.2). Note that recently developed fully automated methods can compute binding sites on the fly [35, 64, 65]. Next, pocket features able to discriminate between druggable and undruggable pockets have to be selected. For classification purpose, different schemes, for example, clustering, regression, machine learning, or simple discrimination functions, are incorporated. Finally, the models have to be calibrated on a preferentially large template dataset to rate druggability. With respect to the nature of the used training data, one can distinguish between methods trained on hit rates [58], on successful versus failed drug targets [41], and on protein–ligand complexes [66].

An important question is how many descriptors are used for the classification. Generally, the published methods in the druggability field agree that there is no linear relationship between a single descriptor and druggability. Nevertheless, the number of used combined parameters per method ranges from a couple to up to several hundreds of classifiers [60, 66]. Hajduk *et al.* [60] examined 13 binding site features with NMR hit rates of 23 targets and found a combination of eight properties to be most discriminative to distinguish between druggable and undruggable targets. Nayal and Honig introduced SCREEN [66], a method to recognize and evaluate drug-binding cavities. Each detected cavity is characterized by a feature vector containing 408 pocket descriptors. Using random forest classifications, a conjunction of 18 descriptors was found to have the highest prediction power on a test set of 99 protein–ligand complexes. A different approach constitutes maximal affinity predicted for passively absorbed oral drug (MAP_{POD}) [41], which investigates the maximal affinity that a small drug-like ligand can achieve in a specific binding pocket based on the available structural and physicochemical environment. A model reduced to important discrete energy terms and properties of the ligands describing oral bioavailability is used. The model has been evaluated on 27 proteins with known drugs and a clear separation between druggable and difficult/undruggable targets could be shown.

Most structure-based methods predict druggability using a set of descriptors incorporated into a scoring function; see, for example, PocketPicker [67], SiteMap [35], Fpocket [68], DoGSiteScorer [65], DLID [54], or DrugPred [69]. To estimate the ability of a pocket to bind small drug-like ligands, PocketPicker [67] uses a 210-dimensional descriptor set encoding size and buriedness of the pocket. Classification of proteins is done via self-organizing maps trained on descriptors from 13,859 partially empty, partially ligand-bound pockets. In 2009, SiteMap [35] was introduced using a rather simple linear regression model for druggability prediction. The function is based on a weighted sum of three descriptors – number of site points (encoding volume), hydrophobicity, and shape of the pocket. Since the authors recognized the difference between bindability and druggability, two functions differing in the coefficients have been generated: SiteScore, for rating ligandability, was calibrated on the PDBbind [43] dataset and DScore was adapted with respect to druggability based on the Cheng et al. [41] dataset. In drug-like density (DLID) [54], the likelihood of a pocket to bind a drug-like ligand is computed based on the number of pockets binding a drug-like ligand in its local structural neighborhood. This local neighborhood is similarly encoded as a linear combination of volume, buriedness, and hydrophobicity of the pockets. The coefficients have been obtained via linear regression on a subset of protein–ligand complexes from the PDB, fulfilling specific quality criteria. Fpocket [68], another fully automated druggability prediction method, introduces an exponential scoring function, containing three physicochemical features normalized by the size of the pocket for druggability calculation: local hydrophobicity density (which encodes the size and spatial contribution of hydrophobic point agglomerations), general hydrophobicity, and normalized polarity. The authors of this method collected the so far largest freely available druggability dataset (DD), consisting of 1070 structures and a nonredundant version of it (NRDD), which was used to train the exponential function via bootstrapping. DoGSiteScorer [65] pursues a similar strategy. A large set of descriptors for self-predicted pockets has been calculated for the DD. Using a discriminate analysis, those features most successful in separating druggable from undruggable features have been extracted and incorporated into a support vector machine (SVM). For druggability prediction, models have been trained on the NRDD, based on selected discriminating pocket features such as depth, fraction of apolar amino acids, and volume. The complete procedure is exemplified in Figure 6.7.3. The success rates on predicting the correct druggability state for all structures from the NRDD and DD dataset are above 88%.

DrugPred [69], a method released in 2011, used partial least squares projection for discriminant feature derivation, resulting in a linear model based on five pocket descriptors. In 2012, Perola et al. [70] introduced a rule-based approach, similar to Lipinski's rule of 5 for orally bioavailable drugs, an intuitive representation of the preferred property space of druggable pockets. Descriptors from pockets of 60 targets of approved drugs have been compared with a diverse set of 440 ligand-binding pockets. A preferred property space based on five rules has been derived: volume ≥ 500 Å^3, depth ≥ 10.4 Å, enclosure ≥ 0.28, percentage charged residues ≤ 26.3, and hydrophobicity ≥ -1.12.

In summary, the findings reveal that combinations of discriminate pocket features are needed, to distinguish druggable from undruggable targets. The methods agree that druggable pockets tend to be larger, more complex and exhibit a higher hydrophobic character than undruggable pockets.

A large collection of different descriptors has been incorporated, and the performance of the methods on retrospective evaluations is quite good; nevertheless, there is still room for improvement. It is particularly necessary to dissociate oneself from the rather global description of a pocket, which may change due to protein flexibility toward a more local representation, which is less dependent on small structural changes. First examples for local considerations are provided in Fpocket [68] where the effect of local environmental changes in accessible surface area is analyzed, and in DoGSiteScorer [65], in which

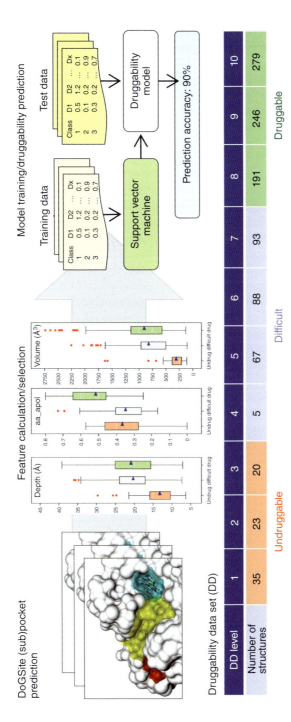

Figure 6.7.3 DoGSiteScorer-based model building and druggability prediction: First, pockets were predicted for all structures of the DD dataset. Descriptors were calculated and discriminative features were selected, based on which a SVM model was trained. Finally, this model can be used for druggability predictions of novel target structures.

pairwise distances between functional groups are incorporated. Improvement in prediction accuracy could also be achieved by adding so far missing features, for example, considering contributions related to bound water molecules, metal ions, or other ions. Very recently, a second class of druggability prediction algorithms based on or tailored for MD simulations have been proposed [71, 72]. The *j*ust *e*xploring *d*ruggability at protein *i*nterfaces (JEDI) [72] algorithm relies on a set of geometric parameters describing the volume, the enclosure, and the hydrophobicity of a binding site and can be collected on the fly during MD simulations. JEDI was shown to be able to detect cryptic druggable binding sites in proteins and to deliver conformations suitable as input for subsequent docking calculations.

Another drawback is the limited availability of large and reliable, as well as unbiased, datasets for method training. Recently, large datasets as the DD (1070 structures, containing druggable, difficult, and undruggable pockets) [68] and the DLD (115 structures, including druggable and less druggable pockets) [69] have been made freely available. Nevertheless, still two main problems remain. One problem is the bias of most datasets toward druggable pockets. The annotation of positive data is relatively easy based on the fact that a drug binding to these pockets is known. The inverse, namely, that the absence of a known drug (to date) implies "undruggability" is however very questionable. This annotation may change in the near future, once a new drug is found against this target. The same holds for the term ligandability, where empty pockets in crystal structures are assumed to be unable to bind a ligand. Thus, a rather forward-looking approach is to base such training on only a positive class against the rest of the pocket universe [70]. The second problem is the ambiguous definition of the term druggability, which makes a clear class affiliation difficult. Unifying and clarifying the term druggability, and with it the class annotation within the different datasets, would be an important step forward.

In conclusion, several computational methods for druggability prediction methods have been introduced over the last decade with very promising results. Many of them have been integrated into the drug discovery pipeline, generating valuable impact for target assessment.

6.7.4 Search for Sequentially Homologous Pockets

As discussed in the previous sections, the flexibility of proteins often constitutes a limiting factor for the analysis of binding sites. This might lead to deviating results when starting from different input structures. Besides conformational variation, mutated amino acids and structural artifacts may also influence the outcome of structure analyzing calculations. Considering an ensemble of pocket conformations or sequentially homologous binding sites rather than a single structure is a popular way to incorporate protein variability. Ensembles generally allow a more detailed description of the protein structure and facilitate to draw additional conclusions about intrinsic binding site features. Hence, the search and selection of suitable alternative structures from a given structure database is an important preprocessing step for all computational approaches based on protein structure.

6.7.4 Search for Sequentially Homologous Pockets

In the easiest case, the structure identification process can be realized by any of the well-established sequence search algorithms, for example, BLAST [73], comparing the entire protein sequence and resulting in a similarity-ranked list of chains from different structures. This approach, however, is limited to pockets that are solely formed by a single chain, as it is the case for CDK2 (see Figure 6.7.4a, 1AQ1 [74]). Furthermore, it does not allow focusing on binding site-specific features. Pockets that are composed of different chains (see Figure 6.7.4b, 5KR2 [75]) usually require the analysis of alternative chain combinations and a consideration of structural information for the investigation and comparison of the mutual chain orientation in different structures. As this is clearly a more elaborate task, the search for alternative binding site conformations is more complicated and therefore more time consuming in these cases.

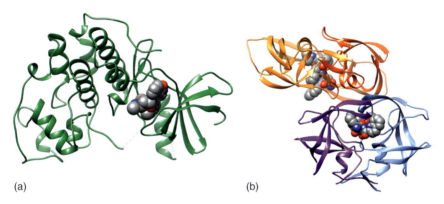

Figure 6.7.4 (a) Binding site of monomeric cyclin-dependent kinase 2 (PDB code 1AQ1) [74]. (b) Four identical subunits of HIV-1 protease forming two symmetrical binding sites (PDB code 5KR2) [75].

A possible strategy to circumvent long search times in large structure databases is the precalculation of pocket similarities, as applied in the Pocketome project [76, 77]. The Pocketome database is a comprehensive collection of superimposed binding site structures that represent different states of the same or sequentially similar pockets. It is based on a preprocessed but routinely updated ensemble compilation procedure that applies a pocket detection algorithm and sequence clustering of PDB chains. For all pockets, the Pocketome database offers a web-based visualization of the resulting binding site ensembles and provides additional information like root mean square deviation (RMSD) values, mutations, ligand overlaps, or interaction patterns.

The ensemble precalculation of the Pocketome database enables a quick overview of accessible alternative pocket structures, but does not allow a case-specific binding site definition or certain search constraints that are necessary to meet the requirements of the desired application. Such more application-oriented tasks can be tackled by the ensemble compilation approach SIENA [78]. Starting from an arbitrary user-defined binding site, SIENA extracts all pocket analogues from a given structure database that fulfill further case-specific search constraints defined by the user. This might, for example, involve a maximum mutation rate in the binding site, RMSD thresholds, detection of structural artifacts, structural features like resolution and deposition year, or ligand presence. Furthermore, SIENA contains different approaches to reduce the resulting binding site ensembles to a better manageable amount of diverse structures.

The underlying search procedure is based on the binding site alignment algorithm ASCONA [79] that is specialized in the processing of alternative conformations rather than identifying structurally similar regions, the objective of most other pocket comparison approaches (see Section 6.7.5). This allows a more accurate identification of flexible regions and therefore also a more precise conformation selection in SIENA. Furthermore, its focus is on the respective binding site and on those features that are relevant for conformational analysis. As a positive side effect, this leads to a reduced runtime that can compete with classical sequence search procedures.

6.7.5 Target Comparison: Virtual Active Site Screening

Similar to methods for active site prediction, as described in Section 6.7.2, methods dedicated to target comparison can be split into sequence- and structure-based approaches. For a long period of time, sequence-based methods have been the most commonly used approaches to protein comparison and function prediction. In these methods, knowledge is transferred based on sequence similarity to proteins with a well-known function. Multiple sequence alignments can either be calculated for the whole sequence of the protein as in BLAST [80] and PFAM [81] or for specific sequence motifs as in PROSITE [82], BLOCKS [83], and PRINT [84]. Other forms of information incorporated into the evaluation of the relationships between proteins are, for example, gene expression data, gene ontology, phylogeny, and coevolution, as used in methods like GoFigure [85], Phydbac [86], SIFTER [87], and FlowerPower [88].

The large and still growing amount of solved protein structures, as well as the fact that structure was found to be more conserved than sequence [89], has increasingly promoted structure-based methods for protein comparison. One strategy to identify structural similarities is the comparison of the overall protein fold. Protein folds can be compared with methods like SCOP [90], CATH [91], or FSSP/Dali [92]. Other methods perform overall structural comparisons in more detail through structural alignments. These can be calculated based either on the complete structure (FATCAT [93], PAST [94], and VAST [95]) or on structural fragments, which are later recombined to a complete alignment (3DCOMB [96] and PROCAT [97]).

Nevertheless, examples showing that proteins can share functionality even without exhibiting overall sequence or structural similarity moved the focus to active site comparison decoding potential local similarities between distantly related structures. Structural methods dedicated to binding site comparison can usually be divided into three main parts [3, 98]. First, selected binding site properties are encoded as molecular recognition features; this reduces the complexity of the comparison problem. Second, similarities between these features are detected mainly based on structural feature alignments or fingerprint comparison. Third, the similarity between two binding sites is quantified with the aid of a scoring function. The common strategies for detecting similarities can be divided into alignment-based and alignment-free methods, as well as some recent intermediate methods.

Alignment-based methods incorporate the encoded features to calculate the best possible superimposition of two binding sites. The most common strategies are geometric matching (ProSurfer [99], SuMo [100], and SiteBase [101]), geometric hashing (SiteEngine [102] and TESS [103]), or clique detection (CSC [104], CavBase [105, 106], eFsite [107], eFseek [108], IsoCleft [109], and ProBis [110]). In SiteBase [101], an example for geometric matching, the algorithm generates possible triangles, based on all triplets of binding site atoms with a pairwise atom distance within a predefined range. These triangles are systematically compared, and matching triangles are subsequently used to superimpose both structures. Matching triangles must have the same corner atom types, and the corresponding side lengths may only differ by a certain tolerance value. Finally, the alignment of two structures is rated based on the number of superimposed atoms of the same type, for example, carbon, nitrogen, or oxygen atom. In SiteEngine [102], geometric hashing is used to efficiently search for matching binding site descriptors. First, each residue surrounding the binding site of interest is represented by a set of pseudo-centers with a certain interaction type, for example, hydrogen-bond donor or acceptor. Triplets of these pseudo-centers are stored using a hash function encoding side lengths and interaction types. Additionally, the types of all pseudo-centers are individually assigned to the combined Connolly surface of all atoms belonging to the respective pseudo-center, so-called surface patches. A query protein is processed in the same way, and the resulting triplets of pseudo-centers are used to retrieve matches from the previously filled hash. Each match represents an alignment that is scored based on the agreement of the interaction type and the geometry of aligned surface patch centers and their local environments. In CavBase [105, 106], clique detection on a product graph is used to determine the best possible alignment of two binding sites. For this purpose, each cavity flanking residue is translated into a pseudo-center and assigned with a special interaction type corresponding to the respective amino acid. For the comparison of two binding sites, a product graph is calculated as follows: A node is inserted for each pair of compatible pseudo-centers, and an edge is placed between two nodes if the corresponding nodes in the original graphs are connected with an edge of nearly equal length. Subsequently, a clique search is performed on this product graph detecting a maximum common subgraph, which is then used to align the binding sites. Finally, the similarity is measured based on the overlap of surface points sharing the same type as well as on the RMSD of the superimposed pseudo-centers.

While an alignment of two binding sites is a useful and easy to interpret result, the drawback is its computational expensive calculation, which is usually in the order of minutes for a pairwise comparison. Therefore, efforts have been undertaken to develop faster alignment-free methods, performing pairwise comparisons in the order of milliseconds. A commonly used approach is the conversion of binding sites into fingerprints, that is, bit strings, encoding distances between specific binding site atoms or features and their distribution in the active site [111–115]. PocketMatch [111], for example, represents each binding site by 90 binned distance histogram sets. Each residue within a 4 Å radius of the ligand is represented by three points, the Cα atom, the Cβ atom, and the centroid of all atoms of the residue's side chain. Each of these points is assigned to one of five possible groups, representing the chemical properties of the corresponding residue. Then, the distances between all pairs of points in one binding site are calculated. Thus, each pair belongs to one of 90 potential sets (six possibilities for point type pairings times 15 possibilities for chemical group pairings). Each distance is stored in the matching distance bin of the set that correspond to the respective combination of types of both points. The similarity of two binding sites is the net average of the number of matching distance elements within these 90 sets divided by the total number of distances in the bigger set. PocketFeature is another alignment-free method comparing so-called microenvironments [116]. A microenvironment is defined by a sphere with a radius of 7.5 Å surrounding the functional center of each residue of a binding site, divided into six concentric shells. For each shell, 80 descriptors are calculated representing the physiochemical properties in this shell, resulting in a vector of 480 property values per microenvironment. The similarity between two binding sites is calculated by comparing the vectors of all pairs of microenvironments that were built for the same type of amino acid (e.g., positively charged, negatively charged, aromatic). The method only considers the most similar pairs of microenvironments per type and sums up the respective similarity scores into a final score for the whole binding site. Note that the relative orientation of the spheres to each other is not explicitly captured in the method. Other prominent alignment-free approaches such as FLAP [117], SiteAlign [118], and FuzCav [119] follow the concept of pharmacophore-based fingerprints. To calculate the FuzCav fingerprint, each residue of the binding site is represented by its Cα atom and annotated by one of six pharmacophore properties, for example, aromatic, aliphatic, and hydrogen-bond donor. The fingerprint contains 4833 integers, each representing the count of unique pharmacophore triplets, which are defined by three properties as well as the three related distances of the Cα atoms. The similarity between two binding sites is based on the number of entries that are greater than zero in both fingerprints normalized by the number of nonzero entries in the fingerprint with less nonzero entries.

Another group of methods follows a slightly different approach by using rotational invariant pocket representations. These methods encode the shape of a binding site as a functional series expansion and thus, a structure can be compactly represented as a vector of coefficients. For example, spherical harmonics [120] or 3D Zernike descriptors [121] are used to represent the structure as a vector of coefficients of the functional series. In the case of spherical harmonic functions, the binding site is described by rays scanning the binding sites surface.

As mentioned before, the problem of alignment-free methods lies in the lack of information about the features that were responsible for the detected similarity. As a consequence, acceleration of alignment-based methods has been investigated to overcome those shortcomings. Recently developed methods are able to perform pairwise alignment-based comparisons in the order of seconds. In

BSAlign [122], the structure of the binding site is represented as a graph. Each residue forms a vertex and is annotated with certain features, like solvent accessibility and physicochemical properties. An edge is inserted between any two vertices if the distance between the respective $C\alpha$ atoms is less than 15 Å and is labeled with the exact distance value as well as the angle between $C\alpha$ and $C\beta$ vectors of the residues. The matching algorithm takes query and target input graphs and transforms them into a single edge-product graph, similar as in CavBase [105], but adding solvent accessibility to the considered features. In contrast to CavBase, BSAlign [122] uses the efficient Cliquer [123] algorithm to find the maximum clique and, therefore, the maximum common subgraph of both input graphs. To achieve further speedup in clique detection, BSAlign [122] reduces the size of the product graph if the number of edges in the graph exceeds a certain threshold by reducing tolerance values for distances, angles, and solvent accessibility. The resulting common subgraph is used to align the respective residues and to superimpose both structures. Finally, the alignment is refined and scored based on the RMSD of superimposed $C\alpha$ atoms.

Recently, multiple approaches aimed on increasing the efficiency of the clique detection-based method of CavBase itself. First, the local clique (LC) method was introduced, which used a faster heuristic clique detection approach together with an extended graph model. By annotating more information to the pseudo-centers representing the binding site, namely, properties encoding the shape of the close-by protein surface as well as the surrounding physicochemical environment, the size of the resulting product graphs, and the runtime could be reduced significantly [124]. In a subsequent development (DivLC), the problem was further simplified by partitioning the input graphs into seven different disjoint graphs, one for each type of pseudo-center. This again achieved an acceleration of one order of magnitude [125]. The new method RAPMAD (RApid Pocket MAtching using Distances) from the same group, however, does no longer rely on clique detection [126]. RAPMAD uses the binding site representation of CavBase and converts it into multiple distance histograms. For this purpose, the pseudo-centers are partitioned as in the DivLC approach, and for each set of pseudo-centers, all distances with respect to two reference points (the centroid and the closest to the centroid) are calculated and weighted with respect to the relative pseudo-center frequency in CavBase. The histograms are used for direct comparison of binding sites with linear complexity, thus, enabling more than 20,000 comparisons per second.

Desaphy et al. [127] published a method using pharmacophore-annotated shape comparison. The binding site is mapped onto a grid by assigning grid points with a property according to nearby protein atoms. A grid point can be inside or outside of the protein. The algorithm annotates, if an inner grid point is close to a hydrogen-bond donor or acceptor or if it is close to an aromatic atom. The total number of pharmacophore-annotated grid points describes the global cavity volume. To align two binding sites, this shape is approximated by smooth Gaussian functions and the optimum volume overlap is calculated. Subsequently, the best shape alignment is scored with respect to the underlying pharmacophore features.

Another method, TrixP [128], compares binding sites based on descriptors encoding pharmacophore and spatial features, using a specialized index for high-dimensional features [129]. In a first step, pharmacophore features of the binding site are determined, which can be of type hydrogen-bond donor, hydrogen-bond acceptor, or hydrophobic region. Triplets of these features, fulfilling specific length and angle constraints, form the later-used triangle descriptors. The descriptor is completed with a set of rays radiating from the center of the triangle describing the local shape of the binding site (see Figure 6.7.5A). The properties of the descriptors, for example, triangle side length, pharmacophore type of the corners and their respective interaction direction are binned and stored in an index. The subsequent matching algorithm generates all site descriptors for a query binding site and uses them to query the index. Only descriptors with matching properties are returned, clustered, and used as basis for the superposition of the respective structures (see Figure 6.7.5B), which results in a large speedup. Each resulting alignment is then scored based on the compliance of pharmacophore features in the binding sites.

Several different methods dedicated to protein binding site comparison have been introduced in this chapter. Nevertheless, the direct comparison between the methods remains difficult. There is no standard benchmark procedure established for the methods, and, thus, the used datasets, as well as the interpretation

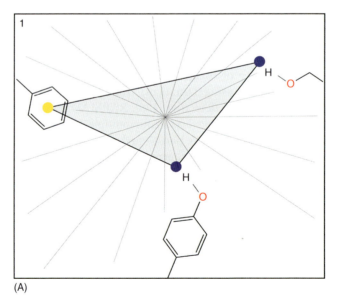

(A)

Figure 6.7.5 Depiction of the structural triangle descriptor used in TrixP. (A) Example of the TrixP descriptor with two hydrogen-bond donors (blue) as well as an apolar point (yellow) as triangle corners. (B) Schematic superposition of two different binding sites based on a matching descriptor with a hydrogen-bond donor (blue), a hydrogen-bond acceptor (red), and an apolar point (yellow) as triangle corners. (a,b) show identical descriptors in two different binding sites and (c) shows the respective superposition of the binding sites based on the matching descriptors.

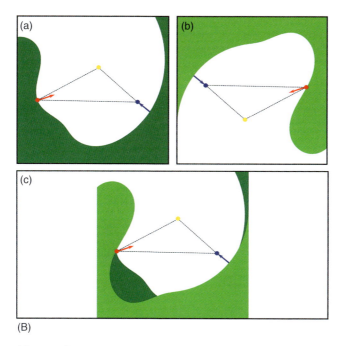

Figure 6.7.5 (Continued)

of the results may differ significantly (see Table 6.7.3). The largest dataset to our knowledge is the scPDB [44] with almost 10,000 structures, used for the evaluation of several methods. The most commonly used validation procedure is a retrospective experiment trying to regain classifications of proteins within a certain dataset. Each query structure is screened against the dataset, and the target proteins are ranked by their similarity scores. The success is rated based on the ability of the method to recover the expected structures, that is, members from the same family on the top ranks.

Another rather small dataset of eight difficult protein pairs – that is, distantly related proteins with known cross-reactivity – has been introduced to evaluate several methods [111, 119, 128]. The objective of this study was to compare the success rates of different methods in detecting similarities between the protein pairs as well as the effective computational times. Note that such an analysis is difficult since the dataset is very small and similarity scores of different methods are in general not directly comparable. Clearly, the attempt to design experiments for the direct comparison of methods for binding site comparison is an important step for the future of the field.

Over the years, many problems concerning speed and accuracy have been solved. Nevertheless, protein flexibility upon ligand binding remains one of the uncertainties during the process. To overcome this problem, recent approaches focus on the comparison of subpockets rather than the whole binding sites with the objective to detect even partial similarities between globally dissimilar binding sites [131, 132]. The described methods proved to be useful in tasks like detecting polypharmacology, protein classification, and function prediction and have a high impact in medicinal chemistry and rational molecular design [5].

Table 6.7.3 Exemplary (non-exhaustive) summary of different evaluation studies of methods for binding site comparison including information about the method category they belong to, the number of structures used for evaluation, and the results as stated in the respective publication.

Tool	Category	Number of structures	Results
SiteBase [101]	Geometric matching	476 phosphate-nucleotide ligand binding sites	Clustering in good agreement with other classifications
SiteEngine [102]	Geometric hashing	4375 structures from the ASTRAL data base [130]	11 out of 15 top ranking hits were family members
CavBase [105]	Clique detection	5248 structures	Top ranks exclusively with family members
PocketMatch [111]	Fingerprint	785 structures from the PDBbind [43]	In good agreement with SCOP classification
FuzCav [119]	Fingerprint	5952 structures from the scPDB (ver. 2008) [44]	75% of all hits were family members
BSAlign [122]	Clique detection	126 structures	9 out of 15 top ranking hits were family members
Desaphy et al. [127]	Shape comparison	9877 structures from the scPDB (ver. 2011)	Mean AUC of ROC curves 0.88
TrixP [128]	Geometric matching	9877 structures from the scPDB (ver. 2011)	Retrieving 84–100% of all included family members

6.7.6 Summary and Outlook

Due to structural genomics projects and advances in structure elucidation techniques, new protein structures, with no information concerning their possible function, are nowadays derived. Thus, computational methods are needed to organize and classify the ever-growing number of solved structures. In the early drug discovery pipeline, often several structures are known that may be responsible for a disease, and it is a challenging task to prioritize which of these targets are worth further investigations. Over the last 20 years, several computational methods, as described in this chapter, have been developed to assist experimental methods and facilitate drug design.

The active site of a protein is its center of action and, the key to its function. Thus, the first step is the detection of active sites on the protein surface and an exact description of their features and boundaries. These specifications are a vital input for subsequent target druggability prediction or target comparison. As described in this chapter, the algorithms and the resulting descriptions have been enhanced during the last years, in terms of specificity and computational efficiency. Thus, many methods allow the prediction or comparison of several thousands of structures with high throughput. Additionally, more training data have become available and the awareness for comparability between methods has been encouraged, leading to increasingly better models and results. The one major aspect not yet completely understood and implemented in computational models is the natural flexibility of proteins, especially upon ligand binding. Although first approaches, incorporating MD simulations or other methods to model structural changes, exist, there is much room for improvement. Another promising opportunity is to extend the level of structural description by the use of conformational ensembles. Furthermore, most approaches today focus on potential interaction sites between proteins and small molecules. Protein–protein interactions have gained much attention in drug discovery. Binding site detection and characterization in this case, especially describing transient protein–protein complexes is certainly an important problem to be addressed in the future.

Essentials

- Algorithms for **active site detection** are based on geometric or energetic features of the protein surface. Geometry-based approaches rely on scanning rays, spheres or Gaussian functions, while energy-based methods use van der Waals energies or docking scores. The detected pockets can be employed for target prioritization and comparison.
- **Druggability predictions** can assist in target prioritization, i.e., ranking of disease-modifying targets based on their potential to be modulated by drug-like compounds. Most computational methods use a combination of binding site descriptors (derived from known (un)druggable targets using machine learning or statistical analysis) to score the druggability of novel targets. Decent volume, depth and hydrophobicity of a pocket are considered important contributors to druggability.
- Since protein flexibility and structural artifacts often affect the accuracy of structure-based methods, various alternative structures can be considered to overcome these issues. **Ensemble generation** approaches assist this strategy by providing pre-calculated or use case-specific collections of sequentially homologous pocket structures, superpositions, and ensemble-related pocket descriptors.
- Methods for **binding site comparison** detect similar arrangements of physicochemical features between the respective sites and can be divided into alignment-based (easier to interpret) and alignment-free (faster) methods. Homologies between active sites are good indicators for potential drug promiscuity and polypharmacology.

Available Software and Web Services (accessed January 2018)

- Open-source protein pocket detection algorithm: Fpocket and MDpocket: http://Fpocket.sourceforge.net/.
- Pocket detection and druggability prediction: DoGSiteScorer is available as part of the modeling support service at http://proteins.plus.

- Active site detection algorithm: MetaPocket 2.0: http://projects.biotec.tu-dresden.de/metapocket/.
- Active site ensemble generation: SIENA is available via the modeling support server at http://proteins.plus.
- Alignment-based pocket comparison server: PROBIS: http://probis.cmm.ki.si/.
- Alignment-free pocket comparison software: SiteAlign and FuzCav: http://bioinfo-pharma.u-strasbg.fr/labwebsite/download.html.

Selected Reading

- Barril, X. (2013) Druggability predictions: methods, limitations, and applications. *Wiley Interdiscip. Rev. Comput. Mol. Sci.*, **3** (4), 327–338.
- Henrich, S., Salo-Ahen, O.M.H., Huang, B., Rippmann, F., Cruciani, G., and Wade, R.C. (2010) Computational approaches to identifying and characterizing protein binding sites for ligand design. *J. Mol. Recognit.*, **23** (2), 209–219.
- Kellenberger, E., Schalon, C., and Rognan, D. (2008) How to measure the similarity between protein ligand-binding sites? *Curr. Comput. Aided Drug Des.*, **4** (3), 209–220.
- Volkamer, A. and Rarey, M. (2014) Exploiting structural information for drug-target assessment. *Future Med. Chem.*, **6** (3), 319–331.

References

[1] Paul, S.M., Mytelka, D.S., Dunwiddie, C.T., Persinger, C.C., Munos, B.H., Lindborg, S.R., and Schacht, A.L. (2010) *Nat. Rev. Drug Discovery*, **9** (3), 203–214.
[2] Egner, U. and Hillig, R.C. (2008) *Expert Opin. Drug Discovery*, 391–401.
[3] Nisius, B., Sha, F., and Gohlke, H. (2012) *J. Biotechnol.*, **159** (3), 123–134.
[4] Volkamer, A., Kuhn, D., Rippmann, F., and Rarey, M. (2013) *Proteins*, **81** (3), 479–489.
[5] Ehrt, C., Brinkjost, T., and Koch, O. (2016) *J. Med. Chem.*, **59** (9), 4121–4151.
[6] Levitt, D.G. and Banaszak, L.J. (1992) *J. Mol. Graphics*, **10** (4), 229–234.
[7] Hendlich, M., Rippmann, F., and Barnickel, G. (1997) *J. Mol. Graphics Modell.*, **15**, 359–363.
[8] Weisel, M., Proschak, E., and Schneider, G. (2007) *Chem. Cent. J.*, **1**, 7.
[9] Volkamer, A., Griewel, A., Grombacher, T., and Rarey, M. (2010) *J. Chem. Inf. Model.*, **50** (11), 2041–2052.
[10] Xie, Z.R. and Hwang, M.J. (2012) *Bioinformatics*, **28** (12), 1579–1585.
[11] Henrich, S., Salo-Ahen, O.M.H., Huang, B., Rippmann, F., Cruciani, G., and Wade, R.C. (2010) *J. Mol. Recognit.*, **23** (2), 209–219.
[12] Liang, J., Edelsbrunner, H., and Woodward, C. (1998) *Protein Sci.*, **7**, 1884–1897.

[13] Aloy, P., Querol, E., Aviles, F.X., and Sternberg, M.J. (2001) *J. Mol. Biol.*, **311** (2), 395–408.
[14] Armon, A., Graur, D., and Ben-Tal, N. (2001) *J. Mol. Biol.*, **307** (1), 447–463.
[15] Pupko, T., Bell, R.E., Mayrose, I., Glaser, F., and Ben-Tal, N. (2002) *Bioinformatics*, **18** (Suppl 1), S71–S77.
[16] Berman, H.M. (2000) *Nucleic Acids Res.*, **28**, 235–242.
[17] Laskowski, R.A. (1995) *J. Mol. Graphics*, **13** (5), 323–330.
[18] Brady, G.P. and Stouten, P.F.W. (2000) *J. Comput.-Aided Mol. Des.*, **14** (4), 383–401.
[19] Edelsbrunner, H. and Mücke, E.P. (1994) *ACM Trans. Graph.*, **13** (1), 43–72.
[20] P. Labute, M. Santavy, *Locating Binding Sites in Protein Structures*, 2001, https://www.chemcomp.com/journal/sitefind.htm, (accessed January 2018).
[21] Peters, K.P., Fauck, J., and Frömmel, C. (1996) *J. Mol. Biol.*, **256** (1), 201–213.
[22] Le Guilloux, V., Schmidtke, P., and Tuffery, P. (2009) *BMC Bioinf.*, **10**, 168.
[23] Delaunay, B. (1934) *Izv. Akad. Nauk SSSR, Otd. Mat. i Estestv. Nauk*, 7.
[24] Zhou, W. and Yan, H. (2014) *Briefings Bioinf.*, **15** (1), 54–64.
[25] An, J., Totrov, M., and Abagyan, R. (2004) *Genome Inform.*, **15** (2), 31–41.
[26] Laurie, A.T.R. and Jackson, R.M. (2005) *Bioinformatics*, **21** (9), 1908–1916.
[27] Huang, N. and Jacobson, M.P. (2010) *PLoS One*, **5** (4).
[28] Ngan, C.-H., Hall, D.R., Zerbe, B., Grove, L.E., Kozakov, D., and Vajda, S. (2012) *Bioinformatics*, **28** (2), 286–287.
[29] Bray, T., Chan, P., Bougouffa, S., Greaves, R., Doig, A.J., and Warwicker, J. (2009) *BMC Bioinf.*, **10**, 379.
[30] Huang, B. and Schroeder, M. (2006) *BMC Struct. Biol.*, **6**, 19.
[31] Capra, J.A., Laskowski, R.A., Thornton, J.M., Singh, M., and Funkhouser, T.A. (2009) *PLoS Comput. Biol.*, **5** (12).
[32] Glaser, F., Rosenberg, Y., Kessel, A., Pupko, T., and Ben-Tal, N. (2005) *Proteins Struct. Funct. Genet.*, **58** (3), 610–617.
[33] Glaser, F., Morris, R.J., Najmanovich, R.J., Laskowski, R.A., and Thornton, J.M. (2006) *Proteins Struct. Funct. Genet.*, **62** (2), 479–488.
[34] Skolnick, J. and Brylinski, M. (2009) *Briefings Bioinf.*, **10** (4), 378–391.
[35] Halgren, T.A. (2009) *J. Chem. Inf. Model.*, **49**, 377–389.
[36] Huang, B. (2009) *OMICS*, **13** (4), 325–330.
[37] Zhang, Z., Li, Y., Lin, B., Schroeder, M., and Huang, B. (2011) *Bioinformatics*, **27** (15), 2083–2088.
[38] Kawabata, T. (2010) *Proteins Struct. Funct. Bioinform.*, **78** (5), 1195–1211.
[39] Yu, J., Zhou, Y., Tanaka, I., and Yao, M. (2009) *Bioinformatics*, **26** (1), 46–52.
[40] Laskowski, R.A., Luscombe, N.M., Swindells, M.B., and Thornton, J.M. (1996) *Protein Sci.*, **5** (12), 2438–2452.
[41] Cheng, A.C., Coleman, R.G., Smyth, K.T., Cao, Q., Soulard, P., Caffrey, D.R., Salzberg, A.C., and Huang, E.S. (2007) *Nat. Biotechnol.*, **25**, 71–75.
[42] Nissink, J.W.M., Murray, C., Hartshorn, M., Verdonk, M.L., Cole, J.C., and Taylor, R. (2002) *Proteins Struct. Funct. Genet.*, **49** (4), 457–471.

- [43] Wang, R., Fang, X., Lu, Y., and Wang, S. (2004) *J. Med. Chem.*, **47** (12), 2977–2980.
- [44] Kellenberger, E., Muller, P., Schalon, C., Bret, G., Foata, N., and Rognan, D. (2006) *J. Chem. Inf. Model.*, **46**, 717–727.
- [45] Xie, Z.R., Liu, C.K., Hsiao, F.C., Yao, A., and Hwang, M.J. (2013) *Nucleic Acids Res.*, **41** (Web Server issue).
- [46] Tripathi, A. and Kellogg, G.E. (2010) *Proteins Struct. Funct. Bioinform.*, **78** (4), 825–842.
- [47] Zhu, H. and Teresa Pisabarro, M. (2009) *Bioinformatics*, **2010** (1), 1–7.
- [48] Schmidtke, P., Bidon-Chanal, A., Luque, F.J., and Barril, X. (2011) *Bioinformatics*, **27** (23), 3276–3285.
- [49] Kokh, D.B., Richter, S., Henrich, S., Czodrowski, P., Rippmann, F., and Wade, R.C. (2013) *J. Chem. Inf. Model.*, **53** (5), 1235–1252.
- [50] Kokh, D.B., Czodrowski, P., Rippmann, F., and Wade, R.C. (2016) *J. Chem. Theory Comput.*, **12** (8), 4100–4113.
- [51] Cimermancic, P., Weinkam, P., Rettenmaier, T.J., Bichmann, L., Keedy, D.A., Woldeyes, R.A., Schneidman-Duhovny, D., Demerdash, O.N., Mitchell, J.C., Wells, J.A., Fraser, J.S., and Sali, A. (2016) *J. Mol. Biol.*, **428** (4), 709–719.
- [52] Brown, D. and Superti-Furga, G. (2003) *Drug Discovery Today*, **8** (23), 1067–1077.
- [53] Hopkins, A.L. and Groom, C.R. (2002) *Nat. Rev. Drug Discovery*, **1** (9), 727–730.
- [54] Sheridan, R.P., Maiorov, V.N., Holloway, M.K., Cornell, W.D., and Gao, Y.-D. (2010) *J. Chem. Inf. Model.*, **50** (11), 2029–2040.
- [55] Ward, R.A. (2010) *J. Mol. Model.*, **16** (12), 1833–1843.
- [56] Edfeldt, F.N.B., Folmer, R.H.A., and Breeze, A.L. (2011) *Drug Discovery Today*, **16** (7–8), 284–287.
- [57] Barril, X. (2013) *Wiley Interdiscip. Rev. Comput. Mol. Sci.*, **3**, 327–338.
- [58] Hajduk, P.J., Huth, J.R., and Tse, C. (2005) *Drug Discovery Today*, **10** (23–24), 1675–1682.
- [59] Pellecchia, M., Bertini, I., Cowburn, D., Dalvit, C., Giralt, E., Jahnke, W., James, T.L., Homans, S.W., Kessler, H., Luchinat, C., Meyer, B., Oschkinat, H., Peng, J., Schwalbe, H., and Siegal, G. (2008) *Nat. Rev. Drug Discovery*, **7** (9), 738–745.
- [60] Hajduk, P.J., Huth, J.R., and Fesik, S.W. (2005) *J. Med. Chem.*, **48** (7), 2518–2525.
- [61] Brenke, R., Kozakov, D., Chuang, G.Y., Beglov, D., Hall, D., Landon, M.R., Mattos, C., and Vajda, S. (2009) *Bioinformatics*, **25** (5), 621–627.
- [62] Zheng, C.J., Han, L.Y., Yap, C.W., Ji, Z.L., Cao, Z.W., and Chen, Y.Z. (2006) *Pharmacol. Rev.*, **58** (2), 259–279.
- [63] Han, L.Y., Zheng, C.J., Xie, B., Jia, J., Ma, X.H., Zhu, F., Lin, H.H., Chen, X., and Chen, Y.Z. (2007) *Drug Discovery Today*, **12** (7–8), 304–313.
- [64] Schmidtke, P., Le Guilloux, V., Maupetit, J., and Tufféry, P. (2010) *Nucleic Acids Res.*, **38** (Web Server issue), W582–W589.
- [65] Volkamer, A., Kuhn, D., Grombacher, T., Rippmann, F., and Rarey, M. (2012) *J. Chem. Inf. Model.*, **52** (2), 360–372.

[66] Nayal, M. and Honig, B. (2006) *Proteins Struct. Funct. Genet.*, **63** (4), 892–906.

[67] Weisel, M., Proschak, E., Kriegl, J.M., and Schneider, G. (2009) *Proteomics*, **9** (2), 451–459.

[68] Schmidtke, P. and Barril, X. (2010) *J. Med. Chem.*, **53** (15), 5858–5867.

[69] Krasowski, A., Muthas, D., Sarkar, A., Schmitt, S., and Brenk, R. (2011) *J. Chem. Inf. Model.*, **51** (11), 2829–2842.

[70] Perola, E., Herman, L., and Weiss, J. (2012) *J. Chem. Inf. Model.*, **52** (4), 1027–1038.

[71] Bakan, A., Nevins, N., Lakdawala, A.S., and Bahar, I. (2012) *J. Chem. Theory Comput.*, **8** (7), 2435–2447.

[72] Cuchillo, R., Pinto-Gil, K., and Michel, J. (2015) *J. Chem. Theory Comput.*, **11** (3), 1292–1307.

[73] Altschul, S.F., Gish, W., Miller, W., Myers, E.W., and Lipman, D.J. (1990) *J. Mol. Biol.*, **215** (3), 403–410.

[74] Lawrie, A.M., Noble, M.E., Tunnah, P., Brown, N.R., Johnson, L.N., and Endicott, J.A. (1997) *Nat. Struct. Biol.*, **4** (10), 796–801.

[75] Liu, Z., Huang, X., Hu, L., Pham, L., Poole, K.M., Tang, Y., Mahon, B.P., Tang, W., Li, K., Goldfarb, N.E., Dunn, B.M., McKenna, R., and Fanucci, G.E. (2016) *J. Biol. Chem.*, **291** (43), 22741–22756.

[76] An, J., Totrov, M., and Abagyan, R. (2005) *Mol. Cell. Proteomics*, **4** (6), 752–761.

[77] Kufareva, I., Ilatovskiy, A.V., and Abagyan, R. (2012) *Nucleic Acids Res.*, **40** (D1), D535–D540.

[78] Bietz, S. and Rarey, M. (2016) *J. Chem. Inf. Model.*, **56** (1), 248–259.

[79] Bietz, S. and Rarey, M. (2015) *J. Chem. Inf. Model.*, **55** (8), 1747–1756.

[80] Altschul, S.F., Madden, T.L., Schäffer, A.A., Zhang, J., Zhang, Z., Miller, W., and Lipman, D.J. (1997) *Nucleic Acids Res.*, **25** (17), 3389–3402.

[81] Finn, R.D., Bateman, A., Clements, J., Coggill, P., Eberhardt, R.Y., Eddy, S.R., Heger, A., Hetherington, K., Holm, L., Mistry, J., Sonnhammer, E.L.L., Tate, J., and Punta, M. (2014) *Nucleic Acids Res.*, **42**, D222–D230.

[82] Sigrist, C.J.A., Cerutti, L., Hulo, N., Gattiker, A., Falquet, L., Pagni, M., Bairoch, A., and Bucher, P. (2002) *Briefings Bioinf.*, **3** (3), 265–274.

[83] Henikoff, J.G., Greene, E.A., Pietrokovski, S., and Henikoff, S. (2000) *Nucleic Acids Res.*, **28** (1), 228–230.

[84] Attwood, T.K. (2002) *Briefings Bioinf.*, **3** (3), 252–263.

[85] Khan, S., Situ, G., Decker, K., and Schmidt, C.J. (2003) *Bioinformatics*, **19** (18), 2484–2485.

[86] Enault, F., Suhre, K., and Claverie, J.-M. (2005) *BMC Bioinf.*, **6** (1), 247.

[87] Ersgelhardt, B.E., Jordan, M.I., Muratore, K.E., and Brersfser, S.E. (2005) *PLoS Comput. Biol.*, **1** (5), 0432–0445.

[88] Krishnamurthy, N., Brown, D., and Sjölander, K. (2007) *BMC Evol. Biol.*, **7** (Suppl 1), S12.

[89] Illergård, K., Ardell, D.H., and Elofsson, A. (2009) *Proteins Struct. Funct. Bioinform.*, **77** (3), 499–508.

[90] Hubbard, T.J.P., Ailey, B., Brenner, S.E., Murzin, A.G., and Chothia, C. (1999) *Nucleic Acids Res.*, **27** (1), 254–256.

[91] Orengo, C.A., Michie, A.D., Jones, S., Jones, D.T., Swindells, M.B., and Thornton, J.M. (1997) *Structure*, **5** (8), 1093–1108.
[92] Holm, L. and Sander, C. (1994) *Nucleic Acids Res.*, **22** (17), 3600–3609.
[93] Ye, Y. and Godzik, A. (2003) *Bioinformatics*, **19** (Suppl 2), ii245-255.
[94] Täubig, H., Buchner, A., and Griebsch, J. (2006) *Nucleic Acids Res.*, **34** (Web Server issue), W20–W23.
[95] Gibrat, J.F., Madej, T., and Bryant, S.H. (1996) *Curr. Opin. Struct. Biol.*, **6** (3), 377–385.
[96] Wang, S., Peng, J., and Xu, J. (2011) *Bioinformatics*, **27** (18), 2537–2545.
[97] Wallace, A.C., Laskowski, R.A., and Thornton, J.M. (1996) *Protein Sci.*, **5** (6), 1001–1013.
[98] Kellenberger, E., Schalon, C., and Rognan, D. (2008) *Curr. Comput. Aided Drug Des.*, **4** (3), 209–220.
[99] Minai, R., Matsuo, Y., Onuki, H., and Hirota, H. (2008) *Proteins Struct. Funct. Genet.*, **72** (1), 367–381.
[100] Jambon, M., Imberty, A., Deléage, G., and Geourjon, C. (2003) *Proteins Struct. Funct. Genet.*, **52** (2), 137–145.
[101] Brakoulias, A. and Jackson, R.M. (2004) *Proteins Struct. Funct. Genet.*, **56** (2), 250–260.
[102] Shulman-Peleg, A., Nussinov, R., and Wolfson, H.J. (2004) *J. Mol. Biol.*, **339**, 607–633.
[103] Wallace, A.C., Borkakoti, N., and Thornton, J.M. (1997) *Protein Sci.*, **6** (11), 2308–2323.
[104] Milik, M., Szalma, S., and Olszewski, K.A. (2003) *Protein Eng.*, **16** (8), 543–552.
[105] Schmitt, S., Kuhn, D., and Klebe, G. (2002) *J. Mol. Biol.*, **323**, 387–406.
[106] Kuhn, D., Weskamp, N., Schmitt, S., Hüllermeier, E., and Klebe, G. (2006) *J. Mol. Biol.*, **359** (4), 1023–1044.
[107] Kinoshita, K., Furui, J., and Nakamura, H. (2002) *J. Struct. Funct. Genomics*, **2** (1), 9–22.
[108] Kinoshita, K., Murakami, Y., and Nakamura, H. (2007) *Nucleic Acids Res.*, **35** (Suppl 2).
[109] Najmanovich, R., Kurbatova, N., and Thornton, J. (2008) *Bioinformatics*, **24** (16), i105-11.
[110] Konc, J. and Janežič, D. (2012) *Nucleic Acids Res.*, **40** (W1), W214–W221.
[111] Yeturu, K. and Chandra, N. (2008) *BMC Bioinf.*, **9** (1), 543.
[112] Binkowski, T.A. and Joachimiak, A. (2008) *BMC Struct. Biol.*, **8**, 45.
[113] Yin, S., Proctor, E.A., Lugovskoy, A.A., and Dokholyan, N.V. (2009) *Proc. Natl. Acad. Sci. U.S.A.*, **106** (39), 16622–16626.
[114] Xiong, B., Wu, J., Burk, D.L., Xue, M., Jiang, H., and Shen, J. (2010) *BMC Bioinf.*, **11**, 47.
[115] Das, S., Kokardekar, A., and Breneman, C.M. (2009) *J. Chem. Inf. Model.*, **49** (12), 2863–2872.
[116] Liu, T. and Altman, R.B. (2011) *PLoS Comput. Biol.*, **7** (12), e1002326.
[117] Baroni, M., Cruciani, G., Sciabola, S., Perruccio, F., and Mason, J.S. (2007) *J. Chem. Inf. Model.*, **47** (2), 279–294.

[118] Schalon, C., Surgand, J.S., Kellenberger, E., and Rognan, D. (2008) *Proteins Struct. Funct. Genet.*, **71** (4), 1755–1778.

[119] Weill, N. and Rognan, D. (2010) *J. Chem. Inf. Model.*, **50** (1), 123–135.

[120] Morris, R.J., Najmanovich, R.J., Kahraman, A., and Thornton, J.M. (2005) *Bioinformatics*, **21** (10), 2347–2355.

[121] Sael, L., Chitale, M., and Kihara, D. (2012) *J. Struct. Funct. Genomics*, **13** (2), 111–123.

[122] Aung, Z. and Tong, J.C. (2008) *Genome Inform.*, **21**, 65–76.

[123] Östergård, P.R.J. (2002) *Discrete Appl. Math.*, **120** (1–3), 197–207.

[124] Krotzky, T., Fober, T., Hullermeier, E., and Klebe, G. (2014) *IEEE/ACM Trans. Comput. Biol. Bioinform.*, **11** (5), 878–890.

[125] Krotzky, T. and Klebe, G. (2015) *Mol. Inform.*, **34** (8), 550–558.

[126] Krotzky, T., Grunwald, C., Egerland, U., and Klebe, G. (2015) *J. Chem. Inf. Model.*, **55** (1), 165–179.

[127] Desaphy, J., Azdimousa, K., Kellenberger, E., and Rognan, D. (2012) *J. Chem. Inf. Model.*, **52** (8), 2287–2299.

[128] von Behren, M.M., Volkamer, A., Henzler, A.M., Schomburg, K.T., Urbaczek, S., and Rarey, M. (2013) *J. Chem. Inf. Model.*, **53** (2), 411–422.

[129] Wu, K. (2005) *J. Phys. Conf. Ser.*, **16**, 556–560.

[130] Chandonia, J.-M., Hon, G., Walker, N.S., Lo Conte, L., Koehl, P., Levitt, M., and Brenner, S.E. (2004) *Nucleic Acids Res.*, **32** (Database issue), D189–D192.

[131] Wood, D.J., de Vlieg, J., Wagener, M., and Ritschel, T. (2012) *J. Chem. Inf. Model.*, **52** (8), 2031–2043.

[132] Kalliokoski, T., Olsson, T.S.G., and Vulpetti, A. (2013) *J. Chem. Inf. Model.*, **53** (1), 131–141.

6.8 Structure-Based Virtual Screening

Adrian Kolodzik, Nadine Schneider, and Matthias Rarey

Universität Hamburg, ZBH – Center for Bioinformatics, Bundesstraße 43, 20146 Hamburg, Germany

Learning Objectives

- To discuss the principles of virtual screening
- To review common docking tools
- To distinguish common scoring functions
- To validate virtual screening results

Outline

6.8.1 Introduction, 313
6.8.2 Docking Algorithms, 315
6.8.3 Scoring, 317
6.8.4 Structure-Based Virtual Screening Workflow, 321
6.8.5 Protein-Based Pharmacophoric Filters, 323
6.8.6 Validation, 323
6.8.7 Summary and Outlook, 326

6.8.1 Introduction

Whenever a structure of a protein of interest is available, computational tools can be used to identify small molecules interfering with the protein's function. The process structure-based virtual screening (SBVS) is a standard technique used in drug discovery as well as in other fields in which identification of bioactive compounds is of interest, for example, agrochemistry or biotechnology. In SBVS projects millions of chemical compounds are docked into target protein structures to find new promising lead compounds. It can be regarded as the computational analog to experimental high-throughput screening (HTS) where millions of new chemical structures are automatically analyzed in activity assays. SBVS methods are less cost intensive and capable of testing very large chemical compound libraries, which might include molecules not yet physically available, in an acceptable time period. The results of SBVS are of similar quality compared with those from experimental HTS [1]. There are several success stories of applied SBVS methods for different protein classes. A short overview is given in Table 6.8.1, and a more detailed description as well as additional examples can be found in [17].

An SBVS approach requires a three-dimensional (3D) structure of the target protein – for example, given by crystallography, NMR spectroscopy, or homology modeling – as well as a virtual library of chemical compounds. A variety of

Table 6.8.1 Successes in structure-based virtual screening.

Protein class	Target protein	Docking tool	Scoring function	Compound library	SBVS hits	Experimental hits[a]	References
Kinases	CK2	DOCK [2]	DOCKScore, SCORE[b]	~400,000 Novartis corporate database	12[c][d]	4	[3]
	Bcr-Abl tyrosine kinase	DOCK	DOCKScore	200,000 commercial available compounds	15[c]	8	[4]
	CHK-1	FlexX-Pharm [5]	Consensus scoring	~200,000 of Astra Zeneca collection	103[d]	36	[6]
Proteases	Dipeptidyl peptidase IV	Glide [7]	GlideScore	~800,000 (~20,000)[e]	4000[c]	51	[8]
	Cysteine protease falcipain-2	Glide, GAsDock [9]		Specs database [10]	81	28	[11]
Nuclear receptors	Thyroid hormone receptor	ICM [12]		~250,000	75	14	[13]
GPCRs	α1A-Adrenergic receptor	Gold [14]		~23,000 Aventis compound library	80[f]	37	[15]

a) Experimental validated hits.
b) Scoring function by Wang et al. [16].
c) Further filtering and clustering steps were applied after the docking procedure.
d) Visual inspection of results before biological testing.
e) Started with 80,000 after filtering protocols and pharmacophore search 20,000 were docked.
f) After postscoring.

different docking algorithms have been developed to place the compounds in the predefined active site of the target protein (reference). The specific orientations of the ligands in the context of a protein are called poses. To find the active conformation of potential lead candidates among the pool of poses, a scoring function is necessary. Additionally, the scoring function should be able to discriminate active from inactive compounds. Until now, more than 60 docking programs and at least 30 scoring functions have been developed, but only some of these programs (AutoDock, DOCK, FlexX, FRED, Glide, GOLD, ICM) have achieved broad acceptance in the chemical computation community [18].

This section summarizes basic knowledge about docking techniques and scoring methods. Section 6.8.2 explains current docking algorithms. Section 6.8.3 introduces several scoring methods used during the docking procedure as well as scoring functions that are applied for rescoring of the docking results. Section 6.8.4 presents an SBVS workflow, and Section 6.8.5 introduces pharmacophoric filters. Finally, in Section 6.8.6 we present validation methods to measure the quality of SBVS results and discuss their applicability.

6.8.2 Docking Algorithms

Since the most commonly active ingredients in drugs are small molecule protein binders, protein–ligand docking is frequently applied during drug development in pharmaceutical research. Although the interest in protein–protein interactions rose in recent years, macromolecular docking like protein–protein and protein–DNA docking are less commonly used and will not be discussed here. For more information about these techniques, please see [19] and [20].

The development of docking programs started more than three decades ago. The first models of the docking problem, which were developed aside from molecular dynamics simulations, were rather simple and assumed rigid molecules [21], which severely restricted their predictive power. In the 1990s a number of more sophisticated programs arose, aiming at modeling the docking problem more accurately. These docking programs use three different strategies to place a flexible ligand into the rigid binding site of the protein:

1. *Stochastic methods* generate random conformations of ligands, which are randomly translated and rotated within the binding site. The resulting pose is then evaluated by a scoring function. To identify good poses, different approaches are available (see next section). Due to random rotations and translations and the probabilistic acceptance of these changes, solutions may vary for subsequent docking runs. Three well-known docking programs using this strategy are AutoDock [22, 23], GOLD [24, 25], and PLANTS [26, 27]. AutoDock uses *simulated annealing* [28] to find a good pose. If a pose has an improved score compared with the former one, it is always accepted. Poses with inferior scores are accepted following the so-called Metropolis criterion [29]. This depends on a temperature parameter, which is stepwise reduced (cooled down). Thus, simulated annealing accepts almost all poses at the beginning and is more stringent at later steps of the docking run. AutoDock Vina [30] is the newest

and enhanced version of AutoDock 4 with improved accuracy in binding mode prediction and lower runtimes. GOLD uses a genetic algorithm (GA) to generate poses. The stochastic GA interprets the conformation of the ligand, the conformation of the protein, and the hydrogen bonds as a single chromosome. Chromosomes are then randomly changed (mutated) and combined (crossover) to produce new solutions. A fitness function evaluates the scores of the docking poses encoded by the chromosomes and favors those with higher fitness as the basis for further mutations. PLANTS uses a heuristic and iterative ant colony optimization (ACO) approach [31] for the generation of poses. Initially a group of poses is generated, scored, and locally optimized. Similar to ants that follow a path of pheromones to a food source, the ACO approach favors translations, rotations, and torsions of poses that have shown good scoring results in previous iterations. Therefore, poses can be iteratively optimized.

2. *Fragmentation-based* methods cut the ligand at rotatable bonds in smaller fragments in order to accurately deal with the conformational flexibility of the ligand. After placing initial fragments, the ligands are subsequently rebuilt in the protein's active site. Each fragment is scored during the construction process. In contrast to stochastic methods like simulated annealing and GA, the construction process is deterministic and therefore allows reproduction of results. FlexX [32] and eHiTS [33] are two typical tools using this strategy. FlexX places an initial fragment into the active site of the protein. Starting from this base fragment, FlexX subsequently rebuilds the ligand. A greedy approach is used in the incremental construction process to limit the number of solutions to several hundred with reasonably high scores. Hence, the algorithm is heuristic in nature. LeadIT [34] is the successor of FlexX and provides improved docking algorithms and more sophisticated scoring combined with a graphical user interface. eHiTS breaks ligands into rigid fragments and connects flexible chains. Rigid fragments are independently docked into the active site. A clique detection algorithm identifies compatible sets of fragment poses. Since the scores of the individually placed fragments are already calculated at this point, the globally best combination of the initially placed rigid fragments can be determined. Finally, the flexible chains are fitted to the rigid fragments to form a rough binding pose, which is then refined by a local energy minimization.

3. *Conformational ensemble* methods generate multiconformer ligand libraries in a first step. These conformations are then placed into the active site of the protein. Examples of docking tools following this strategy are Glide [7, 35], FRED [36, 37], and TrixX [38, 39]. Glide generates conformers by exhaustively enumerating the global energy minima in the torsion-angle space of the ligand. These low energy conformations are used for a fast prescreening in the active site of the protein. Best poses from prescreening are energetically minimized in a first step and then subjected to a Monte Carlo procedure to identify nearby torsional minima. Finally, the best poses are scored by a combination of a scoring function, a molecular mechanics energy calculation, and the calculated ligand strain energy. FRED uses OMEGA [40] as a conformer generator. For each conformer, all rotations and translations are systematically generated with a specified resolution. Poses clashing with the rigid protein or being

role in the protein–ligand binding process and are typically modeled in a scoring function as a sum of individual terms describing the different interactions.

In general, scoring functions can be categorized into three different types:

1. *Force field-based scoring functions* use classical molecular force field terms to estimate the interactions in protein–ligand complexes. Torsional energy, Lennard–Jones potential, and electrostatics are the common energies that are calculated in molecular mechanics force fields like AMBER [53] (designed for proteins) or MMFF [54] (designed for small molecules). Atoms are classified and the resulting atom types are used for calibrating the parameters of the system. These scoring functions are often used to guide the docking algorithm during pose construction. GoldScore [55, 56] is an example for such a force field-based scoring function. It is made up of the following four terms (Eq. 6.8.2):

$$\text{GoldScore} = E_{\text{H-bond,ext}} + E_{\text{vdW,ext}} + E_{\text{vdW,int}} + E_{\text{torison,int}} \qquad (6.8.2)$$

The van der Waals and hydrogen-bond energies between the protein and the ligand are calculated ($E_{\text{H-bond, ext}}$, $E_{\text{vdW,ext}}$) as well as the van der Waals and torsional strain energies of the ligand ($E_{\text{vdW,int}}$, $E_{\text{torsion,int}}$). The strength of force field-based scoring functions lies rather in pose prediction than in ranking or binding affinity prediction due to the missing of terms accounting for entropy and the overestimation of electrostatic interactions and hydrogen bonds.

2. *Knowledge-based scoring functions* take advantage of the continuously growing amount of experimentally resolved protein–ligand complex structures. These scoring functions rely on pairwise atom potentials calculated from a statistical analysis of common interactions/distances occurring in protein–ligand complexes. The final score is calculated as the sum of potential values for all atom pairs within a certain distance cutoff. Here, the disadvantage is that rare interactions, like cation–π or halogen interactions, are not well parameterized. An example for this kind of scoring functions is the potential of mean force (PMF) score (Eq. 6.8.3):

$$\text{PMF} = \sum_{kl} A_{ij}(r) \qquad (6.8.3)$$

A_{ij} is a protein–ligand atom pair interaction free energy at distance r and kl is a ligand–protein atom pair of type ij. The PMF function was developed by Muegge and Martin [57] and was reparameterized in 2006 on a large dataset of protein–ligand crystal structures (7152 protein–ligand complexes from the PDB) [58]. Other examples of this category of scoring functions are DrugScore[PDB] [59] or DrugScore[CSD] [60], which were derived from statistical analysis of protein–ligand complexes from the PDB and small organic molecules in the Cambridge Structural Database [61].

3. *Empirical scoring functions* are the most common type of scoring functions. The binding free energy is calculated as a sum of terms representing different physicochemical effects found in protein–ligand complexes such as hydrogen bonding, hydrophobic effect, or metal interactions. These scoring functions were usually calibrated on experimentally measured binding affinities for complexes with known structure. The first scoring function of this type

was developed by Boehm [46]. Boehm's function was further developed to the widely used ChemScore [47] function, of which the original version is shown as follows (Eq. 6.8.4):

$$\Delta G_{binding} = \Delta G_0 + \Delta G_{hbond} + \Delta G_{metal} + \Delta G_{lipo} + \Delta G_{rot} \quad (6.8.4)$$

$\Delta G_{binding}$ is the estimated free binding energy. Several terms contribute to $\Delta G_{binding}$ such as hydrogen bonding (ΔG_{hbond}), metal interactions (ΔG_{metal}), lipophilic interactions (ΔG_{lipo}), and a term that penalizes the flexibility of the ligand (ΔG_{rot}). The constant term (ΔG_0) arises during calibration on experimentally measured binding. The original ChemScore used a training set of 82 protein–ligand complexes with known binding affinity and reliable binding geometry from the PDB. Another well-known scoring function of this type is the X-Score scoring function [62]. The largest drawback of empirical scoring functions is that they perform better on protein families close to those used in the calibration dataset [63].

In addition to scoring functions fitting to one of the described three categories, alternative ways to estimate the free binding affinity have been developed. Some of these approaches try to model the underlying physics in a more straightforward manner. They do not calibrate the scoring function on protein–ligand structures and measured binding affinity, so that the scoring function is more general. Kellogg et al. have developed the HINT interaction force field [64] that is derived from the partition coefficient of 1-octanol/water (log P). It includes hydrogen bonding, electrostatic interactions, hydrophobic contacts, entropy, and solvation/desolvation. The HYDE [65, 66] scoring function also uses log P values of small molecules for calibration. It consistently describes hydrogen bonds, the hydrophobic effect, and desolvation in protein–ligand complexes.

Another strategy is to train a scoring function especially for one system; these approaches are called tailored or target-specific scoring functions [67]. Here, the scoring function is calibrated on already known ligands or crystal structures of a specific protein target, and therefore the prediction power for this target is increased. The disadvantage is that this approach cannot be used when a new drug design projects starts in which only limited information is available. During the development process, it can be used to rank leads against each other as well as to identify new promising molecules. An example is the ISAC approach [68], which uses the information of activity cliffs to derive pharmacophore hypotheses and target-specific scoring functions.

Another alternative to achieve more reliable results in the scoring of protein–ligand complexes is to use so-called consensus scoring. This strategy combines the results of different existing scoring functions to get a more reliable prediction [69]. Although reliability often increases, consensus scoring typically also reduces the number of identified true actives [70].

Other computationally more demanding approaches, such as molecular mechanics Poisson–Boltzmann/generalized Born surface area (MM-PB/GB-SA), were applied for rescoring of protein–ligand complexes. These scoring functions are more physics based, combining short molecular dynamics simulation runs to account for the flexibility of the system and the change in molecular surface to calculate desolvation effects. Brown and Muchmore [71] have shown promising results with their MM-PBSA method, achieving good correlations between experimental affinity and predicted scores. For the three tested datasets, the Pearson correlation coefficients (R) are in the range of 0.72–0.83. However, these approaches are still computationally too cost intensive to apply them in the context of large-scale virtual screening. The same also applies for free energy calculation methods, like free energy perturbation (FEP) or thermodynamic integration (TI [72]), which have gained more interest recently (for a review, see Chipot and Pohorille [73]). These methods calculate relative energy differences between similar ligands bound to the same protein target to estimate the binding affinity. By modeling the underlying system (protein–ligand complex) more thoroughly, these methods seem to be a promising alternative to simple scoring functions. However, they afford a huge expertise to obtain reliable results [74].

Due to the importance of water molecules in the binding process, novel methods were recently developed to explicitly model and analyze water molecules in the binding pocket. WaterMap [75, 76], for instance, uses molecular dynamics simulations of explicit water molecules to identify regions with high and low water densities in the binding pocket across simulations. It further estimates enthalpy and entropy of these water molecules. Therefore it could assist in the decision which of those water molecules could be replaced, resulting in a gain in binding energy. On the other hand, experimental data of high-resolution X-ray crystallography can be used to statistically analyze water molecules in protein–ligand complexes. One challenge here is to include only water molecules with experimental evidence, meaning measured electron density. A novel descriptor called EDIA was recently developed [77] to quantitatively assess the quality of the electron density of these small molecules.

6.8.4 Structure-Based Virtual Screening Workflow

After the introduction of docking and scoring, the following section describes a typical SBVS workflow. SBVS can be regarded as a sophisticated filtering step during the drug discovery process.

The SBVS workflow involves four basic steps (see Figure 6.8.2). At first, the ligands of the chemical library that will be screened are checked for chemical correctness. In addition, tautomeric forms (see *Methods Volume*, Chapter 3), protonation states, and conformations are generated as required by the utilized docking software.

In a second step, the binding site is defined and prepared by optimizing the protein's internal hydrogen-bonding network. Furthermore, conserved water molecules and important cofactors are usually included.

The third step is the generation of poses by placing the compounds into the binding site of the protein. Because of the huge number of possible translations, rotations, and conformations of each compound, only a limited number of possible poses can be generated. The generation of a pose can be guided using pharmacophore-type filters, which will be discussed in more detail in the following section.

In the fourth step, every pose is scored by a scoring function that predicts the binding affinity of the ligand to the protein, assuming that the pose represents the predominant binding mode. Depending on the best-scored pose, a compound

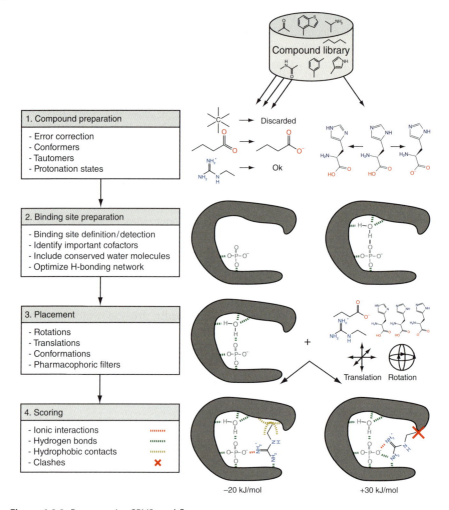

Figure 6.8.2 Best practice SBVS workflow.

is ranked high in the resulting hit list. Furthermore, a cutoff can be applied to the score in order to classify compounds into actives and inactives. Predicted actives can then be filtered and optimized for high binding affinities and improved physicochemical properties. Compounds predicted active and optimized in this way represent promising candidates for synthesis and testing.

Additional filtering steps can be applied before and after virtual screening runs. Filtering beforehand is meaningful due to the high computational demands of SBVS. Usually, these filters limit a molecule's physicochemical properties, like molecular weight, log P, and number of hydrogen-bond donors and acceptors. Depending on the application scenario, also more sophisticated filters like substructures or ligand-based pharmacophores are employed. These filters can drastically reduce the chemical space [78] and therefore speed up the virtual screening run.

Filtering can also be applied after the docking and scoring phase. These post filters usually take the potential binding mode of a compound into account. A common post filter is constructed from protein-based pharmacophoric constraints, which are described in the next section.

6.8.5 Protein-Based Pharmacophoric Filters

A protein-based *pharmacophoric filter* describes the features of a molecule that are responsible for its molecular activity [79]. A feature can be a substructure or a physiochemical property that is required for binding at a certain position of the protein.

If a pharmacophoric filter is carefully designed and supported by experimental data, the knowledge introduced into the docking process by these constraints can significantly increase the quality and the speed of predictions. It is either used during pose generation to reduce the number of poses that have to be scored for a ligand or afterward upon hit list generation. Since each pharmacophoric filter can represent a different binding hypothesis, ligands passing a pharmacophoric filter can be interpreted as a cluster of ligands having similar binding motifs. A drawback of using a pharmacophoric filter is the restriction of solutions to these known binding motifs. Therefore the filter restricts a docking program's ability to discover new binding modes.

6.8.6 Validation

As there are many docking programs that are suitable for different kinds of target proteins, there is a need for methods to evaluate and compare the performance of the different approaches. Redocking experiments evaluate the quality of predicted poses. Screening experiments measure the ability of the docking software for selecting potential actives. A measure of success for redocking is usually the root mean square deviation (RMSD), while enrichment factors (EF) and receiver operating characteristic (ROC) curves [80] evaluate a docking software's performance with respect to sensitivity and specificity. These methods, which are commonly used in publications concerning docking, will be discussed in more detail in the following.

Redocking experiments evaluate a docking program's ability to reproduce binding modes observed by X-ray crystallography. Under the assumption that the crystal structure represents the correct binding mode, a docking run is considered successful, if the deviation of the ligand atom positions in the pose compared with the crystal structure is low.

To perform a redocking experiment, the ligand is separated from the protein–ligand complex. The conformation of the ligand should be randomized to avoid a biased input conformation. Afterward, the ligand is processed using a standard protocol suitable for the applied docking software and docked into the binding site of the protein. A number of metrics are available to describe the deviation of the predicted pose from the observed pose in the crystal structure. The most commonly used metric is the RMSD (Eq. 6.8.5):

$$\text{RMSD}(P, X) = \sqrt{\frac{1}{N} \sum_{\text{Atomsi}} (p_i - x_i)^2} \qquad (6.8.5)$$

Here, N is the number of ligand atoms, and P and X are the atom coordinate vectors of the ligand in the pose and crystal structure, respectively.

Although the RMSD is widely used, it is questionable if it is a suitable criterion for the evaluation of a docking software's quality [81] since it depends on the size and the shape of a molecule. Furthermore, a crystal structure only represents a snapshot of all possible states of a protein–ligand complex. Therefore, deviations of a predicted pose to the co-crystallized ligand are not necessarily an indication of a wrong prediction. To circumvent at least a part of the RMSD's shortcomings as a measure of redocking performance, a number of alternatives have been suggested [82–84]. For example, to circumvent the size dependencies, one can consider redocking experiments successful if all interactions that are predicted for the crystal structure are also predicted for the docking solutions.

An *EF* [85] describes how a docking program performs in identifying true actives compared with a random selection. To calculate an EF, a dataset of known actives and inactives is docked into the binding site of a protein. Then a subset (Compounds$_{\text{selected}}$) of molecules with highest predicted binding affinities (first $x\%$) is chosen from the whole dataset (Compounds$_{\text{total}}$).The number of actives in the subset (Actives$_{\text{selected}}$) is divided by the number of actives in the whole dataset (Actives$_{\text{total}}$) (Eq. 6.8.6):

$$\text{EF} = \frac{\text{Compounds}_{\text{total}}}{\text{Compounds}_{\text{selected}}} \times \frac{\text{Actives}_{\text{selected}}}{\text{Actives}_{\text{total}}} \qquad (6.8.6)$$

An *enrichment plot* illustrates the percentage of identified true actives according to the size of the chosen subset. Figure 6.8.3b shows an enrichment plot to compare two docking tools. Both tools show superior enrichment compared with randomly choosing ligands as active. The x-axis has a logarithmic scale to give a better impression of the quality of the prediction for ligands on the first ranks. Low-percentage values are most important since in many drug design processes, a huge number of ligands are screened, but only a small percentage can be tested *in vitro* or *in vivo*. Consequently, it is especially important to identify truly active compounds among ligands having the highest docking scores. In this regard, docking program A (black) is superior to docking program B (light gray).

A *score histogram* illustrates the ability of a docking software to separate active from inactive compounds. Two curves are depicted that represent docking scores of truly active compounds and docking scores of truly inactive compounds (see Figure 6.8.3a). Truly active compounds that are predicted to be "active" are called *true positives*. Inactive compounds that are predicted to be active are called *false positives*. Correspondingly, truly inactive compounds that are correctly predicted to be inactive are called *true negatives*, and active compounds that are predicted to be inactive are called *false negatives*.

Correctly and incorrectly predicted actives and inactives allow the calculation of the sensitivity and specificity of a docking program. Sensitivity is a synonym

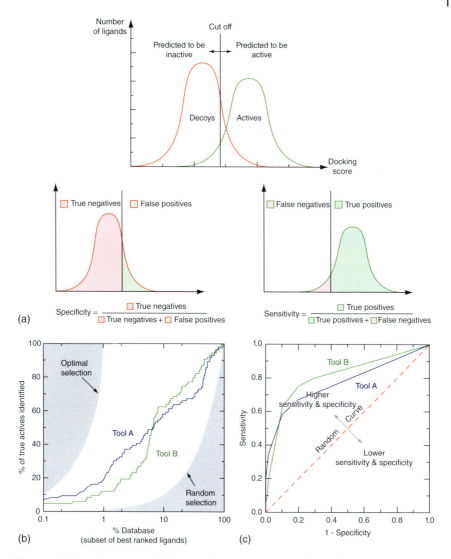

Figure 6.8.3 Score histograms (a) allow an intuitive assessment of a docking program's sensitivity and specificity for a defined cutoff. Enrichment plots (b) and ROC curves (c) are used to assess a docking program's quality. The example shows two hypothetical docking runs for a library of 10,000 compounds containing 100 active ligands.

for "true positive rate," "recall," and "hit rate." A docking tool with high sensitivity correctly predicts most actives. A trivial way to achieve high sensitivity is to predict every ligand to be active. With this approach, one would however not be able to distinguish actives from inactives. Therefore, a high specificity is required in addition to high sensitivity. Specificity is a synonym for "true negative rate" and "correct rejection rate" (compare Figure 6.8.3):

$$\text{Sensitivity} = \frac{\text{True posivitives}}{\text{True positives} + \text{True negatives}} \tag{6.8.7}$$

$$\text{Specificity} = \frac{\text{True negatives}}{\text{True negatives} + \text{False positives}} \qquad (6.8.8)$$

A docking tool with high specificity would identify inactives correctly. Perfect docking software would classify all actives and inactives correctly, resulting in a high sensitivity and specificity. The combination of sensitivity and specificity can be easily illustrated using an ROC curve (see Figure 6.8.3c). The area under the curve (AUC) is an indicator of the overall virtual screening performance. The optimal value here is AUC = 1. For a random selection of actives and inactives, the average AUC is 0.5.

Unfortunately, scoring functions are still not perfect. In a drug discovery project, the number of available compounds is usually large, while the capacities of activity assays are limited. Therefore, it is often acceptable to miss some actives (lower sensitivity) in order to increase the probability that a predicted active is truly active (higher specificity).

Nevertheless, one has to keep in mind that this depends on the underlying application scenario. If, for example, docking is used to predict possible side effects of a drug candidate by docking it into the cavities of a number of human proteins, the opposite strategy might be more suitable. In this case one would not like to miss any possible side effect, but it would be acceptable to predict more side effects than will be observed in the lab or in the clinic.

6.8.7 Summary and Outlook

SBVS has been established as a useful tool in the drug discovery process. Often it is a complement to HTS or even an alternative, achieving hit rates that are 100- to 1000-fold better than those produced in HTS [86]. However, SBVS cannot be used as a black box tool, and expert knowledge about the protein, its functional mechanism, and the essential interactions is indispensable to obtain reasonable results. These results also require the visual inspection and interpretation by a medicinal chemist.

Additionally, there are still many simplifications in the computer-based models used today. Flexibility of the protein target is still an unsolved problem. There exist many approaches that try to model flexibility in different manners, but most of these approaches have increased rates of false positives [43]. Additional problems that are currently investigated are different protonation states and different tautomeric forms of the molecules, which cannot be seen in crystal structures (see *Methods Volume*, Section 2.5). These have a huge impact on a molecule's binding affinity. Methods exist, which address the problem by sampling possible states, but this is a compromise between accuracy and computing time [87–91]. Another challenge is modeling entropy. Although this has been studied for decades, most docking tools only consider the freezing of rotatable bonds during the formation of the protein–ligand complex. The more significant part of the entropy arises from the release of ordered water molecules from the binding pocket as well as from the solvation shell of the ligand. This part of the entropic contribution to the binding energy is mostly neglected. As an approximation the change of the hydrophobic surface of the protein–ligand interface is considered [65, 92].

Despite a number of challenges, there have been major advances in the field of docking, and various successful applications have been published. These successes are especially impressive since the field of docking is still quite young compared with more traditional chemical research. Furthermore, the exponential growth of current computers' performance allows increasingly sophisticated models to be implemented. This will probably not only lead to more accurate predictions of binding affinities but also to docking tools that require less manual intervention.

Essentials

- SBVS is a standard technique used in drug discovery.
- It requires a 3D structure of a protein and predicts potentially active compounds.
- A number of docking tools are available, which place ligands in the active site and score the resulting poses.
- Scoring functions estimate the energies of protein–ligand complexes and take into account ionic interactions, hydrogen bonds, and hydrophobic contacts as well as desolvation effects and entropy.
- A number of validation procedures have been developed to evaluate the performance of the different tools. These include redocking experiments, enrichment factors, and ROC curves.

Available Software and Web Services (accessed January 2018)

- DOCK Blaster – A Free Virtual Screening Server; http://blaster.docking.org/
- Open Data Drug & Drug Target Database; http://drugbank.ca/
- X-Score (scoring function); Comprehensive Cancer Center, University of Michigan; http://sw16.im.med.umich.edu/software/xtool/
- DOCK 6.5 (docking software); University of California, San Francisco; http://dock.compbio.ucsf.edu/Online_Licensing/index.htm
- GOLD (docking software); Cambridge Crystallographic Data Centre (CCDC), Cambridge; http://www.ccdc.cam.ac.uk/products/life_sciences/gold/
- AutoDock 4 and AutoDock Vina (docking software); Molecular Graphics Lab at The Scripps Research Institute, La Jolla; http://vina.scripps.edu/
- Glide (docking software); Schrödinger, New York; http://www.schrodinger.com/products/14/5/
- WaterMap; Schrödinger, New York; https://www.schrodinger.com/watermap
- FRED (docking software); OpenEye Scientific Software, Santa Fe; http://www.eyesopen.com/fred
- LeadIT (docking software); BioSolveIT GmbH, Sankt Augustin; http://www.biosolveit.de/LeadIT/
- ICM (docking software); Molsoft L.L.C., San Diego; http://www.molsoft.com/

- Surflex (docking software); Tripos International, St. Louis; https://www.certara.com/
- PLANTS, University of Tübingen; http://www.tcd.uni-konstanz.de/plants_download/

Selected Reading

- Bissantz, C., Kuhn, B., and Stahl, M. (2010) A medicinal chemist's guide to molecular interactions. *J. Med. Chem.*, **53**, 5061–5084.
- Henzler, A.M. and Rarey, M. (2010) In pursuit of fully flexible protein–ligand docking: modeling the bilateral mechanism of binding. *Mol. Inf.*, **29**, 164–173.
- Klebe, G. (2006) Virtual ligand screening: strategies, perspectives, and limitations. *Drug Disc. Today*, **11**, 580–594.
- Sotriffer, C. (2011) *Virtual Screening*, Weinheim, Germany, Wiley-VCH, 550 pp.

References

[1] Polgar, T., Baki, A., Szendrei, G.I., and Keseru, G.M. (2005) *J. Med. Chem.*, **48**, 7946–7959.
[2] Ewing, T.J.A., Makino, S., Skillman, A.G., and Kuntz, I.D. (2001) *J. Comput.-Aided Mol. Des.*, **15**, 411–428.
[3] Vangrevelinghe, E., Zimmermann, K., Schoepfer, J., Portmann, R., Fabbro, D., and Furet, P. (2003) *J. Med. Chem.*, **46**, 2656–2662.
[4] Peng, H., Huang, N., Qi, J., Xie, P., Xu, C., Wang, J., and Yang, C. (2003) *Bioorg. Med. Chem. Lett.*, **13**, 3693–3699.
[5] Hindle, S.A., Rarey, M., Buning, C., and Lengauer, T. (2002) *J. Comput.-Aided Mol. Des.*, **16**, 129–149.
[6] Lyne, P.D., Kenny, P.W., Cosgrove, D.A., Deng, C., Zabludoff, S., Wendoloski, J.J., and Ashwell, S. (2004) *J. Med. Chem.*, **47**, 1962–1968.
[7] Friesner, R.A., Banks, J.L., Murphy, R.B., Halgren, T.A., Klicic, J.J., Mainz, D.T., Repasky, M.P., Knoll, E.H., Shelley, M., Perry, J.K., Shaw, D.E., Francis, P., and Shenkin, P.S. (2004) *J. Med. Chem.*, **47**, 1739–1749.
[8] Ward, R.A., Perkins, T.D.J., and Stafford, J. (2005) *J. Med. Chem.*, **48**, 6991–6996.
[9] Li, H., Li, C., Gui, C., Luo, X., Chen, K., Shen, J., Wang, X., and Jiang, H. (2004) *Bioorg. Med. Chem. Lett.*, **14**, 4671–4676.
[10] Specs (2008) Chemistry Solutions for Drug Discovery, http://www.specs.net (accessed January 2018).
[11] Li, H., Huang, J., Chen, L., Liu, X., Chen, T., Zhu, J., Lu, W., Shen, X., Li, J., Hilgenfeld, R., and Jiang, H. (2009) *J. Med. Chem.*, **52**, 4936–4940.
[12] Totrov, M. and Abagyan, R. (1997) *Proteins*, **29**, 215–220.
[13] Schapira, M., Raaka, B.M., Das, S., Fan, L., Totrov, M., Zhou, Z., Wilson, S.R., Abagyan, R., and Samuels, H.H. (2003) *PNAS*, **100** (12), 7354–7359.

[14] Jones, G., Willett, P., Glen, R.C., Leach, A.R., and Taylor, R. (1997) *J. Mol. Biol.*, **267**, 727–748.
[15] Evers, A. and Klabunde, T. (2005) *J. Med. Chem.*, **48**, 1088–1097.
[16] Wang, R., Liu, L., Lai, L., and Tang, Y. (1998) *J. Mol. Model.*, **4**, 379–394.
[17] Sotriffer, C. (2011) *Virtual Screening*, Wiley-VCH Verlag GmbH & Co. KGaA, Weinheim, 550 pp.
[18] Moitessier, N., Englebienne, P., Lee, D., Lawandi, J., and Corbeil, C.R. (2008) *Br. J. Pharmacol.*, **153**, 7–26.
[19] van Dijk, M. and Bonvin, A.M.J.J. (2010) *Nucleic Acids Res.*, **38** (17), 5634–5647.
[20] Moreira, I.S., Fernandes, P.A., and Ramos, M.J. (2010) *Comput. Chem.*, **31** (2), 317–342.
[21] Kuntz, I.D., Blaney, J.M., Oatley, S.J., Langridge, R., and Ferrin, T.E. (1982) *J. Mol. Biol.*, **161** (2), 269–288.
[22] Goodsell, D.S. and Olson, A.J. (1990) *Proteins*, **8** (3), 195–202.
[23] Morris, G., Goodsell, D., Halliday, R., Huey, R., Hart, W., Belew, R., and Olson, A.J. (1998) *J. Comput. Chem.*, **19**, 1639.
[24] Jones, G., Willett, P., and Glen, R.C. (1995) *J. Mol. Biol.*, **245**, 43–53.
[25] Verdonk, M.L., Chessari, G., Cole, J.C., Hartshorn, M.J., Murray, C.W., Nissink, J.W.M., Taylor, R.D., and Taylor, R. (2005) *J. Med. Chem.*, **48**, 6504–6515.
[26] Korb, O., Stuetzle, T., and Exner, T.E. (2006) *Lect. Notes Comput. Sci.*, **4150**, 247–258.
[27] Korb, O., Stuetzle, T., and Exner, T.E. (2009) *J. Chem. Inf. Model.*, **49**, 84–96.
[28] Eglese, R. (1990) *Eur. J. Oper. Res.*, **46** (3), 271–281.
[29] Metropolis, N., Rosenbluth, A., Rosenbluth, M., Teller, A., and Teller, E. (1953) *J. Chem. Phys.*, **21** (6), 1087–1092.
[30] Trott, O. and Olson, A.J. (2010) *J. Comput. Chem.*, **31**, 455–461.
[31] Dorigo, M. and Stützle, T. (2004) *Ant Colony Optimization*, MIT Press/Bradford Books, Cambridge, MA. ISBN: 0-262-04219-3
[32] Rarey, M., Kramer, B., Lengauer, T., and Klebe, G. (1996) *J. Mol. Biol.*, **261** (3), 470–489.
[33] Zsoldos, Z., Reid, D., Simon, A., Sadjad, B.S., and Johnson, A.P. (2006) *Curr. Protein Pept. Sci.*, **7** (5), 421–435.
[34] LeadIT, BioSolveIT GmbH, Germany, http://www.biosolveit.de/LeadIT (accessed January 2018).
[35] Halgren, T.A., Murphy, R.B., Friesner, R.A., Beard, H.S., Frye, L.L., Pollard, W.T., and Banks, J.L. (2004) *J. Med. Chem.*, **47**, 1750–1759.
[36] McGann, M.R., Almond, H.R., Nicholls, A., Grant, J.A., and Brown, F.K. (2003) *Biopolymers*, **68**, 76–90.
[37] McGann, M.R. (2011) *J. Chem. Inf. Model.*, **51** (3), 578–596.
[38] Schellhammer, I. and Rarey, M. (2007) *J. Comput.-Aided Mol. Des.*, **21** (5), 223–238.
[39] Henzler, A.M., Urbaczek, S., Hilbig, M., and Rarey, M. (2014) *J. Comput.-Aided Mol. Des.*, **28** (9), 927–939.

[40] Bostrom, J., Greenwood, J.R., and Gottfries, J. (2003) *J. Mol. Graphics Modell.*, **21** (5), 449–462.
[41] Griewel, A., Kayser, O., Schlosser, J., and Rarey, M. (2009) *J. Chem. Inf. Model.*, **49**, 2303–2311.
[42] Schlosser, J. and Rarey, M. (2009) *J. Chem. Inf. Model.*, **49** (4), 800–809.
[43] Henzler, A.M. and Rarey, M. (2010) *Mol. Inf.*, **29** (3), 164–173.
[44] Corbeil, C.R., Englebienne, P., and Moitessier, N. (2007) *J. Chem. Inf. Model.*, **47** (2), 435–449.
[45] Cheng, T., Li, X., Li, Y., and Wang, R. (2009) *J. Chem. Inf. Model.*, **49**, 1079–1093.
[46] Böhm, H.J. (1994) *J. Comput.-Aided Mol. Des.*, **8**, 243–256.
[47] Eldridge, M.D., Murray, C.W., Auton, T.R., Paolini, G.V., and Mee, R.P. (1997) *J. Comput.-Aided Mol. Des.*, **11**, 425–455.
[48] Gilli, G. and Gilli, P. (2009) *The Nature of the Hydrogen Bond*, Oxford University Press.
[49] Meyer, E.A., Castellano, R.K., and Diederich, F. (2003) *Angew. Chem. Int. Ed.*, **42**, 1210–1250.
[50] Panigrahi, S.K. and Desiraju, G.R. (2007) *Proteins*, **67**, 128–141.
[51] Glaser, R., Chen, N., Wu, H., Knotts, N., and Kaupp, M. (2004) *J. Am. Chem. Soc.*, **126**, 4412–4419.
[52] Sarwar, M.G., Dragisic, B., Salsberg, L.J., Gouliaras, C., and Taylor, M.S. (2010) *J. Am. Chem. Soc.*, **132**, 1646–1694.
[53] Cornell, W.D., Cieplak, P., Bayly, C.I., Gould, I.R., Merz, K.M., Ferguson, D.M., Spellmeyer, D.C., Fox, T., Caldwell, J.W., and Kollman, P.A. (1995) *J. Am. Chem. Soc.*, **117** (19), 5179–5197.
[54] Halgren, T.A. (1996) *J. Comput. Chem.*, **17**, 490–519.
[55] Verdonk, M.L., Cole, J.C., Hartshorn, M.J., Murray, C.W., and Taylor, R.D. (2003) *Proteins*, **52**, 609–623.
[56] Mooij, W.T.M. and Verdonk, M.L. (2005) *Proteins Struct. Funct. Bioinf.*, **61**, 272–287.
[57] Muegge, I. and Martin, Y. (1999) *J. Med. Chem.*, **42**, 791–804.
[58] Berman, H.M., Westbrook, J., Feng, Z., Gilliland, G., Bhat, T.N., Weissig, H., Shindyalov, I.N., Bourne, P.E. (2000) *Nucleic Acids Res.*, **28**, 235–242. doi: 10.1093/nar/28.1.235
[59] Gohlke, H., Hendlich, M., and Klebe, G. (2000) *J. Mol. Biol.*, **295**, 337–356.
[60] Velec, H.F.G., Gohlke, H., and Klebe, G. (2005) *J. Med. Chem.*, **48**, 6296–6303.
[61] Groom, C. R., Bruno, I. J., Lightfoot, M. P., Ward, S. C. (2016) *Acta Cryst.*, **B72**, 171–179, DOI: 10.1107/S2052520616003954
[62] Wang, R., Lai, L., and Wang, S. (2002) *J. Comput.-Aided Mol. Des.*, **16**, 11–26.
[63] Perola, E., Walters, W.P., and Charifson, P.S. (2004) *Proteins*, **56** (2), 235–249.
[64] Kellogg, G.E., Burnett, J.C., and Abraham, D.J. (2000) *J. Comput.-Aided Mol. Des.*, **15** (4), 381–393.
[65] Reulecke, I., Lange, G., Albrecht, J., Klein, R., and Rarey, M. (2008) *ChemMedChem*, **3** (6), 885–897.

[66] Schneider, N., Lange, G., Hindle, S., Klein, R., and Rarey, M. (2013) *J. Comput.-Aided Mol. Des.*, **27**, 15–29.
[67] Seifert, M.H.J. (2009) *Drug Discovery Today*, **14**, 562–569.
[68] Seebeck, B., Wagener, M., and Rarey, M. (2011) *ChemMedChem*, **6**, 1630–1639.
[69] Yang, J.M., Chen, Y.F., Shen, T.W., Kristal, B.S., and Hsu, D.F. (2005) *J. Chem. Inf. Model.*, **45**, 1134–1146.
[70] Stahl, M. and Rarey, M. (2001) *J. Med. Chem.*, **44**, 1035–1042.
[71] Brown, S.P. and Muchmore, S.W. (2009) *J. Med. Chem.*, **52**, 3159–3165.
[72] Kirkwood, J.G. (1935) *J. Chem. Phys.*, **3**, 300–313.
[73] Chipot, C. and Pohorille, A. (2007) in *Springer Series in Chemical Physics, Free Energy Calculations: Theory and Applications in Chemistry and Biology*, vol. **86** (eds C. Chipot and A. Pohorille), Springer-Verlag, Berlin, pp. 33–75.
[74] Michel, J. and Essex, J.W. (2010) *J. Comput.-Aided Mol. Des.*, **24**, 639–658.
[75] Abel, R., Young, T., Farid, R., Berne, B.J., and Friesner, R.A. (2008) *J. Am. Chem. Soc.*, **130**, 2817–2831.
[76] Young, T., Abel, R., Kim, B., Berne, B.J., and Friesner, R.A. (2007) *Proc. Natl. Acad. Sci. U.S.A.*, **104**, 808–813.
[77] Nittinger, E., Schneider, N., Lange, G., and Rarey, M. (2015) *J. Chem. Inf. Model.*, **55** (4), 771–783.
[78] Klebe, G. (2006) *Drug Discovery Today*, **11**, 580–594.
[79] Leach, A.R., Gillet, V.J., Lewis, R.A., and Taylor, R. (2010) *J. Med. Chem.*, **53** (2), 539–558.
[80] Triballeau, N., Acher, F., Brabet, I., Pin, J.P., and Bertrand, H.O. (2005) *J. Med. Chem.*, **48** (7), 2534–2547.
[81] Kirchmair, J., Markt, P., Distinto, S., Wolber, G., and Langer, T. (2008) *J. Comput.-Aided Mol. Des.*, **22** (3), 213–228.
[82] Kroemer, R.T., Vulpetti, A., McDonald, J.J., Rohrer, D.C., Trosset, J.-Y., Giordanetto, F., Cotesta, S., McMartin, C., Kihlenand, M., and Stouten, P.F.W. (2004) *J. Chem. Inf. Comput. Sci.*, **44**, 871–881.
[83] Yusuf, D., Davis, A.M., Kleywegtand, G.J., and Schmitt, S. (2008) *J. Chem. Inf. Model.*, **48**, 1411–1422.
[84] Baber, J.C., Thompson, D.C., Cross, J.B., and Humblet, C. (2009) *J. Chem. Inf. Model.*, **49** (8), 1889–1900.
[85] Chen, H., Lyne, P.D., Giordanetto, F., Lovell, T., and Li, J. (2006) *J. Chem. Inf. Model.*, **46** (1), 401–415.
[86] Shoichet, B.K. (2004) *Nature*, **432**, 862–865.
[87] Lippert, T. and Rarey, M. (2009) *J. Cheminf.*, **1** (1), 13.
[88] Bayden, A.S., Fornabaio, M., Scarsdale, J.N., and Kellogg, G.E. (2009) *J. Comput.-Aided Mol. Des.*, **23**, 621–632.
[89] Labute, P. (2009) *Proteins*, **75**, 187–205.
[90] ten Brink, J. and Exner, T.E. (2009) *J. Chem. Inf. Model.*, **49**, 1535–1546.
[91] Bietz, S., Urbaczek, S., Schulz, B., and Rarey, M. (2014) *J. Cheminf.*, **6**, 1–12.
[92] Friesner, R.A., Murphy, R.B., Repasky, M.P., Frye, L.L., Greenwood, J.R., Halgren, T.A., Sanschagrin, P.C., and Mainz, D.T. (2006) *J. Med. Chem.*, **49** (21), 6177–6196.

6.9 Prediction of ADME Properties

Aixia Yan

Beijing University of Chemical Technology, State Key Laboratory of Chemical Resource Engineering, Department of Pharmaceutical Engineering, 15 BeiSanHuan East Road, Beijing 100029, P. R. China

Learning Objectives

- To build an appropriate SPR/QSPR model for the prediction of an ADME property.
- To discuss the relationship between an ADME property and the molecular structural features by means of analyzing the prediction models of aqueous solubility, blood–brain barrier permeability (log BB), and human intestinal absorption (HIA).
- To review the wide range of ADME properties and on models for their prediction.

Outline

6.9.1 Introduction, 333
6.9.2 General Consideration on SPR/QSPR Models, 334
6.9.3 Estimation of Aqueous Solubility (log S), 336
6.9.4 Estimation of Blood–Brain Barrier Permeability (log BB), 342
6.9.5 Estimation of Human Intestinal Absorption (HIA), 346
6.9.6 Other ADME Properties, 349
6.9.7 Summary, 354

6.9.1 Introduction

Absorption, distribution, metabolism, and excretion (ADME) studies are widely used in drug discovery to optimize the balance of properties necessary to convert lead compounds into good medicines, which are crucial for the final clinical success of a drug candidate [1]. It has been estimated that nearly 50% of drugs fail because of unacceptable efficacy, which includes poor bioavailability as a result of ineffective intestinal absorption and undesirable metabolic stability [2].

ADME properties are complicatedly connected with molecular structures, a relationship that needs to be discovered through the information in datasets of known compounds. This is the domain of qualitative structure–property relationship (SPR) modeling, that is, classification, or of quantitative structure–property relationship (QSPR) modeling, that is, property prediction [3]. This approach is an important computational tool employed for predicting various physical, chemical, or biological properties, including ADME properties. The method characterizes the compounds of a dataset by molecular structural

Applied Chemoinformatics: Achievements and Future Opportunities, First Edition.
Edited by Thomas Engel and Johann Gasteiger.
© 2018 Wiley-VCH Verlag GmbH & Co. KGaA. Published 2018 by Wiley-VCH Verlag GmbH & Co. KGaA.

features with a series of numerical values, called molecular descriptors, which can be experimental physicochemical properties of the compound or calculated properties using computational chemistry methods from the molecular structure (see *Methods Volume*, Chapter 10). Then, the relationship between an ADME property and the molecular structural features of a dataset is built by some machine learning method (see *Methods Volume*, Chapter 11). The computational models built using a series of data with known experimental ADME properties are then used as ADME screens, filtering undesired compounds. This can minimize the number of compounds needed to be synthesized.

Some review articles have reported recent achievements and challenges in *in silico* ADME properties prediction models using various machine learning methods [4–9].

This section will review the current state-of-the-art computational models for the ADME property prediction. The methodology and status will be illustrated in more detail with three properties, including aqueous solubility, blood–brain barrier (BBB), and human intestinal absorption (HIA). The number of compounds in the datasets, the types and number of molecular descriptors, the modeling methods, and the performance of the built models will be summarized. In addition, computational models of several other ADME properties will also briefly be introduced. The topic of drug metabolism prediction will not be dealt with here because it is the subject of Section 6.10.

6.9.2 General Consideration on SPR/QSPR Models

In building an SPR/QSPR model, several factors that influence the quality and the reliability of a model need to be considered.

6.9.2.1 Dataset

The quantity and quality of a dataset will strongly influence the quality of an SPR/QSPR model. A dataset will usually be split into a training set and a test set. The training set is used to build a model, and the test set is used to test the model built based on the training set. Sometimes, a validation set working as a test set is used to examine the performance of a model or help to select the optimum parameters of a model. The training set, which determines the applicability domain (AD) of a model, needs to contain large and diverse compounds to cover as large a chemical space as possible. The AD means the response and chemical structure space in which the model can make predictions with a given reliability [3]. The molecular structures and descriptor values of the training set can be used for the assessment of the AD of the model.

6.9.2.2 Molecular Descriptors

A wide variety of molecular descriptors have been developed to represent the molecular structural features [3, 10] (see *Methods Volume*, Chapter 10), which can be used to predict the ADME properties. A molecular structure can be described by fragment codes or by different one-dimensional (1D), 2D, or 3D descriptors. Fragment codes contain information about the presence or absence of a certain substructure. 1D, 2D, and 3D descriptors correspond to its 1D chemical formula (e.g., molecular weight), 2D graphic scheme (e.g., number and types of rings), or 3D molecular structure taking the spatial arrangement of the atoms in the molecule into account (e.g., surface area).

For building an SPR/QSPR model using a specific dataset, the molecular descriptors that are highly correlated with the concerned ADME property need to be selected from the calculated descriptors. The selection of proper molecular descriptors can be done using known experience or knowledge; for example, we know that aqueous solubility is negatively connected with the logarithm of the octanol/water partition coefficient (log P). Thus, log P should be used as a descriptor for predicting solubility. The molecular descriptors can also be selected by trial and error using some statistical methods such as correlation analysis, or other computational tools such as genetic algorithms (GA) [11].

6.9.2.3 Building an SPR/QSPR Model

Building an SPR/QSPR model means to derive a mathematical equation or model that can qualitatively or quantitatively reflect the relationship between the concerned ADME property and the selected molecular descriptors using machine learning methods. In the process of building a model, some different descriptor combinations should be tried, different models should be built, and this process can be iterated until the model with the best prediction performance is obtained. The machine learning methods used for building an SPR/QSPR model include supervised learning methods such as multiple linear regression (MLR) analysis, partial least squares (PLS), decision trees (DTs), random forest (RF), k-nearest neighbors (KNN), artificial neural networks (ANNs), multilayer perceptron (MLP), and support vector machines (SVMs), or unsupervised methods such as self-organizing maps (SOMs) and principal component analysis (PCA) [8] (see *Methods Volume*, Chapter 10).

In the discussion of the models, we will mainly concentrate on the accumulation of a dataset and briefly mention the choice of the data analysis methods. Special attention will be given to the selection of molecular descriptors in order to further the understanding of the factors contributing to the properties modeled. Details on the performance of the models will be accumulated in Tables 6.9.1–6.9.5 and not always discussed in the text.

6.9.2.4 Validating the Model

For an SPR (or classification) model, the performance of a model can be evaluated by the computed values such as predictive accuracy. For a QSPR model, the performance of a model can be evaluated by values such as the square of the correlation coefficient (r^2), the internally cross-validated r^2 (q^2), or the standard deviation (SD) [3]. The stability of a model can also be checked by k-fold cross-validation.

For regulatory purposes, the Organisation for Economic Cooperation and Development (OECD) also recommends that a given QSPR model should follow the "Setubal principles," that is, the model should have a defined endpoint, an unambiguous algorithm, and a defined domain of applicability, be able to appropriate measures of goodness of fit, robustness, and predictiveness, and have a mechanistic interpretation, if possible [12].

6.9.3 Estimation of Aqueous Solubility (log S)

The aqueous solubility of organic compounds is a particularly important property having many applications in pharmaceutical, environmental, and other chemical disciplines. The solubility of a drug is a decisive property that determines its bioavailability and other biological activities. In the drug design process, it is essential to estimate the solubility of a large number of candidates for a drug before the compound is synthesized. A knowledge of aqueous solubility is also necessary for predicting the general environmental distribution of organic pollutants such as highly toxic, carcinogenic, and other undesired compounds. The major approaches to solubility prediction have briefly been introduced in Section 3.3.5 with a more detailed analysis at this place.

From the standpoint of thermodynamics, the dissolving process is the establishment of an equilibrium between the phase of the solute and its saturated aqueous solution. Aqueous solubility is primarily dependent on the intermolecular forces that exist between the solute molecules and the water molecules. The solute–solute, solute–water, and water–water adhesive interactions determine the amount of compound dissolving in water. Additional solute–solute interactions are associated with the lattice energy in the crystalline state.

The solubility of a compound is thus affected by many factors: the state of the solute, the relative aromatic and aliphatic degree of the molecule, the size and shape of the molecule, the polarity of the molecule, the steric effects, and the ability of some groups in participating in hydrogen bonding. In order to predict aqueous solubility accurately, all these factors correlated with solubility should be numerically represented by descriptors derived from the structure of the molecule or from experimental observations.

A compound's solubility is normally represented as log S, where S is the concentration of the compound in mol/L for a saturated aqueous solution in equilibrium with the liquid compound or the most stable form of the crystalline material. In practice, about 85% of drugs have log S values between −1 and −5, and virtually none has a value <−6. Empirically, the log S range of −1 to −5 for most drugs reflects a compromise between the polarity necessary for reasonable aqueous solubility and the hydrophobicity necessary for acceptable membrane transport [13].

An extensive series of studies for the prediction of aqueous solubility has been reported in the literature, as summarized by Lipinski et al. [14], Jorgensen and Duffy [13], and others [5–7].

These methods can be categorized into the following types:

1. Correlation of solubility with experimentally determined physicochemical properties such as melting point and molecular volume

2. Estimation of solubility by group contribution methods
3. Correlation of solubility with descriptors derived from the molecular structure by computational methods

The third approach has been proven to be particularly successfully for the prediction of solubility because it does not need experimental descriptors and can therefore also be applied to collections of virtual compounds.

6.9.3.1 Correlation with Descriptors from Experimental Data

A series of studies has been made by Yalkowsky *et al*. The so-called general solubility equation was used for estimating the solubility of solid nonelectrolytes [15, 16]. The solubility log S (logarithm of solubility expressed as mol/L) was formulated with log P (logarithm of octanol/water partition coefficient) and the melting point (MP) as shown in Eq. (6.9.1). This equation generally works well. For 580 compounds, the prediction resulted in average absolute error (AAE) of about 0.45 log units:

$$\log S = 0.5 - \log P - 0.01(\text{MP} - 25) \tag{6.9.1}$$

It is remarkable that only two descriptors were needed in this method. However, this equation is now mostly only of historical interest as it is of little use in modern drug and combinatorial library design because it requires a knowledge of the compound's experimental MP, which is not available for virtual compounds. Several methods exist for estimating log P [17, 18], but only a few inroads have been made for the estimation of MPs.

6.9.3.2 Group Contribution Method

The group contribution method allows the approximate calculation of solubility by summing up fragmental values associated with substructure units of the compounds. In a group contribution model, the aqueous solubility values are computed in Eq. (6.9.2):

$$\log S = C_0 + \sum_{i=1}^{N} C_i G_i \tag{6.9.2}$$

Here, log S is the logarithm of solubility, C_i is the number of occurrences of a substructure group, i, in a molecule, and G_i is the relative contribution of the fragment, i.

With a group contribution method, Kühne *et al*. developed a solubility model using experimental data on 351 liquids and 343 solids and compared their models with four other group contribution algorithms [19]. The number of fragments and correction terms was about 50 in the best performing models, and an additional term for the MP was added for estimating the solubility of solids. A correlation with $r^2 = 0.95$ and with an *AAE* of 0.38 log S units was obtained.

The group contribution method is, however, restricted to those functional groups it was parameterized for. The functional groups used for building the equation need to cover well a large space of fragments, functional groups, or atom types. Early studies on group contribution schemes contained a paucity of polyfunctional molecules in the datasets, and the number of fragment types was not large enough to treat drug-like molecules well. In 2001, in the work by Klopman and Zhu [20], several new models were proposed with an improvement in the accuracy and the scope of previous models by increasing the size and diversity of the training set.

In 2004, Hou et al. developed a group contribution model [21] by using 76 atom types to classify atoms with different chemical environments, and two correction factors, the number of hydrophobic carbon atoms and the square of the molecular weight, were introduced to account for the inter-/intramolecular hydrophobic interactions and the bulkiness effect.

6.9.3.3 Correlation with Calculated Descriptors

Several research groups have built models using descriptors calculated only from the molecular structure. This approach has been proven to be particularly successfully for the prediction of aqueous solubility without the need for descriptors derived from experimental data. Thus, this approach is also suitable for virtual data screening and for library design.

Recently, several QSPR solubility prediction models based on a fairly large and diverse dataset were generated. In 2000, Huuskonen developed models using MLR analysis and ANN on a dataset of 1297 diverse compounds [22]. The aqueous solubility values were measured at a temperature of 20–25 °C and are expressed as log S. The compounds were described by 24 atom-type E-state indices and six other topological indices.

Using this Huuskonen dataset, several other groups derived prediction models using different types of molecular descriptors and different data analysis methods [21, 23–26]. All these approaches obtained similar prediction performances as shown in Table 6.9.1.

The construction of some of these models is discussed in slightly more detail. The compounds of the Huuskonen dataset were described by two different representation methods: (i) with 18 topological descriptors representing the 2D structure [25] and (ii) with 32 RDF codes representing the 3D structure of a molecule; in addition eight descriptors mainly to reflect hydrogen bonding were used [26] (see below). The dataset was divided into a training set and a test set based on Kohonen's SOM . Several quantitative models for the prediction of the aqueous solubility were developed by using MLR and back-propagation (BPG) neural networks.

6.9.3.3.1 Models with 18 Topological Descriptors

The topological descriptors were calculated as autocorrelation vectors of physicochemical atomic properties with the program ADRIANA.Code [27, 28].

In addition, the following descriptors were also used: the partition coefficient log P because it strongly affects solubility [15, 16] as well as the mean molecular polarizability, the molecular weight and the highest hydrogen bond acceptor potential, the highest hydrogen bond donor potential, the number of hydrogen bond donor groups, and the number of atoms of the elements nitrogen, oxygen, and fluorine. Statistical and correlation analyses allowed the elimination of highly correlated descriptor that eventually left 18 descriptors.

An optimized multilayer neural network of architecture of 18-10-1 gave somehow better results than an MLR analysis (see Table 6.9.1). The prediction results for the test set are shown in Figure 6.9.1.

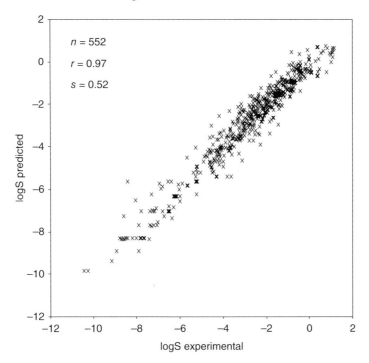

Figure 6.9.1 Predicted versus experimental solubility values of 552 compounds in the test set by a back-propagation neural network with 18 topological descriptors.

6.9.3.3.2 Models with 32 Radial Distribution Function Values and Additional Eight Descriptors

The compounds were described by a set of 32 RDF code values [29] representing the 3D structure of a molecule (see *Methods Volume*, Section 10.3.4.4). The 3D coordinates were obtained using the 3D structure generator CORINA [30]. Additional eight descriptors were also calculated by ADRIANA.Code [28], including mean molecular polarizability, aromatic indicator of a molecule, aliphatic indicator of a molecule, highest hydrogen bond acceptor potential, highest hydrogen bond donor potential, number of hydrogen bond donor groups, and number of atoms of the elements nitrogen and oxygen.

As Table 6.9.1 shows, both approaches using 2D descriptors or using 3D descriptors provided models with similar prediction accuracy. The descriptors for hydrogen bonding and polarizability were dominating, the 2D or 3D descriptors providing similar information.

Collaborators from pharmaceutical industry pointed out that the Huuskonen dataset did not properly reflect structures typically found in drugs. In order to enlarge the AD of the solubility models to structures encompassed in drugs, a larger dataset compiled at Merck KGaA, which included 2084 compounds,

Table 6.9.1 The prediction performance of solubility models based on calculated descriptors.

Author	Year	Number of compounds	Number of descriptors	Type of descriptors	Method	Performance
Huuskonen [22]	2000	$N_{train} = 884$, $N_{test} = 413$, $N_{ext} = 21$	30	E-state	ANN	$r^2_{train} = 0.94$, $SD_{train} = 0.47$; $r^2_{test} = 0.92$, $SD_{test} = 0.60$; $r^2_{ext} = 0.91$, $SD_{ext} = 0.63$
Tetko et al. [23]	2001	$N_{train} = 879$, $N_{test} = 412$, $N_{ext} = 21$	33	E-state, TD	ANN	$r^2_{train} = 0.95$, $SD_{train} = 0.47$; $r^2_{test} = 0.92$, $SD_{test} = 0.60$; $r^2_{ext} = 0.90$, $SD_{ext} = 0.64$
Liu and So [24]	2001	$N_{train} = 1033$, $N_{test} = 258$, $N_{ext} = 21$	7	1D, 2D	ANN	$r^2_{train} = 0.86$, $SD_{train} = 0.70$; $r^2_{test} = 0.86$, $SD_{test} = 0.70$; $r^2_{ext} = 0.79$, $SD_{ext} = 0.91$
Hou et al. [21]	2004	$N_{train} = 878$, $N_{test} = 412$, $N_{ext} = 21$	78	76 atom types, others	MLR	$r_{train} = 0.96$, $SD_{train} = 0.59$; $r_{test} = 0.95$, $SD_{test} = 0.63$; $r_{ext} = 0.94$, $SD_{ext} = 0.64$
Yan and Gasteiger [25]	2003	$N_{train} = 741$, $N_{test} = 552$, $N_{ext} = 21$	18	G1D, TD	ANN	$r^2_{train} = 0.92$, $SD_{train} = 0.51$; $r^2_{test} = 0.94$, $SD_{test} = 0.52$; $r^2_{ext} = 0.83$, $SD_{ext} = 0.80$
Yan and Gasteiger [26]	2003	$N_{train} = 797$, $N_{test} = 496$, $N_{ext} = 21$	40	RDFD, PD	ANN	$r^2_{train} = 0.93$, $SD_{train} = 0.50$; $r^2_{test} = 0.92$, $SD_{test} = 0.59$; $r^2_{ext} = 0.85$, $SD_{ext} = 0.77$
Yan et al. [31]	2004	$N_{train} = 1148$, $N_{test} = 936$, $N_{ext} = 799$	18	G1D, TD	ANN	$r_{train} = 0.93$, $SD_{train} = 0.61$; $r_{test} = 0.92$, $SD_{test} = 0.62$; $r_{ext} = 0.94$, $SD_{ext} = 0.72$

Reference	Model	N	# desc	Descriptors	Method	Statistics
Huuskonen et al. [32]	2008	$N_{train} = 1217$, $N_{test} = 866$, $N_{ext} = 799$	40	RDFD, PD	ANN	$r_{train} = 0.93$, $SD_{train} = 0.60$; $r_{test} = 0.90$, $SD_{test} = 0.73$; $r_{ext} = 0.91$, $SD_{ext} = 0.88$
		$N_{train} = 191$, $N_{test1} = 174$, $N_{test2} = 200$	5	CDPD	MLR	$r^2_{train} = 0.87$, $SD_{train} = 0.51$; $r^2_{test1} = 0.80$, $SD_{test1} = 0.68$; $r^2_{test2} = 0.88$, $SD_{test2} = 0.65$
Raevsky et al. [33]	2015 Model6[a]	$N_{train} = 818$, $N_{test} = 204$	12	PD	RF	$r^2_{train} = 0.97$, $RMSE_{train} = 0.34$; $r^2_{test} = 0.91$, $RMSE_{test} = 0.62$
	Model12[b]	$N_{train} = 2093$, $N_{test} = 522$	12	PD	RF	$r^2_{train} = 0.93$, $RMSE_{train} = 0.61$; $r^2_{test} = 0.84$, $RMSE_{test} = 0.90$
	Model25[c]	$N_{train} = 2911$, $N_{test} = 726$	12	PD	RF	$r^2_{train} = 0.94$, $RMSE_{train} = 0.56$; $r^2_{test} = 0.86$, $RMSE_{test} = 0.84$
	Model26[c]	$N_{train} = 2911$, $N_{test} = 726$	12	PD	SVM	$r^2_{train} = 0.87$, $RMSE_{train} = 0.83$; $r^2_{test} = 0.86$, $RMSE_{test} = 0.85$
	Model27[d]	$N_{train} = 2911$, $N_{test} = 726$	12	PD	RF SVM	$r^2_{train} = 0.92$, $RMSE_{train} = 0.67$; $r^2_{test} = 0.88$, $RMSE_{test} = 0.78$

a) Model based on liquid chemicals.
b) Model based on crystalline chemicals.
c) Models based on liquid and crystalline chemicals.
d) Consensus model of Models 25 and 26.

ANN, artificial neural networks; MLR, multiple linear regression; SVM, support vector machine; RF, random forest; E-state, electrotopological state indices; CD, constitutional descriptors; GD, geometrical descriptors; PD, physicochemical descriptors; TD, topological descriptors; 1D, one-dimensional descriptors; 2D, two-dimensional descriptors; 3D, three-dimensional descriptors; GlD, global descriptors; ShD, shape descriptors; RDFD, radial distribution function descriptors; others, other descriptors.

was used to build models [31]. The structures of the organic compounds were represented by the two methods as before: (i) using 18 topological descriptors [25] and (ii) using 32 RDF codes and eight additional descriptors [26]. After excluding those compounds from the Huuskonen dataset that overlap with the Merck dataset, 799 compounds remained and were used as an external test set. The results are collected in Table 6.9.1. The predictions on the external test set are slightly inferior and indicate that they comprise a somehow different chemical space (as had already initially been pointed out).

In 2008, Huuskonen *et al.* used a training set of 191 drug-like compounds extracted from the AQUASOL database to produce QSPR models on aqueous solubility employing a set of simple structural and physicochemical properties [32] (see Table 6.9.1). Their work suggested that increasing molecular size, rigidity, and lipophilicity decreases solubility, whereas increasing conformational flexibility and the presence of a nonconjugated amine group increases the solubility of drug-like compounds.

In 2015, Raevsky *et al.* [33] collected datasets containing 1022 liquid and 2615 crystalline compounds with solubility values for the intrinsic unionized structures. A series of QSPR models were constructed for the prediction of solubility based on liquid or crystalline chemicals, first individually, and then in combination. Various data analysis methods were used. Either a set of best QSPR models separately for liquids and crystalline molecules or a combined dataset was obtained (see Table 6.9.1).

6.9.3.4 Conclusions on log *S* Prediction

The prediction of aqueous solubility suffers from a lack of high-quality data measured under standardized conditions. The presently used datasets have compiled data from a variety of sources and conditions (e.g., thermodynamic or kinetic values). The modeling approaches in recent years have moved to using descriptors that can be interpreted. Thus, they show the importance of hydrogen bonding as well as of other electronic effects. Use of 3D descriptors did not lead to major improvements in the models. Developing different models for liquids and solids is an interesting approach but provides problems for virtual molecules whose state is not known. In any case, models for the prediction of aqueous solubility of organic compounds now exist that can help in the process of drug design.

6.9.4 Estimation of Blood–Brain Barrier Permeability (log *BB*)

In the discovery and development of a drug active in the central nervous system (CNS) drug, one important property is the ability of a drug to penetrate the BBB [34]. Usually, *BB* is defined as the brain–blood concentration ratio of a compound at steady state, which is commonly expressed by log *BB* with log $BB = \log(C_{\text{brain}}/C_{\text{blood}})$, where C_{brain} and C_{blood} are the equilibrium concentrations of the drug in the brain and the blood, respectively. In addition, compounds that are able to cross the BBB are defined as BBB+, whereas BBB− signifies compounds that have little ability to cross the BBB.

Whether a drug can penetrate the BBB is critical in screening potential CNS active agents and for improving the side effect profile of drugs with peripheral activity. It is an expensive and time-consuming work to obtain the experimental data on the blood–brain distribution ratio of a compound. Therefore, a computational model is very much wanted. In addition, the BBB should also be available for virtual compounds in screening that cannot be measured anyhow. However, finding log *BB* data of high quality (following a uniform standard protocol for experimental determination of the brain/plasma ratio) and of a sufficient quantity is quite difficult. In addition, many factors influence the BBB permeability such as plasma protein binding (PPB), active efflux from the CNS by transporters such as P-glycoprotein (P-gp), and the metabolism of a drug candidate. Most models developed for the BBB assume that the drug penetrates the barrier by passive diffusion [34]. Hence, establishing a useful relationship between the molecular structure and the measured blood–brain partitioning is a challenging task. The prediction of log *BB* using computational models has recently been reviewed [34]. From the standpoint of modeling, the BBB penetration is an interesting property as two types of computational models have been developed for studying the relationship between this property and molecular structural features: an SPR model providing a classification into BBB+ and BBB−, and a QSPR model making quantitative predictions on log *BB*. Several classification models on BBB+/BBB− [35–40] and QSPR models [39–41] on log *BB* have recently been reported (as shown in Tables 6.9.2 and 6.9.3); some are discussed in more detail here.

6.9.4.1 Classification of Blood–Brain Barrier Permeability

In 2007, Zhao *et al.* [35] used a BBB dataset including 1593 compounds (consisting of 1283 BBB+ and 310 BBB− compounds), which had been collected by Adenot and Lahana [36]. An external test set was obtained from Li *et al.* [37]. The molecular structures were represented by 19 simple molecular descriptors or by a fragmentation scheme.

For the classification model, the prediction accuracy (Q) was calculated by the Eq. (6.9.3):

$$Q = \frac{TP + TN}{TP + TN + FP + FN} \times 100\% \quad (6.9.3)$$

where TP stands for true positive and TN means true negative. FP and FN are short for false positive and false negative, respectively.

The results of the classification models of Zhao *et al.* [35] are shown in Table 6.9.2. This work indicated that hydrogen-bonding properties of compounds play a very important role in modeling BBB penetration.

The dataset of Zhao *et al.* [35] was also investigated by Wang *et al.* [38] using Kohonen's SOM and an SVM method to build BBB classification models. A total of 55 descriptors (including 13 global descriptors and 42 2D property autocorrelation descriptors) were calculated using the program ADRIANA.Code [28]. Work on descriptor elimination left five 2D property autocorrelation descriptors. The results of the developed models are shown in Table 6.9.2. The results were

Table 6.9.2 Performance of classification models on blood–brain barrier permeability.

Authors	Year	Number of compounds	Number of descriptors	Type of descriptors	Method	Performance (%)
Zhao et al. [35]	2007	$N_{train} = 1093$ $N_{test} = 500$	4	Abraham descriptors	RP	$Q_{train} = 90.6$ $Q_{test} = 96.8$
		$N_{train} = 1093$ $N_{test} = 500$	69	Fragment schemes	PLS	$Q_{train} = 97.1$ $Q_{test} = 97.2$
		$N_{train} = 1593$ $N_{test} = 397$	2	Abraham descriptors	PLS	$Q_{train} = 91.6$ $Q_{test} = 80.1$
Wang et al. [38]	2009	$N_{train} = 1093$ $N_{test} = 500$	5	TD	SOM	$Q_{train} = 96.6$ $Q_{test} = 97.0$
		$N_{train} = 1093$ $N_{test} = 500$	5		SVM	$Q_{train} = 94.5$ $Q_{test} = 96.8$
		$N_{train} = 1593$ $N_{test} = 396$	5		SOM	$Q_{train} = 95.9$ $Q_{test} = 81.1$
		$N_{train} = 1593$ $N_{test} = 396$	5		SVM	$Q_{train} = 97.2$ $Q_{test} = 76.8$
Brito-Sánchez et al. [39]	2015	$N_{train} = 381$ $N_{test} = 116$	6	2D, 3D	LDA	$Q_{train} = 85.09$ $Q_{test} = 83.33$
Gupta et al. [40]	2015	$N_{train} = 252$ $N_{val} = 54$ $N_{test} = 54$ $N_{ext} = 29$	5	CD, TD, GT	GBT	$Q_{train} = 99.47,$ $Q_{val} = 98.77,$ $Q_{test} = 98.77,$ $Q_{ext} = 86.21$
		$N_{train} = 252$ $N_{val} = 54$ $N_{test} = 54$ $N_{ext} = 29$	5	CD, TD, GT	BDT	$Q_{train} = 98.94,$ $Q_{val} = 95.06,$ $Q_{test} = 97.53,$ $Q_{ext} = 82.76$

RP, recursive partitioning; PLS, partial least squares; SOM, Kohonen's self-organizing maps; SVM, support vector machine; LDA, linear discriminant analysis; GBT, gradient boosted tree; BDT, bagged decision tree; CD, constitutional descriptors; TD, topological descriptors; 2D, two-dimensional descriptors; 3D, three-dimensional descriptors; GD, geometric descriptors; PD, physicochemical descriptors; others, other descriptors.

Table 6.9.3 Performance of QSPR models on logBB.

Authors	Year	Number of compounds	Number of descriptors	Type of descriptors	Method	Performance
Yan et al. [41]	2013	$N_{train} = 198$, $N_{test} = 122$	14	GlD, ShD, RDFD	SVM	$r_{train} = 0.90$, $SD_{train} = 0.61$, $r_{test} = 0.89$, $SD_{test} = 0.56$,
	2013	$N_{train} = 198$, $N_{test} = 122$	14	GlD, ShD, RDFD	ANN	$r_{train} = 0.90$, $SD_{train} = 0.63$, $r_{test} = 0.90$, $SD_{test} = 0.58$
Brito-Sánchez et al. [39]	2015	$N_{train} = 381$, $N_{test} = 116$	10	2D, 3D	MLR	$r^2_{train} = 0.69$, $MAE_{train} = 0.10$, $MAE_{test} = 0.31$
Gupta et al. [40]	2015	$N_{train} = 252$, $N_{val} = 54$, $N_{test} = 54$, $N_{ext} = 29$	3	CD, TD, GD	GBT	$r^2_{train} = 0.957$, $SD_{train} = 0.69$, $r^2_{val} = 0.921$, $SD_{val} = 0.49$, $r^2_{test} = 0.938$, $SD_{test} = 0.51$, $r^2_{ext} = 0.905$, $SD_{ext} = 0.31$
		$N_{train} = 252$, $N_{val} = 54$, $N_{test} = 54$, $N_{ext} = 29$	3	CD, TD, GD	BDT	$r^2_{train} = 0.932$, $SD_{train} = 0.65$, $r^2_{val} = 0.896$, $SD_{val} = 0.45$, $r^2_{test} = 0.913$, $SD_{test} = 0.48$, $r^2_{ext} = 0.913$, $S_{ext} = 0.26$

SVM, support vector machine; ANN, artificial neural networks; MLR, multiple linear regression; GBT, gradient boosted tree; BDT, bagged decision tree; CD, constitutional descriptors; GD, geometric descriptors; PD, physicochemical descriptors; TD, topological descriptors; 2D, two-dimensional descriptors; 3D, three-dimensional descriptors; GlD, global descriptors; ShD, shape descriptors; RDFD, radial distribution function descriptors; others, other descriptors.

slightly better than those obtained by Zhao et al. [35]. However, more important is the fact the selected five autocorrelation descriptors having a good correlation with BBB clearly reflect important electronic effects residing in the charge and electronegativity values. Four of the five descriptors were the squared sigma and pi charges, q_σ^2, q_π^2, as well as the squared pi and lone pair electronegativity, χ_π^2, χ_{lp}^2 [38].

In recent work by Brito-Sánchez et al. [39], BBB passage was modeled using classification and regression schemes on carefully curated data employing six molecular descriptors. The results are shown in Table 6.9.2.

Recently Gupta et al. [40] collected a dataset of 360 compounds. They built models with the relatively new methods of gradient boosted trees (GBTs) and bagging decision trees (BDTs). The classification results were quite good (see Table 6.9.2).

6.9.4.2 QSPR Models on log BB

In 2013, Yan et al. [41] collected a dataset of 320 compounds with log BB values from several publications. Each molecule was represented by global and shape descriptors, 2D autocorrelation descriptors, and RDF descriptors calculated by ADRIANA.Code [28]. Descriptor elimination left 14 mostly 3D RDF descriptors based on charge and electronegativity values. For the best SVM and ANN models, $r = 0.89$ and $r = 0.90$ were obtained for the test set, respectively (see Table 6.9.3).

In the work of Brito-Sánchez et al. [39], an MLR-based model with acceptable explanation of more than 69% of the variance in the experimental log BB was developed.

Gupta et al. [40] also built QSPR models on log BB using the GBTs and BDTs. The three descriptors for the QSPR models comprised XLogP, TPSA, and dipole moment. The results were exceptionally good (see Table 6.9.3).

6.9.4.3 Conclusions on BBB Prediction

The models for BBB permeability have matured to a point that allows the interpretation of the descriptors. They show the importance of polar and hydrophobic effects, as expected; polarity and hydrogen bonding responsible for a drug to remain in the blood, increasing lipophilicity and polarizability, leads to an increase in the penetration into the brain. Progress will certainly be achieved by access to more and better data. Nevertheless, the present models are useful in practice.

6.9.5 Estimation of Human Intestinal Absorption (HIA)

Intestinal absorption can be defined as the transfer of a drug from the site of administering into the blood system. The transfer occurs at the gastrointestinal epithelial membrane, accessing the systemic circulation through the blood vessels and passing to the portal vein [42]. In experiment, HIA is measured by the fraction of absorption, %FA (or %HIA), which is defined as the total mass absorbed divided by the given dose of the drug.

The prediction of HIA was pioneered by the "rule of five" proposed by Lipinski et al. [14]. The rule of five defined several criteria for identifying compounds with potentially poor absorption and permeability: a molecule will be poorly absorbed if the molecular weight > 500; the calculated log P > 5 (CLOGP) or >4.15 (MLOGP); the number of hydrogen bond donors (OH and NH groups) > 5; and the number of hydrogen bond acceptors (N and O atoms) > 10. The method allows – with an acceptable probability – the elimination of molecules that would have improper properties. However, only a rough classification of the molecules is obtained, which is a drawback of the method.

The in silico modeling of HIA has recently been reviewed [6, 42, 43]. The computational models can be performed by building SPR for classification [44–46] or QSPR models for the prediction of HIA values [45–48], as shown in Table 6.9.4.

6.9.5.1 Classification of Human Intestinal Absorption (HIA)

In 2007, Hou et al. [44, 47] studied HIA by classification and QSPR models. The dataset consisted of 578 structurally diverse molecules. Seven molecular descriptors were used with the apparent partition coefficient at pH 6.5 (log D 6.5) and the topological polar surface area (TPSA) being the most important ones (see Table 6.9.4).

In 2011, Suenderhauf et al. collected a dataset of 458 small drug-like compounds that had FDA approval [45]. They calculated 80 1D to 3D physicochemical descriptors and used various methods of feature selection and data analysis methods such as decision tree induction (DTI). The best performance for classification was seen with the DTI method (see Table 6.9.4).

In 2016, Basant et al. [46] used Hou's data [44] of 577 compounds for classification of HIA values. The GBT and BDT were used for the binary classification of HIA. Four 2D molecular descriptors were used: HDon_O (number of hydrogen bonding donors derived from the sum of O—H groups), NAtoms (number of all atoms in the molecule), XLogP (octanol partition coefficient), and TPSA. Both the GBT and BDT classification models rendered excellent classification accuracies (see Table 6.9.4).

6.9.5.2 QSPR Models for the Prediction of Human Intestinal Absorption (HIA)

In 2007, Hou et al. collected data on HIA for 647 drugs and drug-like molecules. After processing, 553 molecules transported by passive diffusion were kept for building quantitative HIA prediction models [47]. The best prediction model (see Table 6.9.4) was obtained by the genetic function approximation (GFA) technique using four molecular descriptors: TPSA, log D 6.5, the number of violations of the Lipinski's rule of five, and the square of the number of hydrogen bond donors.

The same dataset as Hou et al. [47] was used by Yan et al. [48]. 107 descriptors were calculated by ADRIANA.Code and Cerius2. By a GA, nine physicochemical descriptors were selected. All the models (PLS and SVM) had good prediction for high HIA (over 80%) value compounds but poor prediction for low HIA (below 30%) value compounds (see Table 6.9.4). That may have mainly been caused by the

Table 6.9.4 Performance of SPR and QSPR models on human intestinal absorption (HIA).

Authors	Year	Model type	Number of compounds	Number of descriptors	Type of descriptors	Method	Performance
Hou et al. [44]	2007	SPR	$N_{train} = 480$, $N_{test} = 98$	7	CD, TD	SVM	$Q = 97.8\%$ (HIA−), $Q = 94.5\%$ (HIA+)
Suenderhauf et al. [45]	2011	SPR	$N = 376$	80	CD, TD, GD, others	DTI (CHAID)	$Q = 92\%$
Basant et al. [46]	2016	SPR	$N_{train} = 403$, $N_{val} = 87$, $N_{test} = 87$	4	CD, TD	GBT (or BDT)	$Q_{train} = 99.75$, $Q_{val} = 98.85\%$, $Q_{test} = 97.70\%$
Hou et al. [47]	2007	QSPR	$N_{train} = 455$, $N_{test} = 98$	4	CD, others	GFA	$r_{train} = 0.84$, $SD_{train} = 15.50$, $r_{test} = 0.90$
Yan et al. [48]	2008	QSPR	$N_{train} = 380$, $N_{test} = 172$	9	CD, TD, GD, PD	SVM	$r_{train} = 0.81$, $SD_{train} = 12.50$, $r_{test} = 0.88$, $SD_{test} = 9.14$
Suenderhauf et al. [45]	2011	QSPR	$N = 458$	9	CD, TD, GD, others	ANN	$r^2 = 0.6$, $RMSE = 25.82$
Basant et al. [46]	2016	QSPR	$N_{train} = 403$, $N_{val} = 87$, $N_{test} = 87$	3	CD, TD	GBT	$r^2_{train} = 0.972$, $SD_{train} = 30.87$, $r^2_{val} = 0.958$, $SD_{val} = 25.84$, $r^2_{test} = 0.953$, $SD_{test} = 24.23$

SVM, support vector machine; DTI, decision tree induction; CHAID, chi-squared analysis interaction detector; GBT, gradient boosted tree; BDT, bagged decision tree; GFA, genetic function approximation; ANN, artificial neural networks; CD, constitutional descriptors; TD, topological descriptors; 2D, two dimensional; GD, geometric descriptors; PD, physicochemical descriptors; others, other descriptors.

unbalanced distribution of experimental HIA values with 71.7% of compounds having high HIA values over 80% and only 18.9% of compounds having HIA values from 30–80% and 9.4% compounds having HIA values below 30% [48].

Suenderhauf et al. also built QSPR models for the prediction of HIA on 458 molecules [45]. Nine descriptors provided a model with a rather poor prediction of an $r^2 = 0.6$ and an RMSE $= 25.8$ [45].

Basant et al. also built QSPR models for the prediction of HIA with a GBT and a BDT [46]. The GBT model (based on three descriptors: HDon_O, NAtoms, and TPSA) performed slightly better than the BDT model with an r^2 of 0.972, 0.958, and 0.953 for the training, validation, and test set, respectively [46] (Table 6.9.4).

6.9.5.3 Conclusions on HIA Prediction

HIA is correlated with the following factors: violations of the rule of 5 (Nrule5), log P, distribution coefficient at pH 6.5 (log D 6.5), aqueous solubility, charge property, polar surface area, and the hydrogen-bonding ability of a molecule. Several classification and QSPR models for the prediction for HIA values are based on rather unbalanced datasets of experimental HIA values, which are short of poor absorption compounds with HIA < 30%. Better prediction models are subject to finding more drugs with reliable HIA experimental data and more compounds with poor absorption.

6.9.6 Other ADME Properties

In addition to the abovementioned properties of aqueous solubility, BBB, and HIA, there are many other ADME properties. In recent work by Wang and Hou [49], ADME properties were broadly classified into two categories: "physicochemical" and "physiological."

The physicochemical properties include aqueous solubility, log P, log D (logarithm of octanol–water distribution coefficient), and pK_a value, and so on, which are governed by simple physicochemical laws.

The physiological ADME properties can be further grouped into *in vitro* ADME properties (such as Caco-2 permeability and MDCK permeability) and *in vivo* pharmacokinetic properties (such as oral bioavailability, HIA, PPB, volume of distribution, urinary excretion, total body clearance (Cl), and elimination half time ($t_{1/2}$)) [49]. It is clear that the physiological ADME properties are more difficult to be accurately predicted than the physicochemical ADME properties.

In this part, computational models on some of these important ADME properties will be briefly introduced. They include acid dissociation constant (pK_a), Caco-2 cell permeability, human skin permeability, human PPB, human serum albumin (HSA) binding, P-gp substrates/non-substrate, P-gp inhibitors/non-inhibitors, and human oral bioavailability (HOBA). The performance of the classification (or SPR) or QSPR models for prediction of these properties are shown in Table 6.9.5.

The acid dissociation constant (pK_a) measures the strength of an acid, which influences solubility [5], permeability [7], and some other ADME properties such as PPB, cardiotoxicity, and metabolism of a compound [50]. In Section 3.3.7 some models for the prediction of pK_a values were presented. It was emphasized that local site-specific descriptors were considered to be essential for an understanding of the details of dissociation of a proton.

On the other hand, the problem of predicting pK_a values was approached in a collaboration by Fraczkiewicz *et al.* from Simulations Plus and researchers from Bayer by investigating massive amounts of data on pK_a values [50]. The pK_a prediction method from Simulations Plus based on artificial neural network ensembles (ANNE), microstates analysis, and literature data was retrained with a large homogeneous dataset of drug-like molecules from Bayer. The new model was developed with curated sets of about 14,000 literature pK_a values (~11,000 compounds) and about 19,500 pK_a values experimentally determined at Bayer Pharma (~16,000 compounds). The results are shown in Table 6.9.5. For the largest and most difficult external test set with 16,404 pK_a values, the new model achieved an RMSE = 0.67 and $r^2 = 0.93$. The model is commercially available as part of the Simulations Plus ADMET Predictor release 7.0 [50].

Apart from the HIA models discussed above, computational drug permeability models [5–8, 14] can also be developed based on experimentally determined permeability values using different cell culture models. The most commonly used approach is the Caco-2 cell line, which is a human colon carcinoma cell line [5]. Other cell lines include MDCK cells, which originate from canine kidney tissue, and 2/4/A1 cells, which originate from the rat small intestine [5]. The Caco-2 monolayer cell is considered as the best *in vitro* model for an HIA study due to the obvious pertinence between Caco-2 permeability and HIA values [51].

Recently, Wang *et al.* built several QSPR models for the prediction of Caco-2 cell permeability values with a large dataset consisting 1272 compounds [51]. Prediction models with 30 molecular descriptors were obtained (see Table 6.9.5).

Human skin permeability of industrial and household chemicals plays an important role in various fields including toxicology and risk assessment of hazardous materials, transdermal delivery of drugs, and the design of cosmetic products [52]. Recently, Khajeh and Modarress developed several QSPR models for the prediction of human skin permeability with a dataset of 283 diverse compounds [52]. The models built by the nonlinear method of adaptive neuro-fuzzy inference system (ANFIS) performed better than the MLR models (see Table 6.9.5). Their work suggests that hydrophobicity (encoded as log *P*) is the most important factor in transdermal penetration [52].

Many drugs bind with varying degrees of association to human plasma proteins. The most important proteins in terms of drug binding are albumin and α1-acid glycoprotein, followed by lipoproteins [53]. In most cases, the binding of the drug to the plasma proteins is a reversible association due to hydrophobic and electrostatic interactions, and the bound fraction of the drug exists in equilibrium with the free drug [53]. PPB strongly influences the volume of distribution and the half-life of drugs. Extended PPB may be further associated with drug safety issues, low clearance, low brain penetration, and drug–drug interactions [54]. HSA has a central role as a binder, and the affinity of drugs to this protein is considered to dominate PPB and related pharmacokinetic issues [54].

Table 6.9.5 Performance of SPR and QSPR models for the prediction of various ADME properties.

ADME property	Model type	Dataset	Descriptors	Method[a]	Performance	References
pK_a	QSPR	$N_{train} = 25{,}509$ $N_{test} = 8138$ $N_{ext} = 16{,}404$	Simulations Plus ADMET Predictor release 7.0	ANNE	$r^2_{train} = 0.975$, $RMSE_{train} = 0.475$ $r^2_{test} = 0.974$, $RMSE_{test} = 0.479$ $r^2_{ext} = 0.93$, $RMSE_{ext} = 0.67$	[50]
Caco-2 cell permeability	QSPR	$N_{train} = 1017$ $N_{test} = 255$ $N_{ext} = 298$	30 MOE descriptors	Boosting	$r^2_{train} = 0.97$, $RMSE_{train} = 0.12$ $r^2_{test} = 0.81$, $RMSE_{test} = 0.31$ $r^2_{ext} = 0.75$, $RMSE_{ext} = 0.36$	[51]
Human skin permeability	QSPR	$N_{train} = 225$ $N_{test} = 58$	3 Dragon descriptors	ANFIS	$r^2_{train} = 0.899$, $RMSE_{train} = 0.312$ $r^2_{test} = 0.890$, $RMSE_{test} = 0.333$	[52]
Human plasma protein binding	SPR	$N_{train} = 744$ $N_{test} = 186$	7 MOSES descriptors	DTB	$Q_{train} = 99.80\%$ $Q_{test} = 97.58\%$	[53]
Human plasma protein binding	QSPR	$N_{train} = 744$ $N_{test} = 186$	6 MOSES descriptors	DTB	$r^2_{train} = 0.963$, $RMSE_{train} = 7.61$ $r^2_{test} = 0.931$, $RMSE_{test} = 8.65$	[53]

(Continued)

Table 6.9.5 (Continued)

ADME property	Model type	Dataset	Descriptors	Method[a]	Performance	References
Human serum albumin binding	QSPR	$N_{train} = 84$ $N_{val} = 10$	6 Cerius2 descriptors	GFA	$r^2_{train} = 0.83$ $r^2_{val} = 0.82$	[55]
P-Glycoprotein substrate or non-substrate	SPR	$N_{train} = 282$ $N_{test} = 202$	>200 functional groups fingerprints	RF	$Q_{test} = 70\%$	[57]
P-Glycoprotein inhibitor or non-inhibitor	SPR	$N_{train} = 1268$ $N_{test} = 667$	>200 functional groups fingerprints	RF	$Q_{test} = 75\%$	[57]
Human oral bioavailability	QSPR	$N_{train} = 916$, $N_{test} = 80$	Basic molecular properties and structural fingerprint	GFA and MLR	$r_{train} = 0.79$, $RMSE_{train} = 22.3\%$ $r_{test} = 0.71$, $RMSE_{test} = 23.6\%$	[59]

a) ANNE, artificial neural network ensembles; ANFIS, adaptive neuro-fuzzy inference system; DTB, decision tree boost; GFA, genetic function approximation; RF, random forest; MLR, multiple linear regression.

Recently, Basant et al. developed classification and QSPR models for the prediction of human PPB using a decision tree forest (DTF) and a decision tree boost (DTB) method based on a dataset of 930 compounds [53]. The molecules were represented by 2D molecular descriptors using the MOSES program. Seven descriptors for classification and six descriptors for regression QSPR modeling were chosen (see Table 6.9.5). Their work showed that XLogP was the most important descriptor in all the classification and QSPR models [53].

Colmenarejo et al. predicted binding affinities to HSA using a dataset of 94 diverse drugs and drug-like compounds [55]. The GFA method was used to exhaustively search for models and to select the best ones with the training set. Hydrophobicity (as measured by ClogP) was found to be the most important variable determining the binding to HSA.

P-gp (ABCB1) plays a significant role in determining the ADME properties of drugs and drug candidates. Substrates of P-gp are not only subject to multidrug resistance (MDR) in tumor therapy but also associated with poor pharmacokinetic profiles [56, 57]. In contrast, inhibitors of P-gp have been advocated as modulators of MDR [57]. Poongavanam et al. developed classification models from a set of 484 substrates/non-substrates and a set of 1935 inhibitors/non-inhibitors. The molecules were described by a set of over 200 in-house-generated functional-groups-based fingerprints (see Table 6.9.5). The results of their work indicate that the majority of non-substrates contain hydroxyl groups when compared with substrates; compounds that contain alkylaryl ethers, aromatic amines, and tertiary aliphatic amine groups are likely to be P-gp inhibitors [57].

Bioavailability represents the percentage of an oral dose, which is able to produce a pharmacological activity, in other words, the fraction of the oral dose that reaches the arterial blood in an active form [58]. It has been estimated that nearly 50% of drugs fail because of unacceptable efficacy, which includes poor bioavailability as a result of ineffective intestinal absorption and undesirable metabolic stability [2]. Oral bioavailability is related to several factors, such as gastrointestinal transition and absorption, intestinal membrane permeation, and intestinal/hepatic first-pass metabolism. Moreover, many authors have suggested that CYP3A4 in the gut wall and P-gp act in a concerted manner to control the absorption of their substrates [58]. Compared with other ADME-Tox properties, HOBA is particularly important but extremely difficult to predict [49]. The advances in computationally modeling HOBA have recently been reviewed [49].

Tian et al. employed the GFA technique to construct a set of MLR models for oral bioavailability based on a dataset of 996 compounds [59]. The molecules were represented by structural fingerprints as basic descriptors together with several important molecular properties [59]. The results are indicated in Table 6.9.5.

In addition to the ADME properties discussed above in this section, there are many other ADME properties that have to be considered in the drug design process. For example, besides P-gp (ABCB1), there exists some other drug efflux transporters expressed in the intestine and liver, which include bile salt export pump (BSEP) (ABCB11), multidrug resistance proteins (MRP) (1-6, ABCC1-6), breast cancer resistance protein (BCRP) (ABCG2), and all members of the

ATP-binding cassette superfamily. Members of this superfamily use ATP as an energy source, allowing them to pump substrates against a concentration gradient [60].

6.9.7 Summary

The challenge of predicting a variety of all-important ADME properties has been adopted by several groups of researchers. Various approaches using different sets of molecular descriptors and a variety of data modeling techniques have been made. The emphasis has shifted to the use of descriptors that can be interpreted and thus help to understand the factors that contribute to the different properties. Real progress will only be made if more, and in particular, better experimental data become available.

Essentials

- To find the molecular structural features (and/or physicochemical properties) that influence an ADME property, which can be correlated by an SPR/QSPR model with a machine learning method.
- To build an SPR/QSPR prediction model of an ADME property consists of four steps: collection of a dataset, calculation and selection of molecular descriptors, establishment of a mathematical model between the selected descriptors and the ADME property, and model validation.

Available Software and Web Services (accessed January 2018)

- http://www.handwrittentutorials.com/videos.php
- http://www.click2drug.org/index.html#ADME Toxicity
- https://www.researchgate.net/post/How_can_I_predict_the_insilico_ADMET_toxicity_of_a_new_drug
- http://www.acdlabs.com
- http://www.simulations-plus.com/
- http://www.fujitsu.com/jp/group/kyushu/en/
- http://www.cyprotex.com/
- https://www.scm.com/
- http://accelrys.com/
- https://www.chemaxon.com/
- http://www.systems-biology.com/sb.html
- https://preadmet.bmdrc.kr/
- http://www.compudrug.com/
- https://www.schrodinger.com/
- http://www.moldiscovery.com/

Selected Reading

- Bonate, P.L. (2011) *Pharmacokinetic-Pharmacodynamic Modeling and Simulation*, 2nd edn, Springer, New York, 618 pp.
- Gabrielsson, J. and Weiner, D. (2007) *Pharmacokinetic and Pharmacodynamic Data Analysis: Concepts and Applications*, 4th edn, Swedish Pharmaceutical Press, Stockholm, Sweden, 1250 pp.
- Shargel, L. and Yu, A. (2015) *Applied Biopharmaceuticals & Pharmacokinetics*, 7th edn, McGraw-Hill Education/Medical, New York, 928 pp.

References

[1] Beresford, A.P., Selick, H.E., and Tarbit, M.H. (2002) *Drug Discovery Today*, **7**, 109–116.
[2] Kennedy, T. (1997) *Drug Discovery Today*, **2**, 436–444.
[3] Cherkasov, A., Muratov, E.N., Fourches, D. et al. (2014) *J. Med. Chem.*, **57**, 4977–5010.
[4] Gola, J., Obrezanova, O., Champness, E., and Segall, M. (2006) *QSAR Comb. Sci.*, **25**, 1172–1180.
[5] Norinder, U. and Bergstrom, C.A.S. (2006) *ChemMedChem*, **1**, 920–937.
[6] Hou, T.J. and Wang, J.M. (2008) *Expert Opin. Drug Metab. Toxicol.*, **4**, 759–770.
[7] Gleeson, M.P., Hersey, A., and Hannongbua, S. (2011) *Curr. Top. Med. Chem.*, **11**, 358–381.
[8] Maltarollo, V.G., Gertrudes, J.C., Oliveira, P.R., and Honorio, K.M. (2015) *Expert Opin. Drug Metab. Toxicol.*, **11**, 259–271.
[9] Tao, L., Zhang, P., Qin, C., Chen, S.Y., Zhang, C., Chen, Z., Zhu, F., Yang, S.Y., Wei, Y.Q., and Chen, Y.Z. (2015) *Adv. Drug Delivery Rev.*, **86**, 83–100.
[10] Gasteiger, J. (2006) *J. Med. Chem.*, **49**, 6429–6434.
[11] Leardi, R. and Terrile, M. (1992) *J. Chemom.*, **6**, 267–281.
[12] OECD Principles for the Validation, for Regulatory Purposes, of (Quantitative) Structure-Activity Relationship Models, http://www.oecd.org/env/ehs/risk-assessment/37849783.pdf (accessed January 2018).
[13] Jorgensen, W.L. and Duffy, E.M. (2002) *Adv. Drug Delivery Rev.*, **54**, 355–366.
[14] Lipinski, C.A., Lombardo, F., Dominy, B.W., and Feeney, P.J. (1997) *Adv. Drug Delivery Rev.*, **23**, 3–25.
[15] Yalkowsky, S.H. and Valvani, S.C. (1980) *J. Pharm. Sci.*, **69**, 912–922.
[16] Jain, N. and Yalkowsky, S.H. (2001) *J. Pharm. Sci.*, **90**, 234–252.
[17] Grime, K.H., Barton, P., and McGinnity, D.F. (2013) *Mol. Pharmaceutics*, **10**, 1191–1206.
[18] Tetko, I.V. and Poda, G.I. (2013) *Mol. Pharmaceutics*, **10**, 381–406.
[19] Kühne, R., Ebert, R.-U., Kleint, F., Schmidt, G., and Schüürmann, G. (1995) *Chemosphere*, **30**, 2061–2077.

[20] Klopman, G. and Zhu, H. (2001) *J. Chem. Inf. Comput. Sci.*, **41**, 439–445.
[21] Hou, T.J., Xia, K., Zhang, W., and Xu, X.J. (2004) *J. Chem. Inf. Comput. Sci.*, **44**, 266–275.
[22] Huuskonen, J. (2000) *J. Chem. Inf. Comput. Sci.*, **40**, 773–777.
[23] Tetko, I.V., Tanchuk, V.Y., Kasheva, T.N., and Villa, A.E.P. (2001) *J. Chem. Inf. Comput. Sci.*, **41**, 1488–1493.
[24] Liu, R.F. and So, S.S. (2001) *J. Chem. Inf. Comput. Sci.*, **41**, 1633–1639.
[25] Yan, A.X. and Gasteiger, J. (2003) *QSAR Comb. Sci.*, **22**, 821–829.
[26] Yan, A.X. and Gasteiger, J. (2003) *J. Chem. Inf. Comput. Sci.*, **43**, 429–434.
[27] Gasteiger, J. (1988) Empirical methods for the calculation of physicochemical data of Organic compounds, in *Physical Property Prediction in Organic Compounds* (eds C. Jochum, M.G. Hicks, and J. Sunkel), Springer Verlag, Heidelberg, pp. 119–138.
[28] ADRIANA.Code has now been renamed CORINA Symphony Descriptors Community Edition and can be accessed on the web: https://www.mn-am.com/services/corinasymphonydescriptors (accessed January 2018).
[29] Hemmer, M.C., Steinhauer, V., and Gasteiger, J. (1999) *Vibrat. Spectrosc.*, **19**, 151–164.
[30] Sadowski, J. and Gasteiger, J. (1993) *Chem. Rev.*, **93**, 2567–2581. https://www.mn-am.com/products/corina (accessed January 2018).
[31] Yan, A.X., Gasteiger, J., Krug, M., and Anzali, S. (2004) *J. Comput.-Aided Mol. Des.*, **18**, 75–87.
[32] Huuskonen, J., Livingstone, D.J., and Manallack, D.T. (2008) *SAR QSAR Environ. Res.*, **19**, 191–212.
[33] Raevsky, O.A., Polianczyk, D.E., Grigorev, V.Y., Raevskaja, O.E., and Dearden, J.C. (2015) *Mol. Inf.*, **34**, 417–430.
[34] Karelson, M. and Dobchev, D. (2011) *Expert Opin. Drug Discovery*, **6**, 783–796.
[35] Zhao, Y.H., Abraham, M.H., Ibrahim, A., Fish, P.V., Cole, S., Lewis, M.L., de Groot, M.J., and Reynolds, D.P. (2007) *J. Chem. Inf. Model.*, **47**, 170–175.
[36] Adenot, M. and Lahana, R.J. (2004) *J. Chem. Inf. Comput. Sci.*, **44**, 239–248.
[37] Li, H., Yap, C.W., Ung, C.Y., Xue, Y., Cao, Z.W., and Chen, Y.Z. (2005) *J. Chem. Inf. Model.*, **45**, 1376–1384.
[38] Wang, Z., Yan, A.X., and Yuan, Q.P. (2009) *QSAR Comb. Sci.*, **28**, 989–994.
[39] Brito-Sánchez, Y., Marrero-Ponce, Y., Barigye, S.J., Perez, C.M., Le-ThiThu, H., and Cherkasov, A. (2015) *Mol. Inf.*, **34**, 308–330.
[40] Gupta, S., Basant, N., and Singh, K.P. (2015) *SAR QSAR Environ. Res.*, **26**, 1–30.
[41] Yan, A., Liang, H., Chong, Y., Nie, X., and Yu, C. (2013) *SAR QSAR Environ. Res.*, **24**, 61–74.
[42] Silva, F.T. and Trossini, G.H.G. (2014) *Med. Chem.*, **10**, 441–448.
[43] Geerts, T. and Vander Heyden, Y. (2011) *Comb. Chem. High Throughput Screening*, **14**, 339–361.
[44] Hou, T.J., Wang, J.M., and Li, Y. (2007) *J. Chem. Inf. Model.*, **47**, 2408–2415.
[45] Suenderhauf, C., Hammann, F., Maunz, A., Helma, C., and Huwyler, J. (2011) *Mol. Pharmaceutics*, **8**, 213–224.

[46] Basant, N., Gupta, S., and Singh, K.P. (2016) *Comput. Biol. Chem.*, **61**, 178–196.
[47] Hou, T.J., Wang, J.M., Zhang, W., and Xu, X.J. (2007) *J. Chem. Inf. Model.*, **47**, 208–218.
[48] Yan, A., Wang, Z., and Cai, Z. (2008) *Int. J. Mol. Sci.*, **9**, 1961–1976.
[49] Wang, J.M. and Hou, T.J. (2015) *Adv. Drug Delivery Rev.*, **86**, 11–16.
[50] Fraczkiewicz, R., Lobell, M., Goller, A.H., Krenz, U., Schoenneis, R., Clark, R.D., and Hillisch, A. (2015) *J. Chem. Inf. Model.*, **55**, 389–397.
[51] Wang, N.N., Dong, J., Deng, Y.H., Zhu, M.F., Wen, M., Yao, Z.J., Lu, A.P., Wang, J.B., and Cao, D.S. (2016) *J. Chem. Inf. Model.*, **56**, 763–773.
[52] Khajeh, A. and Modarress, H. (2014) *SAR QSAR Environ. Res.*, **25**, 35–50.
[53] Basant, N., Gupta, S., and Singh, K.P. (2016) *SAR QSAR Environ. Res.*, **27**, 67–85.
[54] Vallianatou, T., Lambrinidis, G., and Tsantili-Kakoulidou, A. (2013) *Expert Opin. Drug Discovery*, **8**, 583–595.
[55] Colmenarejo, G., Alvarez-Pedraglio, A., and Lavandera, J.L. (2001) *J. Med. Chem.*, **44**, 4370–4378.
[56] Wang, Z., Chen, Y.Y., Liang, H., Bender, A., Glen, R.C., and Yan, A.X. (2011) *J. Chem. Inf. Model.*, **51**, 1447.
[57] Poongavanam, V., Haider, N., and Ecker, G.F. (2012) *Bioorg. Med. Chem.*, **20**, 5388–5395.
[58] Wang, Z., Yan, A.X., Yuan, Q.P., and Gasteiger, J. (2008) *Eur. J. Med. Chem.*, **43**, 2442–2452.
[59] Tian, S., Li, Y.Y., Wang, J.M., Zhang, J., and Hou, T.J. (2011) *Mol. Pharmaceutics*, **8**, 841–851.
[60] Shugarts, S. and Benet, L.Z. (2009) *Pharm. Res.*, **26**, 2039–2054.

6.10 Prediction of Xenobiotic Metabolism

Anthony Long and Ernest Murray

Lhasa Limited, Granary Wharf House, 2 Canal Wharf, Holbeck, Leeds, LS11 5PS West Yorkshire, UK

Learning Objectives

- To discuss the importance of xenobiotic biotransformation in the life sciences.
- To describe the difference between local and global methods.
- To discuss how scientists use metabolism information in different ways.
- To review how biotransformation information can be used in medicinal chemistry and rational drug design, metabolite identification, the augmentation of existing metabolomic data, and the search for putative adduct-forming metabolites and pathways of toxification.

Outline

6.10.1 Introduction: The Importance of Xenobiotic Biotransformation in the Life Sciences, 359
6.10.2 Biotransformation Types, 362
6.10.3 Brief Review of Methods, 364
6.10.4 User Needs: Scientists Use Metabolism Information in Different Ways, 370
6.10.5 Case Studies, 372

6.10.1 Introduction: The Importance of Xenobiotic Biotransformation in the Life Sciences

We can think about the interaction of foreign chemicals with a living organism in two different but ultimately related ways. In pharmacology we speak of pharmacodynamic effects (what a substance does to the body) – how a medicinal compound interacts with a biological receptor to effect a biochemical adjustment and, hopefully, a downstream therapeutic physiological change. Pharmacokinetics (what a body does to the substance) describes the processes by which compounds are taken into the body – through the skin and gastrointestinal or respiratory tracts (absorption); how they move throughout the body to different organs and tissues (distribution); how they are chemically modified (metabolism); and finally how they are removed from the body (excretion). These four collected phenomena (*absorption, distribution, metabolism,* and *excretion*) are sometimes abbreviated to *ADME* properties. Although people often used the term metabolism to mean any ADME property in general, in this section the terms metabolism and biotransformation will be used interchangeably according to this definition:

Applied Chemoinformatics: Achievements and Future Opportunities, First Edition.
Edited by Thomas Engel and Johann Gasteiger.
© 2018 Wiley-VCH Verlag GmbH & Co. KGaA. Published 2018 by Wiley-VCH Verlag GmbH & Co. KGaA.

"The chemical modifications made by an organism on a xenobiotic [1], and these could include (as well as medicinal compounds): plant and fungal toxins, cosmetics, fragrances and other personal products, food and drink additives, agrochemicals, other industrial chemicals or environmental contaminants."

In the literature of drug metabolism, the term "drug-metabolizing enzyme" is often used – that is, enzymes that specifically metabolize xenobiotics as opposed to enzymes that catalyze the chemical transformations of endogenous biochemistry. Biochemistry is the study of the chemical interconversions associated with life-sustaining processes and involves things like the breaking down of large molecules, for example, lipid degradation, processes associated with growth and maturation and the synthesis of large molecules from small ones such as polypeptides and proteins from amino acids, regulatory pathways such as gene expression, homeostasis, and cell signaling, and energy-producing pathways such as glycolysis, the Krebs tricarboxylic (or citric) acid cycle, oxidative phosphorylation, and so on (see Section 4.2). The need of living organisms to limit the uptake of and to eliminate biomaterial is an evolutionary inevitability as the endless accumulation of physiologically useless molecules, even if they are not toxic, is incompatible with survival of the organism. Multiple mechanisms have evolved to facilitate this such as physical elimination by passive or active excretion, but when the structure of a chemical invader is such that it resists direct removal, chemical modification of the molecule can help overcome barriers to elimination. Xenobiotic metabolism can be thought of as having evolved as a mechanism of defense against such chemical invasion. Most biotransformations are catalyzed by the so-called drug-metabolizing enzymes. These enzymes share, to a variable extent, characteristics that apparently contrast them from the enzymes of endogenous biochemistry:

- They have low substrate specificities and can be highly promiscuous for large varieties of structurally diverse substrates.
- In some cases they demonstrate low chemoselectivities being able to catalyze the formation of more than one type of functional group.
- They tend to have low catalytic rates, which are sometimes compensated for by the presence of very large amounts of the enzyme.
- They have a distinct preference for lipophilic substrates that they convert to more polar substrates, which may be more easily excreted so that loss of water from the body is minimal.
- Many enzymes are inducible by xenobiotics, particularly by some of their own substrates.

These characteristics of drug-metabolizing enzymes support the evolutionary theory of generalized, nonselective defense rather than the development of selective or specific biofunctionality. However, a sharp delineation between the two classes does not exist, and they have many overlapping aspects. For example, they share many common cofactors, such as NADPH, and common electron transfer enzymes, such as cytochrome b_5. Some drugs, because of their chemical similarity to metabolome components, are modified by enzymes of endogenous biochemistry.

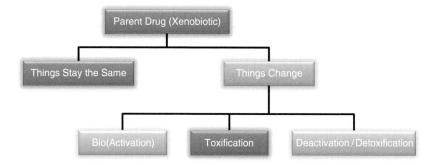

Figure 6.10.1 Generalized scheme of the consequences of metabolic biotransformation.

The pharmacodynamic consequences of drug metabolism can be placed into two broad categories, one of which can be subdivided into three subcategories (Figure 6.10.1).

In the first main category we can state that "things stay the same." This is the case where biotransformation generates one or more metabolites that have a very similar chemical, physicochemical, and ligand binding characteristics to the parent drug and consequently express pharmacology very similar to the parent, for example, as an agonist or antagonist. In the second main category, we can state that "things change." This is the case where biotransformation generates one or more metabolites that have different chemical and physicochemical characteristics and biological affinities and, consequently, could exhibit very diverse pharmacologies. Diversity in this context may indicate that the metabolites are different from each other, different from the parent drug, or a combination of these two effects. In the first subcategory we can state that "things change for the better." Enzymatic modification of a drug can lead to such chemical changes that increase the binding potential of a metabolite relative to the parent drug at a target receptor, and this may result in an increase in the desired activity. This effect can be deployed as a deliberate strategy in drug design. The administered compound, usually because it has a more favorable pharmacokinetic profile, is designed to be metabolized into the active molecule – under these conditions the administered compound is called a prodrug. In the second subcategory we can state that "things change for the worse." Enzymatic modification of a drug can lead to such chemical changes that increase the binding potential of a metabolite relative to the parent drug at off-target receptors or to other biomolecules like DNA. This can lead to unfavorable pharmacology, unwanted side effects, and, in the extreme case, measurable apical toxicity. In the third subcategory we can state that "bioactivity is lost." Differences can lead to lower binding affinity at on- or off-target receptors, therefore leading to decreased activity of any sort. An overall decrease in lipophilicity and increase in water solubility can aid excretion. The effects within these three subcategories can be summarized as activation or bioactivation for the first, toxification for the second although this is also referred to as bioactivation, and deactivation or detoxification for the third. This is a very simplified scheme, and in reality a combination of all of these effects may occur and compete with each other for any particular drug. Paracelsus taught us, almost 500 years ago, that substances considered toxic are harmless in small doses, and conversely an ordinarily harmless substance can be deadly if overconsumed – for any individual xenobiotic the difference between a curative agent and a poison being, essentially, the dose administered. All of this is very true, but also, in considering the *in vivo* consequence of drug biotransformation, whether bioactivation, toxification, or deactivation predominates ultimately depends on the relative rates of all of the pathways involved.

6.10.2 Biotransformation Types

Enzyme-mediated biotransformations are biochemically categorized as either phase I or phase II. Phase I biotransformations are reactions of functionalization or refunctionalization: either a new functional group is introduced into the molecule or an existing functional group is modified. Phase I biotransformation is very diverse and involves oxidations, reductions, hydrolyses, and many other reaction types. Some reduction reactions in particular are associated with the activity of anaerobic gut bacteria. Some common phase I biotransformations are given in Table 6.10.1.

Table 6.10.1 Some common phase I biotransformation types.

Reaction type	Reaction subtype (functional groups affected)
Oxidation	Hydroxylation (carboaliphatic; carboaromatic; heteroaromatic)
	Epoxidation (alkenes; carboaromatics; heteroaromatics)
	Dealkylation and oxidative ring opening at (or alpha to) nitrogen, oxygen, and sulfur
	Oxidative dehalogenation
	Dehydrogenation (oxidation of alcohol; oxidation of nitrogen-containing functional groups; alkene formation; aromatic ring formation)
	Oxidation of aldehyde
	Oxidation at aliphatic or aromatic nitrogen
	Oxidation at aliphatic or aromatic sulfur
	β-Oxidation of carboxylic acids
Reduction	Carbonyl reduction
	Alkene reduction
	Reduction of nitrogen-containing functional groups (nitro compounds; nitroso compounds; hydroxylamines, oximes, and related compounds)
	Azo compounds
	Aliphatic or aromatic N-oxides
	Reductive dehalogenation
Non-redox	Hydrolyses (esters and lactones; amides and lactams; other carboxylic acid derivatives; nitrogen-containing functional groups: imines, oximes, nitriles, amidines, and so on)
	Hydrolytic dehalogenation
	Dehydration and hydration (including vicinal diols from epoxides)
	Decarboxylation

Some characteristics of drug-metabolizing enzymes were outlined earlier, and there are many families. The more important phase I enzymes include flavin monooxygenases (FMO), monoamine oxidases (MAO), molybdenum oxygenases (aldehyde and xanthine oxidases), peroxidases, alcohol and aldehyde dehydrogenases, esterases, and epoxide hydrolases (EH). However, preeminent among the phase I enzymes are the cytochrome P450s – more commonly shortened to CYP. CYPs typically function as monooxygenases, peroxidases, and reductases. They are not just found in mammals but in all animals and indeed across the entire kingdom of living things (over 8.5 million estimated species). They area superfamily of hemoproteins with approximately 1200 gene protein sequences known, but this number continues to increase. These proteins are classified into families, subfamilies, and individual family members, based on degrees of similarity in their primary structures, in other words the amino acid sequences. Of all known CYP genes, 57 of these code for human cytochromes (plus a few pseudogenes), which are divided into 18 families and 43 subfamilies. As an example of an individual isoform, CYP3A4 denotes CYP family 3, subfamily A, family member 4. There are finer variations than family members within the superfamily, and these are referred to as genetic polymorphisms – also called single nucleotide polymorphisms – and they are stable allelic variants of the gene that codes for an individual family member. It is important to understand these phenotypic differences as alternate polymorphic isoforms can influence the rate of metabolism of a particular drug – altering its pharmacokinetic properties. As an example, 5–10% of Caucasians are poor metabolizers with respect to the CYP2D6 isoform. Ignoring this fact when determining the dosing regimen for a drug metabolized either exclusively or predominantly by this isoform can lead to, for example, unwanted drug–drug interactions (DDIs) or accidental overdosing. Some of the more important xenobiotic-metabolizing cytochromes that are expressed in humans are listed in Table 6.10.2.

Phase II biotransformations are reactions of conjugation – a medium molecular weight endogenous molecule (supplied by a cofactor) is attached to the substrate through an existing functional group. Reactions are catalyzed by transferase enzymes. There are five main categories, three of which (glucuronidation, sulfonation, and glutathione conjugation) produce metabolites that are usually

Table 6.10.2 Some of the important xenobiotic-metabolizing CYPs in humans.

Family	Subfamily	Family members
1	A	CYP1A1, CYP1A2
2	A	CYP2A6
	B	CYP2B6
	C	CYP2C8, CYP2C9, CYP2C19
	D	CYP2D6
	E	CYP2E1
3	A	CYP3A4, CYP3A5

more water soluble than their substrates. Acetylation and methylation produce metabolites that are generally slightly more lipophilic than their substrates and so typically less water soluble. All of these reactions with the exception of glutathione conjugation, which interacts with electrophilic centers, involve conjugation with nucleophilic functional groups (like hydroxy or amino) in the substrate. Excretion of xenobiotics is often (but by no means always) mediated by a sequence of biotransformations – a phase I reaction introducing or modifying a functional group followed by a phase II reaction, which introduces a water-solubilizing moiety into the molecule. Some common phase II biotransformations together with their enzymes and cofactors are given in Table 6.10.3.

Table 6.10.3 Some common phase II biotransformation types.

Reaction type	Functional groups affected	Enzyme/cofactor
Glucuronidation	Phenols, alcohols, carboxylic acids, amines, amides and related compounds, aromatic heterocycles, thiols	Uridine 5′-diphospho-glucuronosyltransferase/uridine diphosphoglucuronic acid
Sulfonation	Phenols, alcohols, amines, hydroxylamines	Sulfotransferase/3′-phosphoadenosine-5′-phosphosulfate
Acetylation	Amines, hydrazines, hydrazides, sulfonamides	N-Acetyltransferase/acetyl coenzyme A
Methylation	Catechols and pyrogallols, amines and amides, aromatic heterocycles, thiols	Methyltransferase/S-adenosylmethionine
Glutathione addition	Electrophilic functional groups: (alkyl halides, epoxides, quinones and related compounds, α,β-unsaturated compounds, aromatic compounds, and so on)	Glutathione-S-transferase/reduced glutathione

6.10.3 Brief Review of Methods

Paradoxically, there is a desire to be able to predict the outcome of xenobiotic metabolism accurately; however the phenomenon itself is notably promiscuous in as much as the enzymes involved usually have low substrate chemo-, regio-, and stereoselectivities (there are, however, many exceptions). For example, CYP3A4 (an isoform of cytochrome P450 found in humans, which we met in Section 6.10.2) metabolizes over 50% of known drugs – many thousands of diverse compounds. These drugs are a few decades old, but the enzyme itself is many millions (if not some billions) of years old. The evolutionary characterization of these biomolecules has not involved rapid adaptive responses to environmental change but rather remarkable conservation of the ancestral gene and retention of such enzyme-substrate promiscuity as it enables generalized chemical defense. Furthermore, despite the rapid and impressive sophistication of chromatographic and analytical techniques, notably mass spectrometry (MS) and NMR, it is often impossible (in the absence of reliable synthetic standards) to establish a full constitutional chemical characterization of all metabolites formed in a study due to the often very small amounts of material involved. Prediction of xenobiotic metabolism has proven useful in pharmacokinetics and toxicology, medicinal chemistry, and drug design and as an aid in the structural characterization of metabolites, among other applications. There are many methods and systems available for this task, and they can all be broadly categorized as local methods or global methods.

6.10.3.1 Local Methods

Local methods are those that attempt to predict specific xenobiotic biotransformation in a restricted manner for a single enzyme (or group of enzymes) or for a single reaction (or reaction type). They can be ligand based (i.e., the models consider only the two- or three-dimensional structure of potential substrates), structure based (the models consider, typically, the characteristics of the catalytic site of the specific drug-metabolizing enzyme), or a combination of these. Models may need to be built on data or can be purely theoretical in nature.

6.10.3.1.1 Quantitative Structure–Activity Relationships (QSARs)

Of the classic data-requiring, ligand-based local methods, two-dimensional quantitative structure–activity relationships (QSARs) are, perhaps, the best established. A dataset consists of negative observations (lack of reactivity) and positive observations – chemical modification at a particular reaction center or of a specific biotransformation type is needed. The data is typically split into a training set and a test set for the purpose of developing the QSAR. Given the importance of the CYPs in xenobiotic metabolism, much effort has been expended in building QSARs for this class. Models have been reported for the major CYP isoforms expressed in human hepatocytes including 1A2, 2C9, 2C19, 2D6, and 3A4. Two-dimensional QSARs for cytochrome activity typically contain a lipophilicity term, but other chemical (both steric and electronic) and physicochemical descriptors have also been used. Best results are observed when the chemical space definable for the test set is similar to the training set (the applicability domain) but less reliable for highly diverse sets of structures. Three-dimensional QSAR approaches to modeling CYP activity have employed pharmacophore descriptors, typically derived over multi-conformational space. Comparative molecular field analysis (CoMFA) approaches have also been developed, and this can be a useful technique for sets of compound containing fairly rigid scaffolds that can be reliably overlaid (see Chapter 2). Besides chemical reactivity, QSARs have been developed for the prediction of CYP induction and inhibition, the latter being useful in the assessment of unfavorable DDIs . Other phase I enzymes, such as FMO and EH, have been the subjects of QSAR investigations as well as a limited number of phase II enzymes: uridine 5'-diphospho-glucuronosyltransferase (UGT) and sulfotransferase (SULT) [2].

6.10.3.1.2 Quantum Mechanical and Molecular Modeling Methods

Oxidation of a substrate by CYPs is a complex process, and multiple mechanisms have been characterized. The simplest case is monooxygenation – the formal insertion of an oxygen atom, derived from molecular oxygen, between the carbon and hydrogen atoms in an aliphatic moiety effecting an overall hydroxylation. A simplified version of this catalytic cycle is shown in Figure 6.10.2a.

At the catalytic center of the CYP active site is the iron-heme prosthetic group (Figure 6.10.2b). Heme is a member of the class of cyclic tetrapyrroles known as the protoporphyrin IX system, and the nitrogen atoms of this macrocycle occupy four out of the six available ligand binding sites of the central iron atom. The fifth ligand site is occupied by the sulfur atom from a cysteine residue in the protein (CYPs are sometimes called heme-thiolate proteins) – this is important for catalysis and may even have prebiotic origins in early iron sulfide-containing ores capable of catalyzing redox chemistry. The sixth binding site is occupied by a water molecule, and this is the ligand that is displaced when the enzyme binds molecular oxygen.

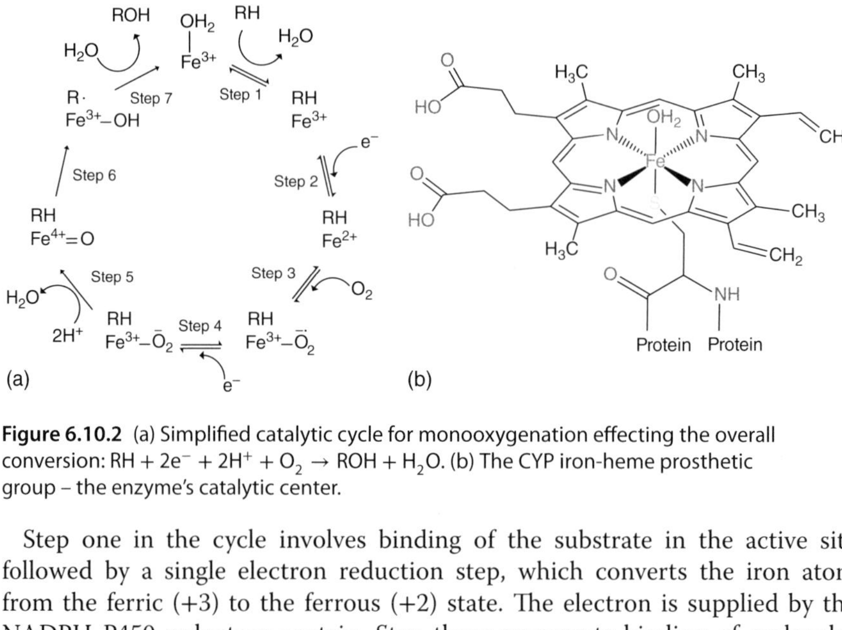

Figure 6.10.2 (a) Simplified catalytic cycle for monooxygenation effecting the overall conversion: $RH + 2e^- + 2H^+ + O_2 \rightarrow ROH + H_2O$. (b) The CYP iron-heme prosthetic group – the enzyme's catalytic center.

Step one in the cycle involves binding of the substrate in the active site followed by a single electron reduction step, which converts the iron atom from the ferric (+3) to the ferrous (+2) state. The electron is supplied by the NADPH P450 reductase protein. Step three represents binding of molecular oxygen to the iron atom at the catalytic center of the heme cofactor followed by step four, the second electron reduction step. Step five represents protonation of bound diatomic oxygen and then scission of the oxygen–oxygen bond with loss of water. Step six represents hydrogen abstraction from the alkyl group to form a short-lived carbon-centered radical (the hydrogen becomes bound to the oxygen atom) onto which the remaining single oxygen atom rebounds to form the hydroxylated product in step seven. This is followed by dissociation of the metabolite from the enzyme [3]. The hydrogen atom abstraction (step six) is sometimes the rate-limiting step of the cycle, and it can be useful to calculate the energy required to remove such a hydrogen atom from the substrate.

Quantum mechanical (QM) methods can be deployed to estimate the energy required for this hydrogen abstraction, and three available systems are very briefly discussed here (see also *Methods Volume*, Section 8.4). As will be seen, most methods these days are in some way hybrid methods combining elements of both the enzyme and the ligand structures. Working with optimized three-dimensional structures, Optibrium's StarDrop P450 prediction module [4] is a ligand-based method. Its algorithm combines hydrogen atom transfer energy calculations generated by the semiempirical AM1 method together with accessibility descriptors to suggest and rank possible sites of metabolism (SoMs) in the study compound. The lipophilicity of the compound is taken into account, and models for the important human hepatic CYPs have been developed. The Meta-Site algorithm [5] is another example of a hybrid method employing molecular interaction fields (MIFs) derived from enzyme structures with QM calculations to identify SoMs for CYP- and FMO-catalyzed reactions. A combination of goodness of fit of the potential substrate in the MIF and the molecular orbital calculations allows all viable SoMs to be ranked. The SMARTCyp algorithm [6] is another example of a QM method for the identification and ranking of SoMs for a number of human CYP isoforms. In this method density functional theory (DFT) is employed to calculate reaction energies for a large number of three-dimensional CYP substrates. Multiple orientations for each substrate are studied in which the carbon–hydrogen bond breaking to be modeled is correctly aligned to the oxygen bound to the heme prosthetic group (although modeling of the whole protein is not required in this technique). It can be shown that carbon–hydrogen bonds in similar proximal chemical environments give similar energies. These environments can be coded in SMARTS line notation (see *Methods Volume*, Chapter 2), allowing much faster pattern matching and energy estimation than running the DFT calculation anew each time. The final scoring and ranking of each SoM is effected using the precomputed activation energies along with a number of topological accessibility descriptors.

The application of structure-based methods to understanding metabolic biotransformation, that is, studying the detailed structure of the enzyme and its interaction with ligands, is challenging. Xenobiotic-metabolizing enzymes tend to have very large active sites that are extremely conformationally flexible and tend to lack specific pharmacophoric binding elements that can be characterized. Also, as many of these enzymes, including the CYPs, are membrane bound to the endoplasmic reticulum, they are difficult to isolate and characterize – the availability of X-ray crystal structures for many of these proteins is a relatively new achievement, and these do not cover the complete conformational space of the enzyme active sites. Nevertheless automated ligand docking has been applied with reasonable success to model substrate binding to the active site and exposition at its catalytic center. Molecular dynamics (MD) simulations have been used to model the conformation changes that occur during ligand binding, and combination techniques involving hybrid quantum mechanical/molecular mechanical (QM/MM) methods have taught us much about the mechanisms of enzyme-mediated metabolic biotransformations. In such hybrid methods, the details of the reaction centers are modeled using QM methods, while the environment (or the whole protein) are modeled using computationally less expensive MM methods.

6.10.3.2 Global Methods

Global methods are those that attempt to predict comprehensively for diverse biological systems taking into account multiple species, many enzymes (both phase I and phase II), and many functional groups and reaction types as well as being able to process for a wide range of small organic molecule types. They are necessarily ligand based rather than structure based and are data-requiring techniques. Two techniques have successfully been applied in this domain, which we discuss in this section – they are machine learning methods and the knowledge-based expert systems approach.

6.10.3.2.1 Machine Learning Methods

As an example of the machine learning approach, we describe the FAst MEtabolizer (FAME) software [7]. FAME uses a collection of random forest models to predict SoMs. Models are built using diverse chemical datasets of more than 20,000 molecules annotated with their experimentally determined SoMs. A large and diverse chemical space is covered within these models and includes drugs, drug-like molecules, endogenous metabolites, and natural products. FAME utilizes six atom descriptors (total partial charge, Gasteiger–Marsili sigma partial charges and sigma electronegativity, pi electronegativity, Sybyl atom type for a specific atom – encoding element type and hybridization state) and one topological descriptor (maximum topological distance between two atoms of a molecule) in its models. Specific models are available for human, rat, and dog metabolism, and prediction can be restricted to phase I or phase II metabolism only. FAME is fast (2–3 s per molecule) and accurate in as much as it is able to identify at least one known SoM among the top 1, top 2, and top 3 highest ranked atom positions in up to 71%, 81%, and 87% of all cases tested, respectively. Results are visualized as a three-dimensional *"hot spot"* diagram with SoMs highlighted by colored spheres, the size of which reflects the probability of metabolism at that site.

6.10.3.2.2 Knowledge-Based Expert Systems

Knowledge-based expert systems for prediction of metabolism have been around for a long time and systems include MetabolExpert [8], Meta-PC [9], MetaDrug [10], TIMES [11], and Meteor Nexus [12]. Expert systems have many features in common. A reasoning (or inference) engine solves problems (or makes predictions) by applying components from a knowledge base in response to single or multiple queries (or hypotheses). For metabolism prediction the knowledge base typically consists of two components: a biotransformation dictionary, which consists of structure–metabolism relationships (SMRs) expressed as generic reaction descriptions (biotransformations), and a rule base, which is used by the program's reasoning engine to discriminate between all possible metabolic outcomes and the most likely ones. The query molecule is matched against the biotransformation dictionary and metabolites generated. The rules are employed to trim this list of reactions according to any processing constraints that the user sets on the system. The results are then organized and displayed as a metabolic tree – the query molecule at the top of the tree generating first-level metabolites (children) and each child metabolite potentially generating second-level metabolites (grandchildren) and so as we go to deeper levels (longer sequences) so the tree broadens. Systems usually have facilities to create exportable and printable reports from such trees. In the Meteor Nexus system, such rules that are used to rank metabolites encode various sorts of information including frequency of occurrence of the reaction in the literature (or dataset) relative to the functional group or substructure activating the particular biotransformation, the depth in the reaction sequence (the longer the sequence gets, the more likely a phase II reaction is to happen), the substrate lipophilicity, the molecular weight of the substrate, and, occasionally, the selected species. In the next section we will briefly describe a hybrid approach in which the Meteor knowledge base is combined with a machine learning method.

6.10.3.2.3 A Composite Knowledge Base/Machine Learning Method

Using the rules and the sorts of information outlined in Section 6.10.3.2.2, previous versions of Meteor Nexus have graded biotransformations into five levels in decreasing order of likelihood: probable, plausible, equivocal, doubted, and improbable. This achieves some sensitivity in prediction as a user can ask, for example, only to see metabolites at the probable and plausible levels; but this approach has some limitations. First, the approach lacks granularity – asking to see metabolites only at the higher levels of likelihood means that some predictions will be missed; moving to lower levels of likelihood means that it is possible to generate too many metabolites (this is a question of balancing sensitivity and specificity, which are discussed in Section 6.10.4.3). Second, the rule base can be thought of as a static model, which addresses the demand: "tell me what you think might normally happen to a molecule with these functional groups/structural characteristics." A dynamic model would better address the question: "what will happen to my specific molecule?" – a more customized approach, which is what most people would wish to know. The new composite knowledge base/machine learning method addresses the granularity issue as well as being fine-tuned to the query structure.

The first part of the analysis under the new system [13] is the same, the query structure is matched against the biotransformation dictionary, and list of possible metabolites is generated – this answers the question: "what reactions could occur?" In the second part of the analysis, the traditional rule base has been replaced with a model that has been machine-learned on a database of experimental metabolic reactions. For each reaction in the database the SoM or SoMs in the reactants have been identified and characterized as an extended atom-centered fingerprint. The fingerprints are hash-coded into a binary data representation, and this information is stored in the model. At runtime, a binary representation of the SoM of the query compound for each biotransformation activated in stage one is generated using the same method, and for each biotransformation a Tanimoto similarity for the query atom-centered fingerprint and the atom-centered fingerprints for appropriate database entries (those expressing the correct biotransformation) is generated.

The eight (default) nearest neighbors from the database are then selected, and a score is generated. The score is a function of the Tanimoto similarities between the nearest neighbors and the query and whether the nearest neighbors have been observed to undergo the biotransformation or not. The score is expressed as a number between 0 and 1000, which should be thought of as a measure of the confidence that the predicted biotransformation will occur. It is important to remember that because a model may generate numerical outputs, this does not necessarily mean that the model predicts quantitatively – this is often, as here, not the case. Because of the extended radius of the fingerprint, much information about the proximal environment of the reaction center is captured. Also there is an option to select nearest neighbors that are close in molecular weight to the query. Both of these features help address the question: "which biotransformations are more likely to occur for my particular molecule?" – in other words a specific rather than a general model. We will see an application of this scoring method in the toxicology case study presented in Section 6.10.5.4.

6.10.4 User Needs: Scientists Use Metabolism Information in Different Ways

A computer system that attempts to predict the metabolic fate of chemicals ideally needs to address several use cases. The optimum output for these different use cases should be very different and deliver variable amounts of relevant information to the user. In this discussion, and as an approximation only, we will use the terms *sensitivity*, in terms of indicating a low rate of false negative predictions, and *specificity*, in terms of indicating a low rate of unconfirmed positive predictions. The term unconfirmed positive is preferred to false positive because *"absence of proof" is not "proof of absence."* In fact, in the vast majority of predictive analyses, many more metabolites are generated than are actually observed and reported, and there are a number of valid, understandable reasons for this, which are indicated later. Swinging the balance toward high sensitivity and lower specificity with the generation of larger trees can be a use case-dependent advantage and not a detriment. This is particularly true when searching for potential metabolic pathways of toxification or when seeking assistance in the chemical characterization of metabolites. This does not infer, of course, that prediction is indiscriminate without a care to unconfirmed positive predictions, and this is why results for individual biotransformations are moderated with confidence ratings or relative rankings. It is possible to list a number of advantages in the generation of larger high sensitivity trees as follows:

- The suggestion of nonobvious alternative routes, comprehensively without the burden of human bias, to observed or putative metabolites.
- The provision of mechanistic insight and the addition of didactic value to the analysis.
- To avoid suspicion in the user – any predictive system (in any knowledge domain) that can only predict what is observed strongly suggests a highly overtrained model.

- The ability to scan for potential pharmacologically/toxicologically important metabolites.
- To provide insight into metabolites that cannot be chromatographically characterized.
- To provide insight into metabolites that cannot be chemically characterized.
- To provide insight into metabolites that may be observable in alternative studies – it is rare for all known metabolites of a parent compound to be observed and reported in a single study.
- To provide insight into the nature of minor or unusual metabolites.
- To provide insight into the nature of metabolism not picked up experimentally due to low metabolite concentrations, poor longevity, or low, *in vivo*, mass balance recovery.

6.10.4.1 Biotransformation in Medicinal Chemistry: Importance in Drug Discovery and Design

The drug design (discovery) use case requires high specificity in metabolism prediction. Medicinal chemists need to address questions like "Where in the molecule are the sites of high metabolic liability?" "Which sites do I need to block or modify?" "Will my analog molecule(s) be metabolized in a similar way?" "How can I improve the metabolic stability of analog compounds and so slow down plasma clearance and maximize exposure?" In this use case it is important to predict only the most likely SoMs rather than all of the possible sites. While most available systems attempt to discriminate between all possible sites, the desirability of increased granularity and specificity of predictions is still high. Predicting for all possible sites (high sensitivity) is not useful in this use case as in reality many of these sites are not affected by metabolism to a significant extent. Having identified metabolically susceptible sites, the medicinal chemist will generally require detailed information regarding neither the metabolic pathways initiated at those sites nor the nature of the metabolites expressed within those pathways. For this reason systems that generate, say, the "top n" metabolic sites and display them as a simple annotated parent structure (sometimes called a "*hot spot*" diagram) are attractive (although often only cytochrome-mediated metabolism is represented) as they answer the questions quickly, concisely, and with the right level of summary information. This is especially true when designing lead compounds from target hits, whereas developing lead compounds toward clinical candidates may require more information as multiparametric optimization (considering the factors that enable the expression of good pharmacokinetics, high efficacy, and low toxicity all at the same time) is the most efficient approach.

6.10.4.2 Biotransformation in Drug Metabolism and Pharmacokinetics (DMPK): Metabolite Identification

The metabolite identification use case requires high sensitivity. It is relatively straightforward to assign obvious mass spectrometric observations to a putative metabolite structure. Addition of 16 mass units is a hydroxylation, oxygenation, or epoxidation; loss of 14 mass units is a demethylation; and so on. Analysis of fragmentation patterns is also useful as this may indicate the region of the molecule that has changed. However when the molecular mass of a metabolite is such that it is not obvious to interpret in terms of a core biotransformation (or combination thereof), an extensive simulated metabolic profile can be useful. Typically the parent compound can be entered into the program and analyzed using relaxed processing constraints, generating a large metabolic tree showing the more likely and less likely pathways together. The system calculates the molecular mass of each metabolite as it is formed and stores this on the final metabolic tree. It is then possible to search or filter the tree for those metabolites that have a specific molecular mass, a molecular mass within a requested range, or satisfy a "greater than" or "less than" molecular mass filter. Pathways that fail on these filtering criteria are collapsed or hidden, leading to an overall simplification of the tree. Because such data reduction is possible and still addresses the fundamental question being asked, the overall number of metabolites, or indeed our confidence in the prediction of those metabolites, does not really matter. High sensitivity for this use case is an advantage and not a detriment – unusual or rarely observed biotransformation combinations would not be expected to yield the most likely predicted metabolites. The technique is particularly useful if a molecule undergoes post-enzymatic rearrangement or when a molecule undergoes extensive breakdown as a result of, for example, a hydrolytic or oxidative dealkylation processes in which the substrate is cleaved at an unknown site somewhere in the middle of the molecule. In the latter case, precomputed mass differences are not useful as an aid to metabolite identification.

6.10.4.3 Biotransformation in Toxicology: The Search for Putative Toxic Metabolites

The toxicity assessment use case requires both high sensitivity and moderate specificity and is perhaps the most difficult to address. High sensitivity is required because we would not wish to miss any low likelihood predicted metabolites that may contain alerting features of toxicological concern – pharmacologically important metabolites are not always the major or most likely ones. Moderate specificity is required because if all possible alerting metabolites are generated indiscriminately, which ones should actually raise our level of concern? Herein lies another paradox – metabolites of toxic concern are sometimes minor ones and in fact can be near impossible to detect in experimental systems. This is often because they do not attain a steady-state concentration, either due to slow rates of formation or poor longevity attributable to intrinsic molecular instability or chemical reactivity.

6.10.5 Case Studies

In this final section we present four case studies, which foreground the usefulness of *in silico* metabolite prediction across four different domains, namely, drug design in medicinal chemistry, metabolite identification in forensic toxicology, the search for novel components in metabolomics, and the search for putative toxic metabolites and pathways of toxification. Three of these case studies come from the published literature, and one is a recent in-house investigation. The main focus is on Lhasa Limited's Meteor software as it is the system with which the authors are, by far, most familiar.

6.10.5.1 *In Silico* Metabolism Prediction in Medicinal Chemistry and Drug Design: Indomethacin Analogs

In a study from 2009 [14], the MetaSite program was used to study the oxidative liabilities of a number of nonsteroidal anti-inflammatory indomethacin derivatives (Figure 6.8.3). Unlike indomethacin (**1**) itself, the phenylethyl amide derivative (**2**) is a potent and selective cyclooxygenase-2 (COX-2) inhibitor and a nonulcerogenic anti-inflammatory agent in the rat. However, unlike (**1**), which has an acceptable half-life in rat and human liver microsome incubations (>90 min), derivative (**2**) underwent extremely rapid clearance ascribable to a shift in metabolism from the methoxyphenyl group to the phenylethyl side chain. MetaSite's 3A4, 2C9, and 2D6 models were used to predict preferred SoMs of the virtual analogs (**3**, **4**, and **5**). These results predicted a shift in metabolism away from the amide side chain and back to the methoxyphenyl group (Figure 6.10.3). Subsequent synthesis and investigation of these three analogs showed extended half-life values relative to (**2**) as indicated in Figure 6.10.3. Metabolism of analogs (**3**), (**4**), and (**5**) was confirmed as being by oxidative demethylation at the methoxyphenyl group. The selective COX-2 inhibition demonstrated for (**2**) was retained in analogs (**3**) and (**4**), so demonstrating the usefulness of a metabolism prediction technique in the design of analog compounds with improved metabolic stabilities.

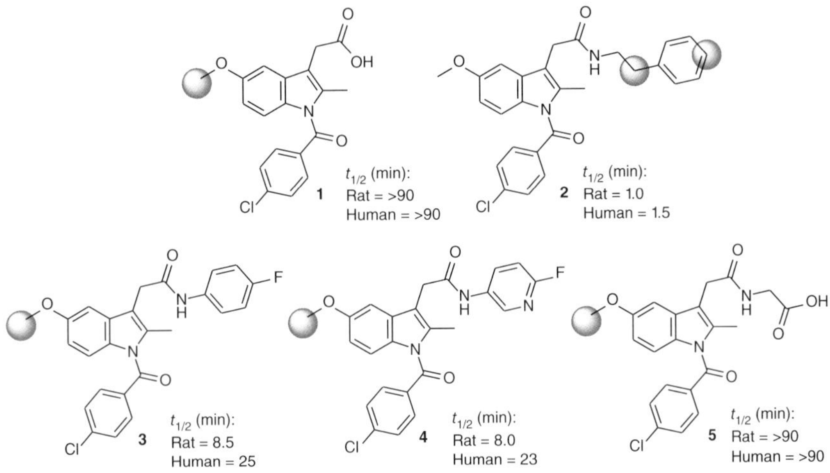

Figure 6.10.3 Indomethacin and a number of its amide derivatives (from Ref. [13]) with half-life values in minutes ($t_{1/2}$) indicated for rat and human liver microsomes. Preferred sites of metabolism for each analog as estimated by the MetaSite program are indicated by a grey circle. (Adapted from Marchant *et al.* 2016 [13].)

6.10.5.2 *In Silico* Metabolism Prediction and Metabolite Identification in Forensic Toxicology: Quetiapine

In a study from 2009 [15], the metabolism of quetiapine (**6**, Figure 6.10.4) was studied using a combination of analytical and *in silico* methods. Quetiapine is an antipsychotic agent with complex pharmacology affecting multiple receptor families (D_3, D_4, 5-HT_{2A}, 5-HT_{2C}, 5-HT_7, α_1-and α_2-adrenergic receptors) with a wide range of IC_{50} values and can be clinically indicated under low to high dosing regimens for conditions ranging from mild sleep and anxiety disorders to schizophrenia and acute manic episodes of bipolar disorder. It also has something of a reputation as a substance of abuse. Although it is extensively metabolized, the chromatographic behavior of its metabolites was not well documented, and synthetic reference standards were unavailable. The experiment used 10 human postmortem urine samples collected at autopsy.

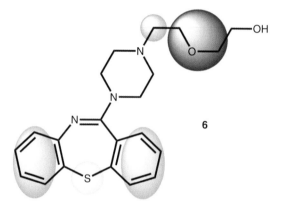

Figure 6.10.4 The structure of quetiapine (**6**). Areas of metabolic liability are indicted by grey colored spheres: *O*-dealkylation, *N*-dealkylation, sulfur oxidation, and carboaromatic hydroxylation.

Thirteen phase I metabolites of quetiapine were detected in urine samples using liquid chromatography/time-of-flight mass spectrometry (LC/ToFMS). The metabolites were associated with fragmentation of the flexible side chain by *O*- and *N*-dealkylation, oxidation at the ring sulfur, and hydroxylation in the aromatic rings (see grey spheres in Figure 6.10.4). Meteor predicted 14 phase I metabolites under default processing constraints, and eight of these were the same as those experimentally determined. The predicted metabolites were associated with metabolism in the grey colored spheres. Meteor did not show products of carboaromatic ring hydroxylation as the likelihood levels fell below the default constraints. In the grey "sulfur" zone Meteor predicted two metabolites: a sulfoxide (monooxygenation at the sulfur atom) and a sulfone (dioxygenation at the sulfur atom). Some of the metabolites here corresponded to sequences of biotransformations in all three zones, complete or partial fragmentation of the side chain, carboaromatic hydroxylation, and oxidation at the sulfur atom. In the absence of synthetic standards though, it was impossible to tell the difference between a sulfone and a sulfoxide with an additional hydroxyl group attached to the aromatic ring as their accurate masses are identical.

So, although useful suggestions, are the predicted sulfone metabolites to be believed? On this occasion the use of a second piece of software – ACD/MS Fragmenter – indicated that the MS fragmentation patterns for the hydroxy sulfoxides and the sulfones were different. Comparison of the experimentally observed fragmentations with those predicted by ACD/MS Fragmenter strongly supported the idea that the hydroxy sulfoxides and not the sulfones were formed on this occasion. The position of hydroxy substitution in the aromatic ring (the regiochemistry) remained undetermined as aromatic rings are not generally susceptible to very much fragmentation in the mass spectrometer. This paper highlights the value of computer methods (indeed, multiple methods) in the determination of metabolite structures found in complex biological matrices and in the absence of reliable synthetic reference standards.

6.10.5.3 *In Silico* Synthesis of Novel Biochemicals: The Search for Novel Metabolome Components

A metabolome can be defined as the complete set of small organic molecules found in an organism or in a biological sample. We do not know how many compounds (biochemicals) exist in the human metabolome, but what is quite clear is that the number is much higher than the number of compounds characterized in biochemical databases. There are around 70,000 chemicals in current biochemical databases, but it has been suggested that the human metabolome may contain some 200,000 lipids alone. There is clearly a need then to supplement our knowledge of the metabolome with other rationally determined virtual structures. In an elegant study from 2013 [16], David Grant and coworkers described the generation of the *In Vivo/In Silico* Metabolites Database (IIMDB). In this study, some 23,000 known compounds from existing databases – mammalian metabolites, drugs, secondary plant metabolites, and glycerophospholipids – were run through Meteor under default processing constraints in an automated, high-throughput batch mode. More than 400,000 phase I and phase II metabolites were generated. The IIMDB consists of the original 23,000 compounds plus their *in silico* generated metabolites. Ninety-five percent of these virtual metabolites could not be found in any existing biochemical database. However over 21,000 of them had entries in the PubMed, HMBD, KEGG, or HumanCyc databases. Most of these were ranked as "biological" using the computer software BioSM [17], a program that identifies biochemical-like molecules in diverse chemical space. This new database then is a useful tool for nontargeted metabolomics studies. What has been achieved here is the conversion of unknown–unknowns into known–unknowns – viable biochemical structures that may exist. As we said at the beginning of the section, the delineation between xenobiotic-metabolizing enzymes and enzymes of endogenous biochemistry is not a sharp one, and the conclusions from this paper do seem to support that assertion.

6.10.5.4 *In Silico* Metabolism and the Search for Putative Pathways of Toxification: 25B-NBOMe

In a study from 2015 [18], the metabolism of 25B-NBOMe (**7**), a potent 5-HT$_{2A}$ receptor agonist and recreational hallucinogen, was investigated.

The 2,5-dimethoxy-N-benzylphenethylamine class of compounds undergo extensive first-pass hepatic metabolism. However, severe and in some cases fatal toxicity has been reported for these compounds, and it is thought that their unpredictable toxicity is idiosyncratic in nature and caused by the formation of toxic metabolites. The study aims to characterize metabolites that may be associated with toxicity. The reported metabolic fate of 25B-NBOMe *in vivo* and *in vitro* is shown in Figure 6.10.5. *In vivo*, the main clearance pathway of 25B-NBOMe is 5′-demethylation followed by rapid phase II conjugation to give the glucuronide (**8**). The phenolic intermediate (**9**) is reported at only very low concentrations in plasma. Incubation of 25B-NBOMe with human and porcine liver microsomes showed evidence of five significant metabolites – (**9–13**) in Figure 6.10.5 – formed by, respectively, *O*-demethylation for metabolites (**9**), (**10**), and (**11**) and *N*-dealkylation for (**12**) and aromatic hydroxylation for (**13**). Phase II conjugates (glucuronides) were also detected.

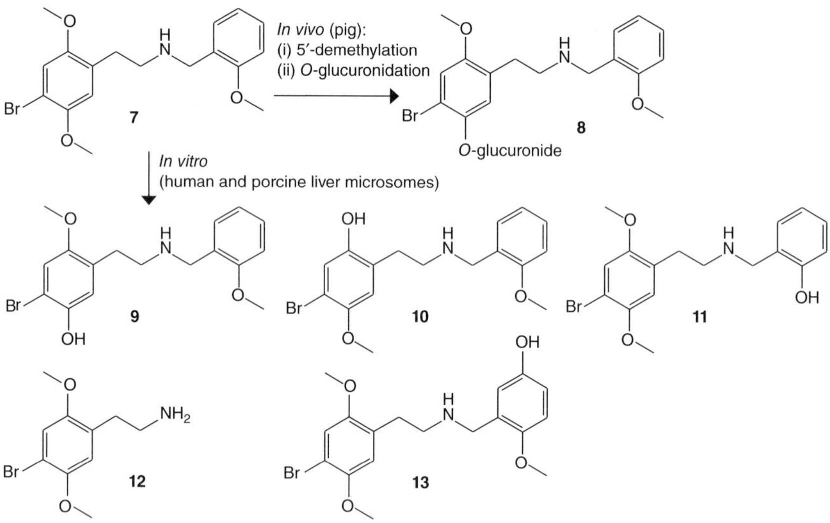

Figure 6.10.5 *In vivo* (pig) and *in vitro* (human and porcine liver microsomes) metabolism of 25B-NBOMe.

The parent compound was analyzed using Derek Nexus and Meteor Nexus. This compound was unknown to both programs – not being part of their respective supporting information datasets. Derek Nexus did not activate any hepatotoxicity structural alerts for the parent compound at the plausible or probable levels. The metabolism was investigated using Meteor Nexus, and the results are compared with the literature results (this is shown in Figure 6.10.6 and Table 6.10.4).

Metabolites were processed using Derek Nexus. For conciseness in this discussion, only the hepatotoxicity endpoint in humans was queried at the probable and plausible levels. Meteor predicts all three observed phenols (**9, 10,** and **11**) as first-generation metabolites, all derived from an *O*-demethylation biotransformation with scores of 731, 629, and 409 using the SoM-based

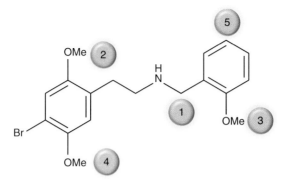

Figure 6.10.6 Top five observed sites of metabolism as predicted by Meteor Nexus. The annotated sites of metabolism (SoMs) are indicated in Table 6.10.4.

Table 6.10.4 First-generation biotransformation predictions from a Meteor Nexus analysis of 25B-NBOMe.

Biotransformation number	Biotransformation name	Score	SoM
243	Oxidative N-dealkylation	743	1
118	Oxidative O-demethylation	731	2
118	Oxidative O-demethylation	629	3
118	Oxidative O-demethylation	409	4
225	4-Hydroxylation of 1,2-disubstituted benzenes	396	5
243	Oxidative N-dealkylation	324	
468	Amide formation from benzylic amines	272	
533	PROTOTYPE – catechol or mercapturic acid formation via arene oxides	122	
533	PROTOTYPE – catechol or mercapturic acid formation via arene oxides	122	
240	3-Hydroxylation of 1,2,4,5-tetrasubstituted benzenes	118	
364	Carbamoyl glucuronides from amines	116	
9	N-Methylation of secondary amines	0	
35	N-Glucuronidation of secondary amines	0	
82	Dihydrodiols via arene oxides	0	
82	Dihydrodiols via arene oxides	0	
82	Dihydrodiols via arene oxides	0	
83	Premercapturic acids via arene oxides	0	
83	Premercapturic acids via arene oxides	0	
83	Premercapturic acids via arene oxides	0	
84	Mercapturic acids via arene oxides	0	
84	Mercapturic acids via arene oxides	0	
84	Mercapturic acids via arene oxides	0	

Table 6.10.4 (Continued)

Biotransformation number	Biotransformation name	Score	SoM
85	2-Halophenols via arene oxides	0	
97	N-Hydroxylation of secondary aliphatic amines	0	
138	Aromatic reductive dehalogenation	0	
169	Non-oxidative aromatic hydroxylative dehalogenation	0	
226	3-Hydroxylation of 1,2-disubstituted benzenes	0	
226	3-Hydroxylation of 1,2-disubstituted benzenes	0	
240	3-Hydroxylation of 1,2,4,5-tetrasubstituted benzenes	0	
564	N-Formylation of primary and secondary amines	0	

fingerprint comparison method described in Section 6.10.3.2.3. None of these metabolites activated a hepatotoxicity alert. Also predicted with scores of 743 and 396, respectively, were biotransformations (N-dealkylation and carboaromatic hydroxylation), leading to the formation of observed metabolites (**12**) and (**13**). Together, these are the top five predicted biotransformations (shown in Table 6.10.4 to no. 225) and correspond well with the observed first-generation metabolites observed in microsome incubations. The next six predictions (shown in Table 6.10.4 to no. 364) correspond to biotransformations that have calculated scores of greater than zero but did not generate observed metabolites in this study. The rest of the predictions (shown in Table 6.10.4) have calculated scores of zero, indicating no literature precedent for this chemistry among the nearest neighbors found for each biotransformation. There are six potential regioisomeric sites for carboaromatic hydroxylation in this molecule. The highest scoring prediction corresponds to the one that is observed – metabolite (**13**). All but one of the others score zero. Two N-dealkylation biotransformations are predicted. The first involving reaction at the benzylic position leading to the formation of metabolite (**12**) scored 743, while the same biotransformation applied to the alterative alpha carbon scored only 324 – a viable prediction – but the corresponding metabolites were not observed on this occasion.

Metabolite (**12**) activated a hepatotoxicity alert at the plausible level. Although this metabolite was observed, the counterpart metabolites (there are also an intermediate carbinolamine and aldehyde) of this N-dealkylation (**14**), shown in Table 6.10.5, and its corresponding alcohol (not shown) were neither observed nor reported. It is common, in dealkylations and other processes, that small or acidic components are not observed, and this is particularly true if the MS experiments are conducted in positive ion scan mode only. Metabolite (**14**) activated a hepatotoxicity alert at the plausible level, and this metabolite is very likely to be present at some concentration. Metabolites (**15**) and (**16**) in Table 6.10.5 are, respectively, the alcoholic and acidic components of the unobserved N-dealkylation biotransformation, and both activate hepatotoxicity alerts. It is interesting to note that while metabolites (**12**) and (**15**) activate

Table 6.10.5 Selected metabolites from a Meteor Nexus analysis of 25B-NBOMe and their corresponding Derek Nexus hepatotoxicity alerts. Alerting substructures are shown in light grey.

Metabolite structure	Meteor Nexus score	Observed	Derek Nexus hepatotoxicity alert	Derek Nexus reasoning level
12	743	Yes	557: Halobenzene	Plausible
14	743	No	551: Salicylic acid or analog	Plausible
15	324	No	557: Halobenzene	Plausible
16	324	No	620: 2-Arylacetic or 3-arylpropionic acid	Plausible

the halobenzene alert for hepatotoxicity, the parent compound (**7**), also a halobenzene, does not. This is because the SAR associated with the alert has a parameter restricting its application to lower molecular weight substrates only. It is important to reemphasize that while these metabolites (**14–16**) are not observed, the predicted scores indicate that this biotransformation is viable and that these metabolites may be present at low concentrations.

The biotransformations scoring greater than zero involved in the prediction of unobserved metabolites clarify some interesting chemistry and point to some putative pathways of toxification (Figure 6.10.7). The unobserved N-dealkylation has already been discussed.

The hydroxy metabolite (**13**) is experimentally observed, and it is possible to generate a *para*-dihydroquinone (**17**) by a second O-demethylation and subsequent oxidation to a potentially adduct-forming quinone (**18**) (Figure 6.10.7, Pathway 1). Other *para*-dihydroquinones are also feasible in the bromine-containing ring. Significantly these are not observed in this study, and on this occasion both experimental evidence and prediction strongly suggest that glucuronidation can effectively compete with these second oxidations, effectively preventing the quinone formation.

The amide (**22**) is also not an observed metabolite, but the presence of the carbinolamine (**19**) is implied by the fact that that the N-dealkylated product (**12**)

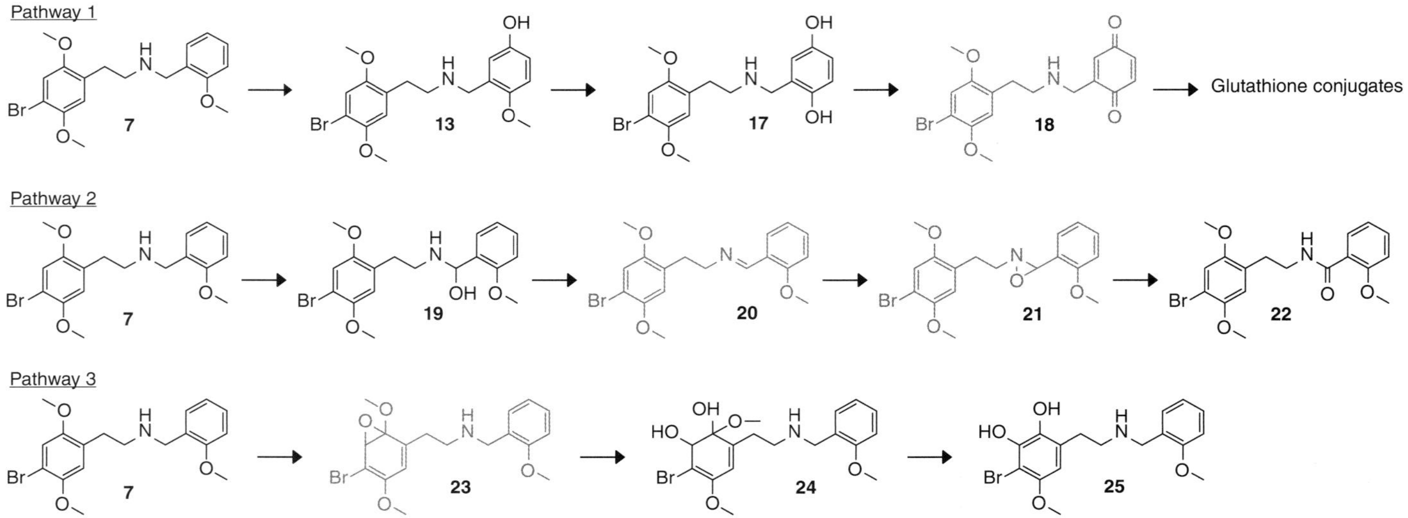

Figure 6.10.7 Putative pathways of toxification suggested by a Meteor Nexus analysis of 25B-NBOMe. Potentially adduct-forming intermediates are depicted in light grey (18, 20, 21, 23).

is observed and (**19**) must be an intermediate in this process. Benzylic amides are thought to form via the intermediacy of the potentially adduct-forming imine (**20**) and oxaziridine (**21**). These are alternative downstream intermediates from the carbinolamine (**19**) for which there is indirect evidence in this study (Figure 6.10.7, Pathway 2).

Meteor Nexus also suggests the formation of the catechol (**25**) – the mechanism here is thought to involve an arene oxide (**23**) formed by *ipso*-substitution in the bromine-containing ring, with subsequent ring opening of the oxirane to afford intermediate (**24**), which loses methanol to give the catechol (**25**) (Figure 6.10.7, Pathway 3).

Not shown here, because the predictions all scored zero, are a number of biotransformations associated with arene oxide formation in the non-bromine-containing ring shown in gray in Table 6.10.4. With the exception of polynuclear aromatic hydrocarbons, arene oxide formation in phenyl rings is not often observed (as ever, there are some exceptions), and the suggestion of such a biotransformation will not cause a medicinal chemist/toxicologist to become alarmed. What we demonstrate here is the utility of an expert system working together with a dynamic scoring system to indicate alerting features in molecules and their metabolites, which should raise a level of concern even though such metabolites may not be readily observed.

Essentials

- Living things have a need to remove chemical invaders (xenobiotics) from their systems. One method for achieving this is by enzymatic chemical modification of substances (biotransformation).
- Metabolic biotransformation is a pharmacokinetic process, but the consequences of biotransformation – the formation of metabolites – can influence the pharmacodynamic effects of the substance.
- These alterations include deactivation (detoxification) and, ultimately, excretion of the substance, activation of the substance into a molecule with enhanced pharmacology (a prodrug), and activation of the substance into a molecule with undesirable pharmacology (a toxicant).
- Some combination of these effects usually operates. The ability to predict the likely biotransformation profile of a molecule is key to understanding its effects when exposed to animals and humans.
- Systems for the *in silico* prediction of metabolism are of two fundamental types. Local systems predict for simple biochemical systems: single enzymes or isoforms, single functional group, and reaction or reaction type. Global systems predict for complex and diverse biochemical systems: multiple enzymes, diverse chemistries, many functional groups and reaction types, and broad chemical space.
- Local systems can be data-requiring, statistical methods (e.g., QSARs) or may be theoretical and calculation based. Methods may use characteristics of substrates

(ligand based) or enzymes (structure based). Hybrid systems are increasingly becoming the norm.
- Global systems are ligand based and include machine learning methods and knowledge-based expert systems. Again, hybrid systems are finding increasing utility.
- Scientists use metabolism information in different ways. Medicinal chemists may only require quite high-level information relating to SoMs and clearance rates in comparison with synthetic analogs, pharmacokineticists and toxicologist need more detailed information about metabolite structures and pathways, and metabolomic scientists need vast amounts of predictive data to augment deficiencies in our understanding of the human metabolome.
- *In silico* metabolism prediction systems then can help in both reductionist and holistic information-requiring domains. We have presented case studies in medicinal chemistry, structure identification in forensic analysis, and metabolomics and in the search for novel pathways of toxification.

Available Software and Web Services (accessed January 2018)

- Cytochrome P450 Homepage: http://drnelson.uthsc.edu/CytochromeP450.html.
- List of steroid ligands of P450s: http://www.icgeb.org/~p450srv/steroid_list.html.
- History of Xenobiotic Metabolism: http://www.issx.org/page/History.
- MetaPrint2D: http://www-metaprint2d.ch.cam.ac.uk/.
- RS-WebPredictor: http://reccr.chem.rpi.edu/Software/RS-WebPredictor/.
- XenoSite: http://swami.wustl.edu/xenosite.
- SyGMa: https://github.com/3D-e-Chem/sygma.
- Meteor Nexus: https://www.lhasalimited.org/products/meteor-nexus.htm.
- Derek Nexus: https://www.lhasalimited.org/products/derek-nexus.htm.
- StarDrop P450 Metabolism: http://www.optibrium.com/stardrop/stardrop-p450-models.php.
- MetaSite: http://www.moldiscovery.com/software/metasite/.

Selected Reading

- Guengerich, F.P. (2017) Intersection of the roles of cytochrome P450 enzymes with xenobiotic and endogenous substrates: relevance to toxicity and drug interactions. *Chem. Res. Toxicol.*, **30**, 2–12. doi: 10.1021/acs.chemrestox.6b00226

- Ioannides, C. (ed.) (2002) *Enzyme Systems that Metabolise Drugs and Other Xenobiotics*, John Wiley & Sons, Ltd., Chichester, 588 pp.
- Khojasteh, S.C., Wong, H., and Hop, C.E.C.A. (2011) *Drug Metabolism and Pharmacokinetics Quick Guide*, Springer, New York, 214 pp.
- Kirchmair, J. (ed.) (2014) *Drug Metabolism Prediction: Methods and Principles in Medicinal Chemistry*, vol. **63**, R. Mannhold, H. Kubinyi and G. Folkers (series editors), Wiley-VCH, Weinheim, 536 pp.
- Kirchmair, J., Goller, A.H., Lang, D., Kunze, J., Testa, B., Wilson, I.D., Glen, R.C., and Schneider, G. (2015) Predicting drug metabolism: experiment and/or computation? *Nat. Rev. Drug Discov.*, **14**, 387–404. doi: 10.1038/nrd4581
- Long, A. (2012) Drug metabolism in silico – the knowledge-based expert system approach: historical perspectives and current strategies. *Drug Discovery Today: Technologies*, **10** (1), e147–e153. doi: 10.1016/j.ddtec.2012.10.006 (Online Only).
- Raunio, H., Kuusisto, M., Juvonen, R.O., and Pentikainen, O.T. (2015) Modeling of interactions between xenobiotics and cytochrome P450 (CYP) enzymes. *Frontiers in Pharmacology*, **6**, 213. doi: 10.3389/fphar.2015.00123 (Open Access).
- Roskar, R. and Lusin, T.T. (2012) Analytical methods for quantification of drug metabolites in biological samples, in *Chromatography – The Most Versatile Method of Chemical Analysis* (ed. L. Calderon), InTech. doi: 10.5772/51676
- Testa, B. (1995) *The metabolism of drugs and other xenobiotics*, in *Biochemistry of Redox Reactions* (eds B. Testa and J. Caldwell), Academic Press, London, 471 pp.
- Thompson, R.A., Isin, E.M., Ogese, M.O., Mett, J.T., and Williams, D.P. (2016) Reactive metabolites: current and emerging risk and hazard assessments. *Chem. Res. Toxicol.*, **29**, 1505–1533. doi: 10.1021/acs.chemrestox.5b00410

References

[1] Testa, B. and Jenner, P. (1976) *Drug Metabolism: Chemical and Biochemical Aspects*, Drugs and the Pharmaceutical Sciences Series, vol. 4, Dekker, New York, 500 pp.
[2] Long, A. and Walker, J.D. (2003) *Environ. Toxicol. Chem.*, **22**, 1894–1899. doi: 10.1897/01-480
[3] Munro, A.W., Girvan, H.M., and McLean, K.J. (2007) *Nat. Prod. Rep.*, **24**, 585–609. doi: 10.1039/B604190F
[4] http://www.optibrium.com/stardrop/stardrop-p450-models.php (accessed January 2018)
[5] Cruciani, G., Carosati, E., De Boeck, B., Ethirajulu, K., Mackie, C., Howe, T., and Vianello, R. (2005) *J. Med. Chem.*, **48**, 6970–6979. doi: 10.1021/jm050529c
[6] Rydberg, P., Gloriam, D.E., Zaretzki, J., Breneman, C., and Olsen, L. (2010) *ASC Med. Chem. Lett.*, **1**, 96–100. doi: 10.1021/ml100016x

[7] Kirchmair, J., Williamson, M.J., Afzal, A.M., Tyzack, J.D., Choy, A.P.K., Howlett, A., Rydberg, P., and Glen, R.C. (2013) *J. Chem. Inf. Model.*, **53**, 2896–2907. doi: 10.1021/ci400503s

[8] Darvas, F. (1987) MetabolExpert, an expert system for predicting metabolism of substances, in *QSAR in Environmental Toxicology - II* (ed. K.L.E. Kaiser), Riedel, Dordrecht, pp. 71–81.

[9] Klopman, G., Dimayuga, M., and Talafous, J. (1994) *J. Chem. Inf. Comput. Sci.*, **34**, 1320–1325. doi: 10.1021/ci00022a014

[10] Kleemann, R., Bureeva, S., Perlina, A., Kaput, J., Verschuren, L., Wielinga, P.Y., Hurt-Camejo, E., Nikolsky, Y., van Ommen, B., and Kooistra, T. (2011) *BMC Syst. Biol.*, **5**, 125. doi: 10.1186/1752-0509-5-125

[11] Mekenyan, O.G., Dimitrov, S.D., Pavlov, T.S., and Veith, G.D. (2004) *Curr. Pharm. Des.*, **10**, 1273–1293.

[12] www.lhasalimited.org/products/meteor-nexus.htm (accessed January 2018)

[13] Marchant, C.A., Rosser, E.M., and Vessey, J.D. (2016) *Mol. Inform.* doi: 10.1002/minf.201600105

[14] Boyer, D., Bauman, J.N., Walker, D.P., Kapinos, B., Karkiamd, K., and Kalgutkar, A.S. (2009) *Drug Metab. Dispos.*, **37**, 999–1008. doi: 10.1124/dmd.108.026112

[15] Pelander, A., Tyrkko, E., and Ojanpera, I. (2009) *Rapid Commun. Mass Spectrom.*, **23**, 506–514. doi: 10.1002/rcm.3901

[16] Menikarachchi, L.C., Hill, D.W., Hamdalla, M.A., Mandoiu, I.I., and Grant, D.F. (2013) *J. Chem. Inf. Model.*, **53**, 2483–2492. doi: 10.1021/ci400368v

[17] http://metabolomics.pharm.uconn.edu/Software.html (accessed January 2018)

[18] Leth-Peterson, S., Gabel-Jensen, C., Gillings, N., Lehel, S., Hansen, H.D., Knudsen, G.M., and Kristensen, J.L. (2016) *Chem. Res. Toxicol.*, **29**, 96–100. doi: 10.1021/acs.chemrestox.5b00450

6.11 Chemoinformatics at the CADD Group of the National Cancer Institute

Megan L. Peach and Marc C. Nicklaus

National Cancer Institute, NIH, NCI-Frederick, 376 Boyles Street, Frederick, MD 21702, USA

Learning Objectives

- To describe the online chemical information services offered by the CADD Group of the NCI.
- To perform chemical structure lookups and conversions.
- To execute optical structure recognition and to create GIFs of chemical structures.
- To predict chemical activity by QSAR models.
- To predict various biological activities.
- To calculate pseudorotation parameters of nucleosides, nucleotides, DNA, and RNA.
- To design synthetically accessible compounds from a building block database.
- To evaluate a variety of database files for download.

Outline

6.11.1 Introduction and History, 385
6.11.2 Chemical Information Services, 386
6.11.3 Tools and Software, 388
6.11.4 Synthesis and Activity Predictions, 391
6.11.5 Downloadable Datasets, 391

6.11.1 Introduction and History

The Computer-Aided Drug Design (CADD) Group is a research unit in the National Cancer Institute (NCI), part of the US National Institutes of Health (NIH), that conducts research on all aspects of chemical information and molecular modeling. The focus of the group is on computational handling of small molecules in the context of drug design. The CADD Group carries out modeling and drug discovery projects internal to NCI, but in addition to this we offer a significant number of freely available chemoinformatics services and tools to the public at large through our web server at https://cactus.nci.nih.gov/. These online resources will be the main focus of this section.

Chemoinformatics has a 60-year history at the NCI. Its starting point is generally agreed to be the congressional authorization, in 1955, of $5 million for the establishment of a drug development program. In the first 30 years of this program, which became operational in 1957, approximately 13,000 new compounds were tested each year. Managing this volume of chemical samples soon made it apparent that only the use of computers could help in dealing with the flood of chemical structure and animal testing data [1]. Beginning in the mid-1960s, therefore, the data were entered into a computer system named the NCI Drug Information System (DIS). While the test results ("biology files") were kept in-house, the chemistry data were processed by Chemical Abstract Services (CAS) under a contract from NCI. The interactive search capability for the NCI data that evolved over the years of this contract proved to be one of the forerunners of the CAS ONLINE system [2].

During the first three decades of the screening program, NCI used *in vivo* animal tumor models to screen compounds for possible antitumor activity. In 1986 this was replaced by assays run against a panel of 60 different human cancer cell lines widely known as the NCI-60 screen [3]. A subset of the database was also tested for evidence of anti-HIV activity. The chemistry data and its associated biological testing data evolved over the years to what has become widely known as the NCI Database. The nonconfidential "open" part of the NCI Database, which currently numbers just over 280,000 records, was used in many early chemoinformatics projects as a benchmark set, to conduct "stress tests" on software, and as a reference of sorts for other large databases at the time [4].

In parallel to the large-scale screening effort that was undertaken by the NCI's Developmental Therapeutics Program (DTP), in 1988 a molecular modeling section was established as part of the (then) Laboratory of Medicinal Chemistry (LMC). Its task was to conduct collaborative scientific work in, and provide support to, drug discovery and development projects at the LMC and other laboratories at NCI. It evolved into the NCI CADD Group and is now a research unit within the Chemical Biology Laboratory (CBL) at the NCI's Center for Cancer Research. Along with drug design, the CADD Group developed both expertise and resources in the field of chemoinformatics in general.

6.11.2 Chemical Information Services

The NCI CADD Group has maintained a set of web-based user resources since 1998. The current home page URL is https://cactus.nci.nih.gov. The server name "cactus" refers to **C**ADD Group **C**hemoinformatics **T**ools and **U**ser **S**ervices. Here we will only outline the major features of the tools and services on cactus; more details can be obtained at the web pages for the individual resources where extensive documentation is provided. Many of these resources are the results of collaborations with groups or individuals outside NCI (or only temporarily at NCI), which are mentioned in the Acknowledgments.

6.11.2.1 NCI Database Browser

After the public release of a significant part (about 125,000 structures) of the NCI Database by DTP in 1994 [5], work began on a service that was to become known as the "NCI Database Browser." The initial coding of this service was done in collaboration with the Computer-Chemie-Centrum at the University of Erlangen-Nuremberg and was (and remains) based on the chemoinformatics toolkit CACTVS [6]. (Note the uppercase and spelling with a "V" for the toolkit, in contrast to "cactus" the web server.) This service was one of the first public and freely available web-based GUIs for a large small-molecule database with advanced capabilities such as full structure and substructure search and is one of the oldest continuously operational Internet resources of this kind. It is currently accessible via the URL https://cactus.nci.nih.gov/ncidb2.2/. This service offers complex queries, hit list display and management, visualization, and data retrieval capabilities of search results (Figure 6.11.1).

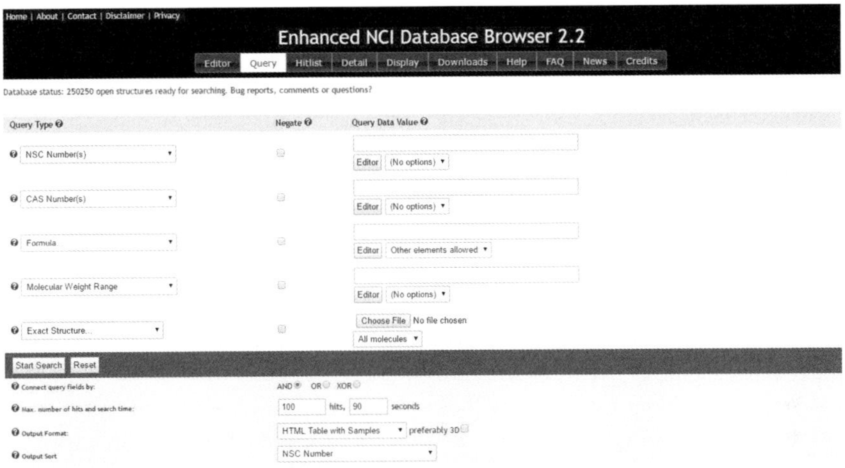

Figure 6.11.1 Web form of the Enhanced NCI Database Browser.

The bulk of the data are the structures (approximately a quarter million) of the open NCI Database, plus their assay results in the cell-based (NCI-60) growth inhibition screens, the yeast anticancer drug screens, and the AIDS antiviral screens, as provided by DTP [7]. Typical molecular structure properties as well as a number of physical and chemical properties such as log P range and drug likeness have been calculated and are searchable. Predictions of 565 different biological activities made by the program PASS [8] for the vast majority of the structures are also available and can be displayed [9].

6.11.2.2 Chemical Identifier Resolver (CIR)

The Chemical Identifier Resolver (CIR) is presently the most used service by far on cactus. It is designed to convert one structure identifier or representation into another. It is human-usable via the web form provided at https://cactus.nci.nih.gov/chemical/structure/ but it is mostly used via programmatic access. This is done with the following URL API appendix scheme: https://cactus.nci.nih.gov/chemical/structure/ "structure identifier"/"representation".

The "structure identifier" must be a single-line structure description since it has to be entered as part of the URL. The identifier can either describe the full chemical structure (e.g., SMILES, InChI [10]), or be a hashed structural key (e.g., InChIKey), or a chemical name (Figure 6.11.2). CIR returns the requested output "representation" with the corresponding MIME-type specification. This

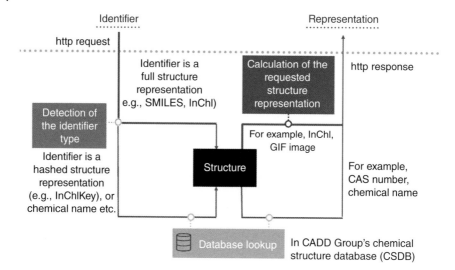

Figure 6.11.2 Possible workflows of CIR queries.

representation can be any of a large number of different file formats, a 2D drawing, a chemical identifier, or a molecular property such as ring count or molecular weight.

The database that is used for CIR lookups is aggregated from about 140 external databases, comprising about 120 million structure records with approximately 110 million unique standard InChIKeys. As per its intention and design, CIR has been embedded programmatically in various external web tools, commercial software products, educational websites, databases, and search tools. One example of an educational tool that uses CIR is CheMagic.org [11].

6.11.2.3 Chemical Structure Lookup Service (CSLS)

The Chemical Structure Lookup Service (CSLS) (https://cactus.nci.nih.gov/lookup/) allows the user to enter a chemical identifier or structure to ask, "In which database is this chemical found?" CSLS has an auto-detect facility that applies some heuristics to determine the type of query text submitted. As an example, if the user submits "740" then CSLS returns hits to the cancer drug methotrexate (which has an ID number of 740 in the NCI Database). If the user then submits the InChIKey of methotrexate, which can be done by clicking a link in the results column, CSLS returns 84 hits, as this well-known molecule occurs in many databases. From a CSLS results page (see Figure 6.11.3), there is also a link to do a Google search directly with the InChI of each returned structure.

6.11.3 Tools and Software

6.11.3.1 Optical Structure Recognition Application (OSRA)

The Optical Structure Recognition Application (OSRA) (https://cactus.nci.nih.gov/osra/) converts graphical representations of chemical structures in journal

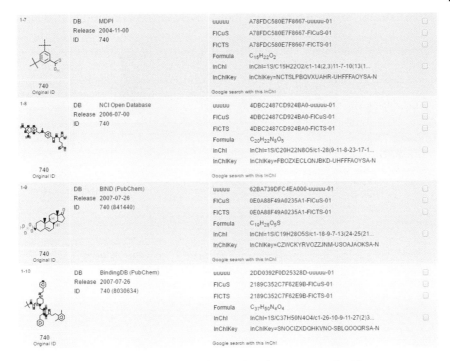

Figure 6.11.3 CSLS results page (excerpt) for search with query string "740."

articles, patent documents, textbooks, trade magazines, and so on, into SMILES or SDF representations [12]. OSRA can read images in any of about 90 graphical formats including GIF, JPEG, PNG, TIFF, and PDF. It is primarily meant to be downloaded as a utility and used in-house, either as a stand-alone executable or integrated as a library into other software. It makes heavy use of open-source libraries such as GraphicsMagick [13]. Further development has continued outside NCI, and a major rewrite, OSRA II, is available at https://sourceforge.net/projects/osra/.

6.11.3.2 Online SMILES Translator

The Online SMILES Translator (https://cactus.nci.nih.gov/translate/) allows the user to submit chemical structure representations containing one or several molecules in SMILES strings, SDF, PDB, MOL, and other formats and request conversion of these structures into Unique SMILES (USMILES) [14], SDF, PDB, or MOL file formats. The input structure can also be drawn with the JSME Structure Editor [15]. For SDF, PDB, or MOL file formats, either 2D or 3D coordinates (calculated with CORINA [16]) can be requested. This service is in its entirety based on CACTVS.

6.11.3.3 GIF Creator for Chemical Structures

The GIF Creator for Chemical Structures (https://cactus.nci.nih.gov/gifcreator/) generates a 2D structure drawing in either GIF or PNG formats from (i) a

SMILES string, (ii) a structure drawn interactively, or (iii) a user-submitted file in a wide variety of chemical structure formats. Numerous options to tailor the 2D drawing are available (see Figure 6.11.4). It is written in the form of a CACTVS script.

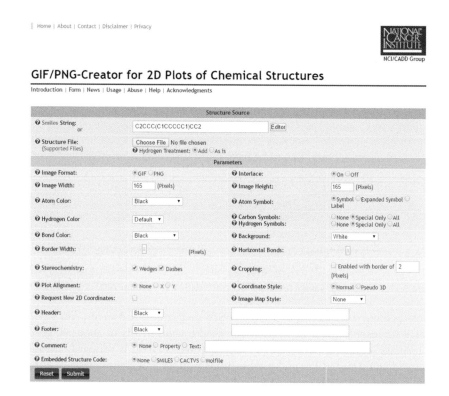

Figure 6.11.4 Web form of the GIF/PNG Creator web service.

6.11.3.4 Pseudorotational Online Service and Interactive Tool (PROSIT)

The Pseudorotational Online Service and Interactive Tool (PROSIT) (https://cactus.nci.nih.gov/prosit/) computes, and displays in tabular format, pseudorotation parameters [17] including the pseudorotational phase angle P, the glycosyl torsion angle χ, and the sugar puckering amplitude v_{max} for user-submitted 3D structures of nucleosides, nucleotides, and their analogs, as well as for DNA and RNA single and double strands [18]. Major chemical file formats such as PDB, SDF, MOL, and xyz are supported. PROSIT was used for a survey of nucleosides contained in the Cambridge Structural Database and nucleotides in high-resolution crystal structures from the Nucleic Acid Database, as well as a study of a parallel-stranded guanine tetraplex and a four-way Holliday junction [19].

6.11.4 Synthesis and Activity Predictions

6.11.4.1 Synthetically Accessible Virtual Inventory (SAVI)

Current databases of commercially available screening samples such as the ChemNavigator iResearch™ Library [20] are approaching the 100 million compound mark. Still, given that the chemical space of drug-like compounds has been estimated to be at least 10^{40} molecules (for 30 or fewer heavy atoms), one does not usually find a newly designed compound from a CADD project in these databases. Synthesis efforts for an arbitrary small organic molecule may be difficult and costly. The Synthetically Accessible Virtual Inventory (SAVI) project therefore aims at turning the question, "Where do I get a new molecule from?" on its head and asks instead, "What can I make reliably and cheaply?" Then, with modern CADD approaches, this set of easily synthesizable molecules can be searched for new potentially bioactive compounds. This project therefore aims at the virtual generation of up to 1 billion molecules that are reliably and cheaply synthesizable and have been filtered by criteria that are seen as desirable in a modern drug development context [21].

Through contributions from international collaborators, SAVI combines a set of reliably available and generally inexpensive commercial starting materials, richly annotated chemical transforms from the LHASA project knowledge base [22–24], and new code implemented in CACTVS to allow parsing of the original LHASA transforms and to handle the reversal of the original retrosynthetic transforms into forward-synthetic transforms.

The SAVI project has the goal of ultimately creating a large database of high-quality screening samples; each annotated with a computer-proposed easy synthetic route, made freely available on the cactus server for download. A first beta set of products of only 14 (out of ~2300 possible) transforms, applied in one-step reactions to a building block database of 377,000 starting materials, has been made available on cactus for download and early testing (see Section 6.11.5). This dataset numbers about 283 million compounds, with more than 99% of them found to be novel structures.

6.11.4.2 Chemical Activity Predictor (CAP)

The Chemical Activity Predictor (CAP) (https://cactus.nci.nih.gov/chemical/apps/cap) uses QSAR models created with the program GUSAR [25] to provide predictions of properties of interest in drug discovery for small molecules. CAP includes the following models: HIV-1-related activities (two models), models for a small number of PubChem assays (five models), environmental toxicity (T.E.S.T.) (five models), acute rat toxicity (four models), and physicochemical properties (nine models) including aqueous solubility. CAP uses CIR to resolve submitted identifiers.

6.11.5 Downloadable Datasets

The CADD Group has created a number of files that have been used internally and felt to be of potential interest for external users as well. The following datasets are therefore available for download at https://cactus.nci.nih.gov/#3:

- SD files of structures from PubChem with assay data included as properties, suitable for building QSAR or other types of models
- NCI Database related datasets including the "raw" data in bulk format that were used in building the NCI Database Browser
- SD file versions of Structured Product Labeling (SPL) index files of substances indexed by the FDA
- Structures as well as biological and bibliographic annotations of HIV-1 integrase inhibitors collected from the literature
- A set of approximately 283 million virtual synthesis products generated in the beta phase of the SAVI project

Essentials

The NCI CADD Group has made chemoinformatics resources freely available since the late 1990s on its web server https://cactus.nci.nih.gov:

- Interactive tools for generation and interconversion of chemical information and representations.
- Services to locate a large number of compounds from many external databases.
- Tools such as optical structure recognition in documents and images, predictive QSAR models, and file format converters.
- Downloadable database files, generated or converted to more user-friendly formats by the CADD Group.

Available Software and Web Services (accessed January 2018)

- NCI CADD Group: https://cactus.nci.nih.gov.
- Chemical Identifier Resolver: https://cactus.nci.nih.gov/chemical/structure.
- Chemical Structure Lookup Service: https://cactus.nci.nih.gov/lookup/.
- OSRA: https://cactus.nci.nih.gov/osra/, https://sourceforge.net/projects/osra/.
- Online SMILES Translator: https://cactus.nci.nih.gov/translate/.
- GIF Creator for Chemical Structures: https://cactus.nci.nih.gov/gifcreator/.
- PROSIT: https://cactus.nci.nih.gov/prosit/.
- Chemical Activity Predictor https://cactus.nci.nih.gov/chemical/apps/cap.
- CACTVS: http://xemistry.com/.

References

[1] Milne, G.W. and Miller, J.A. (1986) *J. Chem. Inf. Comput. Sci.*, **26**, 154–159.

[2] Dittmar, P.G., Farmer, N.A., Fisanick, W., Haines, R.C., and Mockus, J. (1983) *J. Chem. Inf. Comput. Sci.*, **23**, 93–102.

[3] NIH Developmental Therapeutics Program, https://dtp.cancer.gov/discovery_development/nci-60/methodology.htm (accessed January 2018).

[4] Voigt, J.H., Bienfait, B., Wang, S., and Nicklaus, M.C. (2001) *J. Chem. Inf. Comput. Sci.*, **41**, 702–712.

[5] CCL.NET NCI Structure Database, http://www.ccl.net/cgi-bin/ccl/message-new?1994+11+11+006 (accessed January 2018).

[6] Ihlenfeldt, W., Takahashi, Y., Abe, H., and Sasaki, S. (1994) *J. Chem. Inf. Comput. Sci.*, **34**, 109–116.

[7] NIH DTP Bulk Data for Download, https://dtp.cancer.gov/databases_tools/bulk_data.htm (accessed January 2018).

[8] Lagunin, A., Stepanchikova, A., Filimonov, D., and Poroikov, V. (2000) *Bioinformatics*, **16**, 747–748.

[9] Poroikov, V.V., Filimonov, D.A., Ihlenfeldt, W.-D., Gloriozova, T.A., Lagunin, A.A., Borodina, Y.V., Stepanchikova, A.V., and Nicklaus, M.C. (2003) *J. Chem. Inf. Comput. Sci.*, **43**, 228–236.

[10] Heller, S.R., McNaught, A., Pletnev, I., Stein, S., and Tchekhovskoi, D. (2015) *J. Cheminf.*, **7**, 23.

[11] CheMagic, http://chemagic.org/home/ (accessed January 2018).

[12] Filippov, I.V. and Nicklaus, M.C. (2009) *J. Chem. Inf. Model.*, **49**, 740–743.

[13] GraphicsMagick Image Processing System, http://www.graphicsmagick.org/ (accessed January 2018).

[14] Weininger, D., Weininger, A., and Weininger, J.L. (1989) *J. Chem. Inf. Comput. Sci.*, **29**, 97–101.

[15] Bienfait, B. and Ertl, P. (2013) *J. Cheminf.*, **5**, 24.

[16] Gasteiger, J., Rudolph, C., and Sadowski, J. (1990) *Tetrahedron Comput. Methodol.*, **3**, 537–547.

[17] Altona, C. and Sundaralingam, M. (1972) *J. Am. Chem. Soc.*, **94**, 8205–8212.

[18] Sun, G., Voigt, J.H., Filippov, I.V., Marquez, V.E., and Nicklaus, M.C. (2004) *J. Chem. Inf. Comput. Sci.*, **44**, 1752–1762.

[19] Sun, G., Voigt, J.H., Marquez, V.E., and Nicklaus, M.C. (2005) *Nucleosides Nucleotides Nucleic Acids*, **24**, 1029–1032.

[20] ChemNavigator iResearch™ Library, http://www.chemnavigator.com/cnc/products/iRL.asp (accessed January 2018).

[21] Bruns, R.F. and Watson, I.A. (2012) *J. Med. Chem.*, **55**, 9763–9772.

[22] Olsson, T. (1986) *Acta Pharm. Suec.*, **23**, 386–402.

[23] Johnson, A.P., Marshall, C., and Judson, P.N. (1992) *Recl. Trav. Chim. Pays-Bas-J. R. Neth. Chem. Soc.*, **111**, 310–316.

[24] Judson, P.N. and Lea, H. (1996) *Chim. Oggi-Chem. Today*, **14**, 21–24.

[25] Filimonov, D.A., Zakharov, A.V., Lagunin, A.A., and Poroikov, V.V. (2009) *SAR QSAR Environ. Res.*, **20**, 679–709.

6.12 Uncommon Data Sources for QSAR Modeling

Alexander Tropsha

Division of Chemical Biology and Medicinal Chemistry, UNC Eshelman School of Pharmacy, University of North Carolina at Chapel Hill, Chapel Hill, NC 27599, USA

Learning Objectives

- To identify large datasets for QSAR modeling now publicly available in ChEMBL, PubChem, and ToxCast.
- To distinguish that many data are available as metadata in MEDLINE abstracts and conference reports.
- To utilize printed data by lexical and linguistic methods for QSAR modeling.
- To create a workflow that combines data from multiple sources.

Outline

6.12.1 Introduction, 395
6.12.2 Observational Metadata and QSAR Modeling, 397
6.12.3 Pharmacovigilance and QSAR, 398
6.12.4 Conclusions, 401

6.12.1 Introduction

In early days of the field, quantitative structure–activity relationship (QSAR) models have been developed with relatively small sets of experimental data that were either obtained from the own or collaborating medicinal chemistry laboratories or reported in medicinal chemistry journals. Subsequent years have seen an immense growth of biomolecular databases, especially in the public domain. This growth has been enabled by both rapidly decreasing cost of high-throughput chemical synthesis and biological screening technologies that brought them into academic institutions as well as commendable efforts by federal institutions such as NIH (Molecular Libraries Initiative [1] and PubChem (Section 6.5) [2]) and EPA (ToxCast program [3]). Concurrent efforts have gone in Europe as well where, for instance, the European Bioinformatics Institute acquired the company Inpharmatica with their unique database of chemical bioactivity records for hundreds of thousands of bioactive molecules, which enabled the creation of the publicly accessible ChEMBL database [4]. Today, ChEMBL and other similar databases contain thousands of datasets available for cheminformatics and QSAR analysis. For instance, the PubChem database (http://pubchem.ncbi.nlm.nih.gov), developed as the central repository of structure–activity data, contains more than 157 million chemical records; more than 1 million compounds have been tested in about 3000 bioassays, and more than 500,000 were found active. ToxCast (https://www.epa.gov/chemical-research/toxicity-forecasting) contains data on over 1800 compounds tested in more than 700 assays, and ChEMBL (https://www.ebi.ac.uk/chembl/) integrates data on more than 1.5 million distinct chemicals and almost 14 million activity measurements for over 11,000 targets. Concurrently (and not surprisingly), the size of biomedical literature has grown following the same steep trajectory: PubMed contains the bibliographic information on over 20 million records and nearly 10 million of these contain chemical-relevant information. These developments have actually stimulated the growth of QSAR literature as was alluded to in the recent perspective on QSAR review [5] and illustrated in Figure 6.12.1 that compares and contrasts the growth of chemical data and accumulation of publications on QSAR modeling.

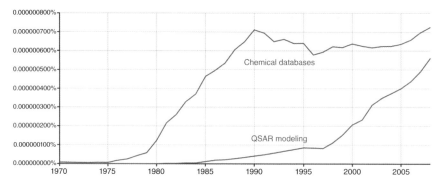

Figure 6.12.1 The growth of publications on QSAR modeling correlates with the accumulation of experimental data. The chart is generated by Google Ngram Viewer (http://books.google.com/ngrams); Y-axis – percentage among all books in the Google Ngram.

The transformative growth of biomolecular data, both in size and diversity, coupled with substantial decrease in the cost of chemical synthesis and especially, biological screening, has created new, and highly stimulating, challenges for QSAR modelers. In recent years, we have seen the growing development of novel approaches capable of multi-objective optimization of large and diverse datasets [6], for example, including recent interest in the application of deep learning [7] and active learning [8] in QSAR. However, as important as methodology development is for QSAR, no models could be developed without both chemical and bioactivity data. The growth of the aforementioned databases such as ChEMBL and PubChem that store chemical data in ready-to-model formats (e.g., sdf or SMILES linked to bioactivity values) has made the task of data exploration and model building fairly straightforward. Indeed, many published models have been derived from such data including even a recent attempt to develop a model of the entire PubChem using deep learning approaches [9]. These traditional data sources are well known to the research community and will not be reviewed here. However, recent years have seen an emergence of the unusual sources of data often processed with different types of text mining approaches. We will briefly comment here on such data and opportunities to develop models from uncommon sources as they highlight the growing diversity of data sources and correspondingly and growing appeal of QSAR to such diverse research fields as text mining and medical informatics.

6.12.2 Observational Metadata and QSAR Modeling

Arguably, one of the first instances of QSAR model development using uncommon input data is provided by the study of chemical hepatotoxicity in collaboration with BioWisdom Ltd [10]. In that study, a dataset of 951 compounds was compiled and reported to produce a wide range of effects in the liver in different species, comprising humans, rodents, and non-rodents. The unique aspect of the study was the source of data used to develop QSAR models; liver effects were obtained as assertional metadata, generated from MEDLINE abstracts and conference reports as well as other electronic sources using a unique combination of lexical and linguistic methods and ontological rules. The assertions comprised thousands of highly accurate and comprehensive observational statements. These statements were represented in triple constructs: concept_*relationship*_concept, for example, cafestol_*suppresses*_bile acid biosynthesis, azathioprine_*induces*_cholestasis, and so on. Each assertion was derived from and evidenced by a variety of electronic data sources. Behind each assertion was a rich vocabulary that rendered that assertion semantically consistent with the other assertions around the same concept. For example, the liver pathology term, cholestasis, can be described across the literature as bile stasis, biliary stasis, cholestasia, and cholestatic injury. Assertional metadata generated in this manner facilitated the semantically consistent integration of disparate observations across historic literature. Each compound in the dataset was eventually assigned a category "0" if no assertion concerning its hepatotoxicity could be identified in the literature or category "1" if such assertion could be found. This assignment enabled the binary classification modeling of hepatotoxicity following the important step of name to structure conversion of each molecule in the dataset.

After creating this unusual dataset, we have analyzed it using conventional chemoinformatics approaches and addressed several questions pertaining to cross-species concordance of liver effects, chemical determinants of liver effects in humans, and prediction of whether a given compound is likely to cause a liver effect in humans. We found that the concordance of liver effects was relatively low (about 39–44%) between different species, raising the possibility that species specificity could depend on specific features of chemical structure. Compounds were clustered by their chemical similarity, and similar compounds were examined for the expected similarity of their species-dependent liver effect profiles. In most cases, similar profiles were observed for members of the same cluster, but some compounds appeared as outliers. The outliers were the subject of focused assertion regeneration from MEDLINE, as well as other data sources. In some cases, additional biological assertions were identified, which were in line with expectations based on compounds' chemical similarity. The assertions were further converted to binary annotations of underlying chemicals (i.e., liver effect vs no liver effect), and binary QSAR models were generated to predict whether a compound would be expected to produce liver effects in humans. Despite the apparent heterogeneity of data, models have shown strong predictive power, with external accuracy in the 64–72% range, assessed by external fivefold cross validation procedures. The external predictive power of binary QSAR models was further confirmed by their application to compounds that were retrieved or studied after the model was developed.

6.12.3 Pharmacovigilance and QSAR

To the best of our knowledge, the study described above was the first application of QSAR modeling and other cheminformatics techniques to observational data generated by the means of automated text mining with limited manual curation, opening up new opportunities for generating and modeling chemical toxicology data. Very recently, we followed this original investigation into the use of uncommon electronic sources of input data for QSAR modeling in another study [11] that looked into a specific adverse drug effect known as Stevens–Johnson Syndrome (SJS).In our opinion, there were two unique elements of the investigation that are worth expanding upon here as we think more studies of this kind should be conducted: first, the uncommon source of data and second, the overall study design. We review both of these components below.

To develop QSAR models, a reference set of drugs was extracted from a database called VigiBase [12] based on their reported correlations with the SJS. VigiBase, the World Health Organization (WHO) global individual case safety report database, is maintained and analyzed by the Uppsala Monitoring Centre. As of the time of the study [11], VigiBase contained 7,014,658 reports from 107 countries, covering approximately 20,000 drugs (i.e., generic substances) and 2000 adverse drug reactions (ADRs) coded according to the WHO Drug Dictionary Enhanced and the WHO Adverse Reactions Terminology, respectively. A drug was defined as *active* if it had higher-than-expected reporting with SJS, as indicated by a positive coefficient in a shrinkage regression model for the reporting of SJS in VigiBase [13]. By considering all 20,000 drugs simultaneously, regression is more conservative than standard disproportionality analysis and minimizes false inclusion of innocent bystanders co-reported with true SJS actives. Drugs were defined as *inactive* if they had no or minimal reporting correlation with SJS, as specified by the following criteria: for drugs with less than 1000 reports in total, inactives must never have been reported with SJS; for drugs with at least 1000 reports in total, inactives must have disproportionately few SJS reports as indicated by a negative information component (IC) 95% credibility interval [14, 15] and never be the sole suspect drug in any SJS report. The *sole suspect* criterion minimizes the risk of including as "inactive" drugs that have weak overall correlation to SJS in the database but may have strong implications for a causal link in one or a few reports. We stress that in this case we used assertions concerning drug adverse reactions as opposed to laboratory measurements, which is analogous to the hepatotoxicity modeling study described above. It is still uncommon to use such data sources in QSAR studies, but it is important to realize that their exploration can bring about highly significant opportunities for impactful model development that would not be realized otherwise.

Figure 6.12.2 presents a general methodological workflow integrating the development, interpretation, and validation of QSAR models for SJS. We shall highlight that despite the unusual first step, the workflow integrated the most rigorous and comprehensive elements and best practices of model development and interpretation.

First, as described above, a diverse set of drugs associated with SJS including positive (active) and negative (inactive) associations was selected from VigiBase

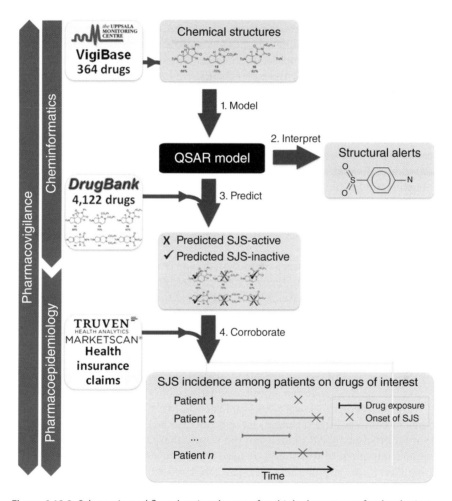

Figure 6.12.2 Schematic workflow showing the use of multiple data sources for developing, interpreting, and validating QSAR models that classify drugs as SJS-active or inactive. VigiBase provided 364 drugs whose chemical structures were used as variables for QSAR modeling. QSAR models provided structural alerts for interpretation and predicted potential SJS actives and inactives in DrugBank. Finally, the predicted actives and inactives were evaluated for evidence of SJS activity or lack thereof in VigiBase, ChemoText, and Micromedex (see text for additional discussion).

as described above. Because of the nature of the initial dataset, drugs were initially listed by name, which required us to translate name to structure (we used ChemSpider, chemspider.com, for this task). This initial transformation was followed by rigorous chemical curation (see *Methods Volume*, Section 12.2). Of the 436 drugs extracted from VigiBase (excluding mixtures and biologics), drug chemical structures were retrieved and curated to ensure that they were correctly represented and standardized prior to model development [16]. After removing salts, metal-containing compounds, large molecules (molecular weight > 2000 Da), and structural duplicates (using ChemAxon v.5.0; Pipeline Pilot Student Edition v.6.1.5), 194 actives and 170 inactives remained for QSAR modeling.

Second, binary QSAR models were developed for the curated set of SJS assertions. We used three different sets of chemical descriptors: Dragon, ISIDA substructural fragments, and MACCS fingerprints. Dragon descriptors (v.5.5, Talete SRL) [17], known for their comprehensive characterization of chemicals structures, include constitutional functional groups, atom-centered fragments, molecular properties, and 2D frequency fingerprints. For each of the three descriptor sets, two classification methods (random forest (RF) [18] and support vector machines (SVMs) [19]) were used to build QSAR models. All models were evaluated by external fivefold cross validation [20] whereby the entire dataset was divided randomly into five equal parts. Each individual part was systematically left out as an external validation set, whereas the remaining 80% of compounds in the dataset were used for model development.

Third, QSAR models were interpreted for important chemical features to detect structural alerts, which are chemical substructures characteristic of SJS-active drugs.

Fourth, these models were used to screen DrugBank [21] for potential SJS-active drugs. Finally, predictions were checked for either the evidence of SJS or lack thereof using VigiBase [12], ChemoText [22], and Micromedex [23].

With the exception of the source of input data, all steps in Figure 6.12.2 are characteristic of any QSAR modeling investigation, but we would like to point to the somewhat uncommon approach taken to validate these models. Recall that the input data were a result of semiautomatic text mining of various literature sources for SJS assertions. A somewhat similar approach was taken for evaluating the predicted novel associations between drugs and SJS; that is, whereas predictions were accomplished using QSAR models built with chemical compound descriptors, we did not use laboratory measurements as input for building QSAR models nor did we use these for model validation. As a unique component of this study, we used three knowledge bases, VigiBase [12], ChemoText [22], and Micromedex [23], that reflect the association between the predicted drug and SJS in various data sources, namely, spontaneous ADR reports, the biomedical literature, and a curated knowledge source, respectively. VigiBase [12], a repository of global spontaneous ADR reports, provided an IC value, which is measured if each drug was linked to a disproportionate number of spontaneous SJS reports; this database was used to create the modeling set.

ChemoText, a chemocentric database of MeSH annotations sourced from PubMed [22], provided the number of human studies co-annotating the drug of interest and "SJS"(also included MeSH synonyms and related terms "erythema multiform," "epidermal necrolysis, toxic"). Micromedex [23], an evidence-based resource referenced by clinicians as an industry standard, was searched for co-mentions of SJS and related hypersensitivity. The latter two databases were used to validate model predictions for drugs not listed as SJS-active in VigiBase. In summary, the study described herein has uniquely applied chemoinformatics concepts to the field of pharmacovigilance to arrive at novel predictions for drugs that should be monitored for possible SJS effects.

6.12.4 Conclusions

There are growing indications and exciting developments that point to the utility of even more unusual sources of data for cheminformatics analysis such as social media. Indeed, there have been recent reports concerning the applications of text mining and cognitive computing to extracting terms and assertions relevant to both therapeutic and adverse effects of drugs from unstructured texts in scientific publications, Facebook and Twitter exchanges, and other internet sources. Examples include the development of the SIDER drug side effect database using FDA black label warnings [24], adverse effect database created by Yahoo scientists based on text mining of Google pages [25], identification of adverse drug effects by mining Twitter [26], and novel target discovery by text mining of biomedical literature [27]. Modern cheminformaticians should be aware of all sources of data concerning biological effects of chemicals and learn to explore to explore these sources. However, it is critically important to curate the raw data properly to avoid the use of erroneous or duplicative data for model generation as ignoring this critical component of the data analysis will likely lead to irreproducible results.

Essentials

- ChEMBL, PubChem, and ToxCast provide large amounts of data on chemical structures and biological activity.
- Reports in printed media and meta data from MEDLINE can be exploited by text mining coupled with lexical and linguistic methods as well as with ontological rules to provide datasets for QSAR modeling.
- A workflow has been established for combining data from multiple sources.
- First studies along these lines have been made for developing models for hepatotoxicity and for the Stevens–Johnson Syndrome.

Available Software and Web Services (accessed January 2018)

- https://www.ebi.ac.uk/chembl/
- https://pubchem.ncbi.nlm.nih.gov/

- https://www.epa.gov/chemical-research/toxicity-forecasting
- https://www.ncbi.nlm.nih.gov/pubmed/
- https://www.nlm.nih.gov/bsd/pmresources.html

Selected Reading

- Low, Y.S., Caster, O., Bergvall, T., Fourches, D., Zang, X., Norén, G.N., Rusyn, I., Edwards, R., and Tropsha, A. (2016) *J. Am. Med. Inform. Assoc.*, **23** (5), 968–978.
- Rodgers, D., Zhu, H., Fourches, D., Rusyn, I., and Tropsha, A. (2010) *Chem. Res. Toxicol.*, **23**, 724–732.

References

[1] Austin, C.P., Brady, L.S., Insel, T.R., and Collins, F.S. (2004) *Science*, **306** (5699), 1138–1139.

[2] Bolton, E.E., Wang, Y., Thiessen, P.A., and Bryant, S.H. (2008) Annual reports in computational chemistry, , vol. 4, American Chemical Society, Washington, DC, pp. 217–241.

[3] Dix, D.J., Houck, K.A., Martin, M.T., Richard, A.M., Setzer, R.W., and Kavlock, R.J. (2007) *Toxicol. Sci.*, **95**, 5–12.

[4] Gaulton, A., Bellis, L.J., Bento, A.P., Chambers, J., Davies, M., Hersey, A., Light, Y., McGlinchey, S., Michalovich, D., Al-Lazikani, B., and Overington, J.P. (2012) *Nucleic Acids Res.*, **40**, D1100-7.

[5] Cherkasov, A., Muratov, E.N., Fourches, D., Varnek, A., Baskin, I.I., Cronin, M., Dearden, J., Gramatica, P., Martin, Y.C., Todeschini, R., Consonni, V., Kuz'min, V.E., Cramer, R., Benigni, R., Yang, C., Rathman, J., Terfloth, L., Gasteiger, J., Richard, A., and Tropsha, A. (2014) *J. Med. Chem.*, **57**, 4977–5010.

[6] Varnek, A., Gaudin, C., Marcou, G., Baskin, I., Pandey, A.K., and Tetko, I.V. (2009) *J. Chem. Inf. Model.*, **49**, 133–144.

[7] Schmidhuber, J. (2014) *Neural Networks*, **61**, 85–117.

[8] Paricharak, S., IJzerman, A.P., Jenkins, J.L., Bender, A., and Nigsch, F. (2016) *J. Chem. Inf. Model.*, **56**, 1622–1630.

[9] Ramsundar, B., Kearnes, S., Riley, P., Webster, D., Konerding, D., and Pande, V. (2015) *Massively Multitask Networks for Drug Discovery* (accessible at https://arxiv.org/pdf/1502.02072.pdf).

[10] Rodgers, A.D., Zhu, H., Fourches, D., Rusyn, I., and Tropsha, A. (2010) *Chem. Res. Toxicol.*, **23**, 724–732.

[11] Low, Y.S., Caster, O., Bergvall, T., Fourches, D., Zang, X., Norén, G.N., Rusyn, I., Edwards, R., and Tropsha, A. (2016) *J. Am. Med. Inform. Assoc.*, **26**, 968–978.

[12] Lindquist, M. (2008) *Drug Inf. J.*, **42**, 409–419.

[13] Caster, O., Norén, G.N., Madigan, D., and Bate, A. (2010) *Stat. Anal. Data Min.*, **3**, 197–208.
[14] Bate, A., Lindquist, M., Edwards, I.R., Olsson, S., Orre, R., Lansner, A., and De Freitas, R.M. (1998) *Eur. J. Clin. Pharmacol.*, **54**, 315–321.
[15] Norén, G.N., Hopstadius, J., and Bate, A. (2013) *Stat. Methods Med. Res.*, **22**, 57–69.
[16] Fourches, D., Muratov, E., and Tropsha, A. (2010) *J. Chem. Inf. Model.*, **50**, 1189–1204.
[17] Todeschini, R. and Consonni, V. (2000) *Handbook of Molecular Descriptors*, Wiley-VCH Verlag GmbH, Weinheim, Germany, 688pp.
[18] Breiman, L. (2001) *Mach. Learn.*, **45**, 5–32.
[19] Vapnik, V.N. (2000) *The Nature of Statistical Learning Theory*, Springer, New York.
[20] Tropsha, A. and Golbraikh, A. (2007) *Curr. Pharm. Des.*, **13**, 3494–3504.
[21] Wishart, D.S., Knox, C., Guo, A.C., Cheng, D., Shrivastava, S., Tzur, D., Gautam, B., and Hassanali, M. (2008) *Nucleic Acids Res.*, **36**, D901–D906.
[22] Baker, N.C. and Hemminger, B.M. (2010) *J. Biomed. Inform.*, **43**, 510–519.
[23] HealthTruven Health Analytics (2012) *Micromedex Healthcare Series. DRUGDEX System*.
[24] Kuhn, M., Campillos, M., Letunic, I., Jensen, L.J., and Bork, P. (2010) *Mol. Syst. Biol.*, **6**, 343.
[25] Yom-Tov, E. and Gabrilovich, E. (2013) *J. Med. Internet Res.*, **15**, e124.
[26] Freifeld, C.C., Brownstein, J.S., Menone, C.M., Bao, W., Filice, R., Kass-Hout, T., and Dasgupta, N. (2014) *Drug Saf.*, **37**, 343–350.
[27] Spangler, S., Myers, J.N., Stanoi, I., Kato, L., Lelescu, A., Labrie, J.J., Parikh, N., Lisewski, A.M., Donehower, L., Chen, Y., Lichtarge, O., Wilkins, A.D., Bachman, B.J., Nagarajan, M., Dayaram, T., Haas, P., Regenbogen, S., Pickering, C.R., and Comer, A. (2014) Automated hypothesis generation based on mining scientific literature. Proceedings of the 20th ACM SIGKDD International Conference on Knowledge Discovery and Data Mining – KDD '14, pp. 1877-1886.

6.13 Future Perspectives of Computational Drug Design

Gisbert Schneider

Swiss Federal Institute of Technology (ETH), Department of Chemistry and Applied Biosciences, Vladimir-Prelog-Weg 4, CH-8093 Zurich, Switzerland

Learning Objectives

- To identify opportunities for chemoinformatics in drug discovery.
- To apply integrated discovery processes for hypothesis testing.
- To compare practicability of computational methods for molecular design.

Outline

6.13.1 Where Do the Medicines of the Future Come from?, 405
6.13.2 Integrating Design, Synthesis, and Testing, 408
6.13.3 Toward Precision Medicine, 409
6.13.4 Learning from Nature: From Complex Templates to Simple Designs, 411
6.13.5 Conclusions, 413

6.13.1 Where Do the Medicines of the Future Come from?

Innovative bioactive agents fuel sustained drug discovery and the development of new medicines. The majority of low molecular weight drug candidates have originated from target-based discovery, often involving high-throughput compound and fragment screening in combination with structure elucidation of the ligand–receptor complexes, as well as natural product-inspired drug discovery and phenotypic screening [1]. In light of high clinical failure rates, with approximately 10% of drug candidates entered in clinical trials ultimately gaining approval by the US Food and Drug Administration (FDA), there is ample room for innovation, especially in the control of safety-related compound properties [2]. This inattention to the metabolic and toxicological aspects of drug design results in poor outcomes in two regards: overlooking potentially promising

Note: Parts of this article have previously been published by the author (*Mol. Inf.* 2014, **33**, 397–402) and are reproduced here in a redacted form, with kind permission from the publisher (Wiley-VCH, Weinheim).

Applied Chemoinformatics: Achievements and Future Opportunities, First Edition.
Edited by Thomas Engel and Johann Gasteiger.
© 2018 Wiley-VCH Verlag GmbH & Co. KGaA. Published 2018 by Wiley-VCH Verlag GmbH & Co. KGaA.

compounds and failing to dismiss those unsuitable for further development early enough in the process. Therefore, it seems advisable to identify and exclude the bad apples early on (the "fail early, fail often" concept) and, at the same time, make smarter, informed choices of the molecules we bring forward in the process. The human mind is easily overstrained and struggles to cope with the combination of the enormous number of chemical entities available for study and the multiple objectives for which we might optimize them. Therefore, future success in chemical biology and pharmaceutical research alike will fundamentally rely on the combination of advanced synthetic and analytical technologies, embedded in a theoretical framework that provides a rationale for the interplay between chemical structure and biological effect [3]. A driving role in this setting falls to bleeding-edge concepts and developments in computer-assisted molecular design, by providing access to a virtually infinite source of novel drug-like compounds and through guiding experimental screening campaigns. In fact, the computational generation of new chemical entities (NCEs) with a desired set of properties lies at the heart of chemoinformatics in future medicinal chemistry.

In this setting, the three cardinal challenges for automated drug design are as follows [4]:

- The assembly of synthetically accessible structures
- The scoring and property prediction of the candidates
- The systematic optimization of promising molecules in adaptive learning cycles (the "active learning" concept)

Numerous methods, algorithms, and heuristics have been proposed to address each of these problems [5]. While the generation of NCEs with attractive chemical scaffolds has become feasible by reaction-driven fragment assembly, and the *in silico* optimization problem may also be considered largely solved, the persistent issue of compound-scoring remains intractable. Scoring entails prioritizing compounds to pick the best from a large pool of possibilities. This process typically includes both ligand- and structure-based (receptor-based) virtual screening of the computationally generated molecules. Not surprisingly, this virtual compound pool contains many more inactive or problematic chemical structures than desirable ones. While compound elimination by appropriate scoring models enables discarding the bulk of the designs ("negative design") with acceptable accuracy, the selection of the best or most promising ones ("positive design") remains error prone. The established techniques employed at this step of the selection process include coarse-grained and application-specific heuristics. These include physicochemical property calculation, quantitative and qualitative structure–activity relationship models, similarity calculations at various levels of detail and with different molecular representations, shape-matching and automated ligand docking, as well as the detection of potentially toxic compounds, and otherwise unwanted chemical structures [6].

More recently, both qualitative and quantitative on- and off-target prediction methods have been added to the molecular designer's tool chest [7]. A common vision of future "precision medicine" is to treat patient groups with specially selected or even custom-tailored drugs to maximize the efficacy of treatment while minimizing adverse drug effects. It has long been noted that a drug typically interacts with several targets, such as proteins, nucleic acids, lipids, and higher-order structures like cells and organs, and that a macromolecular target may accommodate different types of ligands. There is no doubt that future medicinal chemistry can benefit from the tight integration of sophisticated molecular design concepts into the drug discovery process. At the same time, our mind-sets, skill-sets, and lead discovery strategies will have to adapt to fully explore the possibilities of advanced molecular informatics approaches and harvest the benefits for innovative future medicines. Automated computer-assisted *de novo* design of new molecular entities (NMEs) will undoubtedly serve as an enabling technology. Genetics and genomics will likely play a seminal role by finding predictors of drug efficacy, thereby amalgamating "chemoinformatics" with "bioinformatics."

There is still a long way to go before tailored drugs will be available on demand, and in the author's view there are several fundamental issues outstanding that require our attention. These are related to both theoretical (computational, biophysical) and practical (chemical, biological) aspects of *de novo* design. For example, notwithstanding the recent advances in scoring function development achieved by deep machine learning techniques [8, 9], the question of how to efficiently (i.e., for millions of computationally generated designs) compute the entropic contributions to ligand–receptor interaction remains unanswered, thus rendering the *ab initio* prediction of ΔG values impractical at best. Furthermore, despite much progress in computer hardware technology (e.g., by graphics processing unit computing or specialized hardware architecture) and the underlying biophysical models, the computational treatment of the dynamics of commonly employed molecular models falls short of realism [10, 11]. This rather unsatisfactory situation certainly does not mean that such approaches are futile. There are fascinating success stories of atomistic, long-time scale molecular dynamics simulations that provided insight into the mechanism of ligand–receptor interaction and enabled the discovery of NCEs [12, 13]. More rigorous or theoretically more appropriate methods are often simply impractical or conceptually inapplicable to real-world scenarios in drug discovery. Future chemoinformatics will also have to deal with increasingly complex molecular frameworks like macrocycles, peptides, nucleic acids, and "unusual" chemical groups (e.g., inorganic elements, fluorescent moieties) that have rarely been considered for rational drug design. Keeping these and other caveats for future breakthroughs in mind, short-term progress in computer-assisted *de novo* design can be expected from the following;

- *Reliable* prediction of the synthesizability of NCEs and suggestion of short synthesis routes and reactions, directly coupled to integrated synthesis and test platforms
- *Robust* quantitative prediction of the target cross-activities ("polypharmacology") of the computer-generated NMEs
- *Adaptive* structure–activity relationship models ("fitness landscapes") that incorporate information about multiple properties including solubility.

6.13.2 Integrating Design, Synthesis, and Testing

Chemical synthesis of the computer-generated designs, and subsequent biochemical and biological activity determination close one round of the molecular design cycle. While this is commonly achieved by conventional organic synthesis, it is no longer science fiction to sketch a fully integrated drug design automation in benchtop format (Figure 6.13.1). Lab-on-a-chip platforms, integrating synthesis, analytics, and bioactivity determination and control by adaptive chemistry-driven *de novo* design software will play an important role for future drug discovery. The fast iteration of molecular design, synthesis, and testing enables rapid learning. Based on newly measured activity data, improved structure–activity relationship models can be developed that help generate the next generation of molecular structures. The human mind depends on both new existing and new information as it learns; the same holds true for machine intelligences. Lab-on-a-chip systems and microfluidics chemistry are breakthrough technologies that have already led to proof-of-concept applications of this autonomous molecular design process [14]. For example, researchers at F. Hoffmann-La Roche have reported an integrated design platform by which they identified compound **1** (Figure 6.13.2), a potent low-nanomolar inhibitor of beta-secretase [15]. A full synthesis–purification test cycle took approximately one hour per compound with the biochemical on-chip assay running over 30 min. Multistep continuous flow synthesis platforms were also conceived and applied by other companies, for example, at Lundbeck, to obtain piperazine derivatives **2** as chemokine receptor CCR8 ligands mimicking template pharmacophore models [16]. First organ-on-a-chip devices and numerous miniaturized microfluidics-based compound screening systems have been developed, and we can expect this field at the interface between chemical engineering, biology, and medicinal chemistry to revolutionize the way drugs will be discovered and profiled [17].

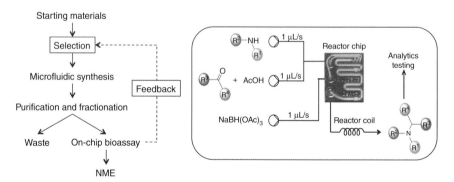

Figure 6.13.1 Schematic of an integrated design, synthesis, and screening platform illustrating the fully automated process with a feedback loop for adaptive compound optimization. An adaptive quantitative structure–activity relationship model guides the compound design and selection process. The diagram on the right illustrates an on-chip microreactor platform as a prototype of fully integrated future design–synthesize–test instruments for drug discovery. The example depicts a module for reductive amination.

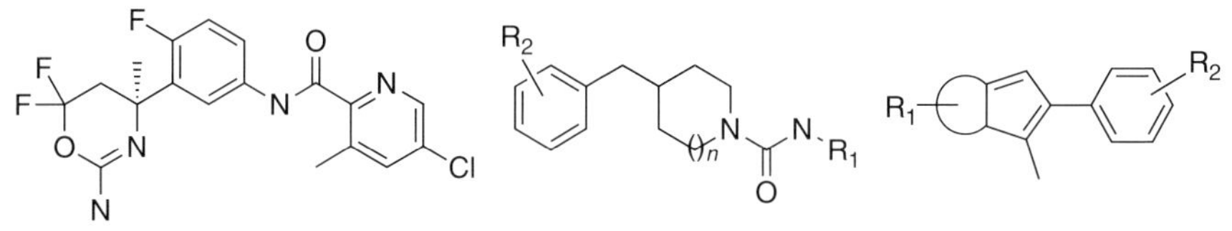

Figure 6.13.2 Examples of computationally *de novo* designed and chemically synthesized bioactive compounds, taken from recent publications.

It goes without saying that the actual synthesis of computer-generated compounds does not necessarily have to be performed in line by an automated platform. Computer-assisted synthesis planning supports skilled chemists to creatively and wisely choose the most appropriate synthesis route and reactions [18]. In this context, integrated design, synthesis, and test platforms offer the unique possibility to not only quickly obtain chemically meaningful hits and tool compounds for medicinal chemistry but also reduce the time needed for hit finding, save precious materials, and allow researchers to explore unusual or rarely used chemistry for drug discovery.

6.13.3 Toward Precision Medicine

Over the past few years, several prediction tools for the identification of the macromolecular targets of small drug-like molecules have been developed [19]. In particular, G-protein-coupled receptor (GPCR) modulators and kinase inhibitors seem to be a domain of interest for ligand-based methods. For example, the imidazopyridine scaffold **3** (Figure 6.13.2) was identified as a GPCR-privileged chemotype by target prediction and subsequent microfluidics-assisted synthesis [20]. Multi-objective design led to the discovery of target panel selective, minimalistic sigma-1 receptor, and dopamine D4 receptor antagonist, which feature nanomolar potencies and high ligand efficiencies. The target panel activity of the designed compounds was predicted by a combination of quantitative structure–activity landscape models (Figure 6.13.3) [21]. Visualizing such "fitness landscapes" can help the chemist decide which next design step may be best to pursue [22]. Other *de novo* design approaches employ *pseudo*-synthesis schemes, estimate synthetic feasibility from molecular structure, or explicitly use virtual organic synthesis reactions [23, 24].

The target profile of a designer molecule can also be "inherited" from the mother compound (template) without explicit macromolecular target prediction. This is achieved by optimizing the pharmacophoric feature similarity between the template and the newly generated chemical structures. The *de novo* generated tetrazole derivative **5** (Figure 6.13.4) is an example. It resulted from ligand-based design, taking the drug Fasudil **4** as design template. Fasudil inhibits death-associated protein kinase 3 (DAPK3), among other targets. The design algorithm assembled NCEs from molecular building blocks using virtual organic reaction schemes, so that the virtual molecules possess similar pharmacophore mimetics, or natural product-inspired synthetic agents, it may be wise to also rely on natural chemistry for future drug discovery [28]. Natural products may be considered as biologically "pre-validated" compounds. Advances in both analytical techniques and synthetic biology, as well as genomics and metabolomics, have already resulted in a large pool of structurally determined natural products, even without the inclusion of the vast number of biopolymers. Chemoinformatics methods are indispensable for the extraction of their drug-relevant molecular features for use in drug design. For example, biology-oriented synthesis (BIOS) synthetic principle explores molecular scaffolds and building blocks from natural products with proven success and high hit rates [29]. Comparisons of the pharmacophores and network pharmacology of natural products and drugs suggest that natural products offer yet-untapped bioactivity space. Furthermore, exhaustive target prediction resulted in statistically supported predictions for a mere 25% of the currently known natural products, indicating unclaimed target space for discovery [30]. Compound 7 (Figure 6.13.5) is the first example of a computationally designed, synthetically accessible mimetic of the structurally complex natural anticancer agent (−)-englerin A (**6**), a sesquiterpene from the plant *Phyllanthus engleri* [31]. Again, the design algorithm first optimized the pharmacophoric feature similarity between the virtually generated compounds and the natural product template by searching the reaction tree spanned by the software. After reranking the top-scoring designs, the (−)-englerin A mimetic **7** was obtained by straightforward three-step synthesis, while the shortest

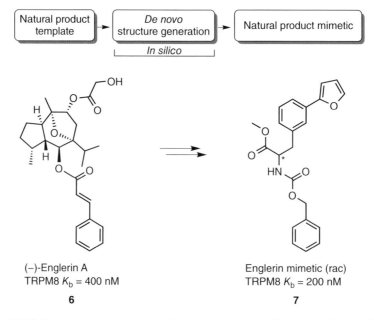

(−)-Englerin A
TRPM8 K_b = 400 nM

6

Englerin mimetic (rac)
TRPM8 K_b = 200 nM

7

Figure 6.13.5 From a complex natural product template to a synthetically easily accessible mimetic by computer-assisted *de novo* design. Morphing of the anticancer natural product (−)-englerin A into an isofunctional compound with a different scaffold enabled the discovery of a novel class of potent and selective inhibitors of transient receptor potential (TRP) M8 calcium channels.

total synthesis of the mother compound requires 14 steps. Importantly, both compounds turned out to be equally potent inhibitors of the intended drug target (transient receptor potential (TRP) M8 calcium channels).

This and other proof-of-concept applications of automated *de novo* structure generation demonstrate the practicability of contemporary molecular design software for natural product research. In addition to the computational design concept potentially providing a partial answer to the age-old question of how to convert complex into simpler chemistry, it offers protection for natural product resources and follows the principles of green chemistry. It goes without saying that more advanced methods with further improved capabilities (speed, accuracy, reliability) will be needed for broad application in medicinal chemistry. For example, the tools currently developed and employed offer insufficient consideration of additional performance and acceptability critical drug properties like biodistribution and metabolism in the design process [32].

6.13.5 Conclusions

Chemoinformatics, as distinguished from other disciplines, bridges the "art of molecule making" with rational industrial design, thereby converting serendipity-driven discovery into model-based research. It is fair to say that virtual screening and *de novo* compound construction can be efficiently managed with the currently existing software tools, which have proven table to come up with NCEs that satisfy one or more objectives. Although these molecular designs may not represent the globally optimum solutions, one can expect them to be locally optimum. Keeping in mind the often-unavoidable factual and conceptual inaccuracies of an experimental setup, one might actually not necessarily want to look for globally optimum states (according to a mathematical model) but strive for practically optimum solutions. The particular challenge for polypharmacological *de novo* design lies in the definition and accuracy of the target profile prediction. Improved scoring functions for all kinds of endpoints will keep chemoinformaticians busy. It seems that chemical database annotation and integration, data ontologies, sources and errors, chemical structure standardization, and so on, and in particular advanced machine learning models will play a decisive role for future progress in drug design. There is still much to be explored with the molecular mechanisms and physical forces that govern ligand–receptor interaction (e.g., agonist vs inverse agonist and antagonist modes of action, allosterism, flexible fit phenomena, the role of water). To an even greater extent, our current understanding of the triggering of physiological effects caused by drug target panel interaction will require models that go far beyond chemogenomic interaction networks. The era of "big data" and the automation of science for drug discovery is in its infancy. Artificial intelligence methods for pattern recognition will play an ever-increasing role in this setting, and consequent transdisciplinary thinking and action will be required. This latter aspect should not be limited to the experimental skills of the researchers involved. It demands an open mind-set, critical thinking, and hard work. Chemoinformatics has always borrowed methodological thinking from engineering and the computer sciences, so that tailored solutions could be found for challenges in chemistry. It will be wise to continue along these lines. There is much to gain from combining mathematical theory and informatics with the applied sciences. However, mutual respect among the theoreticians and practitioners involved will be an indispensable prerequisite for future success. The experience gained during the past two decades has taught valuable lessons in this regard; one has to stay open minded to explore and appreciate innovative ways of working in interdisciplinary teams. New and potentially better methodological approaches will be born from deep reasoning and creative thinking. Loosely speaking, innovation often evolves from a state that one may call the "edge of chaos," a phrase coined by mathematician Doyne Farmer [33]. If this notion is true, how can we achieve such a desirable state of innovation in chemoinformatics? Educating students may be the key to answering this question. The fundamentally interdisciplinary nature of computer-assisted drug design calls for researchers who are both willing and capable of standing their ground between multiple scientific disciplines.

Essentials

- Chemoinformatics methods enable macromolecular target prediction and the automated generation of innovative chemical structures.
- Ligand- and receptor-based methods and concepts complement each other, and there is no universally applicable method.
- Predicting (bio)activity is the most difficult aspect of *in silico* drug design.
- Rapid feedback cycles and automation facilitate computational molecular design and optimization.
- Computer-assisted synthesis design should be integrated into the drug design process.
- Interdisciplinary thinking enables innovation in drug discovery.

Selected Reading

- Eder, J., Sedrani, R., and Wiesmann, C. (2014) The discovery of first-in-class drugs: origins and evolution. *Nat. Rev. Drug Discov.*, **13**, 577–587.
- Hartenfeller, M. and Schneider, G. (2011) Enabling future drug discovery by *de novo* design. *WIREs Comp. Mol. Sci.*, **1**, 742–759.
- Schneider, G. (2018) Automating drug discovery. *Nat. Rev. Drug Discov.*, nrd. 2017, 232.

References

[1] Eder, J., Sedrani, R., and Wiesmann, C. (2014) *Nat. Rev. Drug Discovery*, **13**, 577–587.

[2] Waring, M.J., Arrowsmith, J., Leach, A.R., Leeson, P.D., Mandrell, S., Owen, R.M., Pairaudeau, G., Pennie, W.D., Pickett, S.D., Wang, J., Wallace, O., and Weir, A. (2015) *Nat. Rev. Drug Discovery*, **14**, 475–486.

- [3] Schneider, P. and Schneider, G. (2016) *J. Med. Chem.*, **59**, 4077–4086.
- [4] Schneider, G. and Fechner, U. (2005) *Nat. Rev. Drug Discovery*, **4**, 649–663.
- [5] Schneider, G. (ed.) (2013) *De Novo Molecular Design*, Wiley-VCH Verlag GmbH & Co. KGaA, Weinheim, 576 pp.
- [6] Bickerton, G.R., Paolini, G.V., Besnard, J., Muresan, S., and Hopkins, A.L. (2012) *Nat. Chem.*, **4**, 90–98.
- [7] Ain, Q.U., Aleksandrova, A., Roessler, F.D., and Ballester, P.J. (2015) *Wiley Interdiscip. Rev. Comput. Mol. Sci.*, **5**, 405–424.
- [8] LeCun, Y., Bengio, Y., and Hinton, G. (2015) *Nature*, **521**, 436–444.
- [9] Gawehn, E., Hiss, J.A., and Schneider, G. (2016) *Mol. Inf.*, **35**, 3–14.
- [10] Beier, C. and Zacharias, M. (2010) *Expert Opin. Drug Discovery*, **5**, 347–359.
- [11] Piana, S., Klepeis, J.L., and Shaw, D.E. (2014) *Curr. Opin. Struct. Biol.*, **24C**, 98–105.
- [12] Dror, R.O., Green, H.F., Valant, C., Borhani, D.W., Valcourt, J.R., Pan, A.C., Arlow, D.H., Canals, M., Lane, J.R., Rahmani, R., Baell, J.B., Sexton, P.M., Christopoulos, A., and Shaw, D.E. (2013) *Nature*, **503**, 295–299.
- [13] Kohlhoff, K.J., Shukla, D., Lawrenz, M., Bowman, G.R., Konerding, D.E., Belov, D., Altman, R.B., and Pande, V.S. (2014) *Nat. Chem.*, **6**, 15–21.
- [14] Rodrigues, T., Schneider, P., and Schneider, G. (2014) *Angew. Chem. Int. Ed.*, **53**, 5750–5758.
- [15] Werner, M., Kuratli, C., Martin, R.E., Hochstrasser, R., Wechsler, D., Enderle, T., Alanine, A.I., and Vogel, H. (2014) *Angew. Chem. Int. Ed.*, **53**, 1704–1708.
- [16] Petersen, T.P., Ritzen, A., and Ulven, T. (2009) *Org. Lett.*, **11**, 5134–5137.
- [17] Esch, E.W., Bahinski, A., and Huh, D. (2015) *Nat. Rev. Drug Discovery*, **14**, 248–260.
- [18] Ravitz, O. (2013) *Drug Discovery Today Technol.*, **10**, e443–e449.
- [19] Lavecchia, A. and Cerchia, C. (2016) *Drug Discovery Today*, **21**, 288–298.
- [20] Reutlinger, M., Rodrigues, T., Schneider, P., and Schneider, G. (2014) *Angew. Chem. Int. Ed.*, **53**, 582–585.
- [21] Reutlinger, M., Guba, W., Martin, R.E., Alanine, A.I., Hoffmann, T., Klenner, A., Hiss, J.A., Schneider, P., and Schneider, G. (2011) *Angew. Chem. Int. Ed.*, **50**, 11633–11636.
- [22] Reutlinger, M. and Schneider, G. (2012) *J. Mol. Graphics Modell.*, **34**, 108–117.
- [23] Vinkers, H.M., de Jonge, M.R., Daeyaert, F.F., Heeres, J., Koymans, L.M., van Lenthe, J.H., Lewi, P.J., Timmerman, H., van Aken, K., and Janssen, P.A. (2003) *J. Med. Chem.*, **46**, 2765–2773.
- [24] Gasteiger, J. (2007) *J. Comput.-Aided Mol. Des.*, **21**, 33–52.
- [25] Rodrigues, T., Reker, D., Welin, M., Caldera, M., Brunner, C., Gabernet, G., Schneider, P., Walse, B., and Schneider, G. (2015) *Angew. Chem. Int. Ed.*, **54**, 15079–15083.
- [26] Erlanson, D.A., Fesik, S.W., Hubbard, R.E., Jahnke, W., and Jhoti, H. (2016) *Nat. Rev. Drug Discovery*, **15**, 605–619.
- [27] Hiss, J.A., Hartenfeller, M., and Schneider, G. (2010) *Curr. Pharm. Des.*, **16**, 1656–1665.
- [28] Harvey, A.L., Edrada-Ebel, R., and Quinn, R.J. (2015) *Nat. Rev. Drug Discovery*, **14**, 111–129.

[29] Wetzel, S., Bon, R.S., Kumar, K., and Waldmann, H. (2011) *Angew. Chem. Int. Ed.*, **50**, 10800–10826.

[30] Rodrigues, T., Reker, D., Schneider, P., and Schneider, G. (2016) *Nat. Chem.*, **8**, 531–541.

[31] Friedrich, L., Rodrigues, T., Neuhaus, C.S., Schneider, P., and Schneider, G. (2016) *Angew. Chem. Int. Ed.*, **55**, 6789–6792.

[32] Kirchmair, J., Göller, A.H., Lang, D., Kunze, J., Testa, B., Wilson, I.D., Glen, R.C., and Schneider, G. (2015) *Nat. Rev. Drug Discovery*, **14**, 387–404.

[33] Langton, C.G. (1990) *Physica D*, **42**, 12–37.

7 Computational Approaches in Agricultural Research
Klaus-Jürgen Schleifer

BASF SE, Computational Chemistry and Bioinformatics, A30, 67056, Ludwigshafen, Germany

Learning Objectives

- To utilize chemoinfomatics in agricultural research.
- To apply ligand-based approaches to lead finding.
- To report examples for structure-based approaches in agricultural research.
- To use databases and programs for toxicity prediction.
- To illustrate examples for *in silico* toxicity models.

Outline

7.1 Introduction, 417
7.2 Research Strategies, 418
7.3 Estimation of Adverse Effects, 429
7.4 Conclusion, 435

7.1 Introduction

Stronger guidelines of registration authorities in terms of risk assessment will squeeze out a lot of current products from the market. This is a great opportunity for agrochemical companies to substitute the upcoming gaps with innovative novel compounds. However, in order to fulfill the specific requirements for future registration, new strategies in the R&D process have to be implemented taking into account not only the classical lead identification and optimization process, with a special focus on biological activity, but also the risk assessment of compounds in very early stages. Concomitant with a permanent cost pressure, efficient strategies have to consider inexpensive computational approaches instead of additional extensive lab experiments to support this enormous effort.

This contribution will give an overview on current computational techniques for lead identification and lead optimization based on molecular structure information. Additionally, first *in silico* toxicology approaches for the estimation of specific risk profiles will be discussed having an emerging impact for registration.

7.2 Research Strategies

Two general screening strategies are followed to identify potential lead structures (see also Section 6.1). First, chemicals are directly tested on harmful organisms (e.g., weeds), and relevant phenotype modifications are rated (e.g., bleaching). This *in vivo* approach indicates biological effects without knowledge of the addressed mode of action (*MoA*). Optimization strategies have to consider that several MoAs may be involved and that during synthetic optimization the original MoA might be changed. In addition, all observed effects reflect a combination of target activity and bioavailability of the compounds.

A second strategy, the so-called *mechanism-based* approach, allows specific target activity optimization. A fundamental condition for this procedure is the availability of the molecular target protein and a suitable biochemical assay to study the protein function in the presence of screening compounds. In this case, transfer of activity from the biochemical assay to the biological system is the challenge.

This clearly reflects that – independent of the screening strategy – hits rarely fulfill all necessary criteria for a new lead structure. Therefore, medicinal chemists have to analyze the screening results (usually structural formulas with corresponding biological or biochemical data) in order to derive a first structure–activity relationship (SAR) hypothesis.

Sometimes, 2D analyses are not sufficient to clarify the *real* situation, which is in nature three-dimensional (3D). So, minor chemical variations may completely change the geometry of a molecule (Figure 7.1), while even diverse substances (from a 2D view) may bind to a common binding site (e.g., acetylcholinesterase inhibitors).

Nowadays, molecular modeling packages are applied to calculate relevant conformations of a molecule via an energy function (i.e., force fields [1]; see also *Methods Volume*, Section 8.2) that is adjusted to experimentally derived reference geometries (mostly X-ray structures). *Van der Waals* and Coulomb terms define steric and electrostatic features, and each mismatch to reference values is penalized.

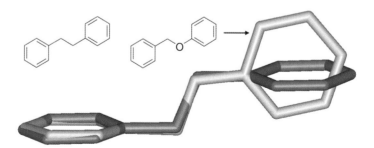

Figure 7.1 Chemical structures and superimposed X-ray coordinates of 1,2-diphenylethane (dark, CSD-code DIBENZ04) and benzyloxybenzene (bright, CSD-code MUYDOZ) indicating the different orientations of one phenyl ring induced by substitution of methylene with an ether function.

7.2.1 Ligand-Based Approaches

In order to identify molecular features crucial for biological activity, all compounds of a common hit cluster have to be superimposed to yield a pharmacophore model. Since this is done in 3D space, relevant conformers of each ligand and critical molecular functions have to be determined. X-ray crystal structures of the ligands (or of congeners) can be helpful to solve the conformational problem since they indicate at least one potential minimum conformation. Even more helpful can be the 3D structure of the physiological endogenous substrate or a postulated transition state of an enzyme reaction (Figure 7.2; see also Section 4.3.5.2).

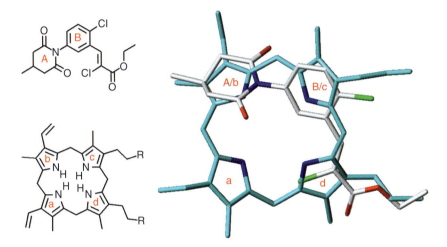

Figure 7.2 Superposition of a Protox inhibitor from pyridinedione type on a calculated protoporphyrinogen-like template (cyan). For reasons of clarity corresponding ring systems are indicated and hydrogen atoms are omitted. Atoms are color coded as follows: carbon gray, nitrogen blue, oxygen red, sulfur yellow, and chlorine green.

Sometimes however, there is no experimental data at all. In this situation a theoretical exploration of relevant conformers has to be performed, taking into consideration all rotational degrees of freedom (e.g., systematic conformational search). The derived conformations are evaluated with respect to their potential energy. Corresponding to Boltzmann's equation, low energy values indicate higher chances to resemble reality. Very often, several distinct conformers are assessed to be energetically similar. In this case the most rigid highly active ligand serves as a template molecule to superimpose all other minimized ligands (i.e., active analogue approach).

Identification of crucial functions – which should be present (at least in part) in all active ligands – takes place via an SAR analysis of all compounds of the cluster. Hypotheses derived from SAR (Figure 7.3) may be experimentally validated by testing compounds with an absent or optimized substitution pattern.

To superimpose all ligands in an appropriate manner, essential groups (e.g., carbonyl groups, aromatic rings, etc.) of energetically favorable conformers are chosen as fit points. The yielded pharmacophore model characterizes the common

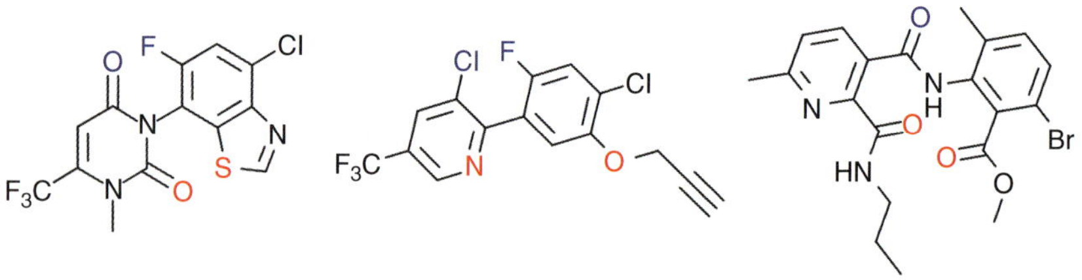

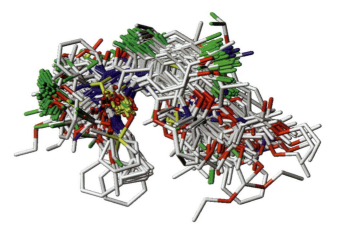

Figure 7.3 Common interaction pattern of potent Protox inhibitors from uracil (left) and pyridine type. Each molecule comprises two ring systems and electron-rich functions on both sides of the linked rings (blue and red colored).

Figure 7.4 Pharmacophore model of 318 Protox inhibitors (color code as indicated at Figure 7.2).

bioactive conformations because similar functional groups (e.g., hydrogen bond acceptors) of all molecules are pointing to the same 3D space (Figure 7.4). Lack of one or several of these functions is usually associated with a drop in activity.

Pharmacophore models may be used to derive ideas for the substitution of one group (e.g., hydroxyl) against another chemical group with similar features (e.g., amine group as hydrogen bond donor and acceptor). This is a helpful indication facilitating planned synthesis strategies or a guided compound purchase. Modeling tools like comparative molecular field analysis (CoMFA) [2], comparative molecular similarity indices analysis (CoMSIA) [3], or PrGen [4] even allow estimation of effects on a quantitative level. These so-called *three-dimensional quantitative structure–activity relationship* (3D-QSAR) studies require the pharmacophore model to determine significantly different interaction patterns that are directly associated with experimental data (e.g., activity). The statistical machinery behind this is mainly based on principal component analyses (PCAs) and partial least squares (PLS) regression. PCA transforms a number of (possibly) correlated variables into a (smaller) number of uncorrelated variables called *principal components*. PLS regression is probably the least restrictive of the various multivariate extensions of the multiple linear regression models. In its simplest form, a linear model specifies the (linear) relationship between a dependent (response) variable Y and a set of predictor variables, the Xs, so that

$$Y = b_0 + b_1 X_1 + b_2 X_2 + \cdots + b_p X_p \tag{7.1}$$

In Eq. (7.1) b_0 is the regression coefficient for the intercept, and the b_i values are the regression coefficients (for variables 1 through p) computed from the data.

The correlation of experimental and calculated activities assesses the quality of 3D-QSAR models. The squared correlation coefficient (r^2) yielded by this statistics is a measure of the goodness of fit. The robustness of the model is tested via cross-validation techniques (leave x% out), indicating the goodness of prediction (q^2). Models with q^2 values >0.4–0.5 are considered to yield reasonable predictions for hypothetical or not yet tested molecules that are structurally comparable to those compounds used to establish the model (Figure 7.5).

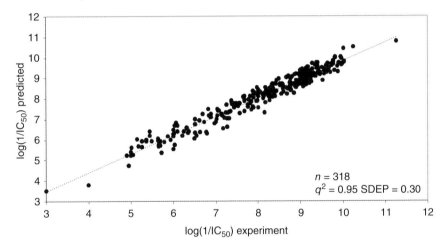

Figure 7.5 Graph indicates the correlation of experimental and predicted IC$_{50}$ values yielded by a "leave-one-out" cross-validation ($q^2 = 0.95$) for the pharmacophore model shown in Figure 7.4.

CoMFA and CoMSIA not only derive a mathematical equation but also generate contour maps (e.g., steric or electrostatic fields) that should or should not be occupied by new compounds with optimized characteristics (Figure 7.6).

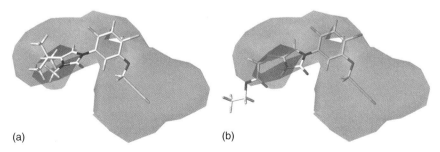

Figure 7.6 Contour map derived from a 3D-QSAR study. Clouds indicate favorable space to be occupied by potent Protox inhibitors. While the highly active imidazolinone derivative (a) fits almost perfectly, the ethylcarboxylate residue of the weaker ligand protrudes the preferred region (b).

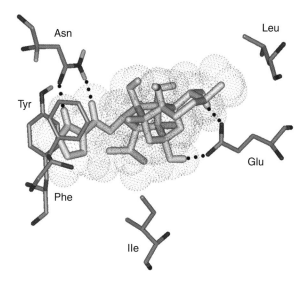

Figure 7.7 Pseudoreceptor model for insecticidal ryanodine derivates constructed with the program PrGen [4]. The binding site model is composed of seven amino acid residues and contains the structure of ryanodine [5]. Hydrogen bond interactions are indicated with dashed lines.

PrGen [4] creates a pseudoreceptor model around the pharmacophore representing an image of the hypothetical binding site (Figure 7.7). Ligand–pseudoreceptor site interactions, solvation, and entropic energy terms are calculated to correlate experimental and computed free binding energies. The binding site construction may take into account experimentally determined amino acid residues of the real binding site or just residues with complementary features to the ligands.

New hypothetical compounds may be introduced in the validated pseudoreceptor model to estimate free binding energies and, thus, to prioritize lab capacities.

A common drawback of ligand-based approaches is the fact that data derived from screening hits may only be interpolated to somehow similar compounds. If any structural information is not present in the training set compounds, transfer to totally new structures is generally not possible [6].

7.2.2 Structure-Based Approaches

New scaffolds for active ingredients are classically obtained by an experimental random screening. Essential for this high-throughput experiment is a multitude of compounds that has to be purchased or synthesized and handled. For capacity reasons, it is desirable to test not all available compounds, but only those with a high chance of success. One helpful strategy to focus a compound library to a particular target is based on the molecular structure of this protein.

At present, highly sophisticated analytical methods like X-ray crystallography, NMR, or cryo-electron microscopy are applied to solve 3D structures of enzymes, ion channels, G-protein-coupled receptors, and other proteins. A collection of more than 125,000 protein coordinates is freely available at the Protein Data Bank (PDB) [7]. In some cases even ligand–protein co-crystal structures are solved. Coordinates derived from co-crystals unambiguously localize the binding site and provide insight into the binding mode of a bound ligand. This allows computational chemists to characterize specific interaction patterns as being crucial for tight binding.

Equipped with this information, the binding site may be used like a lock in order to find the best fitting key by virtually screening diverse compound libraries (i.e., lead identification) or increasing the specific fit of weak binders (lead optimization). Automation of this so-called (protein) *structure-based approach* [8, 9] is typically divided into a docking and a scoring step [10]. While docking yields the pose(s) of a ligand in the complex, scoring is necessary to discriminate good and bad binders by calculating free energies of binding for each generated conformer of a ligand.

In this context it is common to differentiate between *empirical* and *knowledge-based* scoring functions [11]. The term "empirical scoring function" stresses that these quality functions approximate the free energy of binding, $\Delta G_{binding}$, as a sum of weighted interactions that are described by simple geometrical functions, f_i, of the ligand and receptor coordinates r (Eq. 7.2). Most empirical scoring functions are calibrated with a set of experimental binding affinities obtained from protein–ligand complexes, that is, the weights (coefficients) ΔG_i are determined by regression techniques in a supervised fashion. Such functions usually consider individual contributions from hydrogen bonds, ionic interactions, hydrophobic interactions, and binding entropy. As with many empirical approaches, the difficulty with empirical scoring arises from inconsistent calibration data:

$$\Delta G_{binding} = \sum \Delta G_i \cdot f_i(r_{ligand}, r_{receptor}) \qquad (7.2)$$

Knowledge-based scoring functions have their foundation in the inverse formulation of the Boltzmann law, computing an energy function that is also referred to as a "potential of mean force" (PMF). The inverse Boltzmann technique can be applied to derive sets of atom pair potentials (energy functions) favoring preferred contacts and penalizing repulsive interactions. The various approaches differ in the sets of protein–ligand complexes used to obtain these potentials, the form of the energy function, the definition of protein and ligand atom types, the definition of reference states, the distance cutoffs, and the several additional parameters [12].

An extension of docking procedures is *de novo* design [13] with BASF's archetype LUDI [14]. Here, molecular fragments are composed inside a given binding pocket in order to design a perfectly matched new molecule.

Both attempts rely on an accurate binding site characterization, an appropriate ligand/binding site complex generation, and a reliable estimation of the free binding energies. The principle of a docking and scoring procedure is illustrated in Figure 7.8.

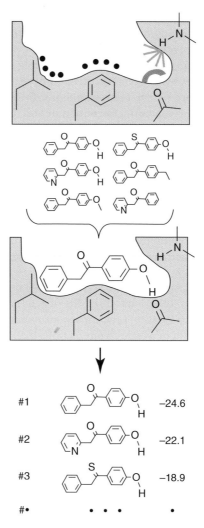

Figure 7.8 Protocol of a classical docking and scoring procedure. The binding site cavity is characterized via, for example, hydrophobic (circles), hydrogen bond donor (lines), and hydrogen bond acceptor properties (circle segment). Each compound of a database (or real library) is flexibly docked into the binding site, and the free binding energy [kJ/mol] for each of the derived poses is estimated by a mathematical scoring function.

In order to demonstrate a docking application, the crystal structure of mitochondrial protoporphyrinogen IX oxidase (Protox) from common tobacco complexed with an acidic phenylpyrazole inhibitor (INH) and a non-covalently bound flavin adenine dinucleotide (FAD) cofactor was chosen (PDB ID code 1SEZ [15]). A salt bridge primarily fixes the inhibitor from the carboxylate group to a highly conserved arginine (Arg98) at the entrance of the binding niche. Further stabilization is due to hydrophobic contacts to Leu356, Leu372, and Phe392 in the core region.

In the first step, INH was extracted from the binding site cavity, and a commercial docking program (FlexX [16]) was applied to determine whether the original binding pose of the X-ray structure could be refound. For this calculation only a volume with a radius of 10 Å around the binding site was considered, not the complete protein.

The program detects 98 favorable docking solutions within an energy range of 10.0 kJ/mol ($\Delta\Delta G$). Except for two poses, all solutions strictly interact with their acidic function to the basic Arg98. However, only 20 of them are really located in the binding niche. The energetically most favorable proposals fix the guanidinium group of Arg98 from the solvent side (Figure 7.9). Another 13 solutions are blocking the gorge to the binding site.

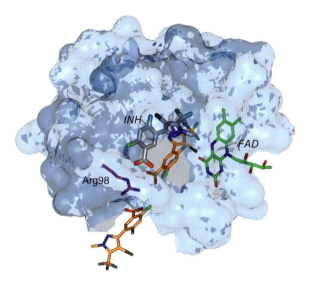

Figure 7.9 X-ray crystallographically determined binding site of Protox [15] including the co-crystallized inhibitor INH (structural formula see Figure 7.10) and a part of the cofactor FAD. Highlighted is Arg98 at the entrance of the binding site cavity interacting with INH and almost all solutions of the FlexX approach via electrostatic and hydrogen bond interactions. Two docking poses representing a cluster of yielded solutions are indicated: one at the outside and one inside the binding site cavity (orange-colored carbon atoms).

In order to rationalize the docking process and to circumvent nonrealistic solutions (i.e., outside the known binding region), two pharmacophore-type constraints may be set. First, an interacting group in the receptor site may be specified (i.e., interaction constraint). During the simulation each docking solution is checked to see whether there is a contact between the ligand and this particular hot spot. If not, the solution is discarded. The second type is a spatial constraint. Here, a spherical volume is defined in the active site and a specified atom or group of atoms from the ligand must lie within this sphere in the docking solution. FlexX-Pharm [17] offers both constraint types that even may be combined.

Taking into consideration only the 20 accurate docking solutions, it must be stated that the original pose of INH is not perfectly found. Although most acidic groups interact with Arg98, the binding mode is different compared with the experimentally solved X-ray structure. Only the more hydrophobic pyrazole ring matches (in some cases) its reference counterpart. Furthermore, two docking solutions are totally different. Their acid function interacts with the terminal amide group and the backbone NH of Asn67, which is opposite to Arg98 (Figure 7.10).

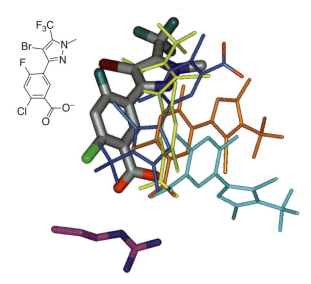

Figure 7.10 Structural formula of INH and comparison of the poses derived from FlexX docking (single colored) and crystallization experiment (thick). Indicated is the crucial Arg98 that stabilizes all poses with the exception of the blue-colored solution, which interacts with the acid group (red-colored oxygen atoms) to the opposite side (i.e., Asn67).

It is worth mentioning that the docking procedure used does not take into account flexibility of the binding site residues. Only the ligand is considered flexible in an energetically restricted range. However, there are programs that allow concerted consideration of flexibility for ligand and binding site residues to simulate induced fit docking (e.g., GLIDE, FlexE).

In contrast to the charged INH inhibitor used for the co-crystallization experiment, all of BASF's in-house compounds presented in the abovementioned 3D-QSAR study are uncharged. Therefore, a second docking study with a neutral uracil derivative (UBTZ, Figure 7.11) should clarify how ligands without acid function bind to this target site.

Applying the default parameters, FlexX produced 122 solutions predominantly located in the binding cavity. The two energetically most favorable solutions are compared with the original pose of INH (Figure 7.11). It is interesting to note that each pose of UBTZ has a direct contact to Arg98. In one pose, a carbonyl oxygen of the uracil and a fluorine of the benzothiazole ring are involved. Alternatively the nitrogen atom of the benzothiazole ring is directed to the positively charged Arg98. This docking solution shows a better total overlap with INH. Interestingly, although chemically diverse, both types of inhibitors (INH and UBTZ) obviously mimic similar binding properties necessary for complex formation.

In a next step we tried to dock the physiological substrate, protoporphyrinogen IX, to the INH and Triton-X100 cleaned enzyme. This attempt failed, although

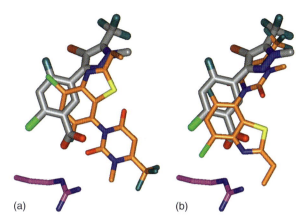

Figure 7.11 Comparison of two docking solutions for BASF's uracil derivative UBTZ with the bound INH. UBTZ interacts with Arg98 over the carbonyl oxygen of uracil and a fluorine of the benzothiazole ring (a) or the nitrogen atom of the benzothiazole ring (b).

the maximum overlap volume and the clash factors were modified in such a way that the narrow binding pocket was apparently relaxed and, subsequently, even the cofactor FAD was (nonphysiologically) removed. Only the product of the enzyme reaction, protoporphyrin IX, which is sterically not as demanding as the substrate, could be docked into the FAD-free binding site cavity (Figure 7.12). The yielded solution interacts loosely with one propionate group to Arg98. The second acid group protrudes into the solvent region. Although the cofactor was not present during the calculation, the final pose indicates carbon atom C20 of protoporphyrin IX in close proximity to the electron-accepting nitrogen atom N5 of the flavin ring. This is in general agreement with the results obtained

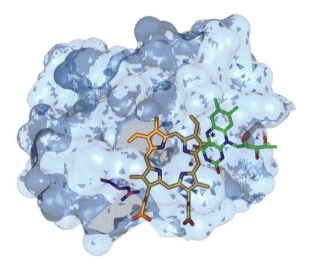

Figure 7.12 Docking solution for protoporphyrin IX in the Protox binding site. One propionic acid is close to Arg98, but does not form an explicit hydrogen bond. Asterisks indicate the proposed reaction centers C20 of protoporphyrin IX and N5 of FAD (see text for details).

by Koch *et al.* [15] and Jordan [18] that propose the initial hydride transfer at C20, followed by hydrogen rearrangements in the whole ring system by enamine–imine tautomerizations.

One possible explanation for the failed docking of the tetrapyrrole derivatives under physiological conditions (i.e., in presence of FAD) might be the reference topology of the binding site. During the co-crystallization experiment, the ligand and the binding pocket adapt to each other perfectly forming a ligand-specific complex structure. Via this induced fit, the narrow cleft is not able to incorporate much larger ligands. Therefore, only the flattened protoporphyrin IX could be introduced, but not in the intuitively expected manner (i.e., completely buried in the binding site cavity with a tight contact to Arg98). In order to circumvent such ligand-specific binding topology for a general docking approach, the ligand-free protein (apoprotein or holoprotein) may be relaxed applying a molecular dynamics (MD) simulation. MD is a computer simulation of physical movements of atoms and molecules. For a given protein, the atoms are allowed to interact for a period of time, giving a view of the motion of the atoms. In the most common version, the trajectories of atoms are determined by numerically solving the Newton's equations of motion for a system of interacting particles, where forces between the particles and potential energy are defined by molecular mechanics force fields [19]. As a result of the simulation, several diverse protein conformers may be chosen as reference input structures for a docking approach.

Besides the binding site flexibility, water molecules may also play a crucial role in protein–ligand binding. Taking the abovementioned Protox as an example, there are two relevant X-ray structures available in the PDB. The first one, used for the modeling study, is from tobacco and has a low resolution of only 2.9 Å. In its binding pocket only one water molecule is to be seen not showing any interaction to the co-crystallized ligand or the cofactor. Therefore this water molecule was not considered in the docking approach. In contrast to that, the human Protox X-ray structure, published 7 years later (pdb-code 3NKS), has a high resolution of 1.9 Å. Here the binding pocket encloses the ligand, the cofactor, and a cluster of about 10 water molecules forming a hydrogen bond network to the ligand, to the cofactor, and among each other [20]. In this case, explicit water molecules would have to be taken into account as fixed anchor points to elucidate reasonable binding poses for the ligands in docking experiments. The relevance of such particular water molecules for lead optimization is described in a recent example for the herbicidal target protein IspD [21]. In this study a water molecule forming hydrogen bonds to the ligand and a residue of the binding pocket was replaced by an additional substituent of a new ligand to form almost the same interaction pattern as the water molecule. Introduction of a nitrile group, as a hydrogen bond acceptor, yielded a fourfold increase of binding affinity. Unexpectedly, a carbonic acid function at the same position, also mimicking an H-bond acceptor, decreased the binding affinity of the ligand by a factor of almost 2000.

These examples indicate typical challenges of structure-based approaches starting from the need of a highly resolved target structure, a multitude of yielded docking poses, and problematic estimations of free binding energies. An enormous advantage of this technique is the unbiased use. Results obtained for a particular target site provide information for new chemical structures without prior expert knowledge or selection. From a technological viewpoint, there is ongoing improvement to gain more realistic docking solutions (e.g., interaction and spatial constraints or a post-processing step). Additionally, the quality of the energy estimation may be increased by tailor-made scoring functions. This requires much experimental data (e.g., co-crystal data and IC_{50} values) for a particular family of targets (e.g., kinases) for the calibration.

To summarize this topic, structure-based methods are extremely helpful to create ideas for new scaffolds and further optimization strategies.

7.3 Estimation of Adverse Effects

Even a compound with highest efficacy can fail because of undesired adverse effects. The new Plant Protection Products Regulation (Regulation EC No. 1107/2009) clearly indicates a strict guideline for relevant toxicological endpoints. Furthermore, there are defined cutoff (restriction) criteria for hazardous properties that will result in a substance being banned even if it can be applied in a safe manner. This comprises compounds proven to be *c*arcinogenic, *m*utagenic, toxic for *r*eproduction (CMR), endocrine disruptors, and persistent (bioaccumulative). Most of this information can only be produced by costly higher tier studies. Especially in early research stages, such experiments would be prohibitive and in contradiction to the 3R philosophy (replacement, refinement, and reduction) to replace animal studies. Therefore cell-based indication studies and/or even cheaper *in silico* techniques will have to be developed, providing first insight into the risk potential of novel compounds.

Efficacy and lack of harmful effects on human health (i.e., toxicological effects) is not sufficient to obtain approval for a novel active ingredient. Legislation clearly demands "no unacceptable effects on the environment." This means that ecotoxicological effects (toxicity in e.g., honeybees, birds, rodents, fishes, etc.) and environmental fate (i.e., behavior of the compound and relevant metabolites in soil, ground water, air, etc.) are of highest relevance. Mainly based on the insufficient number and quality of experimental data, only a few *in silico* tools are available for this purpose (e.g., enviPath, a database and prediction system for the microbial biotransformation of organic environmental contaminants [22]). The ongoing need for effective and registrable products coupled with a growing ecological awareness of the society will be a strong motivator for companies to strengthen their experimental and computational efforts in this particular area (see Chapter 8).

7.3.1 *In Silico* Toxicology

In silico toxicology indicates a variety of computational techniques that relate the structure of a chemical to its toxicity or fate with the advantages of cost-effectiveness, speed compared with traditional testing, and reduction in animal use. Based on experimental data two general techniques are followed for toxicity predictions:

- Rule-based expert systems that rely on a set of chemical structure alerts
- Correlative SAR methods based on statistical analysis.

7.3.2 Programs and Databases

Several software tools are available, allowing statistical estimation of some critical endpoints. The commercial computer program *Deductive Estimation of Risk from Existing Knowledge* (DEREK, Lhasa Inc.) is designed to assist chemists and toxicologists in predicting likely areas of possible toxicological risk for new compounds, based on an analysis of their chemical structures. DEREK indicates whether a specific toxic response may occur, but it does not provide a quantitative estimate of the prediction. The program accepts as input a "target" (the molecule to be analyzed) drawn in the language of structural formulae that is common to all organic chemists. DEREK scans a "rule base" of substructures that are known to have adverse toxicological properties, looking for matches to substructures in the target molecule. "Hits" in the rule base are shown to the user on a graphical display and summarized in tabular form for hardcopy output.

Toxicity prediction by komputer assisted technology (TOPKAT) quantifies electronic, bulk, and shape attributes of a structure in terms of electrotopological state (E-states) values of all possible two-atom fragments, atomic size-adjusted E-states computed from rescaled count of valence electrons, molecular weight, topological shape indices, and symmetry indices. The methodology is an extension of classic QSAR.

Leadscope is used to analyze datasets of chemical structures and related biological or toxicity data. Structures and/or data can be loaded from an SD file and data can be loaded from a text file. Leadscope provides a number of ways to group a set of compounds, including the chemical feature hierarchy (27,000 named substructures), recursive partitioning/simulated annealing (a method for identifying active classes characterized by combinations of structural features), structure-based clustering, and dynamically generated significant scaffolds or substructures. Based on the presence or absence of the substructures, models are generated with training sets of compounds associated with wanted (activity) or unwanted (toxicological endpoint) biological data. Test compounds with unknown biological activity are then classified according to relevant substructures (i.e., descriptors) of the validated models, and activity is estimated. In addition to that, Leadscope automatically calculates the following properties for all imported compounds: aLogP, polar surface area, number of hydrogen bond donors, number of hydrogen bond acceptors, number of rotatable bonds, molecular weight, number of atoms, and Lipinski score.

Another example of a knowledge-based system is computer-automated structure evaluation (CASE) and its successor, MCASE/MC4PC (former Multi-CASE), designed for the specific purpose of organizing biological/toxicological data obtained from the evaluation of diverse chemicals. These programs can automatically identify molecular substructures that have a high probability of being relevant or responsible for the observed biological activity of a learning set composed of a mix of active and inactive molecules of diverse composition. New, untested molecules can then be submitted to the program, and an expert prediction of the potential activity of the new molecule is obtained. Another program is Case Ultra, which was especially developed with an objective to meet the current and most updated regulatory needs for safety evaluation of compounds.

Like DEREK, OncoLogic [23], HazardExpert [24], and ToxTree [25] are other knowledge-based expert systems for the prediction of toxicological endpoints.

Independent of the particular program, the crucial basis for good models and predictions is the quality of data. In best-case scenario, all data are obtained by in-house experiments for the toxicological endpoints of interest. In many cases, however, there will not be enough experiments to develop a general model, only a tailor-made scaffold-related model for a certain endpoint. Therefore, external data might be of value to provide a better coverage of chemical space. Publicly available data sources exist for endpoints like human health cancer and mutagenicity (ISSCAN, CPDB, OASIS Genetox), skin sensitization (local lymph node assay, guinea pig maximization test, and ECETOC skin sensitization), mammalian single/repeated dose toxicity studies (Japan EXCHEM, RepDose), eye irritation (ECETOC), skin irritation (OECD Toolbox), and skin penetration (EDETOX). The OECD (Q)SAR Toolbox, which is a software application intended to be used by governments, chemical industry, and other stakeholders in filling gaps in (eco)toxicity data needed for assessing the hazards of chemicals, allows further estimation of bioaccumulation, acute aquatic toxicity, and estrogen receptor binding.

Parallel to these freely available databases and tools, there are also commercially compiled databases of toxicological information. Typically addressed endpoints are carcinogenicity, genetic toxicity, chronic and sub-chronic toxicity, acute toxicity as well as reproductive and developmental toxicity, mutagenicity, skin/eye irritation, and hepatotoxicity. Providers are companies like Leadscope (Leadscope database), Lhasa Ltd. (VITIC Nexus), TerraBase Inc. (TerraTox), or MDL (RTECS). In order to combine a number of databases and resources, several tools have become available. Resources from TOXNET, the OECD's eChemPortal, and the US EPA Aggregated Computational Toxicology Resource (ACToR) are most likely to be useful for obtaining information about single compounds. ACToR is a freely available collection of databases from more than 200 sources. Data include chemical structure, physicochemical properties, *in vitro* assay, and *in vivo* toxicology data for industrial chemicals with mostly high and medium production volume, pesticides and potential groundwater or drinking water contaminants. Based on the analysis of Judson *et al.* [26] using approximately 10,000 substances (industrial chemicals and pesticide ingredients), it was shown that acute hazard data are available for 59% of the surveyed chemicals, testing information for carcinogenicity for 26%, developmental toxicity for 29%, and reproductive toxicity for only 11%. In order to fill the toxicity data gaps, EPA has designed the ToxCast screening and prioritization program. ToxCast is profiling over 300 well-characterized chemicals (mostly pesticides) in over 400 HTS endpoints. These endpoints include

- Biochemical assays of protein function
- Cell-based transcriptional reporter assays
- Multicell interaction assays
- Transcriptomics on primary cell cultures
- Developmental assays in zebrafish embryos

Almost all the compounds have been tested in traditional toxicology tests, including developmental toxicity, multigeneration studies, and sub-chronic and chronic rodent bioassays. These data, collected in the Toxicity Reference Database (ToxRefDB; http://www.epa.gov/ncct/toxrefdb/), will be used to build computational models to forecast the potential human toxicity of chemicals with the aim to lead more efficient use of animal testing.

7.3.3 *In Silico* Toxicology Models

SAR and QSAR for toxicological endpoints were applied only sporadically in drug discovery in the 1960s and 1970s. This was primarily due to the combination of a lack of detailed understanding of most mechanisms of toxicology, a lack of systematically generated datasets around specific toxicities, and a lack of general directives from regulators for standard tests that should be performed prior to drug testing in humans. With the advent of the *Salmonella* reverse mutation assay (Ames test) in the early 1970s, this picture began to change. In the Ames test [27] frameshift mutations or base-pair substitutions may be detected by exposure of histidine-dependent strains of *Salmonella typhimurium* to a test compound. When these strains are exposed to a mutagen, reverse mutations that restore the functional capability of the bacteria to synthesize histidine enable bacterial colony growth on a medium deficient in histidine ("revertants"). In some cases there is need to activate the compounds via a mammalian metabolizing system, which contains liver microsomes (with S9 mix). Ames positive compounds significantly induce revertant colony growth at least in one out of usually five strains, either in the presence or absence of S9 mix. A compound is judged Ames negative if it does not induce significant colony growth in any reported strain. The adoption of the assay for assessment of the mutagenic potential of chemicals provided a relatively consistent data source previously unknown to toxicology. In 2009 Hansen *et al.* [28] collected 6512 nonconfidential compounds (3503 Ames positives and 3009 Ames negatives) together with their biological activity to form a new benchmark dataset for *in silico* prediction of Ames mutagenicity. The dataset contains 1414 compounds from World Drug Index (i.e., drugs) and has a mean molecular weight of 248 ± 134 (Median MW: 229). With these data the commercial software programs DEREK, MultiCASE, and Pipeline Pilot (Accelrys) as well as four noncommercial machine learning implementations (i.e., support vector machines (SVMs), Gaussian process (GP) classification, random forest, and k-nearest neighbor) were evaluated. Molecular descriptors were extracted from DragonX version 1.2 [29]. The final statistical results indicate the general applicability of all programs (Table 7.1).

All commercial programs have fixed sensitivity levels, whereas the sensitivity of all parametric classifiers (e.g., SVM) can be calculated to arbitrary levels of specificity. A sensitivity value of 0.93 for the SVM model at a 50% false positives rate indicates that for a given test set of 200 compounds with 100 mutagens and 100 non-mutagens, 93 mutagens will be classified as true positives together with 50 false positives. A shift to the 36% false positives level yields 88 true positives together with only 36 false positives. Of the commercial programs, Pipeline Pilot performs best due to the explicit training with the given validation

Table 7.1 Comparison of all classifiers for mutagenicity applied from Hansen et al. [28].

	50% false positives	43% false positives	36% false positives	Model
Sensitivity	0.93 ± 0.01	0.91 ± 0.01	0.88 ± 0.01	SVM
	0.89 ± 0.01	0.86 ± 0.01	0.83 ± 0.02	GP
	0.90 ± 0.02	0.87 ± 0.02	0.84 ± 0.03	Random forest
	0.86 ± 0.02	0.86 ± 0.02	0.81 ± 0.02	k-Nearest neighbor
	—	—	0.84 ± 0.02	Pipeline Pilot
	0.73 ± 0.01	—	—	DEREK
	—	0.78 ± 0.02	—	MultiCASE

True-positive predictions of mutagens (i.e., sensitivity) relative to false-positive rates (50%, 43%, and 36%) for trained support vector machines (SVMs), Gaussian process (GP) classification, random forest, and k-nearest neighbor in comparison to commercial programs.

dataset, whereas DEREK and MultiCASE are based on a fixed set of mainly 2D descriptors (MultiCASE) or a static system of rules derived from a largely unknown dataset and expert knowledge (DEREK). However, the latter two programs provide structure–activity and/or mechanistic information essential for structure optimization and regulatory acceptance. The parametric classifiers outperform the commercial programs and may be applied for larger screening campaigns. On the other hand, these classifiers do not provide hints why a certain compound was predicted to be mutagen, and thus optimization guidance is not given.

In a second case study [30], aquatic toxicity was predicted for a dataset of 983 unique compounds tested in the same laboratory against *Tetrahymena pyriformis* [31]. For model validation, 644 compounds were chosen randomly and the remaining 339 compounds were used as a first external test set (external test I). In addition to that, a second test set (external test II) was used with compounds recently published by the same laboratory. Six independent academic groups developed 15 different types of QSAR models with a particular focus on the predictive power for the external test sets. Each group relied on its own QSAR modeling approaches to generate toxicity models using the same datasets. For all models the applicability domain was calculated yielding a measure whether compounds of the test sets may be predicted or not. In case that totally novel compounds in the test set have no equivalents in the training set, prediction is restricted, and this unique compound will not be considered. Use of applicability domains will lead to lower coverage rates (i.e., <100%) but typically to better predictions of remaining the test sets.

The internal prediction accuracy for the modeling set ranged from 0.76 to 0.93 as measured by the leave-one-out cross-validation correlation coefficient (Q_{abs}^2). The prediction accuracy for the external validation sets I and II ranged from 0.71 to 0.85 (linear regression coefficient R_{absI}^2) and from 0.38 to 0.83 (R_{absII}^2), respectively. Finally, several consensus models were developed by averaging the predicted aquatic toxicity of all 15 models. Results of several individual models

Essentials

- The same types of techniques used in drug discovery are also used in agricultural research.
- Ligand-based approaches have identified crucial functions of agrochemicals.
- Structure-based approaches are increasingly used in agricultural research.
- Docking and scoring steps are separately investigated.
- The estimation of adverse effects, such as different types of toxicity, of plant protection candidates is routinely investigated.
- Various commercial programs are used for toxicity prediction.
- Publicly available databases on toxicity data play an important role in toxicity prediction.

Available Software and Web Services (accessed January 2018)

- http://www.epa.gov/ncct/toxrefdb/

Selected Reading

- Hansen, K., Mika, S., Schroeter, T., Sutter, A., ter Laak, A., Steger-Hartmann, T., Heinrich, N., and Müller, K.R. (2009) *J. Chem. Inf. Model.*, **49**, 2077–2081.
- Schleifer, K.-J. (2000) *J. Comput. Aided Mol. Des.*, **14**, 467–475.
- Schleifer, K.J. (2012) *Computational Approaches in Agricultural Research*, In: P. Jeschke, W. Krämer, U. Schirmer, and M. Witschel (eds), *Modern Methods in Crop Protection Research*, Wiley-VCH, Weinheim, Germany, pp. 21–41. The present Chapter is a somehow updated version of this publication. Reproduced with permission.

References

[1] For references see: https://en.wikipedia.org/wiki/Force_field_(chemistry). (accessed January 2018)

[2] Cramer, R.D. III, Patterson, D.E., and Bunce, J.D. (1988) *J. Am. Chem. Soc.*, **110**, 5959–5967.

[3] Klebe, G., Abraham, U., and Mietzner, T. (1994) *J. Med. Chem.*, **37**, 4130–4146.

[4] Zbinden, P., Dobler, M., Folkers, G., and Vedani, A. (1998) *Quant. Struct.-Act. Relat.*, **17**, 122–130.

[5] Schleifer, K.-J. (2000) *J. Comput.-Aided Mol. Des.*, **14**, 467–475.

[6] Bordás, B., Komives, T., and Lopata, A. (2003) *Pest. Manag. Sci.*, **59**, 393–400.

[7] RCSB PDB Protein Data Base, http://www.rcsb.org (accessed January 2018).
[8] Waszkowycz, B. (2002) *Curr. Opin. Drug Discovery Dev.*, **3**, 407–413.
[9] Taylor, R.D., Jewsbury, P.J., and Essex, J.W. (2002) *J. Comput.-Aided Mol. Des.*, **3**, 151–166.
[10] Kitchen, D.B., Decornez, H., Furr, J.R., and Bajorath, J. (2004) *Nat. Rev. Drug Discovery*, **11**, 935–949.
[11] Gohlke, H. and Klebe, G. (2002) *Angew. Chem. Int. Ed.*, **41**, 2644–2676.
[12] Gohlke, H., Hendlich, M., and Klebe, G. (2000) *J. Mol. Biol.*, **295**, 337–356.
[13] Schneider, G. and Fechner, U. (2005) *Nat. Rev. Drug Discovery*, **4**, 649–663.
[14] Böhm, H.J. (1992) *J. Comput.-Aided Mol. Des.*, **6**, 61–78.
[15] Koch, M., Breithaup, C., Kiefersauer, R., Freigang, J., Huber, R., and Messerschmidt, A. (2004) *EMBO J.*, **23**, 1720–1728.
[16] Rarey, M., Kramer, B., and Lengauer, T. (1995) *Proc. Int. Conf. Intell. Syst. Mol. Biol.*, **3**, 300–308.
[17] Hindle, S.A., Rarey, M., Buning, C., and Lengauer, T. (2002) *J. Comput.-Aided Mol. Des.*, **16**, 129–149.
[18] Jordan, P.M. (1991) in *Biosynthesis of Tetrapyrroles* (ed. P.M. Jordan), Elsevier, New York.
[19] Hansson, T., Oostenbrink, C., and van Gunsteren, W. (2002) *Curr. Opin. Struct. Biol.*, **2**, 190–196.
[20] Qin, X., Tan, Y., Wang, L., Wang, B., Wen, X., Yang, G., Xi, Z., and Shen, Y. (2011) *FASEB J.*, **25**, 653–664.
[21] Witschel, M.C., Höffken, H.W., Seet, M., Parra, L., Mietzner, T., Thater, F., Niggeweg, R., Röhl, F., Illarionov, B., Rohdich, F., Kaiser, J., Fischer, M., Bacher, A., and Diederich, F. (2011) *Angew. Chem.*, **123**, 8077–8081.
[22] Wicker, J., Lorsbach, T., Gütlein, M., Schmid, E., Latino, D., Kramer, S., and Fenner, K. (2016) *Nucleic Acids Res.*, **44**, D502–D508.
[23] Woo, Y.-T. and Lai, D.Y. (2005) *OncoLogic: a mechanism-based expert system for predicting the carcinogenic potential of chemicals*, in *Predictive Toxicology* (ed. C. Helma), CRC Press, Boca Raton FL, USA, pp. 385–413.
[24] Lewis, D.F.V., Bird, M.G., and Jacobs, M.N. (2002) *Hum. Exp. Toxicol.*, **21**, 115–122.
[25] Patlewicz, G., Jeliazkova, N., Safford, R.J., Worth, A.P., and Aleksiev, B. (2008) *SAR QSAR Environ. Res.*, **19**, 495–524.
[26] Judson, R., Richard, A., Dix, D.J., Houck, K., Martin, M., Kavlock, R., Dellarco, V., Henry, T., Holderman, T., Sayre, P., Tan, S., Carpenter, T., and Smith, E. (2009) *Environ. Health Perspect.*, **117**, 685–695.
[27] Ames, B.N., Lee, F.D., and Durston, W.E. (1973) *Proc. Natl. Acad. Sci. U.S.A.*, **70**, 782–786.
[28] Hansen, K., Mika, S., Schroeter, T., Sutter, A., ter Laak, A., Steger-Hartmann, T., Heinrich, N., and Müller, K.R. (2009) *J. Chem. Inf. Model.*, **49**, 2077–2081.
[29] Todeschini, R. and Consonni, V. (2002) *Handbook of Molecular Descriptors*, 1st edn, John Wiley & Sons, Inc., 688 pp.
[30] Zhu, H., Tropsha, A., Fourches, D., Varnek, A., Papa, E., Gramatica, P., Oberg, T., Dao, P., Cherkasov, A., and Tetko, I.V. (2008) *J. Chem. Inf. Model.*, **48**, 766–784.

[31] Schultz, T.W. and Netzeva, T.I. (2004) in *Modeling Environmental Fate and Toxicity*, vol. 4, Chapter 12 (eds M.T. Cronin and D.J. Livingstone), CRC Press, Boca Raton, FL, pp. 265–284.

[32] Varnek, A., Fourches, D., Solov'ev, V.P., Baulin, V.E., Turanov, A.N., Karandashev, V.K., Fara, D., and Katritzky, A.R. (2004) *J. Chem. Inf. Comput. Sci.*, **44**, 1365–1382.

[33] Fjodorova, N., Vracko, M., Novic, M., Roncaglioni, A., and Benfenati, E. (2010) *Chem. Cent. J.*, **4** (Suppl 1), S3.

[34] Fitzpatrick, R.B. (2008) *Med. Ref. Serv. Q.*, **27**, 303–311.

[35] Christmann-Franck, S., Bertrand, H.O., Goupil-Lamy, A., der Garabedian, P.A., Mauffret, O., Hoffmann, R., and Fermandjian, S. (2004) *J. Med. Chem.*, **47**, 6840–6853.

[36] Schlegel, B., Stark, H., Sippl, W., and Höltje, H.D. (2005) *Inflamm. Res.*, **54** (Suppl 1), 50–51.

8 Chemoinformatics in Modern Regulatory Science

Chihae Yang[1,2,3], James F. Rathman[1,2,3], Aleksey Tarkhov[1], Oliver Sacher[1], Thomas Kleinoeder[1], Jie Liu[2], Thomas Magdziarz[1], Aleksandra Mostraq[2], Joerg Marusczyk[1], Darshan Mehta[3], Christof Schwab[1], and Bruno Bienfait[1]

[1] Molecular Networks GmbH, Neumeyerstr. 28, 90411 Nürnberg, Germany
[2] Altamira LLC, 1455 Candlewood Dr., Columbus, OH 43235, USA
[3] The Ohio State University, Department of Chemical and Biomolecular Engineering, 151 W. Woodruff Ave. Columbus, OH 43210, USA

Learning Objectives

- To compare the various roles chemoinformatics plays in modern regulatory science.
- To describe how the threshold of toxicological concern (TTC) and read-across methods are used in risk assessment.
- To design databases suitable for risk assessment of chemicals.
- To apply new types of descriptors to address the new approach methods encouraged for regulatory purposes.
- To revisit the concept of compound similarity and how both chemistry and biology are taken into consideration.
- To explore methods of analyzing and comparing the chemical space of one or more datasets.

Outline

8.1 Introduction, 439
8.2 Data Gap Filling Methods in Risk Assessment, 441
8.3 Database and Knowledge Base, 448
8.4 New Approach Descriptors, 453
8.5 Chemical Space Analysis, 462
8.6 Summary, 464

8.1 Introduction

8.1.1 Science and Technology Progress

The parallel development of computational science and computer technology has enabled countless advances in many fields. Figure 8.1 presents a snapshot of this timeline to illustrate how progress in early computational chemistry of molecular orbital theories was strongly associated with the advances in computing power of that era. A bit later, molecular modeling was enabled by the development of workstations with sophisticated graphics capabilities. The progress of

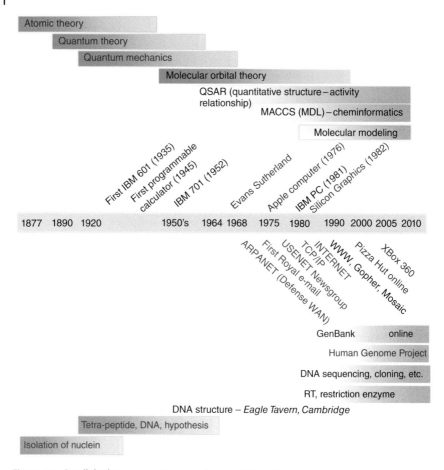

Figure 8.1 Parallel advances in science and computational technology over time.

chemoinformatics and, even more so, bioinformatics was also greatly assisted by the development of database technologies for efficiently storing and managing large amounts of information.

Computational chemistry and chemoinformatics have become integral to the drug discovery process over the past several decades (see Chapter 6). More recently, chemoinformatics has also come to play a role in regulatory science, an area that historically has relied mostly on experimental data and expert opinion. *In silico* approaches are becoming more widely accepted and are viewed as ways to complement and enhance traditional risk assessment and safety evaluation. Modern regulatory science therefore presents new opportunities and challenges for the field of chemoinformatics.

8.1.2 Regulatory Science in Twenty-First Century

There is considerable motivation to predict the toxicity potentials for chemicals in relation to international regulatory initiatives even beyond the traditional application in drug discovery. These initiatives include the Canadian

Domestic Substance List (DSL) [1], European Union's Registration, Evaluation, Authorization, and Restriction of Chemicals (REACH) regulation (Article 13(1)) [2], Cosmetics Directive [3], and Assessment and Control of DNA Reactive (Mutagenic) Impurities in Pharmaceuticals to Limit Potential Carcinogenic Risk from the International Council for Harmonization of Technical Requirements for Pharmaceuticals for Human Use (ICH M7) [4]. There are many reasons why requiring *in vivo* testing results is not practical:

1. Experimental testing, especially *in vivo*, of extremely large numbers of chemicals is prohibitively expensive
2. Lack of availability of isolated chemicals
3. Restrictions on animal testing, such as the ban on animal testing associated with marketing cosmetics ingredients in Europe

In 2007, the National Research Council published the paradigm-changing report entitled "Toxicity Testing in the 21st Century" [5]. This vision and strategy proposed in this report was further elaborated by Collins *et al.* [6] – a shift from primarily *in vivo* animal studies to *in vitro* assays, *in vivo* assays with lower organisms, and computational modeling for toxicity assessment. Since the challenge from the US National Academy of Science for a paradigm shift for regulatory science toward computational methods to reduce animal testing, EPA initiated the ToxCAST™ program in 2007, joining interagency programs such as Tox21 and NTP initiated in 2004 in order to work together to elucidate toxicity pathways through testing protocols involving quantitative high-throughput screening (qHTS) assays and computational toxicology methods [6]. Since then, ToxCast has completed both phases I and II covering almost 1000 chemicals, generating over 500 assays for each chemical. Including Tox21 assays, there are easily over 8000 chemicals tested for screening assays. Many of the assay results were anchored to *in vivo* studies such that the *in vitro* results can be associated with *in vivo* findings. In addition, establishing the *in vitro* assays requires knowledge of potential metabolites of the chemicals in the testing environment. Hence, metabolomics using liquid chromatography–mass spectrometry (LC-MS) has been of much interest along with metabolic predictors.

8.2 Data Gap Filling Methods in Risk Assessment

Risk assessment is an important part of regulatory science that is conducted when the safety of a test substance needs to be addressed in a specific exposure condition. A risk can be evaluated from three vantage points:

1. Intrinsic toxicity (hazard or safety)
2. Exposure of such chemical in the target population, either human or environment
3. Probability of such exposure

Once identified, risks arisen from chemical exposure are characterized and managed. A highly toxic chemical might not be considered high risk if it does not have a sufficiently high probability of reaching the target population. A good example is the radioactivity from a nuclear power plant. Although the toxicity is lethal, the risk to humans and environment is minimal as long as it is contained securely to prevent exposure. A pharmaceutical ingredient (e.g., drug) with a demonstrated effect in rat liver steatosis may pose an unacceptable health risk to healthy humans, but the risk may be deemed acceptable by patients with a disease the drug is treating.

During the risk identification and characterization process, scientists often must compile data from many sources to build a profile. For example, for human health, all toxicity data of target organs from short-term to lifetime dosing, reproductive/developmental studies, carcinogenicity, and even genetic toxicity endpoints are considered. Unless the chemical is a drug, it is quite unlikely that the full toxicity data will be available; hence, quite often researchers are faced with data gaps. Filling these data gaps is therefore an important activity of risk assessment.

Although predictions of toxicity are not yet universally accepted by regulatory programs, there is a strong correlation between acceptance and lack of data. The more mechanistic the prediction method is, the more amenable they become to the regulators who require complete transparency. The prediction methods can include quantitative structure–activity relationships (QSARs) (see Chapter 2), structural knowledge, read-across (RA), and threshold of toxicological concern (TTC).

8.2.1 QSAR and Structural Knowledge

Exploring the relationships between the chemical structure of a molecule and its biological activities is a fundamentally important focus of chemistry and life sciences. It is also covered in other chapters of this book (see Chapters 6, 7, and 10). Structure–activity relationships (SARs) are based on electronic and steric effects [7], which are quantitatively represented by the structures as molecular properties and/or structural features along with physicochemical properties, such as membrane partition coefficients [8]. These molecular descriptors are expressed in mathematical relationships to biological activity or toxicity using statistical methods and are commonly referred to as "statistical approaches" in risk assessment. The models are required to be internally cross-validated during the modeling process and further validated against an appropriate external dataset. The QSAR methods therefore require full documentation on the training dataset, the description of predictors, the algorithm used to fit the quantitative relationship, the prediction power with appropriate statistical metrics, and the external validation statistics to assess model applicability and reliability.

Structural alerts are an expert-driven approach generally based on a causal relationship between a specific chemical structural motif and the biological activity of interest, which is the fundamental basis of an SAR. As compared to the "bottom-up" data-driven process of the QSAR methods, the structural rules are developed from the knowledge of experts based on first principles or mechanistic rationales; hence, this is often called a "top-down" approach. The presence of structural alerts in compounds is reported as "hits" with certain likelihood ratings. Common examples are the structural alerts that describe DNA or protein binders that may lead to genotoxicity or skin sensitization.

The influence of chemoinformatics on QSAR is evident from the vast array of publicly available descriptors (see *Methods Volume*, Chapter 10). For structure knowledge, chemoinformatics enables more efficient developments by systematically improving the process of data mining based on ontology and systems approach. The strengths and weaknesses of QSAR and rule-based systems are compared in Table 8.1. Chemoinformatics can contribute to improve the shortcomings of these methods and help leverage the two approaches.

Table 8.1 Comparison of QSAR and structural rules.

	Strength	Weakness
QSAR	• Data driven, hence specific mechanistic understanding is not necessary • Large applicability domain of global models • Powered by machine learning • Ability to predict negatives	• Limited mechanistic insights • Chance correlations
Structure rules	• Mechanistic rationale possible • Intuitive interpretation • Combining with machine learning, negative rules can be developed	• Limited number of rules • Rules are defined for positives; hence negative prediction is theoretically not feasible

8.2.2 Threshold of Toxicological Concern (TTC)

The Threshold of Toxicological Concern (TTC) is a risk assessment method that can be used to screen substances with few or no toxicological data and for which human exposures are likely to be low. The concept was initially developed at the Center for Food Safety and Applied Nutrition (CFSAN) of the US Food and Drug Administration (FDA) to address the challenges in the safety assessment of food contact substances. The TTC approach has an origin related to the regulatory program of threshold of regulation (TOR) implemented at FDA CFSAN (Title 21 of the US Code of Federal Regulations section 170.39) for food contact substances, involving very low-level potential exposure to a wide range of chemicals migrating to food or substances that posed significant challenges regarding regulatory and testing efficiency. The current TOR exempts from the regulatory requirement any substance used in food contact substances (e.g., food packaging or food processing equipment) that migrates, or that may be expected to migrate, into food, if it becomes a component of food only at levels that are below the TOR, that is, 0.5 µg/kg of diet (0.5 ppb) or equivalent to an intake of 1.5 µg/person/day or 25 ng/kg-body weight/day based on a body weight of 60 kg adult and daily intake of 1500 g each of solid food and liquid. The TOR is intended to be protective for all toxicological endpoints, including carcinogenicity, although US law does not permit known carcinogens to be regulated as food and color additives according to the Delaney clause [9].

In 1996 Munro et al. [10] extended this threshold concept to address the general non-cancer endpoints by analyzing the no observed effect level (NOEL) distributions of a database grouped by structure classes based on the decision tree proposed by Cramer et al. in 1978 [11] (see Section 2.6.3.1) to correlate potency with chemical structure (Table 8.2). The database included 613 diverse chemicals including pesticides, cosmetics, food additives, drugs, and industrial and environmental chemicals with known biological properties. Three Cramer classes are defined: class III (most likely to be toxic), class II (intermediate), and class I (least likely to be a potent toxicant). The fifth percentile NOEL, that is, the 5% quantile, of each class was then derived from the cumulative distribution function.

Table 8.2 logNOEL distributions and Cramer classes.

	Cramer et al. (1978) [11]		Munro et al. (1996) [10]	
	NEL (mg/kg/d)	Histogram	NOEL (mg/kg/d)	Histogram
Class I	50–254 (N = 31)		0.018–7204 (N = 137)	
Class II	5–200 (N = 7)		1–1441 (N = 28)	
Class III	0.03–500 (N = 50)		0.005–3775 (N = 448)	

First Cramer and then Munro, 20 years later, both confirmed that the NOEL distributions (referred to as NEL by Cramer) are correlated with the Cramer classes as shown in Table 8.2, indicating that there are structural features that represent stronger potency. This important observation is analogous to SAR and QSAR such that structural features not only are related to the potential toxicity but also may be correlated to potency.

The fifth percentile values are estimated by either parametric (assuming Gaussian distributions) or nonparametric methods, which are then divided by a safety assessment factor (conventionally, a factor of 100 is applied to consider the animal-to-human extrapolation) to derive corresponding human exposure thresholds. To this day, the resultant human exposure thresholds of 1.8, 0.54, and 0.09 mg/person/day for Cramer classes I, II, and III, respectively, are still being used. Although the TTC method was initially applied to flavoring substances and food additives, the approach has been extended to cosmetics-related chemicals, metabolites of pesticides, and genotoxic impurities of pharmaceuticals by the EU Scientific Committee of Consumer Safety (SCCS), European Food Safety Authority (EFSA), Joint FAO/WHO Expert Committee on Food Additives (JECFA), and ICH M7 Guidance on the Assessment and Control of

DNA Reactive (Mutagenic) Impurities in Pharmaceuticals to Limit Potential Carcinogenic Risk [4].

Combining the results from the development of both cancer and non-cancer TTC approaches, a decision tree was proposed by Kroes *et al*. [12] in 2004 to be used as guidance on when and how the TTC could be applied in food safety evaluation. The framework poses questions on the cohort of concerns to remove chemicals with five structural categories, that is, polyhalogenated dibenzo-p-dioxins, polyhalogenated dibenzofurans, and analogs, aflatoxin-like, azoxy, N-nitroso compounds, and steroids, for which compound-specific data are required in all cases. It also removes substances for which the TTC approach is not appropriate; for example, a nonessential metal or metal-containing compound. A compound can also be screened for genotoxic alerts before being allowed to proceed to apply the threshold for cancer. If the chemical is matched with a lower concern genotoxic structural alert (e.g., not containing azoxy, N-nitroso, or aflatoxin moiety), the cancer threshold (0.15 µg/day) is compared with the exposure of the chemical. The non-cancer portion of the decision tree begins with a question about whether the chemical is an organophosphate (OP) to which a lower class-specific exposure threshold can be applied. The non-OP chemical is then allowed to proceed to the node for the comparison of the exposure with the threshold of the appropriate Cramer class. If the exposure (estimated daily intake) exceeds the threshold, the risk assessment requires compound-specific toxicity data. The decision tree in Figure 8.2 incorporated different TTC threshold values to chemicals with different structural characteristics.

The current TTC method therefore requires tools to aid the Cramer classification, to detect chemical classes for either exclusion or separation when executing the TTC framework. Currently, the OECD Toolbox [13], Toxtree [14], and COSMOS TTC workflow [15] provide Cramer class assignments. The Cramer decision tree of 33 rules is therefore a combination of logical statements and structural recognition, achieved by employing either SMARTS or chemotypes (see *Methods Volume*, Section 10.2.1.4). Various structural classes and alerts (steroids, azoxy, benzodioxins, N-nitroso, etc.) are also identified and separated. However, the current Cramer tree has two important shortcomings. One is due to vague or insufficient descriptions of the original rules, leading to confusing interpretations. The second is due to the limitation of many current chemoinformatics tools in handling certain complex issues such as tautomers (see *Methods Volume*, Section 2.5.2) and reactivity. Ongoing efforts to improve this situation are focused on designing more precise rule definitions and expanding the chemoinformatics methods to include chemical and metabolic reactivity [16]. Also under consideration is the idea of moving away from the Cramer decision tree and classes and instead focusing on building a series of chemotypes that accurately represent the various structural categories and effectively discriminate potency [17, 18].

8.2.3 Read-Across (RA)

RA is a data gap filling technique that has been utilized as an alternative method to address information requirements under various regulatory programs such as

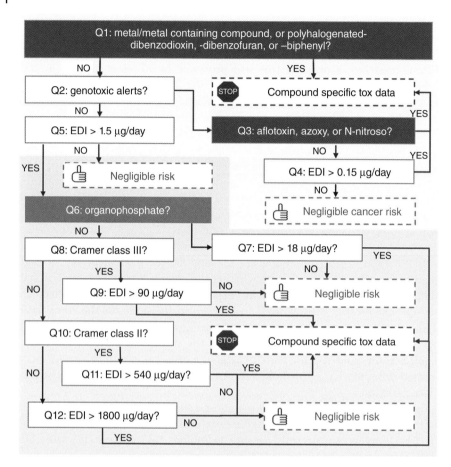

Figure 8.2 The decision framework for the Threshold of Toxicological Concern.

the OECD high production volume (HPV) program as well as the EU's REACH regulation.

In general, RA involves using the endpoint information for one chemical (source compound) to predict the same endpoint for another chemical (target compound), which is considered to be "similar" on the basis of structural similarity or some other relevant measure of similarity (see *Methods Volume*, Section 11.1.3.3). Hence to "read across" is to apply data from a tested chemical for a particular property or effect (cancer, reproductive toxicity, etc.) to a similar untested chemical.

The RA technique is often applied as an analog/category approach, where not every chemical needs to be tested for every endpoint but within groups of similar chemicals assembled for assessment. The analog approach is based on a very limited number of chemicals with a mechanistic perspective based on a well-understood SAR, whereas the category approach groups a larger number of chemicals according to a chemical category. This chemical category is defined as a group of chemicals whose physicochemical properties and human health and/or environmental toxicological and/or environmental fate profiles are expected to be similar or follow a regular pattern as a result of similarities of chemical structures or biological activities. Thus the chemical similarity is judged by the following:

1. Compound structure similarity
2. Chemical reactivity (abiotic and biotic) pattern
3. Physicochemical and molecular properties profile
4. Biological proximity

It is important to note that the similarity of the target molecule is not judged solely by chemical structures.

Although a guidance document of Read-Across Assessment Framework (RAAF) [19] has been recently published from European Chemical Agency (ECHA), describing a systematic approach for consistent evaluation of the scientific aspects of RA studies, there remains a fair amount of confusion and variation across the various international regulatory bodies and agencies when it comes to applying RA. The RAAF defines different scenarios for RA approaches; each scenario is associated with particular aspects ("assessment elements," AEs), which address different scientific considerations deemed crucial to judging the validity and the reliability of RA. Each AE poses questions that lead an assessing expert to select predefined conclusions ("assessment options," AOs), reflecting the strengths and weaknesses of the RA. The approach proposed in the RAAF thus allows one to define a degree of confidence associated with the proposed RA and finally evaluate the reliability of an RA prediction. This procedure is called "quantitative weight of evidence assessment."

Chemoinformatics contributes greatly to the RA method at many critical points. The regulatory community seems to better accept conclusions when the toxicity endpoint data of a target compound can be "read" from experimental data for structures whose chemical and biological properties are highly similar to those of the target. While (Q)SAR methods predict toxicity of a chemical structure from statistical approaches based on large training datasets, the RA approach predicts the target endpoint outcome from a much smaller set of chemicals, selected by the person or group performing the RA based on their relatedness and similarity to the target. While RA is thus perceived by some as being more transparent and expert informed than traditional (Q)SAR approaches, there are of course advantages and disadvantages to both strategies.

Considering possible speciation of the target compound is important in evaluating possible toxicity concerns. Chemoinformatics techniques can be applied to identify tautomers, degradants due to reactivity, and, most importantly, metabolites that may be formed under *in vivo* conditions. The speciation of the target compound in terms of reactivity and metabolic potential, followed by analog searching, results in groups of chemicals closely related in a neighborhood of structure, and reactivity and biological activities. This logical tree of chemical clusters, organized hierarchically to reflect related neighborhoods, is defined in this chapter as the "read-across tree," as depicted in Figure 8.3. This tree assists visualization of the complex relationships and groupings.

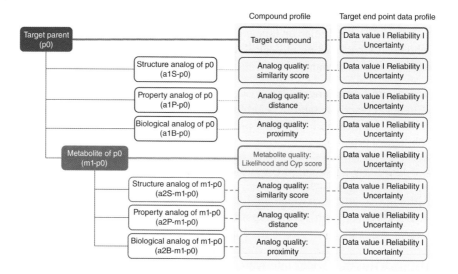

Figure 8.3 Read-across tree, hierarchy of logical relationships.

In RA bookkeeping of these species can get quickly overwhelming and confusing to human end users: hence, development of user-friendly software tools is envisioned.

Comparing these chemical species to compounds for which toxicity data are available is then a key next step. As before, similarities of these species can be evaluated based on structural features, physicochemical or ADME properties, and even biological assays. The term "biological analogs" emphasizes that similarity based only on structure connectivity is not sufficient or relevant in the RA workflow. For example, two compounds that are structurally quite different may be deemed biological analogs if they exhibit similar behavior in biological assays or if they have the same or highly similar metabolites.

Given the daunting lack of sufficient experimental data and increased acceptance of new approaches, the regulatory science community recognizes the need and the power of chemoinformatics in risk assessment. The ECHA encourages the exploration of new approach methodologies (NAMs) in regulatory science [20] that make use of multiple and diverse sources of evidence, including omics, high-throughput screening, and even *in silico* methods. For example, information about the metabolism of a compound can be gleaned from the spectral resolution of LC-MS results from metabolomics. As the NAM framework opens the door to *in silico* approaches, QSAR and other computational methods can sometimes be used to fill data gaps.

8.3 Database and Knowledge Base

8.3.1 Architecture of Structure-Searchable Toxicity Database

Database technology and the content of toxicity databases have made tremendous advancements within the last decade (see *Methods Volume*, Chapter 6).

It is now a standard expectation that a toxicity database should provide chemical structure-based queries in conjunction with toxicological study data. The generic architecture of these databases, implemented within a chemistry-aware relational database management system (RDBMS), is usually a 3-tiered arrangement of a data storage backend (database server), a user interface front end (client), and a middleware layer that enables the communications. The client can be web-based with a web server backend. The backend database can be, for example, Oracle database server, Microsoft SQL Server, PostgreSQL server, or any other relational database.

Since chemical structures are not among the native SQL data types, database-specific chemistry extensions, or cartridges, are necessary. Chemistry extensions add necessary support for chemical structures as data types to the database engine, allowing for storing and indexing molecules in the database. Depending on the database servers, there are numerous chemical cartridges available, including Daylight DayCart Cartridge [21], JChem Cartridge [22], BIOVIA Direct [23], Bingo cartridge [24], OrChem [25], MolSql [26], and RDKit extension [27].

The middleware layer typically implements a database access layer (e.g., Django Python ORM [28] or Java Persistence API [29]) and employs a chemoinformatics platform such as Daylight toolkit [30], ChemAxon IT Platform Toolkit [31], MOSES [32], RDKit [27], or CDK [33]. Together, the chemoinformatics platform, database access layer, and chemistry cartridge provide functionalities for manipulating, storing, searching, and retrieving chemical structures and calculation of molecular property and structural descriptors. The scripting layer provides tools for writing application.

The architecture depicted in Figure 8.4a is common but already a decade old; newer architectures, such as the one illustrated in Figure 8.4b, do not depend on a rigid coupling between layers but rather use a more loosely coupled design for communications and accessing the functionalities via an application programming interface (API), thus allowing flexible exchanges between the systems. This is achieved using representational state transfer (REST), an architectural design and communications strategy for developing web services. For example, the COSMOS DB v2.0 [34] or the eTOX database [35] is not tightly integrated in one RDBMS, but allow users to dynamically query databases of their choice through a data exchange standard specified between them. This type of database architecture is referred to as a federated database system (FDBS) [36]. Although not described any further in this chapter, the move toward FDBS is a noteworthy trend.

8.3.2 Data Model for Chemistry-Centered Toxicity Database

The basic data elements of such a database include chemical structures and their properties, chemical identification, chemical annotation, chemical-level safety evaluation results, study background, study design parameters, study reference, study-level outcome, and study results at dose level for each assay. A high-level summary is listed in Table 8.3.

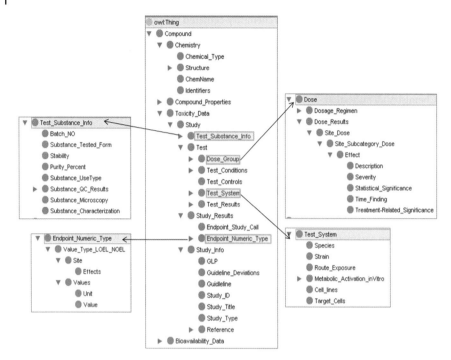

Figure 8.5 Top-level diagram of the data model for the chemistry-centered toxicity database.

8.3.3 Inventories

A chemical inventory is a collection of chemical lists from a specific regulatory entity and may include either chemicals approved by the regulatory program or the chemical testing list from a regulatory science initiative. Examples of regulatory lists include New Drug Application (NDA) from US FDA CDER, generally regarded as safe (GRAS) from US FDA CFSAN, HPV program US EPA, Canadian DSL, Registered Substance list from REACH, and Cosmetics Inventories from European Commission and US Cosmetics Ingredients Review (CIR). Examples of regulatory science testing list inventories include Tox21 [44] in the United States and the Ames/QSAR International Collaborative Study by NIHS Japan [45]. Summary information for several of these inventories is listed in Table 8.4.

The Tox21 inventory consists of ToxCast phase I and II chemicals as well as National Chemical Genomics Center (NCGC), covering 516 and 1000 screening *in vitro* assays, respectively. PAFA is a priority-based assessment of food additives list from US FDA CFSAN. This list is associated with >2000 genetic toxicity study results and toxicity data for repeated dose and developmental and reproductive toxicity studies for over 1000 test substances. The cosmetic inventory was compiled by the COSMOS project by combining the European CosIng list with that of US CIR. These test substances have been reported in Europe and the United States for the use in cosmetics formulations. The eTOX inventory is a nonpublic set of structures that have been compiled from 13 different pharmaceutical companies willing to share their preclinical phase toxicity data [8]. A total of 7228

Table 8.4 Regulatory science-related chemical inventories.

Chemical inventory	Number of original test substances	Number of 2D structures, no duplicates (small fragments intact)	Number of 3D structures, no duplicates, no small fragments, neutralized
Tox21 (US EPA)	8,193	7,993	6,662
PAFA (US FDA CFSAN)	7,198	4,603	3,586
Cosmetics inventory	19,931	4,740	4,459
eTOX active pharmaceutical ingredients	1,856	1,369	1,344

toxicity studies are associated with 1856 active pharmaceutical ingredients. A small subset of the database is made publicly available as a sampler to demonstrate the capabilities of the system and the level of detail to a broader audience [47].

8.4 New Approach Descriptors

This section introduces new types of descriptors applicable to address the previously mentioned NAMs, including metabolomics. Descriptors are generally classified as either structural features or physicochemical/molecular properties (see *Methods Volume*, Chapter 10). These descriptors include categorical variables (e.g., denoting the presence or absence of a particular structural fragment), counts (e.g., number of hydrogen bond donors or acceptors, or the number of occurrences of a particular structural fragment in a molecule), and continuous-valued data (e.g., molecular weight, octanol–water partition coefficient, etc.).

8.4.1 ToxPrint Chemotypes

Chemotypes are predefined structural fragments that can be encoded not only with connectivity and topology but also with properties of atoms, bonds, electronic systems, or molecules [48] (see *Methods Volume*, Section 10.2.1.2). The information is expressed in the extensible markup language (XML)-based chemical subgraphs and reaction markup language (CSRML). This new approach not only can represent molecules or substructures but also can be extended to describe chemical patterns, reaction rules, and reactions. Technical details of the CSRML [20] and reference implementation [49] are available (see also *Methods Volume*, Section 3.2.9).

ToxPrint [50] is a public library of chemotypes specifically designed to provide a broad feature coverage of inventories consisting of tens of thousands of environmental and industrial chemicals, including pesticides, cosmetics ingredients, food additives, and drugs. The current version of ToxPrint (V2.0_r711) offers 729 uniquely defined chemotypes designed to capture salient features important to the safety assessment within FDA and TTC workflows. The object model of CSRML and the query features are documented by Yang et al. [48]. Access to these resources is provided via web sites dedicated to ToxPrint and ChemoTyper [51]. ChemoTyper is a software application that enables browsing of the chemotypes as well as searching the structure sets with chemotypes. ToxPrint has been used to characterize the chemistry of the Tox21 and ToxCast inventories [52].

8.4.1.1 Coverage

The degree to which a given set of chemotypes such as the ToxPrint chemotypes covers a set of compounds can be evaluated by analyzing the following:

1. How the proportion of compounds containing at least one "hit" increases with increasing the number of chemotypes
2. How well the dataset can be described when all available chemotypes are applied

An analysis of the first type is illustrated in Figure 8.6, which was created as follows:

- From ChemoTyper, the structure files and the ToxPrint chemotypes were loaded into the "Match" window.
- The fingerprint files were saved from ChemoTyper. The files contain 0/1 values for the invariant 729 chemotypes.
- From the fingerprint files, a new table containing the sum of the number of hits for each chemotype and the label (chemotype name) was prepared.
- ToxPrint chemotypes were randomly sampled (e.g., 0.5%, 0.7%, 1.0%, 2.5%, 5%, 6.25%, 7.5%, 10%, 20%, 40%, 100%) to search for the matching structures in the dataset.
- The number of matches for the structures containing any selected chemotypes was determined.
- The percent of structures matched in each search is then plotted against the percent of ToxPrint chemotypes engaged.

As illustrated in Figure 8.6, a large proportion of compounds is covered by a relatively small proportion of chemotypes. For example, a subset of approximately 146 ToxPrints (20%) covers more than 80% of the compounds in the Tox21 database and more than 90% in PAFA. As the number of chemotypes used increases, structural matching increases, but levels off after 20% and 40% of the ToxPrints are engaged to describe the 95% of the structures in the Tox21 and PAFA datasets, respectively. The trends observed here are typical. The steepness of the initial slope of the coverage curve is inversely related to the structural diversity of the dataset [53]. Figure 8.6 suggests that the Tox21 dataset may be either less diverse than PAFA or at least there may be some large groups of similar local neighbors in the Tox21 set. A careful analysis of chemotype coverage is an important step in characterizing the chemical space of a dataset.

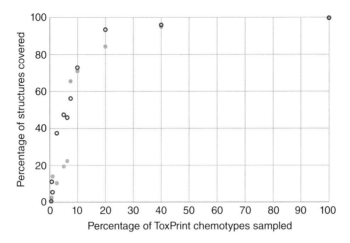

Figure 8.6 Typical coverage plot of ToxPrints against datasets. The solid and open circles represent the PAFA and Tox21 datasets respectively.

8.4.1.2 Information Density

Another useful characterization technique is to examine the information density of the ToxPrint chemotypes. The more precisely annotated a chemotype is, the greater will be its differentiating power when correlating biological properties or comparing different datasets. A feature carrying a high density of relevant information will hit fewer structures than its generic counterpart by being constrained in connectivity or properties. In the extreme case, a single chemotype may be identical to a whole molecule. Although information density may be high, care must be taken when using highly specific features in modeling in order to avoid overtraining. On the other hand, lower information density of highly generic features would not sufficiently differentiate molecules even though these can be useful to quickly group chemicals at a coarse granularity. There is therefore a happy medium where the information density and extensibility can both be acceptable. As an example, consider the following analysis to characterize ToxPrint chemotypes:

- Using the same fingerprint file and count sum for each chemotype, a table of the number of structure hits per chemotype is prepared.
- A histogram of structure hits against the chemotype is displayed in Figure 8.7.

As shown in Figure 8.7, 103 chemotypes hit only 1, 2, or 3 structures in the Tox21 dataset, while approximately the same number (108) do the same in PAFA. These are the "high information density" chemotypes, single features capable of differentiating a compound from all others. At the other extreme, 86 chemotypes hit 400 or more compounds in the Tox21 dataset, while 35 chemotypes hit 400 or more compounds in PAFA. These "low information density" chemotypes may be useful for coarse grouping of compounds but are likely less helpful in understanding more complex differences between molecules. The extremely low information density chemotypes appearing in more than 50% of the compounds

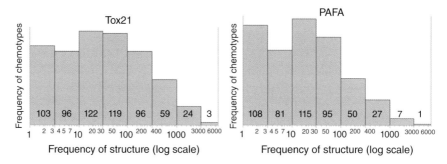

Figure 8.7 Histogram of structural hits matching with the chemotypes.

(3000 or more) include basic chemotypes that encode features such as main group elements, generic carbonyl, and benzene ring.

8.4.1.3 Property Coding

The chemotypes in ToxPrints are also coded with electron system information, in addition to the topological connectivity and subgraphs. Once coded with the physicochemical properties, chemotypes can serve as powerful alerts or reaction rules. The value of this approach will be demonstrated based on the well-known example of the Hammett equation for the acid dissociation coefficients of substituted benzoic acids (Eq. 8.1) (see also Section 3.2.3.3).

$$K_a = \frac{[CO_2^-][H^+]}{[CO_2H]}$$

(8.1)

Any substituent (X) stabilizing the anion or making the proton more labile will increase the acidity of benzoic acid. Hammett constants, which reflect the electron-withdrawing power of the substituent, increase in the order $OH < OCH_3 < CH_3 < H < Cl < Br < I < C(=O)H < C\#N < NO_2$. Table 8.5 summarizes the trend.

This observation is quantitatively explained by introducing the Hammett constant, σ_x, the electronic effect (influence) of substituent X relative to a hydrogen atom on the ionization of benzoic acid. In Figure 8.8, K_x and K_H are the acid dissociation constants of the substituent and the parent (unsubstituted benzoic acid).

This simple but elegant example is at the basis of physical organic chemistry. Along with Hansch and Fujita's addition of membrane partition coefficients, it is still one of the pillars of the modern QSAR approach [8]. The Hammett equation addressed this phenomenon by introducing parameters affecting the substituents at a given carbon atom. It should therefore be possible to define

Table 8.5 pK$_a$ values of benzoic acids.

X(para) e$^-$-withdrawing group	pK$_a$	X(para) e$^-$-donating group	pK$_a$
H	4.20	H	4.20
Cl	3.98	CH$_3$	4.34
Br	3.96	OCH$_3$	4.46
C(=O)H	3.75	N(CH$_3$)$_2$	5.03
C#N	3.55		
(O=)S(=O)CH$_3$	3.52		
NO$_2$	3.41		

chemotypes to substituent effects on the carbon atoms of the benzene ring. Furthermore, these effects can be captured without having to enumerate the electron-withdrawing and -donating groups. The benzoic acid is being used as an initial chemotype (Figure 8.8), which in turn is modified with embedded atomic charges using CSRML.

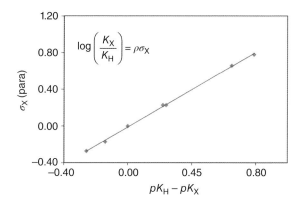

Figure 8.8 Hammett constant and the substitution effect on pK$_a$ values.

The chemotype definition relies on calculations of the summed total σ charge and atomic π partial charge for the (*) carbon atom in each target molecule that matches the para-substituted benzoic acid substructure (see *Methods Volume*, Section 10.1). As shown in Figure 8.9, compounds for which this sum is greater than approximately 0.05 have a lower pK$_a$ value than benzoic acid.

The total σ charge and atomic π partial charge are coded in the CSRML file and can be used for query features. Although a related approach was published by Ertl [54], this is the first time where the generalization was made by a generic chemical representation method capable of carrying both structure connectivity and local properties using publicly available chemotypes and the CSRML reference implementation.

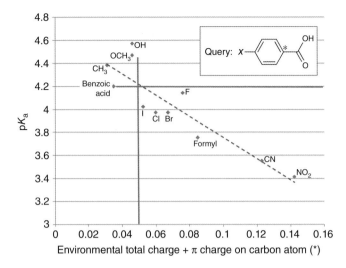

Figure 8.9 Effect of substituents on the charges at the ring carbon atoms (ipso position) and on the pK_a values of substituted benzoic acids.

8.4.2 Liver BioPath Chemotypes

The chemotype concept can be also applied to chemical reactions to represent the chemical species participating in a reaction as chemotype queries. Each species is assigned a role of either reactant or product, and the reaction rule defines a set of transformations that convert reactants into products. Therefore, the rules can be used to generate reactions for matching substrates (reactants) to predict, for example, the metabolites of a chemical query. In Liver BioPath [55], 144 CSRML reaction rules are implemented to cover human liver metabolic transformations based mostly on APIs as summarized in Table 8.6.

Table 8.6 Metabolic transformation reactions in human liver.

Top category	Transformations
Cleavage	Deacetylation, dealkylation, decarboxylation, deformylation, deglycosylation, dehalogenation
Conjugation	Acetylation, glucuronidation, glycination, methylation, phosphorylation, sulfation
Hydrolysis	At carbon–hetero, at carbon–nitrogen, at carbon–oxygen, at hetero–hetero
Hydroxylation	Aliphatic, aromatic, benzylic
Ring closure	Lactam formation
Ring closure	Lactone formation
Oxidation	At carbon, at carbon–nitrogen, at carbon–oxygen, at nitrogen, at sulfur
Rearrangement	Shift or tautomerization
Reduction	At carbon–carbon, at carbon–oxygen, at nitrogen–oxygen, at sulfur–oxygen

These rules have been validated against a metabolic database of human *in vitro* tests (e.g., hepatocytes), from which positive predictivity value (PPV) and negative predictivity value (NPV) as well as odds ratios were calculated. The odds ratio is a quantitative measure of how a rule correlates with the experimental test outcome. This knowledge-based approach quantitatively ranks the likelihood of the generation of metabolites based on observations, hence allowing quantitative data mining well beyond the conventional structural rules developed by expert systems.

An interesting pattern emerges when these transformation rules are applied to chemicals from different chemical structure types, that is, cosmetics, APIs (eTOX), and pesticides and industrial chemicals of Tox21 inventories. While all chemicals (cosmetics, pharmaceuticals, pesticides, industrial) give high hit rates on oxidation reactions (viz, P450 transformations), pharmaceuticals are much more reactive in transformations of cleavage, conjugation, and hydrolysis than the other chemical types.

This simple analysis, illustrated in Figure 8.10, confirms that the transformation profile of chemicals can be useful in understanding the structural diversity and chemical space of the dataset. The concept of metabolic fingerprints as new descriptors is introduced in the next section.

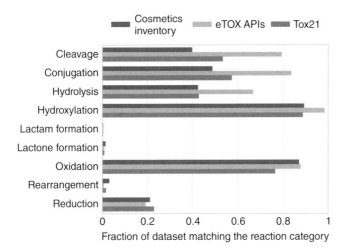

Figure 8.10 Histogram of reactions rules for different chemical inventories.

8.4.3 Dynamic Generation of Annotated Linear Paths

The ToxPrint chemotypes are examples of "static" fingerprints, predefined features that are mapped to a compound dataset. An alternate approach is to generate "dynamic" fingerprints, features extracted "on the fly" from the compound set of interest. There are many methods and sources for generating dynamic fingerprints in public including circular fingerprints [51], extended connectivity fingerprints [56], Lazer [57], PaDEL [58], signature fingerprints [59], and SARpy [60, 61]. These tools extract structural motifs from either SMILES or CTAB and generate fingerprints. Some of them also generate SMARTS or

XML-based fragments on the fly from the input structure set, which can be further defined as alerts.

Figure 8.11 illustrates a process for extracting linear path dynamic fingerprints. The chemical graph is decomposed into linear path diagrams, where each node corresponds to a heavy atom (i.e., not hydrogen atom) in the molecule. Figure 8.11 shows the linear paths from node 1, and a similar path diagram would be extracted for all other nodes in the original graph.

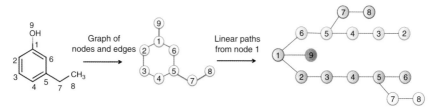

Figure 8.11 Linear fragments using graph theory and a depth-first search algorithm.

Once the linear paths have been constructed, an annotation scheme is applied. Each node can be annotated to encode relevant information such as atom identity, number of heavy atoms connected to the node, number of hydrogen atoms connected to the node, electronic- or bond- or atom-based properties (e.g., partial charge), aromaticity, whether atom is in a ring, and so on. The annotated linear paths are then filtered to identify all unique paths of a particular length or range of lengths, for example, all paths of length 4–8. This set of linear fragments is then used to fingerprint the compound set.

The flexibility in defining the annotation scheme allows one to explore various options to find a scheme that works best for a particular analysis of a dataset. Figure 8.12 shows how the number of fragments generated dynamically varies depending on the annotation scheme for a set of approximately 4400 compounds from the PAFA dataset.

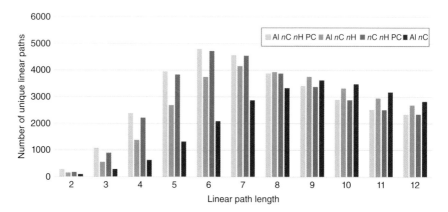

Figure 8.12 Effect of path length and annotation scheme on the number of unique linear paths generated from a set of 4400 compounds from the PAFA database. Annotation options are atom identity (AI), number of heavy-atom connections (nC), number of connected hydrogen atoms (nH), and atom partial charge (PC).

These results illustrate one of the challenges common to dynamic fingerprint generation methods: the rather large number of descriptors that may be generated. Still, the annotated linear path method generates far fewer features in comparison to circular fingerprint and extended connectivity fingerprint methods, for which, depending on the size and structural diversity of the compound dataset, the number of fragments generated may be an order of magnitude or even more higher than for linear paths. Linear path fragments are also simpler and therefore easier to rationalize and interpret to extract mechanistic information. Figure 8.12 also illustrates a common trend: the number of unique paths is generally highest for path lengths in the range 5–7.

Dynamically generated fingerprint descriptors can be used to explore differences between the chemical space covered by different databases, as in Figure 8.13. For this particular example, these results suggest that the Tox21 is structurally more diverse since the number of unique paths is considerably higher than for compounds from the PAFA dataset, especially at path lengths 5 and 6.

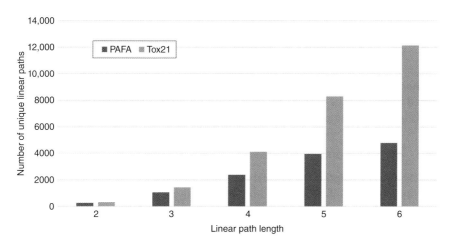

Figure 8.13 Comparison of the number of unique linear paths generated from 4400 compounds from two different datasets: PAFA and Tox21. Annotation scheme used was (AI, nC, nH, PC).

For many applications, the ideal set of structural descriptors may be a combination of dynamically generated features, such as the linear path method described above and features selected from a library such as the ToxPrint chemotypes. Each approach offers advantages for characterizing chemical datasets or for deriving descriptors to be used in QSAR or rule-based prediction methods.

8.4.4 Other Examples of Descriptors

To this point, we have focused on descriptors based on structural features and molecular properties. Although an in-depth discussion is beyond the scope of this chapter, it is important to note that many other types of data can be used to characterize molecules and used as descriptors for quantifying the similarity between molecule or building QSAR models. For example, spectroscopic data

Only compounds unique to a particular dataset were included in these analyses; compounds present in more than one dataset were excluded. Figure 8.14 was constructed as follows:

- Generate 14 physicochemical properties from CORINA Symphony community edition [46]: XlogP, MW, violation of Lipinski Rule of 5, number of hydrogen bond acceptors and donors, number of rotational bonds, water solubility logS, complexity, and ring complexity.
- PCA can be performed with many statistical packages. The covariance was calculated by Euclidean distance of the standardized variables.
- Generate ToxPrint fingerprints of the 2D unique structures from ChemoTyper and select chemotypes appearing at least in five structures. For PCA analysis, the covariance was calculated based on Jaccard distances.
- Each axis represents the scores of the PC and each point represents a chemical structure.

From Figure 8.14a it is clear that there is significant overlap of the four datasets in chemical structure space; however, differences are also apparent. There are regions that contain concentrations of eTox compounds, and other regions highly enriched in COSMOS compounds. The space occupied by PAFA and Tox21 compounds, on the other hand, is mostly overlapped with that of many compounds from the other two inventories. When comparing the inventories based on PCA of molecular properties in Figure 8.14b, a different and somewhat more pronounced separation is observed. There is a region occupied by COSMOS, PAFA, and Tox21 compounds, but very few eTox compounds.

The next step in a more detailed analysis would be to explore various regions of interest and determining which descriptors (e.g., chemotypes or properties) explain the differences observed between the inventories. It is also important to note that Figure 8.14 is in fact three-dimensional (3D) slices through the higher dimensional principal component space. Although 2D and 3D projection plots provide useful and interesting visualizations, a more rigorous and quantitative analysis of the PCA results would often require exploring how the data are distributed in a higher dimensional space. This is computationally more demanding, but the objective and approach remain the same.

8.6 Summary

Regulatory entities focused on safety and risk assessment have recognized the need to include *in silico* approaches to help understand and predict chemical toxicity. This effort involves the design and development of chemical toxicity databases, data mining methods, descriptor identification and knowledge building, and data visualization tools to inform model development and regulatory decision-making.

RA based on analogs requires knowledge of structural features, molecular properties, and biological profiles. Chemotypes allow both structural knowledge and physicochemical property information to be captured. The freely available

ToxPrint chemotypes are tailored for toxicity investigations and are known to provide a broad feature coverage of regulatory inventories consisting of commodity chemicals, cosmetics, food additives, pesticides, and pharmaceuticals. Chemotypes can also be annotated with mechanistic information and potency categories based on NOELs/lowest observed effect levels (LOELs). This provides a powerful chemoinformatics approach for identifying and grouping analogs defined by similarity of structure, properties, potency, and biological mechanisms. The final outcome of the RA is based on quantitative weight of evidence assessment of analog toxicity information as well as *in silico* predictions. This approach should take diverse evidence sources into account, including not only QSAR models and structural alerts but also, when available, experimental results from biological assays and toxicity studies.

Essentials

- Various public projects for gathering toxicity data and for risk assessment
- The content and design of toxicity databases
- The essentials of structural alerts and of RA
- The essentials of TTC
- The usage of chemotypes and especially Toxprint chemotypes for toxicity applications
- The analysis of the chemical space of toxicity databases

Available Software and Web Services (accessed January 2018)

- Strategic Plan for Advancing Regulatory Science at the US Food and Drug Administration (FDA): https://www.fda.gov/ScienceResearch/SpecialTopics/RegulatoryScience/ucm268111.htm
- European Chemical Agency: https://echa.europa.eu/
- COSMOS project: http://www.cosmostox.eu/
- eTOX project: http://www.etoxproject.eu/index.html
- eTOX public sampler: https://etoxsys.eu/etoxsys.v3-demo/#/
- https://chemotyper.org
- https://toxprint.org
- Molecular Networks Biopath database: https://www.mn-am.com/databases/biopath
- https://www.mn-am.com/services/corinasymphonydescriptors
- http://toxtree.sourceforge.net/
- Ambit database (CEFIC LRI) for REACH data from ECHA: http://ambit.sourceforge.net/
- ToxCast: https://www.epa.gov/chemical-research/toxicity-forecaster-toxcasttm-data

Selected Reading

- Cancer TTC: Boobis, A., Brown, P., Cronin, M.T.D., Edwards, J., Galli, C.L., Goodman, J., Jacobs, A., Kirkland, D., Luijten, M., Marsaux, C., Martin, M., Yang, C., and Hollnagel, H.M. (2017) *Critic. Rev. Toxicol.*, **47** (8), 705–727, http://www.tandfonline.com/doi/full/10.1080/10408444.2017.1318822.
- Collins, F.S., Gray, G.M., and Bucher, J.R. (2008) *Science*, **319**, 906–907.
- COSMOS TTC Bioavailability Issues in TTC: Williams, F.M., Rothe, H., Barrett, G., Chiodini, A., Whyte, J., Cronin, M.T.D., Monteiro-Riviere, N.A., Plautz, J., Roper, C., Westerhout, J., Yang, C., and Guy, R.H. (2016) *Regul. Toxicol. Pharm.*, **76**, 174–186.
- Richard, A.M., Judson, R.S., Houck, K.A., Grulke, C.M., Volarath, P., Thillainadarajah, I., Yang, C., Rathman, J., Martin, M.T., Wambaugh, J.F., and Knudsen, T.B. (2016) *Chem. Res. Toxicol.*, **29** (8), 1225–1251.
- Schultz, T.W., Amcoff, P., Berggren, E., Gautier, F., Klaric, M., Knight, D.J., Mahony, C., Schwarz, M., White, A., and Cronin, M.T.D. (2015) *Regul. Toxicol. Pharm.*, **72**, 586–601.
- COSMOS TTC Database and Thresholds: Yang, C., Barlow, S.M., Muldoon Jacobs, K.L., Vitcheva, V., Boobis, A.R., Felter, S.P., Arvidson, K.B., Keller, D., Cronin, M.T.D., Enoch, S., Worth, A., and Hollnagel, H.M. (2017) *Food Chem. Toxicol.*, **109** (Pt 1), 170–193.
- Yang, C., Tarkhov, A., Marusczyk, J., Bienfait, B., Gasteiger, J., Kleinoeder, T., Magdziarz, T., Sacher, O., Schwab, C.H., Schwoebel, J., and Terfloth, L. (2015) *J. Chem. Inf. Mod.*, **55** (3), 510–528.

References

[1] Canadian Domestic Substances list, http://www.ec.gc.ca/lcpe-cepa/default.asp?lang=En&n=5F213FA8-1 (accessed January 2018).

[2] https://echa.europa.eu/regulations/reach/understanding-reach (accessed January 2018)

[3] EU Cosmetics Regulations and Registration, http://www.cirs-reach.com/Cosmetics_Registration/eu_cosmetics_directive_cosmetics_registration.html (accessed 25 January 2018).

[4] US Department of Health and Human Services, Food and Drug Administration (2015), Assessment and Control of DNA Reactive (Mutagenic) Impurities in Pharmaceuticals to Limit Potential Carcinogenic Risk (Guidance for Industry), https://www.fda.gov/downloads/Drugs/GuidanceComplianceRegulatoryInformation/Guidances/UCM347725.pdf (accessed January 2018).

[5] Toxicity Testing in the 21st Century, National Research Council (NRC) (2007), National Academy Press, Washington, DC, 266 pp.

[6] Collins, F.S., Gray, G.M., and Bucher, J.R. (2008) *Science*, **319**, 906–907.

[7] Hammett, L.P. (1970) *Physical Organic Chemistry*, 2nd edn, McGraw-Hill, New York, NY.

[8] Hansch, C., Maloney, P., Fujita, T., and Muir, R.M. (1962) *Nature*, **194** (4824), 178–180.

[9] Janssen, W.F., *The Story of the Laws Behind the Labels*, FDA Consumer (1981), Delaney Clause, http://www.fda.gov/AboutFDA/WhatWeDo/History/Overviews/ucm056044.htm (accessed January 2018).

[10] Munro, I.C., Ford, R.A., Kennepohl, E., and Sprenger, J.G. (1996) *Food Chem. Toxicol.*, **34** (9), 829–867.

[11] Cramer, G.M., Ford, R.A., and Hall, R.L. (1978) *Food Cosmet. Toxicol.*, **16** (3), 255–276.

[12] Kroes, R., Renwick, A.G., Cheeseman, M., Kleiner, J., Mangelsdorf, I., Piersma, A., Schilter, B., Schlatter, J., Van Schothorst, F., Vos, J.G., and Würtzen, G. (2004) *Food Chem. Toxicol.*, **42** (1), 65–83.

[13] The OECD QSAR Toolbox, Organization for Economic Co-operation and Development, http://www.oecd.org/chemicalsafety/risk-assessment/oecd-qsar-toolbox.htm (accessed January 2018).

[14] Toxtree - Toxic Hazard Estzimation by Decision Tree Approach, http://toxtree.sourceforge.net/ (accessed January 2018).

[15] COSMOS Project - Integrated *In Silico* Models for the Prediction of Human Repeated Dose Toxicity of COSMetics to Optimize Safety, http://www.cosmostox.eu/what/databases and http://www.cosmostox.eu/what/ttc/ (accessed January 2018).

[16] Stice, S. and Adams, T.B. (2016) Annual Meeting of the Society of Toxicology: The Application of an Updated Cramer Decision Tree to Food Ingredient Safety Assessment, Abstract 2280.

[17] Yang, C., Arvidson, K., Cheeseman, M., Cronin, M.T.D., Enoch, S., Escher, S., Fioravanzo, E., Jacobs, K., Steger-Hartmann, T., Tluczkiewica, I., Tarkhov, A., Rathman, J., Vitcheva, V., Mostrag, A., and Worth, A. (2016) Annual Meeting of the Society of Toxicology: Development of a Master Database of Non-Cancer Threshold of Toxicological Concern and Potency Categorization Based on ToxPrint Chemotypes, Abstract 2163.

[18] Yang 2017. Yang C, Barlow SM, Muldoon-Jacobs KL, Vitcheva V, Boobis AR, SP Felter, Arvidson KB, Keller D, Cronin MTD, Enoch S, Worth AP, Hollnagel HM. Thresholds of Toxicological Concern for cosmetics-related substances: New database, thresholds, and enrichment of chemical space. *Food and Chemical Toxicology*, **109** (2017) 170–193.

[19] Read-Across Assessment Framework (RAAF), European Chemicals Agency, https://echa.europa.eu/documents/10162/13628/raaf_en.pdf (accessed January 2018).

[20] New Approach Methodologies in Regulatory Science. Proceedings of a scientific workshop, European Chemical Agency, April 2016, https://echa.europa.eu/documents/10162/22816069/scientific_ws_proceedings_en.pdf

[21] DayCart – Chemical Intelligence for the relational database environment, Daylight Chemical Information Systems, Inc., http://www.daylight.com/products/daycart.html (accessed January 2018).

[22] JChem Oracle Cartridge, https://chemaxon.com/products/jchem-engines (accessed January 2018).

[23] BIOVIA Direct, Dassault Systemes, http://accelrys.com/products/collaborative-science/biovia-direct/ (accessed January 2018)
[24] Bingo, Epam Systems, http://lifescience.opensource.epam.com/bingo/index.html (accessed January 2018).
[25] OrChem, http://orchem.sourceforge.net/ (accessed January 2018).
[26] MolSql - Chemistry Cartridge for SQL Server, https://www.scilligence.com/web/dev-suite/ (accessed January 2018)
[27] RDKit: Open-Source Cheminformatics Software, http://www.rdkit.org/ (accessed January 2018).
[28] Django Web Framework, Django Software Foundation, https://www.djangoproject.com/ (accessed January 2018).
[29] Wikipedia, The Free Encyclopedia, "Java Persistence API", https://en.wikipedia.org/wiki/Java_Persistence_API (accessed January 2018).
[30] THOR-Merlin Toolkit – C-language interface for chemical database processing/searching, Daylight Chemical information Systems, Inc., http://www.daylight.com/products/thor_merlin_kit.html (accessed January 2018).
[31] ChemAxon Products, https://www.chemaxon.com/products/ (accessed January 2018).
[32] MOSES – Extensive chemoinformatics platform, Molecular Networks GmbH, https://www.mn-am.com/moses (accessed January 2018).
[33] CDK: Java Libraries for Cheminformatics, https://cdk.github.io/ (accessed January 2018).
[34] https://cosmosdb.eu/cosmosdb.v2/ (accessed January 2018).
[35] eTOX Project Website, http://www.etoxproject.eu/index.html (accessed January 2018).
[36] Wikipedia, The Free Encyclopedia, "Federated Database Systems", https://en.wikipedia.org/wiki/Federated_database_system (accessed January 2018).
[37] Wikipedia, The Free Encyclopedia, "International Chemical Identifier", https://en.wikipedia.org/wiki/International_Chemical_Identifier (accessed January 2018).
[38] SMILES - A Simplified Chemical Language, Daylight Chemical Information Systems, Inc., http://www.daylight.com/dayhtml/doc/theory/theory.smiles.html (accessed January 2018).
[39] Django Mutant, https://pypi.python.org/pypi/django-mutant (accessed January 2018).
[40] Django Dynamo, https://bitbucket.org/schacki/django-dynamo (accessed January 2018).
[41] Wikipedia, The Free Encyclopedia, "Entity-Attribute-Value Model", https://en.wikipedia.org/wiki/Entity-attribute-value_model (accessed January 2018).
[42] Ravagli, C., Pognan, F., and Marc, P. (2016) OntoBrowser: a collaborative tool for curation of ontologies by subject matter experts. *Bioinformatics*, **33**

(1), 148–149; 10.1093/bioinformatics/btw579 and http://opensource.nibr.com/projects/ontobrowser/.

[43] Ontology Lookup Service, http://www.ebi.ac.uk/ols/index (accessed January 2018).

[44] "Toxicology Testing in the 21st Century (Tox21)", U.S. Environmental Protection Agency, https://www.epa.gov/chemical-research/toxicology-testing-21st-century-tox21 (accessed January 2018).

[45] "AMES/QSAR International Collaborative Study", National Institute of Health Sciences, Division of Genetics and Mutagenesis, Japan, http://www.nihs.go.jp/dgm/amesqsar.html (accessed January 2018).

[46] CORINA Symphony Descriptors Community Edition Web Service, Molecular Networks, GmbH, https://www.mn-am.com/services/corinasymphonydescriptors (accessed January 2018).

[47] eTOXsys – Online chemical toxicity database and prediction tools, https://etoxsys.eu/etoxsys.v3-demo/#/ (accessed January 2018).

[48] Yang, C., Tarkhov, A., Marusczyk, J., Bienfait, B., Gasteiger, J., Kleinoeder, T., Magdziarz, T., Sacher, O., Schwab, C.H., Schwoebel, J., and Terfloth, L. (2015) *J. Chem. Inf. Model.*, **55** (3), 510–528.

[49] "ChemoTyper application, a public tool for searching and highlighting chemical chemotypes in molecules.", Altamira LLC and Molecular Networks GmbH, (2013), https://chemotyper.org (accessed January 2018).

[50] "ToxPrint – A public set of chemotypes, including generic structural fragments, Ashby-Tennant genotoxic carcinogen rules, and cancer TTC categories.", Altamira LLC and Molecular Networks GmbH, (2013), https://toxprint.org (accessed January 2018).

[51] Glen, R.C., Bender, A., Arnby, C.H., Carlsson, L., Boyer, S., and Smith, J. (2006) *IDrugs*, **9** (3), 199–204.

[52] Richard, A.M., Judson, R.S., Houck, K.A., Grulke, C.M., Volarath, P., Thillainadarajah, I., Yang, C., Rathman, J., Martin, M.T., Wambaugh, J.F., and Knudsen, T.B. (2016) *Chem. Res. Toxicol.*, **29** (8), 1225–1251.

[53] Yang, C., Richard, A.M., and Cross, K.P. (2006) *Curr. Comput.-Aided Drug Des.*, **2**, 135–150.

[54] Ertl, P. (1997) *QSAR*, **16**, 377–382.

[55] "BioPath – Database on Biochemical Pathways", Molecular Networks GmbH (2018), https://www.mn-am.com/databases/biopath (accessed January 2018).

[56] Rogers, D. and Hahn, M. (2010) *J. Chem. Inf. Model.*, **50** (5), 742–754.

[57] Maunz, A. and Helma, C. (2008) *SAR QSAR Environ. Res.*, **19** (5–6), 413–431.

[58] Yap, C.W. (2011) *J. Comput. Chem.*, **32** (7), 1466–1474.

[59] Faulon, J.L., Visco, D.P. Jr., and Pophale, R.S. (2003) *J. Chem. Inf. Comput. Sci.*, **43** (3), 707–720.

[60] Ferrari, T., Cattaneo, D., Gini, G., Golbamaki, N., Manganaro, A., and Benfenati, E. (2013) *SAR QSAR Environ. Res.*, **24** (5), 365–383.

[61] Ferrari, T., Gini, G., Bakhtyari, N.G., and Benfenati, E. (2011), Mining toxicity structural alerts from SMILES: A new way to derive Structure Activity Relationships, IEEE Symposium on Computational Intelligence and Data Mining (CIDM), 2011, pp. 120–127.

counterpart of chemoinformatics for applications in analytical chemistry. While there is a significant overlap between the computational toolboxes of the two fields, chemometricians also frequently apply data analysis methods that are primarily associated with this field. In fact, the chemometric community has contributed to the collective set of data analysis tools with the development of significant, milestone methodologies. In the following chapter, we give a brief introduction into these techniques, including data preprocessing and validation approaches frequently applied in chemometrics. More details on these methods can be found in the *Methods Volume*, Chapter 11.1.

9.2 Sources of Data: Data Preprocessing

If the data are arranged in a matrix form, the vectors of columns are called variables (features) and the rows are called objects (cases). By convention in chemometrics, the samples are arranged in the rows, whereas the measured quantities, variables, wavelengths, and so on in the columns of the input matrix. In chemoinformatics, similarly, the compounds are usually arranged in the rows by convention. Any data analysis is preceded with a data manipulation step frequently called preprocessing.

> **Good to know**: The term preprocessing should be preferred to the term pretreatment in order not to confuse it with physical sample treatment (preparation) prior to experimental analysis.

Preprocessing not only can lead to a serious information loss but can also separate useful and meaningless modeling. The most frequently applied preprocessing modes are centering, scaling, and transformation. All of the preprocessing techniques imply some assumptions about the variance structure in the data.

Mean centering is to subtract the column averages from each matrix element, for example, to shift the origin of the coordinate system to the center of the point cloud. Centering makes interval-scaled data behave as ratio-scaled data, which is the type of data assumed in most multivariate models. Centering reduces the rank of a model, may increase the fit performance, removes offset, and can avoid numerical errors [1]. Centering can be considered as a projection step, as it removes the postulated offset and leaves the data structure untouched [1].

Standardization means to divide each centered matrix elements with the standard deviations of the columns. If the variables are measured in (significantly) different units, standardization is absolutely necessary.

Scaling or weighting transforms the items for a commensurate scale. The most frequently used weights are the inverses of the standard deviations. In such a way all variables are scaled to unit standard deviation. Standardization allows features with small variations to have the same influence as features with large(r) variations. The term autoscaling is often used instead of standardization. The influence of scaling can be seen in Figure 9.1.

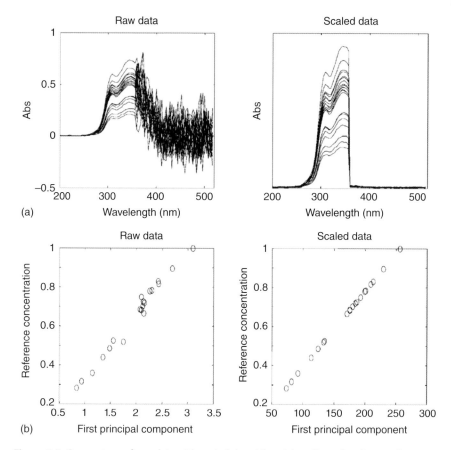

Figure 9.1 Comparison of raw data with scaled data (a) and the effect of scaling on the calibration line (b). (Reproduced from Ref. [1], Copyright © 2003 John Wiley & Sons, Ltd.)

> **Good to know**: Standardization is not necessarily advantageous: the same influence of peaks and noise in spectral data may lead to a loss of interpretability.

In chemometrics, *normalization* means the scaling of all variables to unit length, that is, dividing each matrix element with the column Euclidean distance (it is often confused with standardization).

Range scaling: All variables can be transformed to the [0,1] interval easily (see Eq. 9.1):

$$x_{ij}^{\text{range scaled}} = \frac{(x_{ij} - \text{Min}(x_j))}{(\text{Max}(x_j) - \text{Min}(x_j))} \tag{9.1}$$

where j is the running index for columns $1, 2, \ldots, m$. All columns will necessarily have (at least) one zero and one unity.

> **Good to know:** Range scaling inflates the measurements errors (as standardization also does) and it is sensitive to outliers. No further data preprocessing is recommended after range scaling.

There are many other scaling options (Pareto scaling, vast scaling, level scaling, etc.) (see, e.g., Ref. [2]).

Transformation: Skewed or heteroscedastic data can be transformed to be (approximately) normally distributed. Reciprocal, logarithmic, and power transformations are used most often. A logarithmic transformation makes multiplicative models additive. The exponent of the power function can be adjusted to the error structure. Zeros and negative values cause difficulties for logarithmic transformation. (Negative values are not applicable for square root transformation either.)

> **Good to know:** Centering should be done after transformation. In that case, zeros and negative values will not cause problems.

Data reduction is the unsupervised elimination of constant variables and the elimination of highly correlated variables.

There are many other ways of data preprocessing such as smoothing (which can be realized by Savitzky–Golay filtering; it works online as well), multiplicative signal correction, derivation (once, twice), and handling of missing data (by multiple imputation, substituting the missing data by the (column) average).

Example: Genotoxicity of Mussels

The authors investigated the genotoxicity of mussels by evaluating single-cell gel electrophoresis data in a recent work [3]. Mussels were gathered from the Adriatic Sea, Kotor Bay, to estimate the genotoxicity. Three types of tissues (hemolymph, digestive gland, and gill) were used for the assessment of DNA damage. Three evaluation methods were investigated (tail length, tail intensity, and olive tail moment). The task was to determine which organs and which evaluation methods should be used preferably. As the measurements had resulted in data on different scales, data preprocessing had to be used: standardization, normalization, range scaling, and rank transformation were investigated as alternatives. Data of various sampling sites and seasons were averaged. All three tissues and three evaluation methods could be compared in all combinations using a methodology based on the sum of (absolute) ranking differences (SRD) [4]. Then, the influential factors (types of organs, evaluation methods, and preprocessing options) were revealed by variance analysis (ANOVA). The smaller the SRD values the better. SRD analysis combined with ANOVA provides a unique and unambiguous way of decomposing the effects and determines the best combination of factors in the assessment of comet assay data (Figure 9.2). The rank transformation is far better than any other way of scaling in this particular scenario. This has also been proven by SRD, ANOVA, and cross-validation.

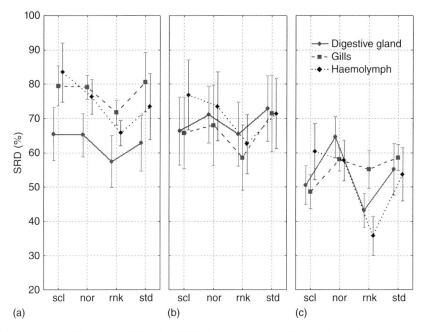

Figure 9.2 Combination of SRD with ANOVA decomposes the effects of factors in an easily perceivable way. Data preprocessing methods: scl, range scaling; nor, normalization to unit length; rnk, rank transformation; std, standardization (autoscaling); type of tissue: digestive gland, circles; gills, boxes; haemolymph, rhombuses; comet assay evaluation methods: tail intensity (a), tail length (b), olive tail moment (c). (Reprinted from Ref. [3], Copyright 2014, with permission from Elsevier.)

9.3 Data Analysis Methods

9.3.1 Qualitative Methods

Qualitative methods of chemometrics are developed and applied for problems of classification (what group of samples do the individual samples belong to?) and pattern recognition (is there any sort of grouping or pattern in the underlying data structure of a set of observations?) [5]. There are two main philosophies upon which these methods are based: unsupervised methods, which aim to reveal the underlying data structure without the potential bias of knowing the group memberships (which samples belong to which group) beforehand, and supervised methods, which aim to produce the best possible separation of the groups and thus to maximize the capability of the classification method to predict the class membership of samples for whose memberships are not known. Therefore, class memberships are taken into account by supervised methods during the development of a classification model, but not by unsupervised methods. Depending on the problem at hand, one group of methods could be more suited for the given purpose, but it is not always an unambiguous choice. In this subsection, we will briefly introduce the most important unsupervised and supervised pattern recognition methods (complemented with some novel, emerging, or simply interesting methods).

9.3.1.1 Unsupervised Pattern Recognition

Classes (or groups) are not even defined for this set of pattern recognition methods. To give a specific example, we will explain principal component analysis (PCA) in more detail.

PCA is one of the most commonly used chemometric methods and probably the single most valuable refined tool in the hands of a chemometrician/data scientist [6]. It is based on dimension reduction and employs matrix decomposition for this purpose. For example, if we have a data matrix of rank r (a data table with m variables and n samples, where $r = \min(m, n)$), PCA will decompose it to matrices of rank 1 (i.e., a specific set of coefficients called loading vectors P'), each multiplied with principal component scores (abstract factors or latent variables (LVs), i.e., score vectors t) (see Figure 9.3).

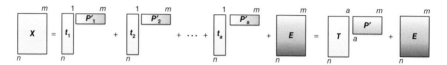

Figure 9.3 Matrix decomposition in principal component analysis. The loading matrix P' contains the coefficients of the original variables in the principal components, while the score matrix T contains the principal component scores (values) of the samples. E is an error matrix in cases of $a < m$. (If $a = m$ the E matrix is empty, contains zeros only).

> **Good to know:** While theoretically the number of loading vectors can be as high as m, in cases where $m \geq n$, any principal component after the nth one will be meaningless ("ill-conditioned problem").

In practice, this means that in the m-dimensional space, we will look for the direction in which the variance of the data is the largest. This will be the direction of the first principal component (PC1), a linear combination of the original m variables defined by the P'_1 loading vector (a unit vector whose elements correspond to the coefficients of the original variables in PC1). Next, the direction with the second largest variance is determined (PC2), with the constraint that it must be orthonormal to PC1 (i.e., it must be a vector of unit length as well). PC3 will have to be orthogonal to both PC1 and PC2 and so on. In the end, the principal components will form an orthonormal basis set.

Once we have determined the principal components, the data can be projected into a subspace of fewer (preferably two or three) dimensions. This practically means plotting the data in 2D or 3D diagrams using the principal components as the axes. Since typically the first few principal components contain the majority of the variance in the data, this simplified interpretation enables the detection of groupings or patterns, which would have been hard to notice in the original m-dimensional data matrix (containing possibly hundreds or even thousands of variables). Therefore, PCA and other related methods can also be called dimension reduction methods. An important question for the application of PCA is the number of principal components that are used for representing the original data matrix. Several approaches are available for determining this number (pseudo-rank, effective rank), including hard cutoffs regarding the explained variance (or the eigenvalue) of the principal components and the scree plot (a plot of the eigenvalues associated with the numbering of principal components in a decreasing order).

Another widely applied tool is cluster analysis or clustering [7]. Cluster analysis is a collective name for a group of methods built upon the same idea: the samples are arranged in groups (or "clusters") based on their distance from each other. The closer the two objects are located, the more similar they are. One of the most commonly applied clustering methods is agglomerative hierarchical cluster analysis (HCA). In this case samples are grouped together into clusters based on their distances, and then the clusters are grouped into successively bigger clusters based on their distances. The result is usually presented in a dendrogram (see Figure 9.4) that gives a detailed visual feedback on the "natural" grouping of the samples. The main problem with clustering is that we have a large number of distance methods (i.e., how do we define the distance of two objects?) and linkage methods ("amalgamation rules", i.e., how do we define the distance of two clusters?) to choose from, which often give diverse results. This could be attributed

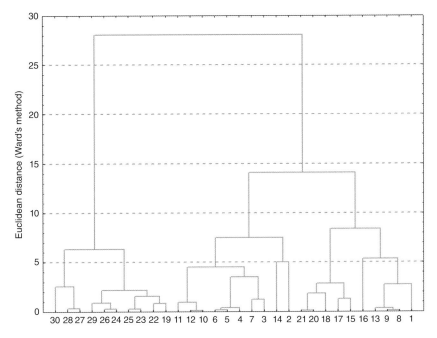

Figure 9.4 Thirty chromatographic columns are grouped according to their various polarity metrics with hierarchical cluster analysis using Ward's method (as the linkage rule) and the Euclidean distance. Arbitrary horizontal lines at about 20 or 10 distance units define two or three clusters, respectively.

to the fact that clustering neglects the information content associated with directions (concentrating only on distances).

> **Good to know**: If the variables are measured on (vastly) different scales, standardization (see Section 9.2) is a must; otherwise the variable(s) with the largest scale(s) dominate(s) the distance(s) and distorts the pattern in the data.

Further examples for widely used unsupervised pattern recognition methods are self-organizing maps (SOMs) (or Kohonen maps) [8]. SOMs are a type of artificial neural networks (ANN) that employ unsupervised learning for producing a map (usually a rectangular or hexagonal grid) of nodes (or neurons) as the simplified (lower-dimensional) representation of the original data structure. After training with input examples (i.e., creating the map), samples can be placed on the map by finding the node that is the closest (more similar) to the given sample.

Example: Classification of Italian Olive Oils

The task was to classify Italian olive oils according to their regional origin. The olive oils originated from nine regions in Italy, namely, northern Apulia (1), Calabria (2), southern Apulia (3), Sicily (4), inner Sardinia (5), coastal Sardinia (6), eastern Liguria (7), western Liguria (8), and Umbria (9), and were characterized by their amount of eight different fatty acids [9, 10]. An SOM of size 15×15 neurons was trained with 250 samples from a total of 572 olive oils producing the SOM shown to the left in Figure 9.5. Sending the rest of 322 samples as test cases into this SOM gave 312 correct predictions, which translates to an excellent correct classification rate of 96.9% [11].

However, another very interesting result can be derived from Figure 9.5: the regions in the SOM reproduce the geography of Italy. The regions from northern

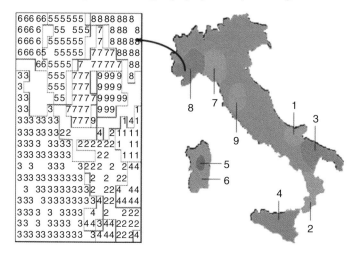

Figure 9.5 Self-organizing map of a dataset of Italian olive oils and its comparison with the map of Italy and the regions of origin of the olive oil samples. (Copyright 1994, with permission from Elsevier.)

Italy and Sardinia are on the top of the SOM, with Sardinia clearly separated from the other regions, and the regions from southern Italy are in the lower part of the SOM. Magic? No, it is in the data! The regions in northern Italy have soils and climates that are distinctly different from those in Sardinia and southern Italy. This underlines the power – and importance – of unsupervised learning: trying to get as much information from the data as they contain.

9.3.1.2 Supervised Pattern Recognition

Contrary to unsupervised methods, supervised classification algorithms are optimized (trained) with a set of observations for which the class memberships are already known (training or learning set). Thus, somewhat different practices have evolved for their use and validation.

One of the most commonly applied supervised classification methods in the domain of chemometrics is linear discriminant analysis (LDA) [12]. The principle of LDA is very similar to that of PCA: both produce linear combinations of the original variables that best represent and explain the original data. Unlike PCA however, LDA explicitly models the differences between the k classes of data using a $(k-1)$-dimensional representation of the m-dimensional space that best separates the classes. Two- (or more-)dimensional plots of these new variables (i.e., canonical variables) can include confidence ellipsoids (or hyperellipsoids) that define the areas associated with a group and lines that separate these areas corresponding to different groups (see Figure 9.6).

While partial least squares (PLS) was originally introduced for regression (see next subsection), it is also applicable to classification. In fact, its use for classification can be thought of as a special case when the dependent (y) variable

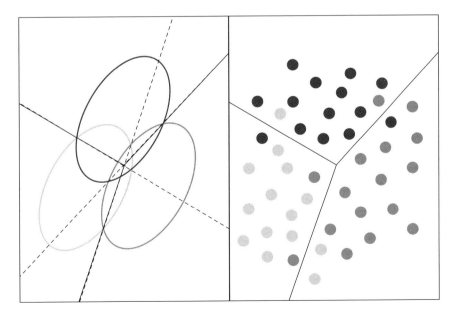

Figure 9.6 LDA plots with confidence ellipsoids and separating lines. Note that the line separating two groups goes by definition through the intersections of the two ellipsoids by definition.

is not continuous, but categorical (corresponding to class memberships). This setup is most commonly referred to as partial least squares discriminant analysis (PLS-DA) [13].

Example: Classification of Historical Hungarian Coins
As the historical sources on the Árpád dynasty (Hungary) are sporadic and deficient, it would be interesting to know whether it is possible to unravel patterns in coinage habits and possible contacts between various kings. The chemical compositions of silver coins from the Árpád dynasty of medieval Hungary were at our disposal determined by X-ray fluorescence spectroscopy. A relatively rarely used supervised method is object target rotation [14]. Nonetheless, the method is suitable for providing detailed and highly specific information on class memberships: for instance, Christie *et al.* have successfully unscrambled a class-in-class situation using this methodology [15]. Three classes of coins were identified, corresponding to three historical periods. The first principal component of the composition of coins reflects the balance between silver and copper in the minting alloys. This principal component score is similarly distributed among the coins from two non-consecutive historical periods (classes 1 and 3), but a sudden change in the Hungarian coin metal composition during the intermediate period – the reign of King Kálmán (class 2) – is an example of coin debasement without subsequent inflation (i.e., independent of the actual silver content of the coins, the merchants' and layman's belief in the value of the coin remained) (Figure 9.7).

In object target rotation, the most central object of each class (period) was selected using the least sum of variance criterion. Then, each object is projected onto the central object, and the assignment of each coin to a given period is expressed in membership percentages. In the cited work, the differences in the compositions of the coins from the historical periods corresponding to classes 1 and 3 were not captured by PCA, but were revealed by object target rotation. In general, "object target rotation can be an alternative to unsupervised and supervised pattern recognition methods such as PCA and partial least squares discriminant analysis, respectively. It is especially useful if the distributions are skewed and/or bimodal. The triumph of the methodology is the efficiency of class assignment combined with the ability to resolve the class-inside-class situation" [15].

Another group of supervised learning methods are based on the concept of recursive partitioning, also called classification and regression trees (CART) [16]. Instead of constructing new variables, these methods implement classification (or regression for a continuous or ordinal y variable) as a set of binary decisions based on the values of the original variables (see Figure 9.8). More advanced implementations of decision trees include bagging trees, random forests, and boosted trees – these methods are also called ensemble methods as they construct more than one decision tree and aggregate their results.

An interesting group of classification methods are support vector machines (SVM). Contrary to most classification algorithms, SVMs project the original data into a space of higher (rather than lower) dimensions (feature space) using a

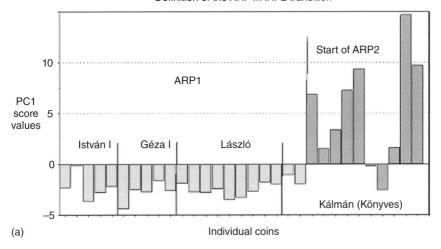

(a)

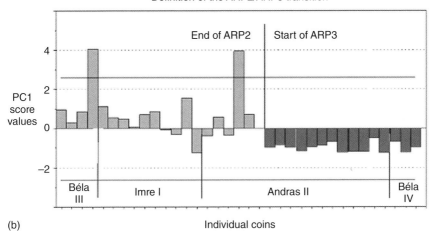

(b)

Figure 9.7 Transition between the three periods as defined by the coins' metal content. (Copyright 2014, with permission from Wiley.)

suitable kernel function. (Most popular functions include polynomial kernels and the Gaussian radial basis function.) In the resulting higher-dimensional space, a linear (or planar, hyperplanar) decision boundary is defined that separates the classes from each other, whereas in the original space the boundary is a non-linear (often complicated) empirical function. There are many variants of SVM, but their common feature is to have two regularization parameters (e.g., C and gamma). (A typical surface in the combination space of the regularization parameters is presented in Figure 9.9.) An easily understandable tutorial review in Ref. [17] summarizes the most characteristic features of SVM.

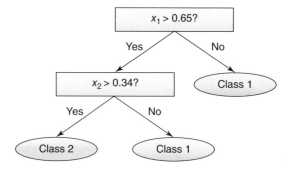

Figure 9.8 A simple example of a classification tree. Each junction corresponds to a binary decision based on the value of a variable, while each leaf is a possible outcome of the classification. (Note that not necessarily just one leaf can classify the sample into a given group.)

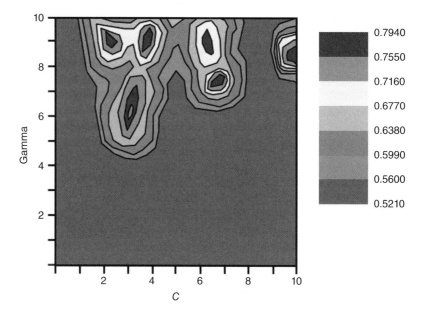

Figure 9.9 Regularization parameter space of a support vector machine classification model. Color coding corresponds to the classification performance (the higher the better). Many combinations may produce the same model goodness.

> **Good to know:** SVM models are sensitive to overfitting, that is, it is relatively easy to achieve 100% correct classification rate on the training set. Careful validation is advised (see later).

Another popular example of machine learning methods is the k-nearest neighbors (kNN) algorithm [18]. Elegantly simple, it assigns objects to classes based on the k (user-defined) closest training examples in the multidimensional space of the (original) variables. Of course this method also needs the optimization of a few parameters beforehand – including the distance metric that is used and the value of k. Recent developments to the algorithm include the N-nearest neighbors (N3) and binned nearest neighbors (BNN) algorithms by Todeschini and colleagues [19].

While receiver operating characteristic (ROC) curves are inherently not classification methods, it is worth taking a short note of them as they are widely used in chemoinformatics to evaluate binary classifiers. ROC curves iterate over a list of observations ranked by a classifier value (e.g., a probability score) whose class memberships (by default, "positive" or "negative") are known. At each iteration, the following question is answered: if I draw a cutoff at the current value of the classifier and predict objects with an equal or higher value to be positives and the rest as negatives, what will be the true positive rate (TPR) (ratio of true positives to all positives) and the false positive rate (FPR) (ratio of false positives – negatives that have been falsely classified as positives – to all negatives)? This will define a point on the ROC curve (see Figure 9.10). The area under the curve ($0 \leq AUC \leq 1$) will be larger as the performance of the classifier method gets better (AUC = 0.5, i.e., the diagonal corresponds to random classification). While ROC curves are traditionally used for the evaluation of binary classifiers, there have been efforts to extend the concept to the domain of multi-class classification, including a recent study by the authors of this chapter [20].

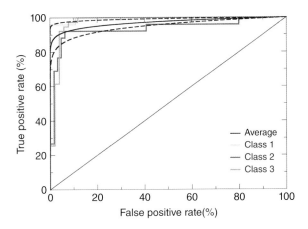

Figure 9.10 An example of n-class ROC curves. Grey curves correspond to individual classes (1–3), while black is the average curve calculated with the Hanley formula. Dashed lines indicate ±1 standard deviation from the average. (Reprinted from Ref. [20], Copyright 2016, with permission from Elsevier.)

9.3.2 Quantitative Methods

The other major domain of analytical chemistry is quantitative analysis. Modern analytical devices can generate huge datasets, often containing thousands of spectral data (Fourier transform infrared (FT-IR)/near-infrared (NIR), mass spectrometry (MS), nuclear magnetic resonance (NMR), etc.). The connection between the detected signals and the exact concentration values is not clear and simple in every case. Thus chemometric analysis of the examined samples plays an important role in the evaluation process. In this subsection, we describe one of the most important families of chemometric methods: regression techniques in quantitative analysis.

All regression methods are based on a simple concept: we explore a connection (linear or nonlinear) between one or several independent variables (X) and one (or more, but usually one) dependent variable (y) (see Eqs. 9.2–9.4). The easiest case is when we have only one dependent and one independent variable: this is called univariate regression. In this case the regression equation is the following: $y = b_0 + b_1 x_1 \pm e$, where b_0 is the intercept, b_1 is the slope, and e is the random noise or residual error. If the number of independent variables is more than one, the following equation can be applied:

$$y = b_1 x_1 + b_2 x_2 + \cdots + b_m x_m \pm e \tag{9.2}$$

or shortly

$$y = \sum_{j=1}^{m} b_j x_j \pm e \tag{9.3}$$

with matrix notations

$$y = Xb \pm e \tag{9.4}$$

where b is the vector of regression coefficients, X is the matrix of independent variables, and y is the vector of dependent variables. These three equations are the various formulations of multiple linear regression (MLR), also called ordinary least squares (OLS). MLR can be used with more than one dependent variable as well. MLR is a frequently used technique in quantitative structure–activity relationship (QSAR) modeling [21]. If the number of variables (m) is more than the number of samples (n), there is no unambiguous solution of the equations. In the case when $m = n$, one and only one solution exists, but for the random errors no degrees of freedom remain. Thus, the best-case scenario is if we have more samples than variables. In this case, more solutions exist for b, but an exact solution can be acquired if we minimize the length of the residual vector (e) (Eqs. 9.5 and 9.6):

$$e = y - Xb = 0 \tag{9.5}$$

$$b = (X'X)^{-1} X' y \tag{9.6}$$

The regression coefficients can be calculated if we multiply y and X with the transpose of X (X') and use the inverse of the $X'X$ matrix. This workaround (Eq. 9.6) is necessary as the original X matrix cannot be inverted in the general case (it is not quadratic). The resulting b vector is generally an approximate solution. The latter equation is called the Moore–Penrose inverse (X^+), when X has linearly independent columns (and thus matrix $X^{*'} X^*$ is invertible, and X^* is the conjugate inverse) and X^+ can be computed as $(X'X)^{-1} X'$. In practice, the columns in X are not linearly independent, that is, in X^+ the small eigenvalues are omitted.

Before the use of regression methods, we should check whether there are any correlations between the variables. This step is important mostly because linear methods cannot be used for nonlinear regression problems. If there are no (or indeed low) correlations between the dependent and independent variables, nonlinear regression techniques can still be applied. Some of the typical methods for correlation analysis are Pearson's r (parametric), Spearman's ρ (nonparametric), Kendall's τ (nonparametric, without ties), and the γ (nonparametric) correlation coefficients [22].

> **Good to know**: Correlation analysis also helps to filter out the highly correlated x variables that carry (closely) the same information. However, the correlation of two variables does not mean in general that we have to omit half of our dataset.

On the other hand, if some of the samples in the dataset have larger error (or we know that some of them are less important than the others for some reason), we can simply downscale the weights of those samples. Thus, they will not influence the analysis as much as the others [23]. (For this purpose we can use predefined functions as well.)

Another important phase of regression model building is variable selection. In the past decades a large number of methods were published. Most of them are using regression parameters such as R^2, $R^2_{adjusted}$, Mallows C_p, and so on for the evaluation of models. The aim of these methods is to reduce the dimension of the original dataset with the selection of the most important variables. Thus, we can improve the model (reduce the chance of overfitting), get a better interpretation, or decrease the measurement costs. One can simply do an MLR forward or backward stepwise manner, where the variables are selected according to the increase of the aforementioned parameters or preferably the Fisher criterion (F value). The use of variable intervals is a good way to select variables, especially in the case of spectral datasets, where the variables (i.e., neighboring wavelengths) are usually highly correlated with each other.

Partial least squares regression (PLSR) is an especially suitable technique for landscape X matrices (where the number of variables is higher than the number of samples). In analytical chemistry, we generally have more wavelengths than samples $m \gg n$. Here we present the most frequently used techniques for variable selection based on Ref. [24].

> **Good to know**: Data preprocessing should be used before the variable selection procedure! Outlier detection should also be applied before variable selection.

Some simple methods for variable selection are filtering based on the regression coefficients (**b**) and the PLS loading vectors. In both cases, one can exclude those parts of the spectra (or any other dataset), where the values of the examined parameters are close to zero. (Closeness here means that the regression coefficients (the **b** values) are not significantly different from zero according to a t-test.) Furthermore, jackknifing (see later in Section 9.4) is also a simple option for variable selection, where the uncertainty of **b** and that of the loading vectors can be calculated by means of cross-validation. The least significant variables can be excluded, until the model improvement stops. Another opportunity is the use of variable importance (VIP) values. The formula of VIP calculation contains the sum of squares of the explained variance, the number of variables, and the weights of the variables.

Some more complex methods, such as genetic algorithms, least absolute shrinkage and selection operator (LASSO), and ridge regression (RR), can also be used for variable selection. Interval partial least squares (iPLS) are based on the original PLS regression (see later). The basic concept is that the dataset (spectra, chromatograms, etc.) can be divided into several equal parts (e.g., 10, 20, 40, or 60). Then PLS is used for these segments of the data, and – based on the root mean square error of cross-validation (RMSECV), R^2, and its cross-validated counterpart Q^2 values – one can select the best models.

> **Good to know:** We can use forward selection or backward selection for the inclusion of the different variable segments in the case of iPLS. Synergy PLS utilizes more segments at a time.

In the case of genetic (or evolutionary) algorithms, the original variables are combined and "evolved" through several "generations," where these combinations (or "chromosomes") are evaluated with an appropriate fitness function (such as R^2 or Q^2) and the best ones are propagated to the next generation. This process is repeated until a stop criterion is reached (it can be a maximum number of generations), after which the best model(s) is (are) selected.

> **Good to know:** Intervals are highly recommended for the use of genetic algorithms instead of individual variables (wavelengths).

Example: Quantitative Determination of Coenzyme Q10
The quantitative determination of coenzyme Q10 in dietary supplements was carried out in a recent study with FT-NIR spectroscopy and chemometrics [25]. Rácz et al. have built a calibration model with partial least squares regression (PLSR) for a dataset of 50 dietary supplements. The combination of FT-NIRS and multivariate calibration methods is a very fast and simple way to replace the commonly used tedious high performance liquid chromatography–ultraviolet (HPLC-UV) method. (In contrast to the chromatographic techniques, sample pretreatment and reagents are not required and virtually no waste is produced.) The calibration models could be improved by different variable selection techniques such as interval PLS or a genetic algorithm. The original idea of Rajalahti et al. (using a selection of single wavelengths) [26] was extended to define interval selectivity ratio, a novel variable selection approach.

LASSO and RR are similar methods, the most important difference lying in the solution of the normal equations. More precisely, when the *b* regression coefficients are calculated, a new meta-parameter (or regularization parameter) is introduced. This parameter can be the Euclidean norm or the one-norm

(Manhattan distance) of the regression coefficients for RR and LASSO, respectively. The equations and other options for Tikhonov regularization (a collective name for methods operating with a regularization parameter) are summarized in details in Ref. [27]. Both methods are applicable to variable selection; however, LASSO usually gives better results and a more easily acceptable selection. RR is implemented in several statistical software; LASSO can be computed, for example, in MATLAB.

MLR was discussed earlier, but there are other linear regression methods as well. Well-known and widely used examples include principal component regression (PCR) and PLSR. Both of these methods are frequently used for spectroscopic datasets. Random forests and boosted trees (see previous subsection) are also suitable for regression analysis. (In this case, the output variable will be continuous as opposed to classification where it is discrete.) PCR is based on PCA (see previous subsection). First the principal component matrix is calculated, and then it is used instead of the X matrix in the basic equation of regression models (Eq. 9.3). The nonlinear iterative partial least squares (NIPALS) algorithm can be used for the calculation of the final model. In the past few years, the popularity of PCR has decreased in contrast to PLSR. The latter one is based on a modified NIPALS algorithm with weights, and in the latter case we have an inner and an outer relation between independent (X) and dependent (Y) variables.

Good to know: The weights in the PLSR algorithm are necessary, because it ensures the orthogonality of the PLS latent variables.

In the modified NIPALS algorithm, a regression is established between the X and Y variables and PLS LVs (components). Figure 9.11 shows the relations of the original variables. As we can see the X and Y matrices can be decomposed to loading (P' and Q'), score (T and U), and error matrices (E and F) as in the case of PCA. This is the so-called outer relation. But we have also the inner relation, which is the following: $u_i = b_i t_i$.

Finally the combined relation can be written in the following form: $Y = TBQ' + F$. After the calculation of LVs (PLS components), we have to

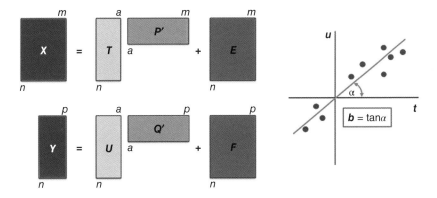

Figure 9.11 Schematic representation of PLS regression.

decide how many components we need to be retained in the model. There are several methods for LV selection. The most popular ones are the use of predictive error of sum of squares (PRESS) and RMSECV values, which are thought to express the predictive ability of the model. On both plots, appropriate number of components is to be selected at the global minimum.

The most famous article about PLSR was written in 1986 [28], but the popularity of this method is still undeniable nowadays. Although in the field of QSAR analysis MLR is more common, PCR and PLSR are preferred in the field of analytical chemistry (multivariate calibration).

Example: Comparison of Different Quantitative Modeling Methods
In the early days of chemometrics, there were fervent discussions about the advantages and disadvantages of PCR, PLSR, and RR. Statisticians prefer(red) RR as it has a continuous regularization (meta) parameter. However, possibilities exist for variable selection (e.g., LASSO) that are better than in RR. Frank and Friedman [29] conducted extended simulations for various situations (well- and ill-conditioned, highly collinear datasets, having unequal true coefficient values, various signal-to-noise ratios, etc.). The algorithmic descriptions imply that PCR, PLS, and RR are very different procedures, leading to quite different model estimates. The authors provided a heuristic comparison suggesting that all techniques are, in fact, quite similar and can be considered as "shrinkage" methods. According to Frank and Friedman's comparison, RR works somewhat better than PLS and PCR and much better than OLS and variable subset selection (VSS). However, they have studied situations, where X variables are independent (but more or less correlated), that is, the rank of X is equal to the number of X variables. LV methods such as PLS are increasingly used in chemistry and other branches of science and technology, where the real rank of X is much lower than the number of X variables (or the number of samples) in a range of important applications. In fact, a great strength of PLS and PCR is the estimation of this real rank, which is given by the number of LVs [30].

There are also some regression tools based on machine learning methods (like SVM). SVMs are discussed earlier (see previous subsection); the main difference in this case is the use of a continuous reference (dependent) Y variable(s) instead of a grouping variable (as in pattern recognition).

In this section, we have presented the most popular methods to build regression models, but none of these models are applicable without the use of proper validation techniques. The validation of any model is an essential part of the work; thus in the following part we discuss the most frequently used validation practices in analytical chemistry.

9.4 Validation

Overfitting arises frequently as highly effective modeling techniques are developed. Overfitting means that the noise is also fitted (not only the systematic information of the data). Hence, the assessment of the predictive performance on future samples (i.e., validation) has gained increasing importance. There is no single best way to determine the predictive performance of a model, though some options such as leave-one-out cross-validation have become a kind of standard. Therefore, the recommended procedures are gathered in the following part, where we also mention some debated issues.

Statisticians recommend [31]: "If possible, an independent sample should be obtained to test the adequacy of the prediction equation. Alternatively, the data set may be divided into three parts; one part to be used for model selection [model building or variable selection], the second part for the calibration of parameters in the chosen model and the last part for testing the adequacy of predictions." In the machine learning field (ANN, SVM, etc.), this is the practice or at least the advocated practice. In many cases the insufficient number of samples leads to the division of the data into two parts. If the calibration of parameters is done using the same part of the data, substantial biases arise.

Some chemists also advocate a separation of an external part for testing [32], while others think opposite: "hold-out sample is far inferior" (as compared with leave-one-out cross-validation) [33] or "hold-out samples are downward biased. … small independent hold-out samples are all but worthless" [34]. Cross-validation is probably the most widely used method for estimating prediction error [35]: "… cross-validation would not suffer much bias … five- or ten-fold cross-validation will overestimate the prediction error. Whether this bias is a drawback in practice depends on the objective. On the other hand leave-one-out cross validation has low bias but can have high variance. Overall, five- or tenfold cross-validation are recommended as a good compromise" [35].

Realization of cross-validation can be completed in different ways. Eigenvector Ltd. summarizes some possibilities [36]:

- *Venetian blinds*: Each test set is determined by selecting every s-th object in the dataset, starting at objects numbered 1 through s.
- *Contiguous blocks*: Each test set is determined by selecting contiguous blocks of n/s objects in the dataset, starting at object number 1.
- *Random subsets*: s different test sets are determined through random selection of n/s objects in the dataset, such that no single object is in more than one test set. This procedure is repeated r times, where r is the number of iterations.
- *Leave-one-out (jackknifing)*: Each single object in the dataset is used as a test set once and only once (it can be considered as an extreme case of leave-many-(M)-out cross-validation).

Recently Bro *et al.* summarized the viable options for cross-validation in a scientifically sophisticated way [37]: Cross-validation can be carried out (i) row-wise, (ii) using a special diagonal pattern suggested by Wold, (iii) row- and column-wise to ensure that each data point was not used at both the prediction and the assessment stages, thus avoiding problems with overfitting as suggested by Eastment and Krzanowski (EK), and (iv) as Eigenvector Ltd. uses an algorithm to solve the missing data problem and two more hybrid methods were suggested by Bro *et al.* based on "expectation maximization" (EM). The authors performed detailed simulations to reveal which method can be recommended for general use. It was expected that higher noise and higher correlation would cause high error for cross-validatory rank estimation. The expectation proved to be valid for the correlation part, but high noise levels are critical only if correlation is also high, in which case all methods fail. Perhaps the best results could be achieved by the eigenvectors algorithm and by EM EK combination.

Strictly speaking a test sample (singly split, hold-out sample) is not a cross-validation technique as the test set has not been used for modeling and the training (learning) sample has not been used for testing, that is, the validation has not realized a "crossing."

Other resampling methods also exist, sometimes (erroneously!) they are also called cross-validation:

- *Randomization test* (Y-scrambling): The response values (dependent variable, Y) for objects (samples) are permuted several times in the first step. (The randomization is equivalent with permuting the running indices.) Afterward the model is tested on the scrambled Y sets. The performance of such permuted models should be deteriorated. During these permutations no exclusion of data is needed from the model building process. Figure 9.12 shows a typical example of a randomization test.
- *Simulation test*: A model can also be tested by randomizing the independent (X) variables. The entire process of model building (variable selection), calibration (setting the parameters of the model), and testing the predictive performance can be carried out similarly to the previous modeling (non-randomized case). While seemingly good models can be achieved on

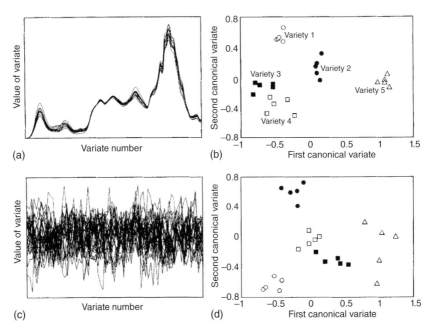

Figure 9.12 Canonical correlation analysis distinguishes five classes (five plant variants b) from real NIR spectra (a). However, a randomization test clearly shows that five arbitrary classes (d) can be found also for random vectors (c). In the present case, canonical modeling could not pass the randomization test, and a serious decrease in the number of included variables (wavelengths) is necessary. (Reprinted from Ref. [5], Copyright 1997, with permission from Elsevier.)

the training set, the model performance should deteriorate on the test set. (Otherwise, the model is likely to be overfitted.)
- *Leave-many-out (LMO)*: In M-fold cross-validation, no repetition is allowed in sampling, that is, the available data (number of rows, N) is split into M disjoint subsets having (approximately) the same size. The model is then recalculated M times, leaving out each time one subset, which is used for validation. The average of the resulting correlation coefficients will be the correlation coefficient of cross-validation, or Q^2.
- *Bootstrap*: Repetitive selection of any objects is allowed in the case of bootstrapping [38, 39]. The training set is composed from randomly drawn data from the original dataset. This kind of resampling can be carried out (many) thousands of times and an empirical distribution can be determined. A bootstrap estimation is biased because of the (possible) repetitions; it overestimates the errors systematically, but the bias can be corrected.
- *Repeated double cross-validation*: Double cross-validation has been introduced to avoid selection bias. It is characterized by two loops; the model building (determination of the optimum number of principal components or variable selection) happens in the inner loop, and the outer loop is used for testing the prediction performance. Leave-one-out (or several-fold) cross-validation is applied in both loops. The whole process can be repeated (approaching bootstrap or Monte Carlo cross-validation) (Figure 9.13). The details can be found in Ref. [40].

The debate about (multiple) cross-validation and external test set validation is starting to be equilibrated: "In conclusion, the two modelling approaches compared here are philosophically different, and neither should be considered

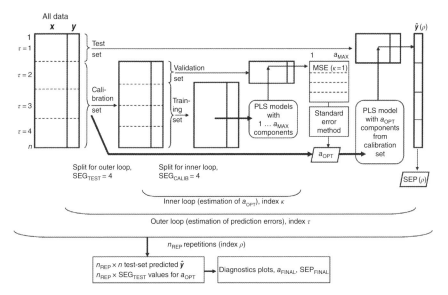

Figure 9.13 The scheme of repeated double cross-validation. (Reproduced from Ref. [40], Copyright © 2009 John Wiley & Sons, Ltd.)

as right or wrong" and "To avoid the limitation of using only a single external set, we ... always verify our models on two/three different prediction sets" [41]. In fact, two or more external sets approximate leave-several-out cross-validation. Recent examinations on real datasets show unambiguously that an external validation on one test set delivers models that are not significantly superior to random choice [42].

9.5 Applications

Table 9.1 collects a small selection of various and colorful applications of chemometric methods in different fields of analytical chemistry. In addition to the references, we have collected the most important details of the cited works, such as the samples that were studied and the analytical and chemometric methods that were applied.

Further interesting applications from the field of food analysis can be found in the work of Beruetta et al. [60].

9.6 Outlook and Prospects

There are many more important and frequently used methods in chemometrics and in analytical chemistry. To provide some recent or simply interesting examples, we briefly introduce the reader to multivariate curve resolution (MCR), ANN, and ant colony optimization (ACO). The latter two are great examples for methods whose main concepts are inspired by natural systems or phenomena.

MCR has a similar algorithm to PCA, and it cannot be solved without constraints (nonnegativity, closure, etc.) because of the rotational ambiguity. Contrary to PCA, MCR is able to find the chemical rank and can theoretically decompose spectra of pure constituents if some conditions are fulfilled. One realization for MCR is alternating least squares (ALS) [61].

ANN were designed to mimic the functions of the real neural network in the brain. Inspired by brain operation, perhaps the best nonlinear fitting method was created. The method develops a relationship between the input (independent) variable(s) and the output (dependent) variable(s). While there are many forms of neural networks, both for unsupervised and for supervised learning, the most frequently used one is a feed-forward ANN, (also referred to as error back-propagation ANN). Generally, one hidden layer is introduced consisting of units called "nodes." The original descriptors (variables) are weighted and transformed between the input/output and the hidden layers. In general, it is advisable to keep the number of hidden nodes at a minimum to avoid overfitting. (Although this method is referred to as nonparametric, the aforementioned weights and also the bias (or offset) are adjustable parameters of ANN.) While ANN can be used as a simple regression technique – especially if the dataset has nonlinearity – there are other important usages, for example, variable selection [62–64].

ACO can be applied for complicated combinatorial optimization problems. The idea is based on real ant colonies, where members of the colony leave

Table 9.1 A selection of applications of chemometric methods in contemporary analytical chemistry literature.

Field of science	Samples	Analytical method	Chemometric methods	Reference
Materials science	Nuclear materials, cultural heritage, and so on	Atomic spectroscopy	PCA and other pattern recognition techniques	[43]
Military science	Chemical and biological warfare agent simulants	Laser-induced breakdown spectroscopy	Linear correlation, PCA, SIMCA	[44]
Pharmaceutical science (quality control)	Herbal medicines	Chromatography (TLC, GC, HPLC) and electrophoretic methods	Spectral correlative chromatogram (SCC), similarity estimations, PCA, kNN, and so on	[45]
Pharmaceutical science (counterfeits)	Pharmaceutical tablets	Raman spectroscopy	Support vector machine (SVM)	[46]
Food science	Edible oils, fats	FT-IR, FT-NIR, FT-Raman spectroscopy	LDA, canonical variate analysis	[47]
Food science	Transgenic tomatoes	Visible/near-infrared spectroscopy	PCA, PLS-DA, discriminant analysis	[48]
Pharmaceutical science	Analysis of nystatin and metronidazole	NIR spectroscopy	PLS regression	[49]
Food science	Spanish honey	Inductively coupled plasma atomic emission spectroscopy (ICP-AES)	Cluster analysis (CA), PCA	[50]
Material science	Geological samples, explosives, and so on	Laser-induced breakdown spectroscopy	PLS, PCA, SIMCA, LDA	[51]

(Continued)

Table 9.1 (Continued)

Field of science	Samples	Analytical method	Chemometric methods	Reference
Pharmaceutical science	Counterfeit medicines	ATR-FT-IR spectroscopy	KNN, PCA, SIMCA, CART	[52]
Pharmaceutical science	Urine samples	^{1}H NMR spectroscopy	PLS, PCA	[53]
Pharmaceutical science	Solid, liquid, and biotechnological pharmaceutical forms	NIR spectroscopy	PLS, PCR, ANN, SVM, MLR	[54]
Pharmaceutical science (QSRR analysis)	Drugs, metabolites and banned compounds	Different LC systems	Multiple linear regression (MLR)	[55]
Environmental science	River water (pollution indicators, e.g., pH, conductivity, etc.)	Non-discussed in details	CA, PCA, FA, DA	[56]
Food science	Italian Barbera wines	Electronic nose, amperometric electronic tongue	PCA, LDA, CART	[57]
Food science	Honey authenticity	ICP-MS	CA, PCA	[58]
Food science	Virgin olive oil	^{1}H NMR spectroscopy	LDA, SIMCA, PLS-DA, CART	[59]

pheromone trails to guide the others to a food source. The method can be used successfully, for example, for spectral wavelength selection or clustering tasks. ACO is a metaheuristic algorithm like genetic algorithms or simulated annealing. In the ACO process several "agents" (like ants) construct iteratively different solutions for an exact problem [65].

While there are well-established methodologies and frameworks in chemometrics, there are always new problems, new challenges to be addressed. In particular, as the amount of the data generated by (hyphenated) analytical techniques increases, chemistry-related data analysis starts to approach the domain of big data. Nevertheless, extensions of the current data analysis methodologies are already being developed and successfully applied in related fields. Machine learning algorithms are succeeded with deep learning methods that can operate with extremely complex models (e.g., neural networks with many thousands or even millions of nodes). However, our care should be of similar proportion when we apply these methods, as they are more difficult to grasp (and to control). Thus, developing suitable validation practices will be of paramount importance as we travel along this road.

Essentials

- Modern chromatographic and spectrometric techniques abundantly provide digitalized data.
- Transforming data into information has become one of the most important tasks in analytical chemistry.
- Chemometrics has developed many methods that are also used in chemoinformatics to such an extent that these two fields are now highly overlapping.
- Preprocessing of data such as mean centering, standardization, normalization, transformation, and data reduction is often necessary before applying a data analysis method.
- There are unsupervised and supervised methods.
- There are qualitative methods for classification and quantitative methods for model building and regression.
- The validation of a model is essential.
- Several types of cross-validations are the most popular ones in model validation.
- Chemometric methods have been applied to a wide range of problems from analytical chemistry from analyzing cultural heritage data through analytical data of herbal medicines to data on Italian Barbera wines.

Available Software and Web Services (accessed January 2018)

- http://www.click2drug.org/index.html#ADME%20Toxicity
- https://www.researchgate.net/post/How_can_I_predict_the_insilico_ADMET_toxicity_of_a_new_drug
- http://www.acdlabs.com
- http://www.simulations-plus.com/

- http://www.fujitsu.com/jp/group/kyushu/en/
- http://www.cyprotex.com/
- https://www.scm.com/
- http://accelrys.com/
- https://www.chemaxon.com/
- http://www.systems-biology.com/sb.html
- https://preadmet.bmdrc.kr/
- http://www.compudrug.com/
- https://www.schrodinger.com/
- http://www.moldiscovery.com/
- http://www.handwrittentutorials.com/videos.php

Selected Reading

- Andersen, C.M. and Bro, R. (2010) Variable selection in regression – a tutorial. *J. Chemom.*, **24**, 728–737.
- Defernez, M. and Kemsley, E.K. (1997) The use and misuse of chemometrics for treating classification problems. *TrAC Trends Anal. Chem.*, **16**, 216–221.
- Devos, O., Ruckebusch, C., Durand, A., Duponchel, L., and Huvenne, J.-P. (2009) Support vector machines (SVM) in near infrared (NIR) spectroscopy: focus on parameters optimization and model interpretation. *Chem. Int. Lab. Sys.*, **96**, 27–33.
- Drab, K. and Daszykowski, M. (2014) Clustering in analytical chemistry. *J. AOAC Int.*, **97**, 29–38.
- Geladi, P. and Kowalski, B.R. (1986) Partial least-squares regression – a tutorial. *Anal. Chim. Acta*, **185**, 1–17.
- Rácz, A., Bajusz, D., and Héberger, K. (2015) Consistency of QSAR models: correct split of training and test sets; ranking of models and performance parameters. *SAR QSAR Environ. Res.*, **26**, 683–700.
- Wold, S., Esbensen, K., and Geladi, P. (1987) Principal component analysis. *Chem. Int. Lab. Sys.*, **2**, 37–52.
- Zupan, J. and Gasteiger, J. (1999) *Neural Networks in Chemistry and Drug Design*, Wiley-VCH, Weinheim, 380 pp.

References

[1] Bro, R. and Smilde, A.K. (2003) *J. Chemom.*, **17**, 16–33.
[2] van den Berg, R.A., Hoefsloot, H.C.J., Westerhuis, J.A., Smilde, A.K., and van der Werf, M.J. (2006) *BMC Genomics*, **7**, 142.
[3] Héberger, K., Kolarević, S., Kračun-Kolarević, M., Sunjog, K., Gačić, Z., Kljajić, Z., Mitrić, M., and Vuković-Gačić, B. (2014) *Mutat. Res. Genet. Toxicol. Environ. Mutagen.*, **771**, 15–22.
[4] Héberger, K. (2010) *TrAC Trends Anal. Chem.*, **29**, 101–109.

- [5] Defernez, M. and Kemsley, E.K. (1997) *TrAC Trends Anal. Chem.*, **16**, 216–221.
- [6] Wold, S., Esbensen, K., and Geladi, P. (1987) *Chemom. Intell. Lab. Syst.*, **2**, 37–52.
- [7] Drab, K. and Daszykowski, M. (2014) *J. AOAC Int.*, **97**, 29–38.
- [8] Kohonen, T. (1982) *Biol. Cybern.*, **43**, 59–69.
- [9] Forina, M. and Armanino, C. (1982) *Ann. Chim (Rome)*, **72**, 127–143.
- [10] Forina, M. and Tiscornia, E. (1982) *Ann. Chim. (Rome)*, **72**, 144.
- [11] Zupan, J., Novic, M., Li, X., and Gasteiger, J. (1994) *Anal. Chim. Acta*, **292**, 219–234.
- [12] Hastie, T., Tibshirani, R., and Friedman, J. (2001) Chapter 4.3 Linear discriminant analysis, in *Elements of Statistical Learning. Data Mining, Inference, Prediction*, Springer, New York, NY, pp. 84–95.
- [13] Barker, M. and Rayens, W. (2003) *J. Chemom.*, **17**, 166–173.
- [14] Kvalheim, O.M. (1987) *Chemom. Intell. Lab. Syst.*, **2**, 283–290.
- [15] Christie, O.H.J., Rácz, A., Elek, J., and Héberger, K. (2014) *J. Chemom.*, **28**, 287–292.
- [16] Breiman, L., Friedman, J., Stone, C.J., and Olshen, R.A. (1984) *Classification and Regression Trees*, Chapman and Hall/CRC press, 368 pp.
- [17] Brereton, R.G. and Lloyd, G.R. (2010) *Analyst*, **135**, 230–267.
- [18] Altman, N.S. (1992) *Am. Stat.*, **46**, 175–185.
- [19] Todeschini, R., Ballabio, D., Cassotti, M., and Consonni, V. (2015) *J. Chem. Inf. Model.*, **55**, 2365–2374.
- [20] Rácz, A., Bajusz, D., Fodor, M., and Héberger, K. (2016) *Chemom. Intell. Lab. Syst.*, **151**, 34–43.
- [21] Gramatica, P., Chirico, N., Papa, E., Cassani, S., and Kovarich, S. (2013) *J. Comput. Chem.*, **34**, 2121–2132.
- [22] Asuero, A.G., Sayago, A., and González, A.G. (2007) *Crit. Rev. Anal. Chem.*, **36**, 41–59.
- [23] de Levie, R. (1986) *J. Chem. Educ.*, **63**, 10.
- [24] Andersen, C.M. and Bro, R. (2010) *J. Chemom.*, **24**, 728–737.
- [25] Rácz, A., Vass, A., Héberger, K., and Fodor, M. (2015) *Anal. Bioanal. Chem.*, **407**, 2887–2898.
- [26] Rajalahti, T., Arneberg, R., Berven, F.S., Myhr, K.-M., Ulvik, R.J., and Kvalheim, O.M. (2009) *Chemom. Intell. Lab. Syst.*, **95**, 35–48.
- [27] Kalivas, J.H. (2012) *J. Chemom.*, **26**, 218–230.
- [28] Geladi, P. and Kowalski, B.R. (1986) *Anal. Chim. Acta*, **185**, 1–17.
- [29] Frank, I.E. and Friedman, J.H. (1993) *Technometrics*, **35**, 109–135.
- [30] Wold, S. (1993) *Technometrics*, **35**, 136–139.
- [31] Miller, A. (1990) *Subset Selection in Regression*, Chapman and Hall, London, 240 pp.
- [32] Esbensen, K.H. and Geladi, P. (2010) *J. Chemom.*, **24**, 168–187.
- [33] Hawkins, D.M., Basak, S.C., and Mills, D. (2003) *J. Chem. Inf. Comput. Sci.*, **43**, 579–586.
- [34] Hawkins, D.M. (2003) *J. Chem. Inf. Comput. Sci.*, **44**, 1–12.

[35] Hastie, T., Tibshirani, R., and Friedman, J.H. (2009) Chapter 7.10 Cross-validation, in *The Elements of Statistical Learning: Data Mining, Inference, and Prediction*, Springer, New York, pp. 214–217.

[36] Eigenvector Research staff and associates. *Using Cross-Validation*, http://wiki.eigenvector.com/index.php?title=Using_Cross-Validation (January 2018).

[37] Bro, R., Kjeldahl, K., Smilde, A.K., and Kiers, H.A.L. (2008) *Anal. Bioanal. Chem.*, **390**, 1241–1251.

[38] Efron, B. and Tibshirani, R. (1993) *An Introduction to the Bootstrap*, Chapman and Hall, New York.

[39] Wehrens, R., Putter, H., and Buydens, L.M.C. (2000) *Chemom. Intell. Lab. Syst.*, **54**, 35–52.

[40] Filzmoser, P., Liebmann, B., and Varmuza, K. (2009) *J. Chemom.*, **23**, 160–171.

[41] Gramatica, P. (2014) *Mol. Inform.*, **33**, 311–314.

[42] Rácz, A., Bajusz, D., and Héberger, K. (2015) *SAR QSAR Environ. Res.*, **26**, 683–700.

[43] Carter, S., Fisher, A., Garcia, R., Gibson, B., Lancaster, S., Marshall, J., and Whiteside, I. (2015) *J. Anal. At. Spectrom.*, **30**, 2249–2294.

[44] Munson, C.A., De Lucia, F.C., Piehler, T., McNesby, K.L., and Miziolek, A.W. (2005) *Spectrochim. Acta B*, **60**, 1217–1224.

[45] Liang, Y.-Z., Xie, P., and Chan, K. (2004) *J. Chromatogr. B*, **812**, 53–70.

[46] Roggo, Y., Degardin, K., and Margot, P. (2010) *Talanta*, **81**, 988–995.

[47] Yang, H., Irudayaraj, J., and Paradkar, M.M. (2005) *Food Chem.*, **93**, 25–32.

[48] Xie, L., Ying, Y., Ying, T., Yu, H., and Fu, X. (2007) *Anal. Chim. Acta*, **584**, 379–384.

[49] Baratieri, S.C., Barbosa, J.M., Freitas, M.P., and Martins, J.A. (2006) *J. Pharm. Biomed. Anal.*, **40**, 51–55.

[50] Fernández-Torres, R., Pérez-Bernal, J.L., Bello-López, M.Á., Callejón-Mochón, M., Jiménez-Sánchez, J.C., and Guiraúm-Pérez, A. (2005) *Talanta*, **65**, 686–691.

[51] Hahn, D.W. and Omenetto, N. (2012) *Appl. Spectrosc.*, **66**, 347–419.

[52] Custers, D., Cauwenbergh, T., Bothy, J.L., Courselle, P., De Beer, J.O., Apers, S., and Deconinck, E. (2015) *J. Pharm. Biomed. Anal.*, **112**, 181–189.

[53] Clayton, T.A., Lindon, J.C., Cloarec, O., Antti, H., Charuel, C., Hanton, G., Provost, J.P., Le Net, J.L., Baker, D., Walley, R.J., Everett, J.R., and Nicholson, J.K. (2006) *Nature*, **440**, 1073–1077.

[54] Roggo, Y., Chalus, P., Maurer, L., Lema-Martinez, C., Edmond, A., and Jent, N. (2007) *J. Pharm. Biomed. Anal.*, **44**, 683–700.

[55] Goryński, K., Bojko, B., Nowaczyk, A., Buciński, A., Pawliszyn, J., and Kaliszan, R. (2013) *Anal. Chim. Acta*, **797**, 13–19.

[56] Kowalkowski, T., Zbytniewski, R., Szpejna, J., and Buszewski, B. (2006) *Water Res.*, **40**, 744–752.

[57] Buratti, S., Benedetti, S., Scampicchio, M., and Pangerod, E.C. (2004) *Anal. Chim. Acta*, **525**, 133–139.

[58] Chudzinska, M. and Baralkiewicz, D. (2010) *Food Chem. Toxicol.*, **48**, 284–290.

[59] Alonso-Salces, R.M., Héberger, K., Holland, M.V., Moreno-Rojas, J.M., Mariani, C., Bellan, G., Reniero, F., and Guillou, C. (2010) *Food Chem.*, **118**, 956–965.
[60] Berrueta, L.A., Alonso-Salces, R.M., and Héberger, K. (2007) *J. Chromatogr. A*, **1158**, 196–214.
[61] Jaumot, J., de Juan, A., and Tauler, R. (2015) *Chemom. Intell. Lab. Syst.*, **140**, 1–12.
[62] Zupan, J. and Gasteiger, J. (1991) *Anal. Chim. Acta*, **248**, 1–30.
[63] Despagne, F. and Massart, D.L. (1998) *Analyst*, **123**, 157R–178R.
[64] Zupan, J. and Gasteiger, J. (1999) *Neural Networks in Chemistry and Drug Design*, Wiley-VCH, Weinheim, 380 pp.
[65] Shmygelska, A. and Hoos, H.H. (2005) *BMC Bioinformatics*, **6**, 30.

10 Chemoinformatics in Food Science

Andrea Peña-Castillo[1], Oscar Méndez-Lucio[1], John R. Owen[2], Karina Martínez-Mayorga[3], and José L. Medina-Franco[1]

[1] Facultad de Química, Departamento de Farmacia, Universidad Nacional Autónoma de México, Avenida Universidad 3000, Mexico City 04510, Mexico
[2] ECIT Institute, High-Performance Computing Research Group, Northern Ireland Science Park, Queens Road, Belfast, BT3 9DT, Northern Ireland
[3] Instituto de Química, Departamento de Fisicoquímica, Universidad Nacional Autónoma de México, Avenida Universidad 3000, Mexico City 04510, Mexico

Learning Objectives

- To identify common chemoinformatics methods applied in food science.
- To describe the current status, successful applications, and limitations of *foodinformatics* as an emerging research field.
- To review molecular libraries of food chemicals.
- To use analysis and visualization applications of the chemical space of food chemicals.
- To discuss structure–property relationship approaches used in food science.
- To report examples of data mining to identify food chemicals with a desired property.

Outline

10.1 Introduction, 501
10.2 Scope of Chemoinformatics in Food Chemistry, 502
10.3 Molecular Databases of Food Chemicals, 503
10.4 Chemical Space of Food Chemicals, 506
10.5 Structure–Property Relationships, 510
10.6 Computational Screening and Data Mining of Food Chemicals Libraries, 513
10.7 Conclusion, 521

10.1 Introduction

The field of chemoinformatics has been extensively developed for drug discovery applications, and it is a central piece in the pharmaceutical industry [1, 2]. However, as discussed in detail in this chapter, many of the principles used in drug development projects can also be applied in food chemistry. A limited number but rich reviews of chemoinformatic approaches implemented in food sciences have been published. For instance, in 2009 Martínez-Mayorga and Medina-Franco presented an overview of representative chemoinformatic methods employed in food chemistry [3]. More recently, Iwaniak *et al.* reviewed the application of chemometrics and chemoinformatics in the analysis of biological active peptides from food sources. In that work, the authors discussed

Applied Chemoinformatics: Achievements and Future Opportunities, First Edition.
Edited by Thomas Engel and Johann Gasteiger.
© 2018 Wiley-VCH Verlag GmbH & Co. KGaA. Published 2018 by Wiley-VCH Verlag GmbH & Co. KGaA.

the principles of artificial neural networks, principal component analysis, partial least squares, and quantitative structure–activity relationship (QSAR) methods to analyze bioactive peptides from food [4]. The book *Foodinformatics: Applications of Chemical Information to Food Chemistry* was published recently [5]. The goal of that book was to discuss basic concepts of chemical information commonly used in the pharmaceutical industry but in the context of food science. The principles of foodinformatics and specific examples are further discussed in this chapter.

The goal of this chapter is to discuss the main topics of chemoinformatics in the field of food science. The manuscript also covers future developments and problems that still need to be solved. After this introduction, it presents a brief discussion of the general scope of chemoinformatics in food chemistry. The next section presents examples of molecular databases of food chemicals, which is followed by a discussion of chemical space. The next two sections discuss structure–property relationships (SPRs) and data mining of libraries of food chemicals, respectively. The last main parts of the chapter collect Essentials, Selected Reading, Available Software and Web Services.

10.2 Scope of Chemoinformatics in Food Chemistry

The study of the information contained in chemical structures is at the heart of chemoinformatics. In principle, any collection of chemicals can be studied for that purpose. As mentioned above, chemoinformatics has found several major uses in the pharmaceutical industry, but other applications include agricultural, environmental, and food chemistries.

Food, defined as any substance consumed to provide nutritional support to the body, is a complex matrix generally composed of dozens of constituents such as water, fats, proteins, vitamins, and minerals and is derived from animal or plant sources. In addition, additives are incorporated into food; the Food and Drug Administration (FDA) in the United States indicates that "any substance that is reasonably expected to become a component of food is a food additive…" Food additives are used for different purposes such as antioxidants, antifoaming, bulking, coloring, flavors, humectants, preservatives, stabilizers, and so on. How do these properties relate to the chemical structures? This is an interesting question to address. Molecules can be categorized as Generally Recognized As Safe (GRAS). Before 1958, molecules were considered GRAS based on their common use in food, which requires a substantial history of consumption for food use by a significant number of consumers. After 1958, the general recognition of safety through scientific procedures requires the same quantity and quality of scientific evidence as is required to obtain approval of the substance as a food additive and ordinarily is based upon published studies, which may be corroborated by unpublished studies and other data and information. As discussed throughout this chapter, the GRAS list is becoming a reference database for chemoinformatic studies.

To be able to perform chemoinformatic studies, it requires to have well-characterized chemical structures. For food chemicals this is not the norm. In fact, in several cases food additives are used as mixtures, for example, as oil extracts. There are at least four particularly relevant features in food chemicals that need to be considered when considering chemoinformatic studies for food chemicals.

- A premise is made that food chemicals are safe for human consumption. Information contained in food chemical databases could be used as reference in chemoinformatic studies, just like marketed drugs are. Notably, food additives are considered safe for human consumption only at a specific concentration and for a given purpose. Comparisons of molecular properties of food chemicals to databases routinely used in pharma-related fields are starting to emerge, but clearly, much work is yet to be done and it is expected that such information will impact and bring new ideas on safety based on molecular properties.
- Concerning safety, a chemical added to food will be considered safe at a given concentration, at the food matrix used for its commercialization, and the process it will undertake before consumption. For example, temperature, humidity, and pH need to be considered for live shelf. Finally, if the food product will be used at high temperatures such as in fried or oven processes, decomposition to toxic components needs to be avoided.
- Other than pure water, food contains several components and in different concentrations; thus synergy and simultaneous interactions with different biological targets routinely occur.
- Customer preferences depend, among others factors, on cultural use. Therefore, the scope of the study should consider the regional nature of the customer preferences.

The evolution of use of food additives parallels the historical need to preserve food supplies. Modern food industry makes use of additives with different purposes such as conservation, taste, patentability, innovation, nutritional enrichment, and so on. Just as Dr. Harvey Washington Wiley, the first commissioner of the FDA, realized the need for food safety and standardization back in 1902, awareness of the need and feasibility of data management of food chemicals is becoming apparent in recent years. In that sense, chemoinformatic methodologies could be used to organize, store, handle, and share this information. However, important features that make the food industry regional and unique will remain.

10.3 Molecular Databases of Food Chemicals

Many public and commercial databases are available for drug discovery, but not so many for food science. Maybe one of the reasons is that food science is too broad that it is difficult to compile all possible compounds in a single database. For instance, a complete Food Science database could include information regarding additives, flavorings, odorants, functional ingredients, and so on. Since compiling such an amount of information is not an easy task, it is more common to find databases containing subsets of all this information [6]. Table 10.1 lists some of

Table 10.1 Examples of compounds and related databases of interest in food science.

Database	Website (January 2018)	Size and comments
Flavors and Scents		
BitterDB	bitterdb.agri.huji.ac.il/bitterdb/	More than 600 bitter compounds
SuperScent	bioinf-applied.charite.de/superscent/	Approximately 2,300 scents
SuperSweet	bioinf-applied.charite.de/sweet/	More than 8,000 sweet molecules
RIFM/FEMA Fragrance and Flavor database	www.rifm.org/index.php	More than 5,100 materials
International Organization of the Flavor Industry (IOFI) database	www.iofi.org	Over 3,500 flavoring substances
Flavor-base database of flavoring materials and food additives	www.leffingwell.com/flavbase.htm	
Volatile compounds in food database	www.vcf-online.nl/VcfHome.cfm	More than 8,200 compounds with odor information
Database of essential oils	www.leffingwell.com/baciseso.htm	More than 4,100 quantitative analyses of essential oils
Allured's Flavor and Fragrance Materials	dir.perfumerflavorist.com/main/login.html;jsessionid=9EC896163AA3A88037DD0BC0E2CE6F65	No structural information is provided
Flavornet database	www.flavornet.org	Freely available
	acree.foodscience.cornell.edu/flavornet.html	More than 730 flavor molecules
Good Scents Company Information System	www.thegoodscentscompany.com/index.html	Public domain of different chemicals related to food
FooDB	http://foodb.ca	More than 26,000 chemicals that can be found in food
Phenol-explorer	www.phenol-explorer.eu/	More than 500 different polyphenols in over 400 foods
Lipids and carbohydrates		
LipidBank	www.lipidbank.jp/	Information on more than 7,000 lipids
LIPIDMAPS	www.lipidmaps.org/	Contains over 40,000 unique lipid structures
GlycomeDB	www.glycome-db.org/	A carbohydrate structure metadatabase

(*Continued*)

Table 10.1 (Continued)

Database	Website (January 2018)	Size and comments
Functional Glycomics Gateway	www.functionalglycomics.org	Contains different information related to glycans, for example, structures, binding proteins
Regulatory information		
Joint FAO/WHO Expert Committee on Food Additives (JECFA) Database	apps.who.int/food-additives-contaminants-jecfa-database/search.aspx	Evaluations of flavors, food additives JECFA has performed
Food additives	https://webgate.ec.europa.eu/foods_system/	Food additives approved for use in food in the EU and their conditions of use
EC Flavor Register	eur-lex.europa.eu/JOHtml.do?uri=OJ:L:2012:267:SOM:EN:HTML	More than 2,500 flavoring substances that can be used in food

the available databases, which contain information that is related to food science. Additional information is listed in the section Available Software and Web Services at the end of this chapter. In Table 10.1 databases were divided into three categories:

1) Food additives databases that comprise information on flavorings, scents, odorants, colorants, or preservatives.
2) Biomolecules databases containing information on molecules such as lipids, carbohydrates, vitamins, or other biomolecules that could be used as food components.
3) Regulatory databases that are usually created and maintained by regulatory agencies to keep an updated record of chemicals approved as food components. Other food-related databases and resources have recently been reviewed [5].

Food additives are an important part of food industry since they are used to preserve or add color, flavor, and texture. There are several databases containing information regarding food additives, for example, the Research Institute for Fragrance Materials (RIFM)/FEMA Fragrance and Flavor database, which contains information on more than 5000 materials. This commercial database represents one of the most comprehensive resources, since it contains information regarding the chemical structure (e.g., CAS numbers and SMILES representations), physicochemical properties, synonyms, and in some cases even health and environmental studies. Other databases are centered on more specific information such as the BitterDB and the SuperSweet databases that contain information on bitter and sweet flavoring agents, respectively. An interesting feature of these public databases is that they provide in addition to the basic structure and physicochemical data also biological information on the receptor protein, that is, the taste receptor (TAS) type 1 or 2 for sweet or bitter compounds, respectively. In a similar way, databases focused on scents are available such as SuperScent or Flavornet. Both open source databases contain molecules that affect flavor; in the former, compounds are classified based on functional groups and scaffolds, whereas in the latter they are organized based on odor classes. Other databases containing food additives can be found in Table 10.1.

Other important sources of food chemical information are regulatory agencies. Usually, these agencies maintain and update databases of chemicals approved for being used as food additives. Generally recognized as safe (GRAS) compounds can be regarded as the most popular set of molecules compiled by a regulatory agency to be used as food additives. This set of compounds can be found in different databases such as the FEMA GRAS database (http://www.femaflavor.org/). It is worth mentioning that the list of GRAS compounds is updated every 1 or 2 years, and the most recent update (GRAS 27) was released in August 2015. Interestingly, not all chemicals approved by other agencies can be considered as GRAS compounds since every agency implements different requirements for approval. An alternative database used for regulatory purposes is "Everything Added to Food in the United States" (EAFUS) databases maintained by the US FDA Center for Food Safety and Applied Nutrition (CFSAN), which lists 3968 substances regulated by FDA, color additives, some GRAS compounds and prior-sanctioned substances (see URLs in Available Software and Web Services section and Table 10.1). Similar databases can be found for other countries or regions such as the food additives database and the EC Flavor Register for the European Union, as well as the Joint Food and Agriculture Organization/World Organization Expert Committee on Food Additives (JECFA) database. Something to keep in mind is that most databases from regulatory agencies are focused on legal aspects, and only few of them provide structural or physicochemical information on the compounds.

10.4 Chemical Space of Food Chemicals

10.4.1 General Considerations

The concept of chemical space is intuitive and has a strong analogy with the cosmic universe [7]. Several definitions of chemical space and visualization methods have been proposed in the literature [8, 9]. Regardless of a precise definition, there is a general agreement that the chemical space highly depends on the structure representation and parameters to define the space (see below).

Chemoinformatic analysis of the chemical space of compound libraries provides a systematic and consistent way to compare the space with other reference collections. Indeed, it is a first major step in the characterization of a molecular library in terms of contents and structural diversity [10]. Chemoinformatic characterizations are usually performed following two general approaches that work best if they are applied in combination:

- Quantitative analysis of all dimensions used to define the space.
- Visual representation of the chemical space along with a qualitative interpretation. To this end, there are several approaches that have been reviewed elsewhere [8, 9].

As mentioned above, the study of the chemical space either qualitatively or quantitatively highly depends on the structure representation (see also *Methods Volume*, Chapter 10). Common representations used to analyze chemical space can be classified into three major groups. These representations are schematically depicted in Figure 10.1 and are briefly described below:

- Whole molecular properties and atom counts; common properties of pharmaceutical relevance are molecular weight, topological surface area, number of hydrogen-bond donors and acceptors, number of carbon, nitrogen, oxygen atoms; and number of aliphatic and aromatic rings, to name a few. The choice of these properties or basically any other continuous property (for instance, energy, charge distribution, etc.) depends on the specific project.
- Molecular scaffolds; frequently obtained by removing the side chains of the chemical structures (such as the Murcko scaffolds) or further disconnecting the rings (scaffold tree).
- Structural fingerprints; including fingerprints where the number of features are independent of the structure such as the dictionary-based MACCS keys [11] or the fingerprints where the number of features depend on the structure such as the extended connectivity fingerprints.

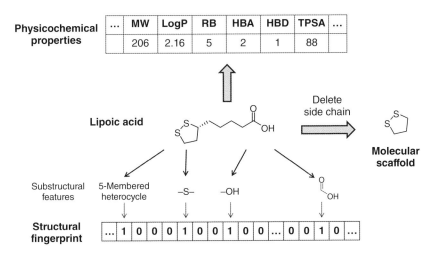

Figure 10.1 Common structure representations of chemical structures to analyze chemical space. The representation of lipoic acid is used as an example.

The choice of the structure representation is at the core of basically any chemoinformatic application. It has been proposed that a comprehensive characterization of chemical compounds should involve more than one criterion since each structure representation has its own advantages and disadvantages. For instance, physicochemical properties and feature counts are straightforward to interpret with the downside that they may be unspecific, that is, a pair of different compounds can have the same molecular weight. Molecular scaffolds are also intuitive and straightforward to understand by chemists. A drawback is that the information of the side chains is missing. Finally, structural fingerprints capture information of the entire structure with the downside that they are more difficult to interpret [12].

10.4.2 Chemical Space Analysis of Food Chemical Databases

Both quantitative and visual approaches using different molecular representations have been employed to analyze the chemical space of databases of food chemicals. Two representative case studies are described in this section; these are not intended to be an exhaustive account of all analysis reported in the literature, but should illustrate the combination of different approaches.

10.4.2.1 Chemical Space of GRAS Compounds

A comprehensive analysis of the recently updated FEMA GRAS list with 2244 flavoring substances (discrete chemical entities only) has been reported [13]. The set of GRAS compounds was compared to two separate collections of natural products from two different vendors containing 2449 and 467 compounds, respectively. GRAS chemicals were also compared with 1713 approved drugs obtained from DrugBank [14]. In addition, a molecular library with 10,000 molecules typically used in general screening, for example, high-throughput screening campaigns, was included in the comparison. Compound databases were assessed using multiple criteria, including molecular properties of pharmaceutical relevance, rings, atom counts, and two types of structural fingerprints of different design: MACCS keys (166-bits) as implemented in the software Molecular Operating Environment (MOE) – Chemical Computing Group, and radial fingerprints as implemented in the software Canvas (Schrödinger). The radial fingerprints used are equivalent to the extended connectivity fingerprints available in other software [15]. A number of data mining and visualization methods were implemented to portray and analyze the data. The main methods were box plots (used to characterize the distribution of the properties and feature counts), principal component analysis (employed to generate a visual representation of the chemical space), and self-organizing maps. Major conclusions of that study (focused on the characterization of the FEMA GRAS list) were as follows:

- The lipophilicity profile of GRAS compounds, a key property to predict human bioavailability, was similar to that of approved drugs.
- Overall, the molecular size of the GRAS flavoring substances was smaller than the molecules of other reference databases analyzed.
- Several GRAS chemicals overlap with an ample region of the property space of drugs.
- The GRAS list had a high structural diversity that was comparable to that of most approved drugs, natural products, and libraries of screening compounds.

That study represented one of the first analyses published toward the use of the distinctive features of flavoring chemicals contained in the GRAS list and natural products to systematically search for compounds with potential health-related benefits [13].

10.4.2.2 Visual Representation of Chemical Space

Reymond and his research group has recently presented a comprehensive analysis of the chemical space of flavoring molecules that were retrieved from four public databases, namely, SuperScent, Flavornet, BitterDB, and SuperSweet [16]. Of note, the SuperSweet set analyzed contained 342 compounds, including many glycosides with demonstrated or potential sweet taste. BitterDB had 606 compounds with bitter taste, containing several alkaloids. In order to explore major differences (or coincidences) in physicochemical properties of flavoring molecules with other types of compounds, the SuperScent, Flavornet, BitterDB, and SuperSweet databases were compared with drug-like compounds with reported biological activity (available in ChEMBL (https://www.ebi.ac.uk/chembl/)), commercial compounds typically used for bioactive screening (retrieved in the ZINC database (http://zinc.docking.org/)), and virtual compounds, that is, theoretically possible molecules up to 13 atoms (obtained from the chemical universe database, GDB-13). The databases were compared by using properties such as log P (as a measure of polarity), number of heavy atoms, number of heteroatoms (in particular, oxygen, nitrogen, and sulfur atoms), and number of cycles (as a measure of structural rigidity). Visualization of the chemical space was done with principal component analysis of 42 molecular quantum numbers. This set of descriptors were previously developed by the same authors to analyze a number of databases [17] and are composed of integer value descriptors of atom counts, bonds, polar groups, and topological features. A visual representation of the chemical space was also generated with principal component analysis of simplified molecular-input line-entry system fingerprint (SMIfp) that counts the occurrences of characters occurring in the SMILES representation of molecules [18].

Analysis and classification of the molecules in the chemical space provided an overall understanding of the different molecular classes. In particular, it was concluded that flavor molecules are located in a defined region of chemical space as represented by the descriptors considered in this study. The distinctive position is due to the relative small size with limited number of polar functional groups. Furthermore, the analysis of the chemical space of flavoring molecules provided a conceptual framework to understand the chemical diversity of taste and smell suggesting approaches to identify new flavors [16]. The authors of that work also concluded that, at least in principle, a detailed SAR analysis of the chemical space could help in revealing the general principles underlying the genetic diversity of the olfactory system [16].

A visual representation of the chemical space of food-related chemicals and other reference compound collections is shown in Figure 10.2. Specifically, the plot shows the chemical space of 2133 EAFUS compounds and 1477 GRAS chemicals. The reference collections are 1798 drugs approved for clinical use obtained from DrugBank and 549 compounds tested as inhibitors of DNA methyltransferase 1 (DNMT1), which is an epigenetic target (see below) [19]. Chemical compounds were represented using MACCS keys (166-bits) as generated with MOE. The visual representation of the chemical space was performed with the generative topographic mapping (GTM) algorithm [20, 21] implemented in MATLAB. In order to produce this plot, the same parameters as previously reported were applied [21] (in particular using the following settings: Parameter: R_G ceil(sqrt(N/10)) and Parameter: L_G ceil(sqrt(N/10)). Interestingly,

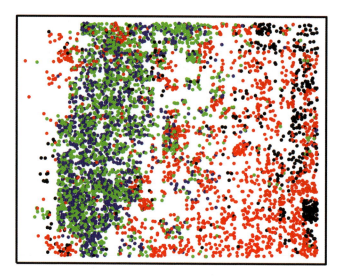

Figure 10.2 Generative topographic mapping (GTM) visualization of the chemical space of 1477 generally recognized as safe (GRAS) compounds (blue), 2133 Everything Added to Food in the United States (EAFUS) (green), 1798 approved drugs from DrugBank (red), and 549 compounds tested as DNMT1 inhibitors (black). Molecules are represented using MACCS keys fingerprints (166-bits). The figure was generated using compound databases prepared by Mariana González-Medina.

in this visual representation of the chemical space, there is an overall separation in chemical space of GRAS and EAFUS compounds (left-hand side of the plot) with approved drugs and DNMT1 inhibitors (right-hand side of the plot). In addition, the visual representation of the chemical space based on MACCS keys/GTM suggests that

- GRAS and EAFUS have a large overlap in chemical space, indicating that the chemical structures are structurally similar (as captured by MACCS keys).
- Compounds tested as DNMT1 inhibitors overlap with the chemical space of drugs and cluster in a restricted space.
- There is quite a limited but intriguing overlap in the chemical space of GRAS/EAFUS with drugs/DNMT1 inhibitors, that is, green and blue data points at the far right-hand side of the plot. Likewise, there are black points on the left-hand side of the GTM plot.

Of course direct/standard similarity searching studies can be conducted with the compounds in these datasets to identify structurally similar molecules (for instance, using MACCS keys and the Tanimoto coefficient) as reported previously [22].

10.5 Structure–Property Relationships

Understanding the SPR of chemical compounds is a common practice of virtually all chemists including, of course, food chemists [23]. In the food industry there is a broad range of properties of interest. A number of chemometrics and chemoinformatics methods are being used in the food sciences. As mentioned above, Iwaniak *et al.* recently reviewed the application of artificial neural networks, principal component analysis, and partial least squares to analyze the SPR of food derived bioactive peptides [4]. Siebert has reviewed applications of chemometrics in brewing [24].

Tan and Siebert have reported QSAR studies of alcohol, ester, aldehyde, and ketone flavor thresholds in beer [25]. Several chapters in the book *Food-Informatics* cover examples of studies aimed at exploring the SPR of food chemicals with different biological activities such as modulators of peroxisome proliferator-activated receptors (PPARs) gamma, dipeptidyl peptidase (DPP)-IV, and cancer epigenetics [5]. In this section we discuss briefly two case studies that illustrate the application of chemical information to elucidate SPRs in food science.

10.5.1 Structure–Flavor Relationships and Flavor Cliffs

In order to explore the complex associations between flavor perception and the chemical structure of molecules contained in a large flavor database, Martínez-Mayorga *et al.* developed a fingerprint-based representation for an initial set of 4181 molecules obtained from the Flavor-Base Pro© 2010 database [26]. In that database the flavor descriptions were composite descriptions that were collected from different sources over the course of more than 40 years. A key aspect of that study was to establish a consistent and valid flavor descriptor that is schematically illustrated in Figure 10.3. In order to achieve that goal, the flavor descriptors were referenced against a detailed and authoritative sensory lexicon (American Society for Testing and Materials (ASTM) publication DS 66) that included 662 flavor attributes. Using clustering and principal component analysis, it established an association between the descriptors and identified those descriptors that are mutually correlated.

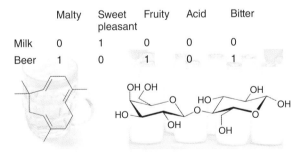

Figure 10.3 Schematic fingerprint-based representation of flavor descriptors. The figure shows five representative flavor descriptors. The descriptors that are commonly obtained from sensory analysis can be encoded in a binary fingerprint. Molecules contained in foods such as humulene and lactose found in beer and milk, respectively, could be related to the flavor descriptors. (Figure courtesy of Karla Nohemí Palma Cruz.)

As part of the analysis of flavor similarity of the molecules, a clustering approach was implemented to rapidly select chemical structures with high or low flavor similarity. Also, the principles of activity landscape modeling [27] were adapted to systematically find associations between structure similarity and flavor similarity. To that end, the structure–activity (property) similarity maps [28], broadly used to explore the SAR of data sets in drug development, were easily adapted to find such associations and identify flavor cliffs, that is, a pair of chemical compounds with high structure similarity but very different flavor (low flavor similarity) as illustrated in Figure 10.4. By analogy with activity cliffs, flavor cliffs can be valuable to guide the design of novel flavor chemicals.

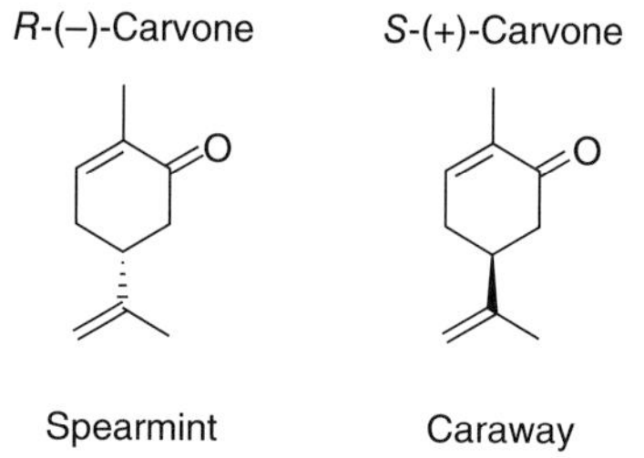

Figure 10.4 Example of a typical flavor cliff: pair of compounds with high structure similarity but very different flavor.

One of the general outcomes of the work of Martínez-Mayorga et al. was to show that common chemoinformatics methods can be adapted to systematically analyze flavor information stored in molecular databases. The study also opened an avenue for conducting additional studies to explore structure–flavor relationships using, for instance, alternative structural representations, selecting groups of compounds from that database in a computational manner, and implementing different classifications methods [26].

10.5.2 Quantitative Structure–Odor Relationships

In an independent study, Kermen et al. explored the association between chemical structure and odor [29] finding that "molecular complexity provides a framework to explain the subjective experience of smells." The authors of that work evaluated the quantitative relationship between the structural complexity of 411 odorants selected from the standardized Arctander atlas [30], and the number of olfactory notes they evoked for experts and nonexperts (where an olfactory "note" is a smell being described as "green," "woody," "tobacco," etc.). The Arctander atlas includes descriptions by experts referring to various chemosensory qualities. The 74 olfactory notes selected by Chastrette et al. [31] were used as a reference list to measure the number of olfactory notes evoked by each odorant. Molecular complexity was measured using an index that takes into account bond connectivity, diversity of non-hydrogen atoms, and symmetry. The complexity values were taken from PubChem where the molecular complexity is computed using the equation developed by Hendrickson et al. [32]. It was concluded that there is a strong association between molecular complexity and the olfactory system. In particular, it was found that the more structurally complex a monomolecular odorant is, the more numerous the olfactory notes it evokes. It was also found that odorants with low structural complexity were also regarded as more unpleasant. It was concluded that there is a strong association between molecular complexity and the olfactory system [29]. This study represents an additional contribution of computational methods toward an understanding between the chemical structures and olfactory receptors at the molecular level [33]. It remains to explore the putative structure–odor relationships with additional metrics of molecular complexity. Of note, González-Medina *et al.* [34] recently quantified the molecular complexity of GRAS compounds using multiple metrics including the fraction of chiral centers, the fraction of sp^3 carbon atoms, and molecular globularity. Interestingly, it was found that GRAS chemicals are more complex than commercial screening compound and approved drugs. One of the hypotheses that emerged from that study was that molecular complexity could also be associated with compound safety. It remains to further assess this hypothesis with different molecular descriptors to quantify molecular complexity and different molecular databases [34].

10.6 Computational Screening and Data Mining of Food Chemicals Libraries

Another promising feature of chemoinformatics in food science is to systematically identify chemical compounds with a desired property [16]. For instance, it is well known that many food chemicals, obtained from various sources, can interfere with a number of biological targets. Typical examples include [16] curcumin isolated from turmeric, genistein obtained from soybean, resveratrol from grapes, and polyphenols from green tea, berries, and cocoa. Likewise, there are a number of compounds present in natural products or food that are known to be related to unwanted effects. A systematic exploration of the associations of food chemicals with biomolecular targets would require an in-depth knowledge of interactions at the molecular level. Despite the fact that a direct approach would be to experimentally test the affinity and biochemical effect of food chemicals with pharmaceutical relevant targets (including the off- or anti-targets), this approach is rather unfeasible not only because of the high cost but also because of the availability of enough samples of the target proteins to perform the experimental tests. Therefore, chemoinformatic approaches play a pivotal role to systematically dissect the food-related chemogenomic space [16].

Using one or more structure-based or ligand-based computational approaches, it is possible to filter compound datasets with a large number of compounds with promising potential to have the desired activity. The compounds need to be experimentally screened to validate the hypothesis. Experimentally validated hits are quite helpful to provide feedback to the computational technique and refine the method in subsequent iterations. This approach called virtual screening is becoming more and more used with success [35].

In drug discovery applications a broad variety of chemical databases are employed such as collections of molecules used in high-throughput screening (e.g., commercial screening libraries), combinatorial libraries, focused or target-oriented synthesis collections, or libraries of natural products. In computer-aided drug repurposing campaigns, databases of approved drugs are used [36]. Essentially the same methods and principles of virtual screening used for drug discovery projects can be implemented to screen databases of food chemicals. Such methods have been extensively discussed in the literature, and the interested reader is directed to such manuscripts [35, 37]. As in any virtual screening campaign, an experimental validation of the virtual screening hits is required.

Despite the fact that virtual screening of databases of food chemicals is not a common practice in food science (at least at present), there are cases reported in the literature that are briefly discussed in this section. Table 10.2 summarizes representative studies that illustrate the efforts of the scientific community for systematically identifying food chemicals with biological activity.

Figure 10.5 shows the chemical structures of representative food and flavor-related chemicals identified from chemoinformatic-driven approaches.

Table 10.2 Examples of virtual screening of food chemical databases.

Study	Chemoinformatic approach	Main outcome	References
Anticonvulsant drug candidates for the treatment of P-glycoprotein-mediated refractory epilepsy	Virtual screening using 2D classifiers, topological models, and molecular docking	EDTA, thioctic acid, sorbitol and, mannitol were active in a maximal electroshock seizure test	[38]
Anticonvulsant effect of sweeteners food preservatives	Virtual screening using linear discriminant functions and ensemble classifiers, molecular docking	Well-known sweeteners and preservatives were active in a maximal electroshock seizure test	[39]
GRAS flavoring substances as modulators of histone deacetylases	Similarity-based virtual screening using MACCS keys/Tanimoto	Nonanoic acid and 2-decenoic acid inhibited histone deacetylase 1 at the micromolar level	[40]
Food chemicals as potential modulators of DNA methyltransferase	Analysis of chemical space and similarity searching	Putative compounds as epigenetic modulators were identified	[22]

10.6.1 Anticonvulsant Effect of Sweeteners and Pharmaceutical and Food Preservatives

With the aim of identifying anticonvulsant drug candidates for the treatment of P-glycoprotein-mediated refractory epilepsy, a virtual screening of DrugBank and ZINC databases was performed using several filters that included the following [38, 41]:

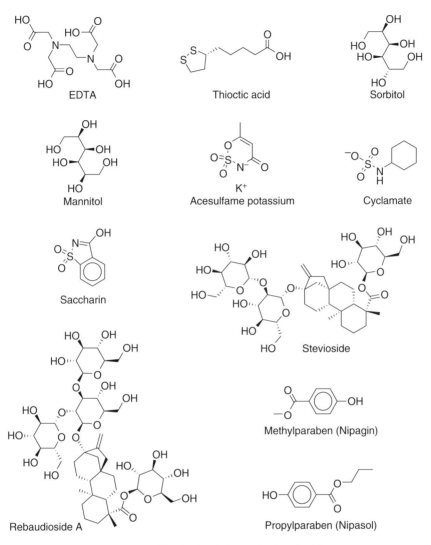

Figure 10.5 Chemical structures of food-related chemicals identified from computational-driven approaches mentioned in Section 10.6.

- Group of 2D classifiers able to differentiating P-glycoprotein substrates from non-substrates.
- Topological model able to identifying anticonvulsants active in a maximal electroshock seizure (MES) test.
- Molecular docking with a 3D structure of human P-glycoprotein. The 3D structure was derived from a homology model built from the crystal structure of P-glycoprotein of mouse.

After applying all virtual screening filters, 10 compounds with diverse structures were acquired for pharmacological testing. Additional criteria to select compounds were structural diversity, accessibility, price, and previous use as either drugs or food additives. All selected compounds showed activity in an MES test. EDTA that is highly used as additive in cosmetics and food and thioctic (or α-lipoic) acid, which is a dietary supplement (Figure 10.5), were among the compounds that showed activity in this test.

Two additional validated hits were sorbitol and mannitol (Figure 10.5). Both are permitted as food additives in both America and Europe. Interestingly, the same authors of that work previously identified other nonnutritive sweeteners as compounds with anticonvulsant activity, namely, acesulfame potassium, cyclamate, and saccharin (Figure 10.5) [39]. The activity of all these compounds further supported the so-called sweetener hypothesis proposed by the same authors. The hypothesis establishes a relationship between the receptor that unchains sweet response in mouth and some of the molecular targets of antiepileptic medications (presumably, metabotropic glutamate receptors). More recently, isolated stevioside and rebaudioside A, two nonnutritive sweeteners approved by the JECFA, were tested in the MES test giving positive results [42].

A validated discriminant function able to differentiate/classify anticonvulsant and non-anticonvulsant compounds in an MES test was applied to virtual screen 10,250 compounds from the Merck Index 13th (excluding inorganic compounds) [43]. The discriminant function was constructed using 2D descriptors calculated with the Dragon software. The hit compounds were tested in mice in the MES test using intraperitoneal administration. One of the experimentally validated hits was methylparaben (Nipagin) (Figure 10.5) that is a preservative widely used in food, cosmetics, and pharmaceutics. Based on these findings, propylparaben (Nipasol), another preservative structurally related to methylparaben was also tested in the same assay. Both methyl and propylparaben were found active in the MES test in mice at doses of 30, 100, and 300 mg/kg.

10.6.2 Mining Food Chemicals as Potential Epigenetic Modulators

Epigenetics is the study of the elements that participate in the regulation of the nucleosome-chromatin as determinants of gene expression. Epigenetic information is crucial for eukaryotic organisms as it impacts a broad range of biological processes from gene regulation to disease pathogenesis [44–46]. These epigenetic patterns include interacting components at transcriptional level, DNA methylation and histone modifications, and at the posttranscriptional level, RNA interference [47]. Interestingly, there is a strong connection between epigenetic and metabolism as there are many enzymes, substrates, and cofactors that are common in metabolic and epigenetic pathways/targets as elegantly shown by Del Rio and Da Costa [48]. Indeed, it is known that there are a number of natural and dietary components that have been identified as being able to interfere with epigenetic and metabolic mechanisms. A rich analysis of such compounds is discussed elsewhere [16]. Since there is a strong link between metabolic and epigenetic mechanisms, it is common that small molecules have polypharmacological effects, targeting those enzymes that use common substrates and cofactors such as S-adenosylmethionine (SAM), α-ketoglutarate, flavin adenine dinucleotide (FAD), acetyl coenzyme A, and nicotinamide adenine dinucleotide (NAD).

The term "nutriepigenomics" describes the study of nutrients and their effects on human health through epigenetic modifications defined as stable heritable patterns of gene expression occurring without changes in the DNA sequence. Nutrigenomics is a discipline within nutritional genomics that explores the effects of food on gene expression [49]. Chemoinformatics is used to identify food chemicals with the potential ability to modulate epigenetic targets, and it is part of the emerging field of Epi-informatics [50].

10.6.2.1 GRAS as Modulators of Histone Deacetylases (HDACs)

One example of the computational-guided identification of bioactive compounds in food chemicals is represented by the similarity-based virtual screening of molecules in the FEMA GRAS list that were compared to approved antidepressant drugs [40]. In that work, the chemical structures of 4600 GRAS flavoring substances were compared with 32 approved antidepressants using MACCS keys (166 bits) and the Tanimoto coefficient. The GRAS flavoring substances considered discrete chemical entities only, and the list was expanded to include all possible stereoisomers. The candidate compounds, that is, hits from similarity searching, were selected as the nearest neighbors of the approved antidepressants following four steps:

- For each of the 32 antidepressant molecules, the similarity to each compound in the expanded GRAS dataset was calculated.
- For each compound in the GRAS set, the maximum of the 32 Tanimoto similarities was computed.
- The GRAS compounds were sorted by maximum similarity.
- The compounds with the largest similarities were selected.

Valproic acid was the antidepressant drug most similar to the FEMA GRAS list, followed by atomoxetine and maprotiline (Figure 10.6). Based on the hypothesis that the inhibition of histone deacetylase-1 (HDAC1) may be associated with the efficacy of valproic acid in the treatment of bipolar disorder, the HDAC1 inhibitory activity of selected GRAS flavoring substances with high structural similarity to valproic acid was evaluated *in vitro*. The GRAS compounds nonanoic acid and 2-decenoic acid (Figure 10.6) inhibited HDAC1 at the micromolar level, with a potency comparable to that of valproic acid under the same assay conditions. The results of that study showed that despite the fact that GRAS compounds likely do not exhibit strong enzymatic inhibitory effects at the concentrations typically employed in foods and beverages, GRAS chemicals are able to bind, albeit weakly, to a major epigenetic target. More studies on bioavailability, toxicity at higher concentrations and off-target effects are warranted. Note, however, that GRAS flavor molecules are regarded as safe when used at or below the levels approved for foods and beverages. That work was a proof of concept that similarity searching followed by experimental evaluation can lead to the identification of GRAS chemicals with possible biological activity [40].

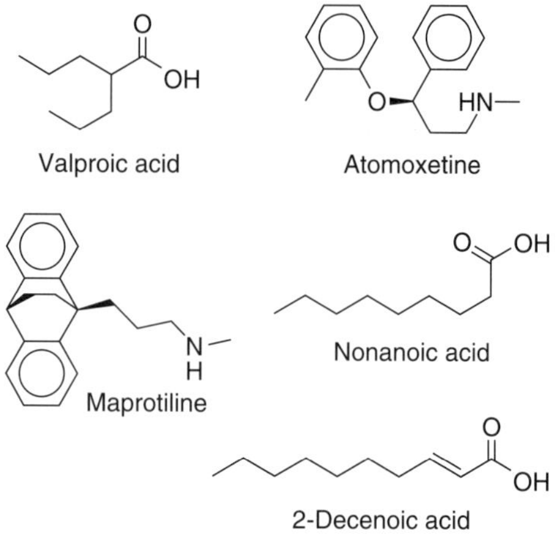

Figure 10.6 Chemical structures of compounds associated with inhibition of histone deacetylases and discussed in Section 10.6.2.1.

10.6.2.2 Mining Food Chemicals as Potential Modulators of DNA Methyltransferases

DNA methylation is an epigenetic modification that occurs after replication. In mammals, it is essential for normal embryonic development and plays important roles in the regulation of gene expression, X chromosome inactivation, chromatin modification, silencing of endogenous retroviruses, and aberrant silencing of tumor-suppressor genes in cancer [51]. In cancer, epigenetic silencing through methylation leads to aberrant silencing of genes with tumor-suppressor functions. This process depends on DNA methyltransferases (DNMTs), enzymes that catalyze the addition of methyl groups to the 5′ carbon of the cytosine residues [52]. DNA methylation occurs by five different enzymes of which DNMT1 is involved in several processes during mammalian development such as stem cell biology, cell proliferation, organ development, senescence, and oncogenesis [53]. Since exposure to external factors seem to have an important participation in abnormal methylation patterns, it has been suggested that a regular ingest of DNA demethylating agents could have a chemopreventive effect. Interestingly, many bioactive dietary components have shown promising results in direct or indirect inhibition of DNMT activity in cancer prevention and therapy [54]. For instance, (−)-epigallocatechin gallate (EGCG) is a major element of green tea with a documented activity as DNMT inhibitor [54]. Since most of the DNA demethylating agents of natural origin have been identified by serendipity, it is hypothesized that computational studies can be used to identify systematically food chemicals and natural products as potential epigenetic modulators [55].

One of the first steps toward the systematic identification of epigenetic modulators of food chemical sources is to characterize the chemical space of food chemicals in terms of structural diversity, profile of physicochemical properties and scaffold content and diversity (see above). With this goal, a curated collection of 2133 EAFUS compounds and 1477 GRAS chemicals was compared to 1798 drugs approved for clinical use, and 549 compounds tested as inhibitors of DNMT1 [19]. The GRAS and EAFUS collections were profiled in terms of six physicochemical properties commonly used in pharmaceutical applications (e.g., molecular weight, number of rotatable bonds, number of hydrogen-bond donors, number of hydrogen-bond acceptors, topological surface area, and log octanol/water partition coefficient). To illustrate this point, the distributions of physicochemical properties are depicted as box plots in Figure 10.7. Although a detailed description of the distribution of the properties is beyond this chapter, it was observed that the distribution of the molecular weight of EAFUS and GRAS was similar; furthermore, compounds in both collections are, in general, smaller than approved drugs and inhibitors of DNMT1. Remarkably, EAFUS

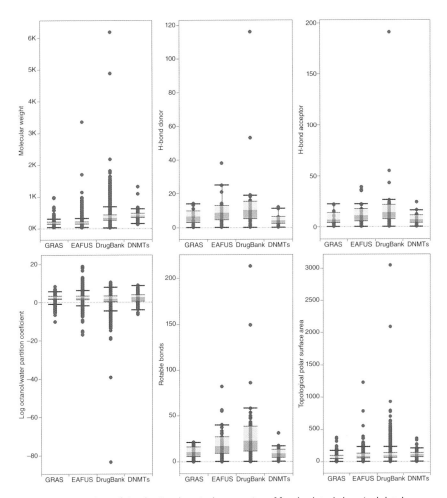

Figure 10.7 Box plots of six physicochemical properties of food-related chemical databases (GRAS and EAFUS), approved drugs (DrugBank), and inhibitors of DNA methyltransferases (DNMTs).

and GRAS compounds have similar distributions of log P as compared with approved drugs. It has been discussed that the distribution of octanol/water partition coefficient is one of the most important physicochemical properties [56]. Interestingly, despite the fact that EAFUS and GRAS have in general similar size, EAFUS compounds are more flexible than GRAS as measured by the number of rotatable bonds. In general, the distribution of rotatable bonds in Figure 10.7 shows that EAFUS compounds are slightly less flexible than chemicals in DrugBank and that GRAS chemicals have flexibility comparable with that of compounds tested as DNMT1 inhibitors. A similar analysis can be performed with the remaining physicochemical properties. This assessment is relevant because it evaluates the overlap in chemical space of food-related chemicals with bioactive compounds (drugs approved for clinical use). It further represents one of the first steps to identify potential modulators of epigenetic targets of food origin.

The scaffolds of the set of compounds tested as inhibitors of DNMT1, approved drugs, GRAS, and EAFUS were calculated with the program DataWarrior [57] computing the Murcko scaffolds. The most frequent scaffolds for each dataset are presented in Figure 10.8. Other than benzene that is a fairly frequent scaffold

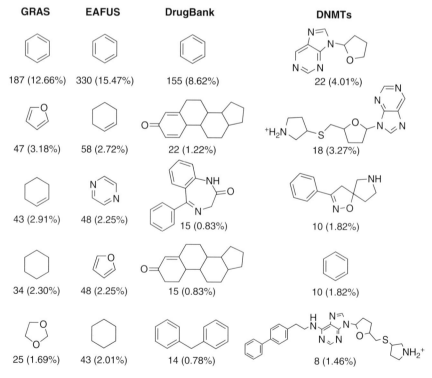

Figure 10.8 Top five most frequent molecular scaffolds identified in 1477 generally recognized as safe (GRAS) compounds and 2133 Everything Added to Food in the United States (EAFUS) chemicals. For reference, the most populated scaffolds in 1798 approved drugs (from DrugBank), and 549 compounds tested as inhibitors of DNA methyltransferase 1 are shown. For each scaffold, the frequency (and percentage) is indicated above the structure diagram.

in many datasets [58–60], the most frequent scaffolds in EAFUS and GRAS collections are small: the main scaffold is a five- or six-membered ring. In contrast, the most frequent scaffolds of approved drugs and DNMT inhibitors are more complex, containing two or more rings. The larger complexity of compounds has been associated with an increased potential specificity. Interestingly a similarity-based fingerprint analysis revealed that there are GRAS compounds with structures identical to compounds tested as DNMT1 inhibitors.

10.7 Conclusion

The application of chemoinformatics methods in food science is constantly growing. A clear example is the organization, storage, handling, dissemination, and mining of molecular databases related to food chemicals. In principle, the availability of a comprehensive food science database including information of additives, flavorings, odorants, and functional ingredients would be desirable. However, currently it is more common to find databases containing subsets with all this information. Another general application of chemoinformatics in food science is the analysis of the chemical space of food chemicals. These analyses have been conducted by quantitative and visual approaches using different molecular representations. The analysis of the chemical space with several chemometrics and chemoinformatics methods is contributing to the exploration of SPRs. A notable instance is the uncovering of complex associations between the chemical structure and the flavor or odor perception of food chemicals. Some of the major future developments of chemoinformatics in food science are the systematic identification of chemical compounds with a desired property and the full characterization of the food-related chemogenomic space.

Essentials

- Chemoinformatics in food science has a broad range of applications many of which are directly associated with the chemical structures of food chemicals.
- Storing, organizing, and mining chemical information of food chemicals in molecular databases is one of the large areas of application and opportunities of chemical information in the food industry.
- The number of chemical databases in food chemistry is limited. Building, curating, and maintaining public large databases of food chemicals is a major area of opportunity at the interface between chemoinformatics and food chemistry.
- Another area of opportunity is to develop (or select among existing ones) molecular descriptors relevant to food industry. For instance, there is a need to define and validate food chemical-oriented fingerprints and molecular representations relevant in sensory analysis.
- Explore, if at all possible, a "food-relevant" chemical space: in contrast to drug discovery where a number of drug-likeness rules have been proposed, a set of boundaries for physicochemical properties relevant for food chemicals have not yet been developed.

- In analogy to "drug-likeness," it is attractive to develop indices to measure food-related chemical likeness, for example, to design new chemical flavors.
- Analysis of the chemical space of food chemicals and bioactive compounds is being employed to evaluate the potential of finding overlapping regions of chemical space and potential bioactive molecules from food chemicals.
- Traditional and novel chemoinformatic approaches have been used to elucidate structure–property relationships (SPRs) in food science. Examples are neural networks, principal component analysis, machine learning, and property landscape modeling.
- Novel concepts such as flavor cliffs have been defined, that is, pairs of compounds with high structure similarity but very different flavor.
- Chemoinformatic studies are being pursued to address complex SPR problems such as the association between chemical structure and odor perception.
- Regarding SPR studies relevant in food sciences, there is a need to further explore potential associations between molecular complexity, odor perception, and compound safety.

Available Software and Web Services (accessed January 2018)

- DataWarrior, a free cheminformatics program for data visualization and analysis; Actelion Pharmaceuticals Ltd., www.openmolecules.org/datawarrior/.
- R: The R Project for Statistical Computing (a free software environment for statistical computing and graphics); The R Foundation, www.r-project.org/.
- MayaChemTools (free collection of Perl scripts to support day-to-day computational discovery needs); Manish Sud, www.mayachemtools.org/.
- Fema flavor, www.femaflavor.org.
- Biopep, www.uwm.edu.pl/biochemia/index.php/en/biopep.
- The Food Chemicals Codex, http://www.foodchemicalscodex.org/.
- FDA "Everything Added to Food in the United States" (EAFUS), www.accessdata.fda.gov/scripts/fcn/fcnNavigation.cfm?filter=&sortColumn=&rpt=eafusListing&displayAll=false#1.
- Prediction of Activity Spectra for Substances, www.pharmaexpert.ru/passonline/index.php.

Selected Reading

- Iwaniak, A., Minkiewicz, P., Darewicz, M., Protasiewicz, M., and Mogut, D. (2015) *J. Funct. Foods*, **16**, 334–351.
- Kermen, F., Chakirian, A., Sezille, C., Joussain, P., Le Goff, G., Ziessel, A., Chastrette, M., Mandairon, N., Didier, A., Rouby, C., and Bensafi, M. (2011) *Sci. Rep.*, **1**, 206.

- Martínez-Mayorga, K. and Medina-Franco, J.L. (2014) *Foodinformatics: Applications of Chemical Information to Food Chemistry*, Springer, New York, 251 pp.
- Talevi, C., Bellera, L., Castro, E.A., and Bruno-Blanch, L.E. (2007) *J. Comp.-Aided Mol. Des.*, **21**, 527–538.

References

[1] Gasteiger, J. (2016) *Molecules*, **21**, 151.
[2] Engel, T. (2006) *J. Chem. Inf. Model.*, **46**, 2267–2277.
[3] Martínez-Mayorga, K. and Medina-Franco, J.L. (2009) Chemoinformatics–applications in food chemistry, in *Advances in Food and Nutrition Research*, vol. 58 (ed. S. Taylor), Academic Press, Burlington, pp. 33–56.
[4] Iwaniak, A., Minkiewicz, P., Darewicz, M., Protasiewicz, M., and Mogut, D. (2015) *J. Funct. Foods*, **16**, 334–351.
[5] Martínez-Mayorga, K. and Medina-Franco, J.L. (2014) *FoodInformatics: Applications of Chemical Information to Food Chemistry*, Springer, New York, 251 pp.
[6] Minkiewicz, P., Miciński, J., Darewicz, M., and Bucholska, J. (2013) *Food Rev. Int.*, **29**, 321–351.
[7] Lipinski, C. and Hopkins, A. (2004) *Nature*, **432**, 855–861.
[8] Medina-Franco, J.L., Martínez-Mayorga, K., Giulianotti, M.A., Houghten, R.A., and Pinilla, C. (2008) *Curr. Comput. Aided Drug Des.*, **4**, 322–333.
[9] Osolodkin, D.I., Radchenko, E.V., Orlov, A.A., Voronkov, A.E., Palyulin, V.A., and Zefirov, N.S. (2015) *Exp. Opin. Drug Discovery*, **10**, 959–973.
[10] Medina-Franco, J.L. (2012) *Drug Dev. Res.*, **73**, 430–438.
[11] Durant, J.L., Leland, B.A., Henry, D.R., and Nourse, J.G. (2002) *J. Chem. Inf. Comput. Sci.*, **42**, 1273–1280.
[12] Singh, N., Guha, R., Giulianotti, M.A., Pinilla, C., Houghten, R.A., and Medina-Franco, J.L. (2009) *J. Chem. Inf. Model.*, **49**, 1010–1024.
[13] Medina-Franco, J.L., Martínez-Mayorga, K., Peppard, T.L., and Del Rio, A. (2012) *PLoS One*, **7**, e50798.
[14] Law, V., Knox, C., Djoumbou, Y., Jewison, T., Guo, A.C., Liu, Y., Maciejewski, A., Arndt, D., Wilson, M., Neveu, V., Tang, A., Gabriel, G., Ly, C., Adamjee, S., Dame, Z.T., Han, B., Zhou, Y., and Wishart, D.S. (2014) *Nucleic Acids Res.*, **42**, D1091–D1097.
[15] Rogers, D. and Hahn, M. (2010) *J. Chem. Inf. Model.*, **50**, 742–754.
[16] Ruddigkeit, L. and Reymond, J.-L. (2014) The chemical space of flavours, in *FoodInformatics: Applications of Chemical Information to Food Chemistry* (eds K. Martínez-Mayorga and J.L. Medina-Franco), New York, Springer, pp. 83–96.
[17] Reymond, J.-L. (2015) *Acc. Chem. Res.*, **48**, 722–730.
[18] Schwartz, J., Awale, M., and Reymond, J.-L. (2013) *J. Chem. Inf. Model.*, **53**, 1979–1989.

[19] Fernandez-De Gortari, E. and Medina-Franco, J.L. (2015) *RSC Adv.*, **5**, 87465–87476.
[20] Nabney, I.T. (2002) *NETLAB. Algorithms for Pattern Recognition*, Springer, London, 438 pp.
[21] Owen, J.R., Nabney, I.T., Medina-Franco, J.L., and López-Vallejo, F. (2011) *J. Chem. Inf. Model.*, **51**, 1552–1563.
[22] Prieto-Martínez, F., Peña-Castillo, A., Méndez-Lucio, O., Fernández-De Gortari, E., and Medina-Franco, J.L. (2016) Molecular modeling and chemoinformatics to advance the development of modulators of epigenetic targets: a focus on DNA methyltransferases, in *Advances in Protein Chemistry and Structural Biology*, vol. 105 (ed. C. Christov), Academic Press, New York, pp. 1–26.
[23] Medina-Franco, J.L., Navarrete-Vázquez, G., and Méndez-Lucio, O. (2015) *Future Med. Chem.*, **7**, 1197–1211.
[24] Siebert, K.J. (2001) *J. Am. Soc. Brew. Chem.*, **59**, 147–156.
[25] Tan, Y. and Siebert, K.J. (2004) *J. Agric. Food. Chem.*, **52**, 3057–3064.
[26] Martínez-Mayorga, K., Peppard, T.L., Yongye, A.B., Santos, R., Giulianotti, M., and Medina-Franco, J.L. (2011) *J. Chemom.*, **25**, 550–560.
[27] Maggiora, G.M. (2006) *J. Chem. Inf. Model.*, **46**, 1535.
[28] Medina-Franco, J.L. (2012) *J. Chem. Inf. Model.*, **52**, 2485–2493.
[29] Kermen, F., Chakirian, A., Sezille, C., Joussain, P., Le Goff, G., Ziessel, A., Chastrette, M., Mandairon, N., Didier, A., Rouby, C., and Bensafi, M. (2011) *Sci. Rep.*, **1**, 206.
[30] Arctander, S. (1994) *Perfume and Flavor Materials of Natural Origin*, Allured Pub. Corp., Carol Stream, IL, 736 pp.
[31] Chastrette, M., Elmouaffek, A., and Sauvegrain, P. (1988) *Chem. Senses*, **13**, 295–305.
[32] Hendrickson, J.B., Huang, P., and Toczko, A.G. (1987) *J. Chem. Inf. Comput. Sci.*, **27**, 63–67.
[33] Tromelin, A., Sanz, G., Briand, L., Pernollet, J.-C., and Guichard, E. (2006) 3D-QSAR study of ligands for a human olfactory receptor, in *Developments in Food Science*, vol. 43 (eds W. Bredie and M. Petersen), Elsevier, pp. 13–16.
[34] González-Medina, M., Prieto-Martínez, F.D., Naveja, J.J., Méndez-Lucio, O., El-Elimat, T., Pearce, C.J., Oberlies, N.H., Figueroa, M., and Medina-Franco, J.L. (2016) *Future Med. Chem.*, **8**, 1399–1412.
[35] Lavecchia, A. and Di Giovanni, C. (2013) *Curr. Med. Chem.*, **20**, 2839–2860.
[36] Méndez-Lucio, O., Tran, J., Medina-Franco, J.L., Meurice, N., and Muller, M. (2014) *ChemMedChem*, **9**, 560–565.
[37] Westermaier, Y., Barril, X., and Scapozza, L. (2015) *Methods*, **71**, 44–57.
[38] Di Ianni, M.E., Enrique, A.V., Palestro, P.H., Gavernet, L., Talevi, A., and Bruno-Blanch, L.E. (2012) *J. Chem. Inf. Model.*, **52**, 3325–3330.
[39] Talevi, A., Enrique, A.V., and Bruno-Blanch, L.E. (2012) *Bioorg. Med. Chem. Lett.*, **22**, 4072–4074.
[40] Martínez-Mayorga, K., Peppard, T.L., López-Vallejo, F., Yongye, A.B., and Medina-Franco, J.L. (2013) *J. Agric. Food. Chem.*, **61**, 7507–7514.

[41] Mauricio, E.D.I., Andrea, V.E., Maria, E.D.V., Blanca, A., María, A.R., Luisa, R., Eduardo, A.C., Luis, E.B.-B., and Alan, T. (2015) *Comb. Chem. High Throughput Screen.*, **18**, 335–345.

[42] Di Ianni, M.E., Del Valle, M.E., Enrique, A.V., Rosella, M.A., Bruno, F., Bruno-Blanch, L.E., and Talevi, A. (2015) *Assay Drug Dev. Technol.*, **13**, 313–318.

[43] Talevi, A., Bellera, C.L., Castro, E.A., and Bruno-Blanch, L.E. (2007) *J. Comput.-Aided Mol. Des.*, **21**, 527–538.

[44] Arguelles, A.O., Meruvu, S., Bowman, J.D., and Choudhury, M. (2016) *Drug Discovery Today*, **21**, 499–509.

[45] Cadet, J.L., Mccoy, M.T., and Jayanthi, S. (2016) *Chem. Biol. Drug Des.*, **99**, 502–511.

[46] Feinberg, A.P., Koldobskiy, M.A., and Gondor, A. (2016) *Nat. Rev. Genet.*, **17**, 284–299.

[47] Remely, M., Stefanska, B., Lovrecic, L., Magnet, U., and Haslberger, A.G. (2015) *Curr. Opin. Clin. Nutr. Metab. Care*, **18**, 328–333.

[48] Rio, A.D. and Costa, F.B.D. (2014) Molecular approaches to explore natural and food compound modulators in cancer epigenetics and metabolism, in *FoodInformatics: Applications of Chemical Information to Food Chemistry* (eds K. Martínez-Mayorga and J.L. Medina-Franco), New York, Springer, pp. 131–149.

[49] Vergeres, G. (2013) *Trends Food Sci. Technol.*, **31**, 6–12.

[50] Martínez-Mayorga, K. and Montes, C.P. (2016) The role of nutrition in epigenetics and recent advances of *in silico* studies, in *Epi-Informatics: Discovery and Development of Small Molecule Epigenetic Drugs and Probes* (ed. J.L. Medina-Franco), London, United Kingdom, Academic Press, pp. 385–398.

[51] Robertson, K.D. (2001) *Oncogene*, **20**, 3139–3155.

[52] Lyko, F., Brown, R., and Natl, J. (2005) *Cancer Inst.*, **97**, 1498–1506.

[53] Benetatos, L. and Vartholomatos, G. (2016) *Ann. Hematol.*, **95**, 1571–1582.

[54] Fang, M.Z., Wang, Y.M., Ai, N., Hou, Z., Sun, Y., Lu, H., Welsh, W., and Yang, C.S. (2003) *Cancer Res.*, **63**, 7563–7570.

[55] Medina-Franco, J.L., López-Vallejo, F., Kuck, D., and Lyko, F. (2011) *Mol. Divers.*, **15**, 293–304.

[56] Ganesan, A. (2008) *Curr. Opin. Chem. Biol.*, **12**, 306–317.

[57] Sander, T., Freyss, J., Von Korff, M., and Rufener, C. (2015) *J. Chem. Inf. Model.*, **55**, 460–473.

[58] Medina-Franco, J.L., Martínez-Mayorga, K., Bender, A., and Scior, T. (2009) *QSAR Comb. Sci.*, **28**, 1551–1560.

[59] López-Vallejo, F., Giulianotti, M.A., Houghten, R.A., and Medina-Franco, J.L. (2012) *Drug Discovery Today*, **17**, 718–726.

[60] Yongye, A.B., Waddell, J., and Medina-Franco, J.L. (2012) *Chem. Biol. Drug Des.*, **80**, 717–724.

11 Computational Approaches to Cosmetics Products Discovery

Soheila Anzali[1], Frank Pflücker[2], Lilia Heider[2], and Alfred Jonczyk[2]

[1] *InnoSA GmbH, Georg-Dascher-Str. 2, 64846 Groß-Zimmern, Germany*
[2] *Merck KGaA, Frankfurter Strasse 250, 64293 Darmstadt, Germany*

Learning Objectives

- To describe the needs and scientific research efforts in cosmetics applications.
- To evaluate the use of chemoinformatics, bioinformatics, and computational methods in the field of cosmetics and to prove the prediction of properties of molecules by *in vivo* studies.
- To perform rational planning of further experiments with information of gene expression analyses.
- To discuss the relevance of computational applications for a fast development of new products for cosmetics use.

Outline

11.1 Introduction: Cosmetics Demands on Computational Approaches, 527
11.2 Case I: The Multifunctional Role of Ectoine as a Natural Cell Protectant (Product: Ectoine, "Cell Protection Factor", and Moisturizer), 528
11.3 Case II: A Smart Cyclopeptide Mimics the RGD Containing Cell Adhesion Proteins at the Right Site (Product: Cyclopeptide-5: Antiaging), 533
11.4 Conclusions: Cases I and II, 542

11.1 Introduction: Cosmetics Demands on Computational Approaches

The development of a new product in cosmetics is an extremely laborious and time-consuming process. Chemoinformatics and bioinformatics methods as well as computational techniques, such as molecular modeling, virtual screening, and gene expression analyses, can particularly facilitate this process and can be beneficially used in the search for lead compounds and their further optimization.

Like the drug discovery process, the cosmetic ingredient development process can start by searching for targets relevant for skin treatment in the cosmetics area. Additionally, the selection rules of so-called beauty compounds (those that show the desirable physicochemical properties in cosmetics) have to be involved in such a process.

Applied Chemoinformatics: Achievements and Future Opportunities, First Edition.
Edited by Thomas Engel and Johann Gasteiger.
© 2018 Wiley-VCH Verlag GmbH & Co. KGaA. Published 2018 by Wiley-VCH Verlag GmbH & Co. KGaA.

The aim of these studies is to present the power of computational methods (bio- and chemoinformatics and quantum mechanical calculations) to accelerate and improve the development of new cosmetics ingredients. The following two examples shall help to demonstrate the potential of the described methods for finding additional values of ingredients in cosmetics research.

11.2 Case I: The Multifunctional Role of Ectoine as a Natural Cell Protectant (Product: Ectoine, "Cell Protection Factor", and Moisturizer)

The protective properties of ectoine formerly described only for extremophilic microorganisms are transferred to the human skin. Our present data give evidence that the compatible solute ectoine protects the cellular membrane from damage caused by surfactants. Trans-epidermal water loss (TEWL) measurements *in vivo* suggest that the barrier function of the skin is strengthened after topical application of an oil/water (O/W) emulsion containing ectoine. Ectoine functions as, for example, a superior moisturizer with long-term efficacy but offers also further benefits to maintain a healthy-looking skin. These findings indicating ectoine as a strong water structure-forming solute are explained *in silico* by means of molecular dynamics (MD) simulations. Spherical clusters containing (i) water, (ii) water with ectoine, and (iii) water with glycerol are created as model systems. The stronger the water-binding activity of the solute, the higher the quantity of water molecules remaining in the cluster. At high temperatures, water clusters around ectoine molecules remain stable for a long period of time, whereas mixtures of water and glycerol break down and water molecules diffuse out of the spheres. Based on these findings we suggest that not only hydrogen bond properties of solutes are responsible for maintaining the specific form of water structure. Moreover, the particular electrostatic potential of ectoine as an amphoteric molecule with zwitterionic character is the major cause for its strong affinity to water. Because of its outstanding water-binding activity, ectoine might be especially useful in preventing water loss in dry atopic skin and in recovering viability and preventing skin aging.

Ectoine is a small organic molecule occurring widely in aerobic, chemoheterotrophic, and halophilic organisms that enable them to survive under very extreme conditions. These organisms protect their biopolymers (biomembranes, proteins, enzymes, and nucleic acids) against dehydration caused by high temperature, salt concentration, and low water activity by substantial ectoine synthesis and enrichment within the cell.

The organic osmolytes ectoines (see Figure 11.1) are amphoteric, water-binding organic molecules. They are generally compatible with the cellular metabolism without adversely affecting the biopolymers or physiological processes and are so-called compatible solutes [1].

The protective function of the compatible solutes in a low-water environment may be explained by the "preferential exclusion model": the solutes are excluded from the immediate hydration shell of, for example, a protein due

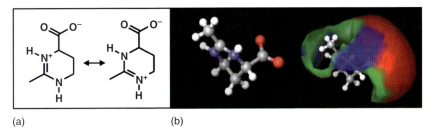

(a)   (b)

Figure 11.1 Molecule structure of ectoine with the two mesomeric forms (a) and its hydrophilic surface colored according to the corresponding atomic partial charges (b).

to an unfavorable interaction with the protein surface. The consequence is preferential hydration of the protein, thus promoting its native conformation. Because compatible solutes do not interact directly with the proteins' surface, the catalytic activity remains unaffected [2, 3].

Yu and Nagaoka reported very interesting results coming from a few molecular dynamics (MD) simulations performed for water–ectoine mixture models around chymotrypsin inhibitor 2. Based on their statement, ectoine maintains water at the surface due to slowing down of the water diffusion around a protein, where it is most needed, while it does not directly interact with macromolecules themselves. Thus, ectoine plays an indirect role in the alteration of the solvent properties and the modification of the stability of proteins [4, 5].

Ectoine minimizes the denaturation that occurs on removal of water molecules by making the unfolding of proteins less favorable [6].

Compatible solutes are amphiphilic in nature and are capable of "wetting" hydrophobic proteins, hence improving their hydration capability [7]. The structure-forming and structure-breaking properties of compatible solutes indirectly influence the hydration shells and hence the activities of the proteins involved [8].

In this way, halophilic organisms and other bacteria utilize ectoine to protect their cytoplasmic biomolecules against heat, freezing, dryness, and osmotic stress [9].

Ectoine can be isolated from halophilic bacteria on a large scale and thus is available as active ingredient for skin care [10].

The protective properties of ectoine formerly described only for microorganisms could be transferred to human skin, which is situated at the interface of the organism and its environment and therefore exposed to a variety of environmental assaults. The *stratum corneum* in particular provides a barrier to the evaporation of water from the viable epidermis. Many factors work to compromise this barrier and increase the rate of water loss from the skin. Exposure to extreme environmental conditions including cold, dry winter weather, frequent washing with soap, and hot water or the exposure to surfactants may cause skin dryness. Besides dryness, the cumulative effect of external factors like radiation, wind, and temperature extremes leads to accelerated skin aging [11, 12].

Various investigations underline the outstanding antiaging properties of ectoine: Epidermal dendritic Langerhans cells are the most important

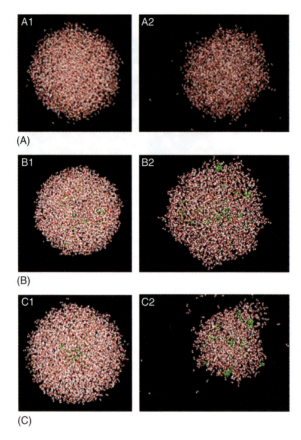

Figure 11.3 Molecular dynamics simulation of different models containing (A) water, (B) water and ectoine, and (C) water and glycerol. The pictures are taken at the beginning of the simulation ($t = 0$, A1, B1, C1) and after 200 (A2), 1000 (B2) and 500 ps (C2) at a constant temperature of 370 K. Water clusters around ectoine molecules remain stable for a long period of time, whereas the cluster of water and glycerol breaks down and water molecules diffuse out of the spheres. The pictures represent the number of water molecules counted during the dynamic simulation as shown in Table 11.1. The solutes are colored in green.

Table 11.1 Results of the molecular dynamics simulation, with the number of water molecules at 370 K.

	A	B	C
t (ps)	Water	Water + ectoine	Water + glycerol
0	3618	3139	3429
200	3026	3138	2339
500	n.c.	3112	1288
1000	n.c.	3103	n.c.

water–ectoine. In this experimental setup the E_{pot} value can be adopted as the stored energy or the energy of position in such a system.

In the pure water and the water–glycerol complexes, the E_{pot} values decrease dramatically during the simulation time, while the E_{pot} value of the water–ectoine sphere remains constant even throughout a longer simulation time (Figure 11.3, Table 11.1). The E_{pot} value of the water–ectoine sphere remains constant at the level indicated in the diagram (data not shown). It is remarkable that the E_{pot} value of regular water molecules *per se* is higher than that of the water–ectoine mixture indicating the strong organizing and complexing properties (kosmotropic) of ectoine.

In case of the dynamic simulation and their animations as well as of the statistical analysis, it can be demonstrated that the water diffusion out of spheres was limited and decreased enormously by adding ectoine molecules into the sphere (Figure 11.3a,b). See also the stick presentation of water and ectoine atoms in Figure 11.2. Even a fivefold-longer simulation time shows a stable water structure due to ectoine properties that is much superior compared to water itself and outstanding compared to a water–glycerol complex (see Figure 11.3a–c).

We propose that not only hydrogen bond properties of solutes are responsible for maintaining the water structure. Moreover, the particular electrostatic potential of compatible solutes like ectoine as an amphoteric molecule with zwitterionic character is the major cause for its *love affairs* with water.

11.3 Case II: A Smart Cyclopeptide Mimics the RGD Containing Cell Adhesion Proteins at the Right Site (Product: Cyclopeptide-5: Antiaging)

Large proteins and small peptides are well known as attractive ingredients for cosmetics. On skin, peptides regulate the activity of many biological processes. Several single natural amino acids originating from biodegradation processes of peptides are deemed to produce positive effects on skin, hair, and nails. For example, arginine (Arg) is an essential amino acid needed when the body is under stress or wounded, and glycine forms glutathione, one of the body's major antioxidants and free radical scavengers.

Peptide stability and optimization of delivery can be seen as a difficult task in cosmetics. The conformational flexibilities of peptides often cause a lower binding affinity to their targets due to entropic effects. This often results also in lower selectivity of such peptidic ligands.

There is growing interest for peptides in topical applications and especially in cosmetics driven by the need of more stable peptides, which are easy to formulate, safe, and efficient with a well-defined target profile.

Therefore, a peptide has been developed, which is the first homodetic cyclic peptide with a new use in cosmetic applications. Cyclopeptide-5 mimics a rigid and defined conformation of ligands with an RGD motif (Figure 11.4), which is found in certain extracellular matrix (ECM) proteins. Cyclic peptides seem to

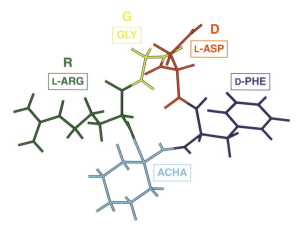

Figure 11.4 Cyclopeptide-5 consists of five amino acids. RGD, which is the one letter code for arginine (R), glycine (G), and aspartate (D) sequence. D-Phe represents D-phenylalanine and ACHA is the short name for aminocyclohexane carboxylic acid.

be more resistant against protease activity than linear analogues. Due to this assumption, it was possible to design peptides, which are found to be more stable also on skin for use in topical applications.

Cyclopeptide-5 is a novel biomimetic peptide containing the sequence Arg-Gly-Asp-DPhe-ACHA (1-aminocyclohexane carboxylic acid). It is a selective ligand for integrins alpha-vβ3 (αvβ3), alpha-vβ5 (αvβ5), and alpha-vβ6 (αvβ6).

Integrins are a large family of divalent cation-dependent heterodimeric transmembrane adhesion receptor proteins. They are composed of non-covalently linked α- and β-chains, generally with large extracellular domains and short cytoplasmatic tails (Figure 11.5).

They are responsible for physiological and pathological processes like cell migration, cell–cell communication, and cell–extracellular matrix interactions. Small fragments of natural integrin ligands containing the binding epitope

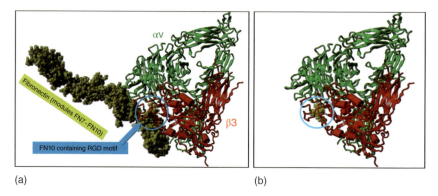

Figure 11.5 Cartoon of the extracellular part (Ectodomain) of αvβ3 integrin crystal structure in the complex with (a) fibronectin with RGD loop in cyan circle and (b) cyclopeptide-5 as a mimic of the RGD loop in fibronectin.

in the right conformation can also promote the function of natural integrin ligands, such as fibronectin, fibulin, collagen, elastin [23, 24]. If the fragments are deposited and anchored, they will promote cell attachment; in solution they will compete with natural ligand binding.

The specificity of Arg-Gly-Asp (RGD)-derived peptides to bind to certain integrins depends on their backbone conformation, the orientations of the charged side chains of Arg and Asp residues, and the hydrophobic moieties flanking the Arg and Asp residues. A key finding remains that the amino acid sequence Arg-Gly-Asp (RGD) in fibronectin serves as a primary cell attachment cue [25].

The ECM is the extracellular part of tissue that provides structural support to the cells and has various important functions. The ECM is the defining feature of connective tissue.

During aging, which is dependent on genetic and environmental factors, tissue repair, and synthesis of proteins that constitute the ECM in skin is slowing down.

The ability of cyclopeptide-5 to compete with vitronectin and fibronectin binding to αvβ3 and αvβ5 integrins was confirmed by carrying out *in vitro* tests [26]. These binding competition studies with isolated integrins αvβ3 and αvβ5 and αvβ6 to their natural ligands like vitronectin or fibronectin showed strong binding competition of cyclopeptide-5 (in-house data, αvβ3: IC50 (nM) = 2.3 ± 0.8; αvβ5: IC50 (nM) = 700 ± 0.8, also active in αvβ6 (nM)).

Recent non-published studies have shown the antiaging effect of cyclopeptide-5 (in-house data, publication in preparation).

This section illustrates part of our results, namely, an analysis of the effects of cyclopeptide-5 in a human three-dimensional skin model as determined by cDNA microarray profiling. Gene expression microarray technology is an extremely powerful tool and has widely been used in many areas of pharmaceutical research. However, its use in cosmetic applications is relatively new.

Gene expression in human skin fibroblasts and keratinocytes treated with cyclopeptide-5 was analyzed by cDNA microarrays. Genes that were identified as deregulated by cyclopeptide-5 treatment were further investigated using GeneGo's MetaCore™ [27]. MetaCore™ comprises a database and algorithms for functional analyses of experimental data like microarray gene expression, metabolomics, and proteomics data, with the following characteristics:

- The content of its database is manually curated from the scientific literature.
- The database includes both metabolic and signaling pathways and different types of interactions between molecular entities, such as genes, proteins, and small molecules.
- After uploading experimental data, these can be analyzed and visualized in the context of pathways, networks, and maps.
- Searching for genes, enzymes, compounds, and reactions is possible.

These results provide strong indication that the treatment of human skin with cyclopeptide-5 has anti-aging, anti-inflammation, and hair growth effects.

11.3.1 Methods

A human full thickness skin model was utilized, which comprises both keratinocytes and fibroblasts [28]. The cells were neither pooled nor genetically modified. Gene expression analysis was carried out with PIQOR™-skin two-color cDNA microarrays, which represent 1308 genes that are involved in target pathways related to stress, inflammation, pigmentation, depigmentation, moisturization, antiaging, and hair follicle development in humans [29]. For instance, probes for genes involved in processes of cell cycle, apoptosis, DNA repair, oxidative metabolism, angiogenesis, cell adhesion, cell–matrix interactions, and cell signaling are included. The gene probes were spotted in quadruplicate on the arrays.

The treatment of the human skin model with cyclopeptide-5 was performed in 0.5 μM concentration (4 days incubation, $n = 3$ per condition). Buffer-treated skin equivalents served as controls in parallel, using a dye-swap scheme for the replications.

After local background subtraction, the data were normalized using the VSN algorithm [30]. Differentially expressed genes were identified using a moderated t-test after fitting linear models to the data of each gene, where the data from replicate spots were used with quality-based weights [31]. Genes with a p-value < 0.01 and an intensity ratio >1.3 between treated cells and controls were considered as significantly deregulated.

The significantly deregulated genes in skin samples treated with cyclopeptide-5 were used for further analyses. It was tested whether this set of deregulated genes is enriched in certain gene sets in MetaCore™, including canonical pathway maps, GeneGo cellular processes, gene ontology biological processes, and disease categories. The degree of "relevance" of different categories is defined by the p-values from the enrichment analysis, so that the lower p-value obtains higher priority.

11.3.2 Results and Discussion

After statistical filtering ($p < 0.01$ and at least 1.3-fold de-regulation), 82 deregulated genes were identified. A subset of these is shown in Table 11.2.

The effects of cyclopeptide-5 on the gene expression profile in human primary epidermal keratinocytes and human fibroblasts were categorized into cellular processes, networks and pathways maps, and description of single genes relevant for cosmetic applications.

11.3.2.1 Cellular Processes and the Most Relevant Pathways Maps

The following cellular and molecular processes were most significantly enriched in genes that were found to be deregulated by cyclopeptide-5: anatomical structure development, response to external stimulus tissue development, positive regulation of cell proliferation, response to wounding, cell proliferation, regulation of cell proliferation, response to extracellular stimulus, cell migration, and extracellular structure organization.

The cDNA microarray analysis of skin treated with cyclopeptide-5 revealed many deregulated genes related to cell organization and communication, also

Table 11.2 Subset of up- and down-regulated genes potentially connected to the effects of cyclopeptide-5 in cosmetic applications. These are presented as log 2 (ratio), where a value >0 represents an up-regulation and <0 a down-regulation, respectively.

Official symbol	Name	Compound versus control, log 2(ratio) value	P-value
AP-1	Jun: transcription factor AP-1	−0.58	0.000000264
BMP7	Bone morphogenetic protein 7	−0.89	0.0019
CCNB2	Cyclin B2 G2/mitotic	0.46	0.000027
CCNB2	Cyclin B2 G2/mitotic specific cyclin B2	0.47	0.0000266
CLU	Clusterin	0.44	0.0048
COL4A1	Collagen alpha 1(IV) chain	0.51	0.0000005
CPE	Carboxypeptidase H	−0.38	0.00407
CYR61	Insulin-like growth factor-binding protein 10	0.43	0.00011
DCN	Decorin	0.40	0.0000934
ENG	Endoglin	0.54	0.000213
FBLN1_4	Fibulin-1	0.52	0.000016
FBLN2	Fibulin-2	0.38	0.00029
FGF1	Acidic fibroblast growth factor	0.51	0.0016
FLOT2	Flotillin-2	0.45	0.0052
GPX3	Plasma glutathione peroxidase	0.53	0.00542
IGFBP4	Insulin-like growth factor-binding protein 4	0.49	0.00089
IL17A	Interleukin-17	−0.67	0.00097
INHA	Inhibin alpha chain	−0.61	0.00411
ITGA5	Integrin alpha-5	0.38	0.000067
ITGB1	Integrin beta-1	0.38	0.0014
KRT10	Cytokeratin 10	−0.39	0.00551
LAMA4	Laminin alpha-4	0.60	0.00000058
LCN2	Lipocalin-2	−0.55	0.000000403
LTBP1	Latent transforming growth factor beta binding protein	0.56	0.000065
MC2R	Melanocortin-2 receptor	−0.38	0.000623
MC3R	Melanocortin-3 receptor	−0.44	0.00268
MGST1	Glutathione s-transferase	0.41	0.0000591
MMP1_1	Matrix metalloproteinase-1	0.87	0.00000003
MMP13	Collagenase 3	−0.47	0.0089
MMP23A/B	Matrix metalloproteinase-23	−0.64	0.00000059
MMP7	Matrilysin	−0.40	0.001
MMP3	Stromelysin-1	0.83	0.000018
NID	Nidogen	0.48	0.000093

(Continued)

Table 11.2 (Continued)

Official symbol	Name	Compound versus control, log 2(ratio) value	P-value
POLE	Dna polymerase II subunit a	−0.43	0.0033
QSOX1	Bone-derived growth factor	0.43	0.0015
S100A7	Psoriasin	−0.51	0.00004
S100A8	Calgranulin A	−0.49	0.000037
S100A9	Calgranulin B	−0.42	0.0000036
SEPRASE	Fibroblast activation protein alpha	0.55	0.00000071
SPRR3	Cornifin-beta	0.74	0.000012
TERT	Telomerase reverse transcriptase	−0.58	0.00176
TIMP1	Fibroblast collagenase inhibitor	0.73	0.0000029
TK1	Thymidine kinase	0.54	0.000000711
TNC	Tenascin	0.88	0.0000063
TOP2A	DNA topoisomerase II	0.62	0.000024
TUBB	Tubulin beta	0.66	0.0005

mediated by integrins. The most relevant network object, based on network analysis and pathway maps, represents cell adhesion and ECM remodeling (Figure 11.6).

The ECM remodeling is involved in normal physiological processes, such as embryonic development, reproduction, proliferation, cell motility, adhesion, wound healing, and angiogenesis, as well as in disease processes, such as arthritis and tumor metastasis. The adhesion of cells to the ECM is a dynamic process, mediated by a series of matrix-associated and cell-surface molecules that interact with each other in a spatially and temporally regulated manner. These interactions play a major role in tissue formation, cellular migration, and induction of adhesion-mediated transmembrane signals.

11.3.2.2 Relevant Results for Cosmetics Applications

The effects of cyclopeptide-5 on the gene level give us new insights into the potential action of the integrin ligand cyclopeptide-5 in a variety of skin and hair regeneration disorders. In the following, the potential relevance of gene expression changes induced by cyclopeptide-5 (Table 11.2) for different applications in cosmetics is outlined: anti-wrinkles/anti-aging, anti-inflammation, hair growth/regrowth – follicle development.

Photoaging is the most common form of skin damage and is associated with skin carcinoma. UV irradiation inhibits the TGF-β1-induced type I procollagen gene expression in cultured human skin fibroblasts [32]. TGF-β is a multifunctional protein that controls proliferation, differentiation, and other functions in many cell types. TGF-β/Smad pathway is the major regulator of type I procollagen synthesis in human skin.

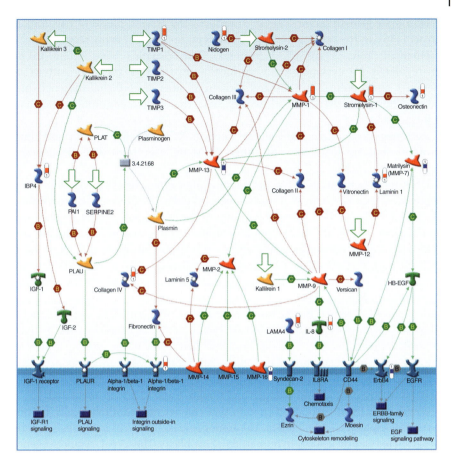

Figure 11.6 The canonical map of cell adhesion and ECM remodeling. Effects of cyclopeptide-5 treatment (0.5 μM) on gene expression in skin cells are visualized on the map as thermometer-like figures. Upward thermometers with red color indicate up-regulation and downward (blue) ones indicate down-regulation.

In the present gene expression experiment, latent TGF-β binding protein (LTBP1), TGF-β1, and TGF-β3 show a trend of being up-regulated after treatment of skin cells with cyclopeptide-5.

Most integrins recognize several proteins of the ECM, including laminin, fibronectin, collagen (types I, II, and IV), elastin, fibulin, osteonectin, hyaluronan, and nidogen [33]. Cyclopeptide-5 is a highly active αvβ3/5 and low active αvβ6 ligand.

Also integrin β1 (ITGB1) and integrin alpha 5 (fibronectin receptor alpha, ITGA5) are up-regulated in the skin treated by cyclopeptide-5. In skin and hair follicle biology, β1 integrins, and their ligands are of particular interest [34]. Integrin α5β1 mediates fibronectin-induced epithelial cell proliferation through activation of the EGFR. Fibronectins are proteins that connect cells with collagen fibers in the ECM, allowing cells to move through the ECM.

Fibroblast growth factor-7 (FGF-7) induces the anagen phase of hair follicle growth [35]. In the skin treated with cyclopeptide-5, FGF-7 is up-regulated.

Fibulin-1 and -2 are ECM proteins. Both proteins were found to be up-regulated by cyclopeptide-5. A locally restricted expression pattern of fibulin-1 and fibulin-2 mRNA and protein at sites of epithelial-mesenchymal interactions was detected in two tissues, the developing tooth and hair follicles [36].

Cyclopeptide-5 might bind at the integrin RGD site like fibulin and thus, promote the up-regulation of other ECM-RGD-proteins like fibronectin, procollagen, laminin, and so on.

DNA topoisomerases (TOPs) are a family of enzymes that is involved in DNA replication and metabolism. These enzymes regulate the helical structure of the double-stranded DNA by breaking strands of the DNA helix [37]. TOP-2 belongs to the genes, which are down-regulated in the human anagen hair follicle bulge [38]. Hair loss is known as a side effect of TOP inhibitors. In this study, both TGF-β1 and TGF-β3 show a trend of up-regulation. TGF-β1 and -β3 activate certain TGF-β receptor Ser/Thr kinases and, thus, might lead via transcription factors like SMADs, AP-1, and so on to the regulation of targets like bone morphogenic proteins (BMPs: down-regulation), nidogen (NID: up-regulation), fibulins (up-regulation), antigen Ki-67 (trend of up-regulation), and TOP2A (up-regulation) [39–42]. In addition, the TGF-β/BMP signaling pathway is known for the regulation of hair follicle development [43]. In the skin treated with cyclopeptide-5, TOP-2 is strongly up-regulated, BMP-2 shows a trend of being down-regulated, and nidogen and fibulin-1 and -2 are up-regulated.

Cyr61 is an ECM-associated protein that can act on endothelial cells, fibroblasts, macrophages, and platelets. The angiogenic factor Cyr61 activates a genetic program for wound healing in human skin fibroblasts. The Cyr61-regulated genes fall into several groups known to participate in processes important for cutaneous wound healing, like ECM remodeling (MMP-1, MMP-3, TIMP-1, uPA, and PAI-1) and cell–matrix interactions (Col11, Col12, and integrins β3 and β5) [44].

Tenascins are ECM glycoproteins. They are abundant in the ECM of developing vertebrate embryos, and they reappear around healing wounds. In the skin treated with cyclopeptide-5, tenascin is strongly up-regulated [45].

In addition, S100 calcium binding proteins A7, A8, and A9 (S100A7, S100A8, and S100A9) are down-regulated genes after treatment of skin cells with cyclopeptide-5.

S100A7 may be involved in epidermal differentiation and inflammation and might therefore be important for the pathogenesis of psoriasis and other diseases. The S100A7 protein, also known as psoriasin, has important functions as a mediator and regulator in skin differentiation and disease (psoriasis) and in breast cancer and as a chemotactic factor for inflammatory cells [46].

In addition, Lener *et al.* investigated genes involved in the natural aging process of the human skin. They found that in total 105 genes change their expression more than 1.7-fold during the aging process in the human skin. S100A7 and S100A9 have been described as genes up-regulated in old skin [25, 47].

S100 proteins are the largest subgroup of Ca^{2+} binding proteins with an EF-hand structural motif. A unique feature of this protein family is that individual members are localized in specific cellular compartments. For example, various S100 proteins are expressed in very restricted regions of the hair follicle [48].

Extracellular proteases are crucial regulators of the cell function. The family of matrix metalloproteinases (MMPs) has classically been described in the context of ECM remodeling, which occurs throughout life in diverse processes that range from tissue morphogenesis to wound healing. Recent evidence has implicated MMPs in the regulation of other functions, including survival, angiogenesis, and inflammation.

MMPs are secreted from keratinocytes and fibroblasts and break down collagen and other proteins that comprise the dermal ECM. Imperfect repair of the dermal damage impairs the functional and structural integrity of the ECM. Repeated sun exposure causes accumulation of dermal damage that eventually results in characteristic wrinkling of photodamaged skin [49]. In the skin, the primary role of MMP enzymes is to recycle the skin matrix, particularly the structural proteins collagen and elastin.

Tissue inhibitor of metalloproteinases (TIMP-1) is strongly up-regulated after treatment of skin cells with cyclopeptide-5. TIMP-1 has been described as a cell survival factor.

TIMP-1 is one representative of the natural MMP inhibitor family, encompassing four members. Its expression is decreased with fibroblast senescence, both *ex vivo* and *in vivo*, thus contributing to increased catabolic activity within dermis. TIMP-1 displays multiple biological functions. It inhibits most MMPs.

The up-regulation of TIMP-1 indicates that cyclopeptide-5 activates up-regulation of dermal fibroblast collagen production and down-regulation of collagen degradation.

Decreasing the expression or activity of matrix metalloproteases has an effect on the biological collagen catabolic process toward skin treatment of aging skin and psoriasis.

The down-regulated MMP genes after treatment of skin cells with cyclopeptide-5 are MMP-7 (type I, II, IV, and V gelatins, fibronectin, and proteoglycan), MMP-13 (degrades type II collagen more efficiently than types I and III), and MMP-23 (may play a specialized role in reproductive processes), whereas MMP-1 (degrades type I, II, and II collagen) and MMP-3 (degrades collagen types II, III, IV, IX, and X, proteoglycans, fibronectin, laminin, and elastin) are in contrast strongly up-regulated. Taking into account the results of TIMP-1, up-regulation of MMP-1 and 3 is an unexpected outcome.

It is likely that up-regulation of TIMP-1 and down-regulation of MMPs result in an increase of collagen fibrils. Also laminin subunit α-4, laminin subunit α-3, laminin subunit β-1, laminin subunit γ-1, and collagen-α (IV) are up-regulated. Laminin is thought to mediate the attachment, migration, and organization of cells into tissues during embryonic development by interacting with other ECM components.

11.4 Conclusions: Cases I and II

In conclusion, our recent studies demonstrate the outstanding role of the compatible osmolyte ectoine in preventing water loss caused by surfactant-induced barrier damage. Ectoine functions as a more potent moisturizer than glycerol and additionally features long-term moisturizing efficacy. These *in vivo* findings could be explained *in silico* by means of MD simulations. Water clusters around ectoine molecules remain stable for a long period of time, whereas mixtures of water and glycerol are disintegrated by diffusion of water molecules out of the spheres.

Because of its strong water-binding activity, ectoine might be especially useful in the prevention of dehydration in dry atopic skin and the recovery of viability and the prevention of skin aging.

An integrin-targeted pentapeptide has been developed. Cyclopeptide-5 is a synthetic peptide, which mimics the Arg-Gly-Asp (RGD) sequence binding site of, for example, fibronectin for the integrins αvβ3, αvβ5, and αvβ6. Fibronectin is a major molecule in the dermis with essential adhesive properties.

The stability and the efficacy of this peptide in cosmetic applications are enhanced by its cyclic conformation. The rigidity of structure through a cyclic form makes it more selective to the desired target, the integrin RGD binding site. Cyclopeptide-5 is an active αvβ3/5/6 integrin ligand. Human full-thickness skin models were treated with cyclopeptide-5, and the effects on the gene expression level were analyzed. The above-mentioned tests and analyses allowed us to postulate that the activity of cyclopeptide-5 is based on its ability to successfully mimic the biological function of cell adhesion proteins of the ECM, like fibronectin, by directing cell behavior toward modulated cell adhesion and ECM remodeling, promoting the signaling pathways and targets involved in skin aging and hair growth.

We also described the possibility of the identified pathways starting from integrin receptors by activation of TGF-β1 and -β3 resulting in regulation of targets BMPs, fibuline, and antigens Ki-67 and TOP2A via transcription factors like SMADs.

In summary, we hypothesize that cyclopeptide-5 slows down skin aging through two distinct pathways: by stimulation of the synthesis of structural proteins like collagens, fibronectins, fibulin, and laminins and simultaneously by up-regulation of TIMP-1 and down-regulation of several MMPs, which disturb collagen and elastin functions. Both features contribute to the maintenance of ECM integrity of the dermis.

The process of development of "beauty compounds" is an extremely detailed process and needs technical refinements. Computer techniques, such as molecular modeling and virtual screening, can particularly facilitate this process and can be beneficially used in the search of new templates and frameworks as lead compounds for further optimization. A good example has been introduced by Ni *et al.* for compounds inhibiting tyrosinase catalytic activity. These compounds are an important class of cosmetic and dermatological agents, which show high potential as depigmentation agents used for skin lightening. A multistep protocol employed for the identification of novel tyrosinase inhibitors incorporated the Shape Signatures computational algorithm for rapid screening of chemical libraries. Shape Signatures excels at scaffold hopping across different chemical families, which enables identification of new actives whose molecular structure is distinct from other known actives. Using this approach, they identified a novel class of depigmentation agents that demonstrated promise for skin lightening product development [50].

Another important use of chemoinformatics is addressing the prediction of safety aspects of cosmetics ingredients.

There is a desire to obtain information about the safety of a cosmetic ingredient directly from its chemical structure. Currently computational, or *in silico*, methods to predict toxicity include the use of strategies for grouping (also termed category formation), read-across within groups, (quantitative) structure–activity relationships ((Q)SARs), and expert (knowledge-based) systems. These are supported by methods to incorporate threshold of toxicological concern (TTC) and kinetics-based extrapolations for concentrations that may arise at the organ level (such as physiologically based pharmacokinetic (PBPK) models).

The COSMOS Project (Integrated *In Silico* Models for the Prediction of Human Repeated Dose Toxicity of COSMetics to Optimize Safety, www.cosmostox.eu) was a unique collaboration addressing the safety assessment needs of the cosmetics industry, without the use of animals.

The COSMOS database framework with different access levels has been created to capture repeat dose toxicity data, adopting a strategy for data quality assessment and quality control, both of chemical structures and toxicity data. A comprehensive inventory of cosmetics ingredients has been compiled with well-defined, unique chemical structures. To assist TTC development, a dataset with repeat dose toxicity data for cosmetic ingredients has been derived with 558 unique chemical structures containing NOEL/NOAEL values. The analysis of the chemical space has shown that the COSMOS TTC dataset is a good representation of the COSMOS Cosmetics Inventory in terms of physicochemical property ranges, structural features, and chemical use categories. Computational workflows, linked to Adverse Outcome Pathways (AOPs), have been developed to identify fragments and properties associated with particular toxicity mechanisms and to form categories and allow for read-across to predict toxicity [51].

The project ran from January 2011 to December 2015.

Essentials

- The physicochemical properties of ectoine, as a strong water structure-forming solute, and the related *in vivo* results have been explained *in silico* by means of molecular dynamic simulations.
- A human full-thickness skin model has been utilized, which comprises both keratinocytes and fibroblasts. Cells are not pooled or genetically modified.
- Human full-thickness skin models were treated with cyclopeptide-5, and the effects on the gene expression level were analyzed.

- With the unique design of cyclopeptide-5 as a highly selective peptide, it is possible to mimic natural processes of skin communication and repair.
- The selection criteria for further studies on cyclopeptide-5 were based on the outcome of the results of the gene expression analyses.
- Cyclopeptide-5 has a direct impact on the process of improving the appearance of aging skin and inflammation. *In vivo* measurements of skin smoothness show that cyclopeptide-5 has a skin smoothing activity within a short period of application time.
- Both cases described examples with their strong background of computational prediction are launched products and are successful on the market.

Available Software and Web Services (accessed January 2018)

- http://www.happi.com/issues/2010-10/view_features/cosmetics-companies-get-active: This article with the title "Cosmetics companies get active" describes the computational methods, which can be used in a Cosmetics Discovery process. http://journals.plos.org/plosone/article?id=10.1371/journal.pone.0112788
- http://www.mdpi.com/1420-3049/20/12/19880/htm: The current work summarizes prospective pharmacophore-based studies conducted in the field of steroid biology, with a special focus on short-chain dehydrogenase/reductase (SDRs), and highlights success stories reported in this area. Endocrine-disrupting chemicals (EDCs) interfering with hormone synthesis, metabolism, and/or hormonal regulation. EDCs include substances used in agriculture, industrial production, dyes, food preservatives, or body care products and cosmetics.
- http://www.cheminformatics-nutrition.recerca.urv.cat/expertesa/en_index/: The activity of the research group in Cheminformatics and Nutrition focuses on using computational tools to (i) predict which natural molecules have one specific bioactivity and (ii) find new uses for specific molecules. Moreover, this group is able to experimentally confirm the predicted bioactivity by means of *in vitro*, *in vivo*, and *ex vivo* experiments.
- http://www.cosmostox.eu/home/welcome/: COSMOS was one of seven projects forming the SEURAT-1 cluster, SEURAT being a European research initiative with the long-term goal of achieving "Safety Evaluation Ultimately Replacing Animal Testing."
- http://www.cyprotex.com: Cyprotex specializes in ADME-Tox and Biosciences including both *in vitro* (laboratory experiments) and *in silico* (computer modeling) approaches.
- https://www.ijert.org/view-pdf/8966/toxicity-prediction-of-cosmetics-product-through-bioinformatics-tool-test: Toxicity Prediction of Cosmetics Product through Bioinformatics Tool: T.E.S.T.

References

[1] Galinski, E.A. (1993) *Experientia*, **49**, 487–496.
[2] Galinski, E.A. (1997) *Comp. Biochem. Physiol. A: Mol. Integr. Physiol.*, **117A**, 357–365.
[3] Kolp, S., Pietsch, M., Galinski, E.A., and Guetschow, M. (2006) *Biochim. Biophys. Acta*, **1764**, 1234–1242.
[4] Goeller, K. and Galinski, E.A. (1999) *J. Mol. Catal. B: Enzym.*, **7**, 37–45.
[5] Yu, I. and Nagaoka, M. (2004) *Chem. Phys. Lett.*, **388**, 316–321.
[6] Crowe, J.H., Carpenter, J.F., Crowe, L.M., and Anchordoguy, T.J. (1990) *Cryobiology*, **27**, 219–231.
[7] Schobert, B. and Tschesche, H. (1978) *Biochim. Biophys. Acta*, **541**, 270–277.
[8] Wiggins, P.M. (1990) *Microbiol. Rev.*, **54**, 432–449.
[9] Lippert, K. and Galinski, E.A. (1992) *Appl. Microbiol. Biotechnol.*, **37**, 61–65.
[10] Lentzen, G. and Schwarz, G. (2006) *Appl. Microbiol. Biotechnol.*, **72**, 623–634.
[11] Orth, D.S. and Appa, Y. (2000) *Glycerine: A natural ingredient for moisturizing skin*, in *Dry Skin and Moisturizers: Chemistry and Function* (eds M. Loden and H.I. Maibach), CRC Press, FL, pp. 213–228.
[12] Rabe, J.H., Mamelak, A.J., McElgunn, P.J.S., Morison, W.L., and Sauder, D.N. (2006) *J. Am. Acad. Dermatol.*, **55**, 1–19.
[13] Toyoda, M. and Bhawan, J. (1997) *J. Dermatol. Sci.*, **14**, 87–100.
[14] Grewe, M. (2001) *Exp. Dermatol.*, **26**, 608–612.
[15] Bushan, M., Cumberbatch, M., Dearman, R.J., Andrew, S.M., Kimber, I., and Griffiths, C.E. (2002) *Br. J. Dermatol.*, **146** (1), 32–40.
[16] Pfluecker, F., Bunger, J., Hitzel, S., and Vitte, J. (2005) *SÖFW J.*, **131**, 20–30.
[17] Buenger, J. and Driller, H.J. (2004) *Skin Pharmacol. Physiol.*, **17**, 232–237.
[18] Grether-Beck, S., Timmer, A., and Felsner, I. (2005) *J. Invest. Dermatol.*, **125**, 545–553.
[19] Yasui, H. and Sakurai, H. (2003) *Exp. Dermatol.*, **12**, 298–300.
[20] Tolmasoff, J.M., Ono, T., and Cutler, R.G. (1980) *Proc. Natl. Acad. Sci. U.S.A.*, **77**, 2777–2781.
[21] Loden, M. (2003) *Clin. Dermatol.*, **21**, 145–157.
[22] Impact (2005) *Version 4.0*, Schrödinger, LLC, New York, NY.
[23] Lin, E., Ratnikov, B.I., Tsai, P.M., Carron, C.P., Myers, D.M., Barbas, C.F., and Smith, J.W. (1997) *J. Biol. Chem.*, **272**, 23912–23920.
[24] Hynes, R.O. (1992) *Cell*, **69**, 11–25.
[25] Schaffner, P. and Dard, M.M. (2003) *Cell Mol. Life Sci.*, **60**, 119–132.
[26] Sagnella, S., Anderson, E., Sanabria, N., Marchant, R.E., and Kottke-Marchant, K. (2005) *Tissue Eng.*, **11**, 226–236.
[27] Metacore™ Version 4.3, https://portal.genego.com/ (accessed January 2018).
[28] Mewes, K.R., Raus, M., Bernd, A., Zöller, N.N., Sättler, A., and Graf, R. (2007) *Skin Pharmacol. Physiol.*, **20**, 85–95.
[29] PIQOR™ Skin Microarray PIQOR of Miltenyi Biotec GmbH, http://www.miltenyibiotec.com/download/ (accessed January 2018).

[30] Huber, W., von Heydebreck, A., Sültmann, H., Poustka, A., and Vingron, M. (2002) *Bioinformatics*, **18** (Suppl. 1), 96–104.
[31] Smyth, G.K., Michau, J., and Scott, H.S. (2005) *Bioinformatics*, **21** (9), 2067–2075.
[32] Quan, T., He, T., Kang, S., Voorhees, J.J., and Fisher, G.J. (2004) *Am. J. Pathol.*, **165**, 741–751.
[33] Lee, J.W. and Juliano, R. (2004) *Mol. Cells*, **17**, 188–202.
[34] Kloepper, J.E. (2008) *Exp. Cell. Res.*, **314**, 498–508.
[35] Paus, R. (2001) *Physiol. Rev.*, **81**, 449–494.
[36] Zhang, H.Y., Timpl, R., Sasaki, T., Chu, M.L., and Ekblom, P. (1996) *Dev. Dyn.*, **205**, 348–364.
[37] Wang, J.C. (1996) *Annu. Rev. Biochem.*, **65**, 635–692.
[38] Ohyama, M., Terunuma, A., and Tock, C.L. (2006) *J. Clin. Invest.*, **116**, 249–260.
[39] Runyan, C.E., Poncelet, A.C., and Schnaper, H.W. (2006) *Cell Signal*, **18**, 2077–2088.
[40] Magan, N., Szremska, A.P., Isaacs, R.J., and Stowell, K.M. (2003) *Biochem. J*, **374**, 723–729.
[41] Grassel, S., Sicot, F.X., Gotta, S., and Chu, M.L. (1999) *Eur. J. Biochem.*, **263**, 471–477.
[42] Sugiura, T. (1999) *Biochem. J*, **338**, 433–440.
[43] Kobielak, K., Pasolli, H.A., and Alonso, L. (2003) *J. Cell Biol.*, **163**, 609–623.
[44] Chih-Chiun, C., Mo, F.-E., and Lau, L.F. (2001) *J. Biol. Chem.*, **276**, 47329–47337.
[45] Latjnhouwers, M.A., Bergers, M., Van Bergen, B.H., Spruijt, K.I., Andriessen, M.P., and Schalkwijk, J. (1996) *J. Pathol.*, **178**, 30–35.
[46] Kulski, J.K., Lim, C.P., Dunn, D.S., and Bellgard, M. (2003) *J. Mol. Evol.*, **56**, 397–406.
[47] Lener, T., Moll, P.R., Rinnerthaler, M., Bauer, J., Aberger, F., and Richter, K. (2006) *Exp. Gerontol.*, **41**, 387–397.
[48] Kizawa, K. and Ito, M. (2005) *Methods Mol. Biol.*, **289**, 209–222.
[49] Fisher, G.J., Kang, S., Varani, J., Bata-Csorgo, Z., Wan, Y., Datta, S., and Voorhees, J.J. (2002) *Arch. Dermatol.*, **138**, 1462–1470.
[50] Ni, A., Welsh, W.J., Santhanam, U., Hu, H., and Lyga, J. (2014) *PLoS One*, **9** (11).
[51] Anzali, S., Berthold, M.R., Fioravanzo, E., Neagu, D., Péry, A.R.R., Worth, A.P., Yang, C., Cronin, M.T.D., and Richarz, A.N. (2012) *IFSCC Magazine*, **15**, 249–255.

12 Applications in Materials Science

Tu C. Le[1] and David A. Winkler[2,3,4,5]

[1] School of Engineering, RMIT University, Swanston Street, Melbourne, Victoria 3001, Australia
[2] Department of Biochemistry and genetics, La Trobe Institute for Molecular Science, La Trobe University, Kingsbury Drive, Bundoora 30186, Australia
[3] School of Medicinal Chemistry, Monash Institute of Pharmaceutical Sciences, Royal Parade, Parkville 3052, Australia
[4] School of Chemical and Physical Sciences, Flinders University, Sturt Rd, Bedford Park 5042, Australia
[5] Biomedical Manufacturing, CSIRO Manufacturing, Bayview Avenue, Clayton 3168, Australia

Learning Objectives

- To describe how chemoinformatics is applied in materials science.
- To review challenges and able to avoid the pitfalls in materials modeling.
- To identify the types of materials properties that can be modeled and predicted.
- To compare the special problems in QSAR modeling of materials such as polymers, catalysts, and nanomaterials with small molecule drugs.
- To gain practical experience in modeling materials properties by the way of real world examples from the scientific literature.

Outline

12.1 Introduction, 547
12.2 Why Materials Are Harder to Model than Molecules, 548
12.3 Why Are Chemoinformatics Methods Important Now? 548
12.4 How Do You Describe Materials Mathematically? 549
12.5 How Well do Chemoinformatics Methods Work on Materials? 551
12.6 What Are the Pitfalls when Modeling Materials? 551
12.7 How Do You Make Good Models and Avoid the Pitfalls? 553
12.8 Materials Examples, 554
12.9 Biomaterials Examples, 561
12.10 Perspectives, 566

12.1 Introduction

Many of the methods for modeling relationships between chemical structure and biological activity or physicochemical properties such as solubility are so-called platform technologies. This means that they are quite general and can be applied to many other types of molecules and properties. One of the exciting changes that have occurred in the chemoinformatics field has been the

Applied Chemoinformatics: Achievements and Future Opportunities, First Edition.
Edited by Thomas Engel and Johann Gasteiger.
© 2018 Wiley-VCH Verlag GmbH & Co. KGaA. Published 2018 by Wiley-VCH Verlag GmbH & Co. KGaA.

realization that computational molecular modeling and design methods developed for drug discovery over the past 50 years can be applied equally well to the discovery and optimization of materials more generally. The application of chemoinformatics methods like quantitative structure–property relationships (QSPRs) modeling to materials science was comprehensively reviewed recently [1]. Two of the hottest areas of research and commercial application in materials science in the past 10–15 years are nanotechnology [2] and regenerative medicine [3]. Although they may seem quite different fields, they do have a lot in common. The differences are largely whether the materials developed are used in biological or non-biological applications. It is obvious that scientists are gaining some degree of mastery over the purposeful design of materials for both of these important areas. We are now designing novel porous materials like metal–organic frameworks (MOFs) to solve environmental problems such as the capture of CO_2 and its conversion into "green" fuel for transport [4]. The unique properties that many materials acquire when in a nanoparticulate form are also being recognized, and products containing nanoparticles are becoming common. Additive manufacturing (3D printing) is finding exciting new applications for novel materials; new polymers are making lighting, communication, and entertainment cheaper and more energy efficient; and we are also seeing the appearance of new cheaper solar cells. In the healthcare area, we are entering an unprecedented area where we can literally reprogram cells to do our bidding and can design new biocompatible materials that can not only replace diseased or damaged parts of the body but also "instruct" cells to speed recovery.

12.2 Why Materials Are Harder to Model than Molecules

One of the most important differences between discrete molecules like drugs and most materials is that most materials are not single precisely defined entities like molecules. Invariably, polymers consist of distributions of chain length, and more complex polymers like block and graft copolymers are statistical distributions of closely related polymers. If the polymers are cross-linked, then the cross-linking is quite stochastic, and it is impossible to know exactly which monomer is connected to which other monomer. Nanomaterials such as nanoparticles similarly are usually distributions of sizes, shapes, and surface chemistries when they are manufactured, and they can also change their properties dynamically over time either by dissolving, by aggregating, or by interacting with other environmental materials like proteins. Most materials are also much larger than discrete molecules like drugs making conversion of their structures or distributions of structures into mathematical descriptors potentially very difficult. However, as difficult as these problems are, as we will illustrate, chemoinformatics methods are capable of generating very useful models and predictions for very complex materials and properties.

12.3 Why Are Chemoinformatics Methods Important Now?

Around 25 years ago, the pharmaceutical industry was subject to a "disruptive technology" in the form of high-throughput combinatorial and robotic chemical synthesis and characterization methods. The driving force was the realization that drug-like chemical space (the number of possible chemical compounds that could theoretically be made using the laws of chemistry) is impossibly large, comparable to the number of atoms in the universe [5]. The premise was that, if drug molecules could be synthesized and screened for biological activity orders of magnitude faster than had previously been the case, then more drugs would result. Sadly, this has not occurred for a number of reasons, primarily a lack of chemical diversity in the libraries generated, issues with late failures of drug leads in clinical trials, unexpected toxicity or insufficient potency, and so on. However, the automation of chemical synthesis and biological screening has benefited the pharmaceutical industry in a myriad of ways that the materials research community is starting to learn from and to adopt these methods. Although still in its infancy, high speed robotized materials synthesis and characterization methods are emerging (e.g., CSIRO's rapid automated materials and processing (RAMP) Centre in Melbourne, Australia). This means that large sets of data for complex materials are starting to emerge, and researchers and companies are looking for methods to analyze and understand these datasets, so they can inform the discovery and optimization of advanced materials. Coupled to this is the emergence of a number of important areas when big data is a major issue some of which, like omics' data, chemoinformatics methods can address in novel ways. If these accelerated technologies are to fulfill their promise, robust methods for extracting useful information and models from the very large data sets they generate are essential.

12.4 How Do You Describe Materials Mathematically?

Because computational modeling and design involves creating simplified mathematical models of molecules and finding their relationship with materials properties (physical or biological), it is essential to convert rather complex molecules or materials into a set of numbers that capture their most physicochemically relevant properties. These sets of numbers are called molecular descriptors, and it is now possible to generate many thousands of them for any particular molecule or material. The choice of descriptors strongly depends on the type of materials being modeled, the property being modeled or predicted, and, for bioactive complex materials, the way in which the material interacts with biology. For example, in the case of nanoparticles or polymers, it is often the properties at the surface that dominate their interactions with cells and tissues.

Simple material compositional and process parameters have been shown to generate useful models, especially for catalysts. Molecular descriptors are also commonly used as these describe the microscopic properties of the materials. However, in the case of polymers, monomers or repeating units are often used to generate molecular descriptors for the models because it is impossible to represent the entire polymer chain in terms of mathematical descriptors (notwithstanding the lack of information on the distributions of chain lengths, block sizes, cross-links, etc.). Furthermore, their chain length and polydispersity are often not well characterized. Measured or calculated physicochemical properties of the materials can also be used as descriptors.

Examples of these are octanol–water partition coefficients, dipole moment, polarizability, or highest occupied molecular orbital–lowest unoccupied molecular orbital (HOMO–LUMO) energy gap. Descriptors describing the size, shape, or porosity of the material components are also useful. Developing descriptors for nanomaterials can be challenging due to their structural diversity, as Figure 12.1 illustrates. Quite often models are built for a single class of nanomaterials using physicochemical properties as descriptors. It has become clear that the development of better ways of capturing, by mathematical descriptors, the complex properties of materials in the most general sense is an extremely important research need now and into the future.

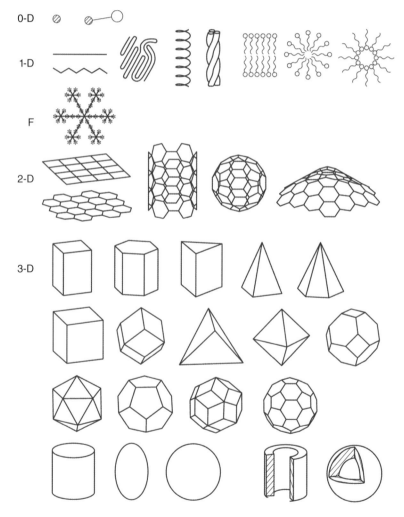

Figure 12.1 Structural diversity of the nanoworld: zero-dimensional (point), one-dimensional (linear), fractal, two-dimensional, and three-dimensional nanoparticle fragments. (Reprinted with permission from Ref. [6]. Copyright 2003 Maik Nauka Interperiodica.)

12.5 How Well do Chemoinformatics Methods Work on Materials?

Because chemoinformatics methods are based on mathematical relationships between descriptors and biological or physicochemical properties, they are intrinsically adaptable to a wide range of materials. Chemoinformatics models are essentially complex, often nonlinear mathematical relationships that link the relevant materials physicochemical parameters, as expressed by mathematical descriptors, with useful materials properties such as hardness, catalytic efficiency, biocompatibility, transparency, conductivity, solubility, cell adhesion, or a myriad of other properties.

The bottom line is that methods capable of generating a statistically valid and predictive mathematical relationship between drug molecule properties and, for example, protein receptor responses can equally be used to generate relationships between microscopic or processing materials parameters and useful macroscopic properties. This realization is relatively recent, and there have been fewer chemoinformatics studies of materials than drug candidates. We recently reviewed the state of the art on materials modeling [1].

The problems with generating good, robust, and predictive chemoinformatics models of materials responses are largely due to the difficulties in characterizing more complex molecules in a mathematically useful way. If this can be done, a myriad of chemoinformatics methods like neural networks, decision trees, and kernel-based methods like support vector and relevance vectors machines can be used to generate good and useful models (see *Methods Volume*, Chapter 11).

12.6 What Are the Pitfalls when Modeling Materials?

All modeling methods, chemoinformatics methods included, need to be applied carefully as it is relatively easy to make mistakes and generate seemingly good models that have no interpretative or predictive value. The errors are not specific to materials and impact equally when modeling small organic molecules like drugs. The main difference in modeling materials is the difficulty in knowing the precise nature of the rather complex material (polymer, nanomaterials, catalyst, porous material, gel) so that its properties can be captured accurately by mathematical descriptors.

The main pitfalls in generating any chemoinformatics models using the quantitative structure–activity relationship (QSAR) method (see *Methods Volume*, Chapter 12) or its property-based variant QSPR are as follows:

- **Uninformative descriptors:** If mathematical descriptors capture properties of materials that are irrelevant to their biological or physicochemical impact, or if descriptors vary only slightly across the dataset, it is extremely unlikely that useful models will result. This is much more likely to be an issue with distributions of complex materials than with single small organic molecules.
- **Insufficient variation** in the useful materials property (dependent variable) being modeled. If all molecules or materials in the datasets generate roughly

the same property then, clearly, there is no SAR to be found. Materials properties should vary by at least a factor of 10 across the dataset, preferably by much more.
- **Under- and overfitting:** Often the number of materials in the dataset is limited by the cost or difficulty of synthesis or biological testing. It is important that chemoinformatics models do not contain too many parameters that can be fitted to the data. In an extreme case it is possible to generate models that predict the dataset perfectly but are incapable of predicting the properties of anything else (model overfitting). This occurs when the numbers of parameters in the model approaches the number of data points or compounds in the dataset. Sparse models where the number of fitted parameters in the models are very much less than the number of data points are likely to make better predictions of the properties of new materials the model has not seen. However, if models are too simple they will not capture the full SAR and again make poor predictions. For example, if a linear model is generated from data that are intrinsically nonlinear (say, parabolic), this model will make poor predictions. Getting the complexity of chemoinformatics models correct can be achieved by regularizing the model. This penalizes overly complex models and finds a balance between model complexity and the accuracy of prediction of the dataset used to generate the models.
- **Poor selection of relevant descriptors and chance correlations:** It is now relatively simple to generate many thousands of mathematical descriptors for a given molecule or material. Selection of the most relevant subset of these needs to be done carefully and in a context-dependent way. If many small subsets of descriptors are repeatedly selected from a very large pool of possibilities, it is likely that a chance correlation will occur, generating a seemingly good model, with very little predictivity. Some time ago, Topliss wrote several important papers that show how chance correlations can occur and how to avoid them [7, 8]. Generally, a small pool of relevant descriptors should be chosen only once from the larger pool. There are some ingenious mathematical methods for making sparse, context-dependent selections of relevant descriptors, including Bayesian sparse feature selection methods [9], decision trees, and methods based on information theory or evolutionary methods.
- **Ill-posed regressions:** Regression is the most important operation in generating chemoinformatics models, but is intrinsically unstable or "ill-posed." This means that small changes to the input parameters can cause large swings in the model outputs. Such models make less reliable predictions and are not robust to the effects of noisy or missing data. Regularization, described above, can convert it into a stable, "well-posed" problem where the models remain relatively constant when generated from noisy data.
- **Predicting outside the model's domain of applicability:** Chemoinformatics models make the best predictions when interpolating to new materials that lie within the multidimensional descriptors space defined by the materials used to train them. This is called the model's domain of applicability. If the model is used to extrapolate to materials outside this domain, the predictions rapidly become less reliable; the further the materials are from the domain of the model. This can be an issue when using a model generated from a relatively

small number of materials to virtually screen a large number of possible new materials.
- **Incorrect processing of outliers:** Datasets always contain errors, either intrinsic experimental uncertainty in the biological response data or in the structure or distribution of materials in the dataset. However, materials can be poorly predicted if they contain unusual molecular or property features (e.g., a functional group that appears in only one material), and sometimes biological data contains simple errors (e.g., incorrect position of a decimal point). It is valid to remove a few materials from a model that are poorly predicted, but the reasons for the omission must be clearly understood and stated in any publications. It is not usually valid or correct to remove data points that just "don't fit" the model and to generate an improved model from the remainder.

12.7 How Do You Make Good Models and Avoid the Pitfalls?

Clearly, to generate good, robust models linking materials properties to their biological impact, you need to avoid the pitfalls identified above. Ideally as much information as possible about the materials should be collected so that the distribution of the structure, size, shape, cross-linking, *etc.* of the materials are understood. Sometimes it is not possible to measure all of the required physical or physicochemical parameters for sets of materials, due to complexity, cost, time, or other limitations. In these cases, inclusion of process variables can help capture some of the missing data and also provides useful interpretative information showing which process variables such as temperature, time, solvent mix, and so on have a major effect on materials properties or which are relatively unimportant. Design of experiments can also help here, to ensure that all of the likely materials or processing variability is sampled by the model in a minimum number of experiments.

To build a successful model, we need to choose descriptors that are relevant to the materials properties being modeled. This is particularly difficult for materials like polymers and nanomaterials, and physiochemical as well as structural descriptors are usually required. For example, when polymers are represented by monomer units, calculated molecular descriptors and physical properties such as glass transition temperature or air–water contact angle might be considered as material descriptors. Additionally, useful materials descriptors can often be obscure or arcane microscopic properties derived from quantum chemical calculations or topological properties of the materials components. Although such descriptors can generate successful and useful models, it is hard to understand how the microscopic properties influence the macroscopic (measured) properties in a mechanistic way. This also makes it hard to "reverse engineer" the model to optimize materials directly.

Recognition of the causes of chance correlations, overfitting, and overtraining (for neural networks), makes avoiding these pitfalls relatively straightforward. By ensuring the number of fitted variables in the model does not exceed 25–50% of the number of data points, overfitting can be avoided. A more elegant method is to use Bayesian methods with sparse priors to choose the best subset of descriptors and automatically optimize model complexity [9, 10]. The statistics from the models can also provide warning that there is a problem. If the statistical parameters (e.g., r^2 and standard error) for the training set and independent test set are similar, the model is probably valid. If the training set statistics are very good (high r^2 and low standard error) and substantially different to those of the test set, overfitting should be suspected. In most models the standard error is a better measure of model predictive power than is the r^2 value [11]. Scrambling the biological data (randomly reallocating the dependent variables of materials in the data set) is also a good way to assess whether a given model is valid. Models derived from scrambled data should have r^2 values close to zero, very substantially lower than the r^2 values for the models.

12.8 Materials Examples

Chemoinformatics methods are starting to be applied to materials generally, stimulated by the recognition of the vast size of materials spaces and the increasing automation of materials synthesis, characterization, and testing discussed above. Table 12.1 summarizes some of the main areas where chemoinformatics studies have been conducted in materials. Interested readers should refer to a recent Chemical Reviews paper for additional information [1].

Table 12.1 Common types of materials chemoinformatics studies [1].

Material type	Property modeled	Example reference
Fullerenes and nanotubes	Solubility	[12]
Homogeneous catalysts	Turnover number and frequency	[13]
Heterogeneous catalysts	Crystallinity and population of phases	[14]
Electrocatalysts	Catalyst performance	[15]
Polymers	Glass transition temperature; refractive index	[16, 17]
Ionic liquids	Melting points; conductivity and viscosity	[18, 19]
Supercritical solvents	Solubility of dyes	[20]
Ceramics	Permittivity and oxygen diffusion	[21]

12.8.1 Inorganic Materials and Nanomaterials

Nanomaterials have been the subject of numerous QSAR/QSPR studies over the last decade. Much of the published QSPR modeling work has concerned the biological effects and safety of nanoparticles, their use in selectively targeting cancer cells for diagnostic or therapeutic purposes [22, 23], and the design of self-assembling, soft nanoparticles as drug delivery systems [24]. Some studies have focused on the solubilities of fullerenes in different solvents. In these studies, descriptors were calculated for the organic solvent molecules. These included constitutional, topological, geometrical, electrostatic, and quantum chemical descriptors because of their ease of calculation using CODESSA [25] or DRAGON [26] and quantum chemical software.

An example of fullerene solubility, log S, model was published by Danauskas and Jurs for C_{60} in 96 solvents [27]. They divided the data into three sets, a training set (76 solvents), a cross-validation set for stopping network training (10 solvents), and an external test (prediction) set (10 solvents). Four types of quantum chemical and topological descriptors were used, individually and in combination. Three types of models were generated: multiple linear regression; three-layer, feed-forward neural network with the descriptors the multivariate linear regression (MLR) model; and neural network where a genetic algorithm was used to select descriptors. The linear MLR solubility model employed nine descriptors chosen from a reduced pool of 85 descriptors and exhibited a training set root-mean-square error (RMSE) of 0.42 log S and a test set prediction error of 0.50 log S. The best neural network model employing the same nine descriptors and three nodes in the hidden layer generated models had RMSE values of 0.30 log S for the training set prediction, 0.45 log S for the cross-validation set, and 0.52 log S for the test set predictions. The best solubility model resulting from the use of genetic algorithms to select descriptors for the artificial neural network (ANN) model had a training set RMSE of 0.26 log S, cross-validation RMSE of 0.25 log S, and test set RMSE of 0.35 log S. In terms of test set prediction accuracy, the MLR and neural network model with the same descriptors have equivalent accuracies (0.50 and 0.52 log S), suggesting that the relationship between solvent structure and fullerene solubility was linear. The improved prediction accuracy of the neural network model that used a genetic algorithm to select descriptors (0.35 log S) suggests that using a method like MLR to select model descriptors is not optimal.

As nanomaterials are finding increasing use in diagnostics and therapeutics, there have been many reports on modeling their biological responses. We have also contributed to this area by publishing our sparse machine learning models to predict the cancer cell uptake of a library of surface modified gold nanoparticles [23]. The QSPR models were built using two sets of data corresponding to amide ligands (mono ligand set) or the amide ligands plus folate (dual-ligand set) on the surface of the nanoparticles. The dataset was divided into eight 30 data point subsets corresponding to the cellular uptake in four cancer cell lines. The structures of different types of surface chemistry were used to generate molecular descriptors that capture their biological relevant properties. The DRAGON software was used to calculate this initial pool of descriptors (482) that mathematically encoded properties such as geometry, partial charges, existence of molecular fragments, or distribution of atoms and atomic mass. QSPR models were then generated using three sparse machine learning methods: multiple linear regression with expectation maximization, nonlinear Bayesian regularized artificial neural networks with Gaussian prior or Laplacian prior. Using these approaches, we generated models with high predictivity for cancer cell uptake. For example, for the cervical cancer cell uptake of nanoparticles with dual ligands (one organic ligand plus folate), the sparsity of the models was increased to reduce the number of descriptors from the initial pool of 482 descriptors. The quality of the models degraded significantly when the number of descriptors was reduced to below 13, indicating that relevant information was being removed from the models. The performance of the optimally sparse linear and nonlinear models is shown in Table 12.2, and their abilities to predict the training (80%) and test (20%) sets are illustrated in Figure 12.2. As can be seen, both models could predict very well the cervical cancer cell uptake, accounting for more than 90% of the variance in the data.

Table 12.2 Statistical results of the best multiple linear regression with expectation maximization and Bayesian-regularized artificial neural networks with Gaussian prior models for the cervical cancer cellular (Hela) uptake of dual-ligand nanoparticles (N_{eff} is number of effective weights (adjustable parameters) in the model).

Method	N_{eff}	r^2	Training set SEE × 10^{-11} g/cell	r^2	Test set SEP × 10^{-11} g/cell
MLREM	14	0.98	0.58	0.93	0.81
BRANNGP	15	0.96	0.28	0.94	0.76

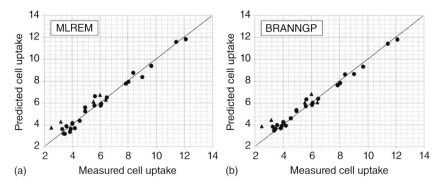

Figure 12.2 Measured and predicted Hela cell uptake (×10^{-11} g/cell) of nanoparticles with dual ligands on their surfaces. (a) The linear model and (b) the nonlinear model. The training set is denoted by circles, and the test set by triangles.

Chemoinformatics has also been shown to be useful in developing improved solar cells to address the rapid growth in energy demand and the urgent need for clean energy. Very recently, Yosipof et al. reported models with good prediction statistics for three important photovoltaic properties of two libraries of metal-oxide-based (TiO$_2$|Cu—O and TiO$_2$|Cu$_2$O) photovoltaic cells [28]. Seven descriptors were used to generate the models:

1. Thickness of the TiO$_2$ window layer
2. Thickness of the Cu—O or Cu$_2$O absorber layer

3. Ratio between the absorber layer and the total (absorber + window) layers
4. Distance of the cell from the center of the depositing plume of the absorber layer
5. Band gap of the absorber layers in electron volts
6. Measured resistance of the absorber layer
7. Maximum theoretically calculated photocurrent

The three device properties that were modeled were the short-circuit photocurrent density, the open-circuit photovoltage, and the internal quantum efficiency. k-Nearest neighbor (kNN) and genetic programming-based symbolic regression methods were employed to build the models. kNN is well known, but the symbolic regression genetic programming method was relatively unusual and worthy of comment. The method generates regression equations from the descriptors by searching the space of mathematical expressions that could be applied to the descriptors to find the model that best fits a given dataset. Methods employing genetic algorithms are being increasingly used for materials property modeling [29]. The kNN approach had relatively higher cross-validation values and external prediction r^2 and q^2 values than the genetic programming approach for both libraries of photovoltaic cells. Feature selection carried out by these methods on the $TiO_2|Cu_2O$ library suggested that the thickness of the TiO_2 window and Cu_2O absorber layers were important factors in determining the photovoltaic properties. To further evaluate the relative importance of these two descriptors, the authors generated linear regression models for the three photovoltaic properties using these descriptors and their interaction terms. The results showed that only the thickness of Cu_2O was a significant predictor for the short-circuit photocurrent density and the internal quantum efficiency, whereas both thicknesses were found important contributors to the open-circuit photovoltage. The work suggested that such models could make good, quantitative predictions and could be used with experimental design to developing novel solar cells in relatively small target materials spaces.

12.8.2 Polymers

A range of polymer properties have been the subject of QSPR studies, including thermophysical properties such as glass transition temperature; thermal decomposition temperature; Flory Huggins parameters; electrical and optical properties such as dielectric constant, electrical conductivity, and refractive index; transport properties such as gas and aqueous diffusion and intrinsic viscosity; mechanical properties such as impact resistance; and so on. The polymer property that has been most widely modeled by QSPR is glass transition temperature, T_g, which can be difficult to determine experimentally because the phase transition may occur over a relatively wide temperature range and depends on measurement technique, duration, and pressure. However studies have shown that it is relatively easy to predict this property using QSPR. The following example of polymer glass transition temperature is taken from Bertinetto *et al.* [16]. The dataset that consists of 615 methacrylic polymers (340 homopolymers and 275 random copolymers) is one of the largest datasets used for T_g models generation. Descriptors were calculated using labeled directed positional acyclic graphs (DPAGs).

Using these descriptors, the recursive neural networks could predict the T_g values to the standard errors of 6–12 K and 6–24 K for the training and test sets, respectively. Figure 12.3 shows the ability of the models to predict the T_g values for a subset of 275 random copolymers.

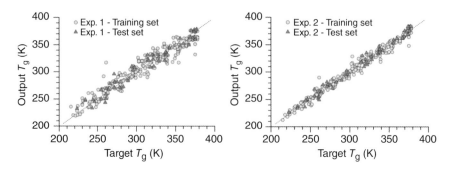

Figure 12.3 Plots of measured versus predicted T_g value for a dataset consisting of 275 random copolymers in experiment 1 where the training error tolerance (TET) is set at 60 K and experiment 2 where TET is 30 K. (Reprinted with permission from Bertinetto et al. [16].)

One interesting study was published by Sumpter and Noid [30]. In this study, nine polymer properties were modeled. These were the molar volume, heat capacity, change in heat capacity at the glass transition temperature, cohesive energy, solubility, glass transition temperature, refractive index, thermal conductivity, and dielectric constant. Unlike most of the other QSPR models for polymers, this study trained the multilayer feed-forward neural network to predict several properties simultaneously by using multiple output nodes. Eighteen topological indices for the repeat units in the polymer were used to generate the models. There were 357 different polymers used to build the QSPR models, which had an average prediction error less than 3%. Neural network models were derived for each physical property separately, resulting in even higher accuracy of prediction.

12.8.3 Catalysts

QSPR models can be used to find optimum reaction conditions, examine the effects of different factors on the catalytic reactions, create virtual catalyst libraries, design new catalysts with better performance, or extract general principles from high-dimensional data resulting from catalysis high-throughput experimentation. In most cases, process variables such as synthesis conditions and catalyst composition were used as descriptors in the QSPR models.

The following example on homogenous catalyst modeling is taken from Burello et al. [13]. 412 Heck cross-coupling reactions were analyzed to generate a QSPR model predicting catalyst performance (turnover number and turnover frequency). Descriptors employed were dipole moment, HOMO and LUMO energies, atom charges, and structural parameters that were relevant to the Heck reaction. Descriptor correlations and principal component analysis (PCA) were used to eliminate descriptors and reduce the dimensionality of the problem to increase robustness and predictivity of the models. Linear regression, ANNs, and classification tree methods were used to model the SPRs. The best models had prediction confidence levels as high as 93%. Palladium loading was the most relevant descriptor for both turnover number and turnover frequency. The models were then used to predict the performance of 60,000 combinations of virtual catalysts and reaction conditions *in silico*. Figure 12.4 shows a contour plot of the predicted turnover number for these 60,000 virtual cross-coupling reactions versus the first two principal components. In this way, a simple and fast selection of the most promising trends in catalysts and reaction conditions can be made.

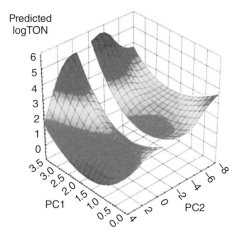

Figure 12.4 Predicted turnover numbers for 60,000 virtual cross-coupling reactions are plotted versus the first two PCs calculated for all the reaction descriptors. The first PC is correlated mainly with the Pd loading and the electronic descriptors of the organic residue on the alkene, while the second represents the ligand's electronic descriptors. (Reprinted with permission from Burello *et al.* [13].)

An example of QSPR models for heterogeneous catalysts is taken from Moliner *et al.* [14]. They reported a high-throughput synthesis study of zeolites using factorial design. Larger datasets, such as those generated by high-throughput experiments, are ideally suited to QSPR modeling. The data for this study were generated by a $3^2\ 4^2$ factorial design, so they consisted of 144 points. These data were partitioned into a training set (100 samples), neural net validation set (20 samples), and a test set (24 samples). The inputs to the neural network were the concentrations of reagents used in the synthesis, and the properties being modeled were the crystallinity and relative amounts of two different zeolite phases. They investigated a range of neural network topologies, ranging from sparse, three-layer networks to more complex four-layer networks. The sparsest models, with the least complex neural network architecture, were able to predict the crystallinity of the two phases at least as accurately as the more complex network architectures. The best QSPR models could predict the training set percent crystallinity to within 5% and the test set crystallinity to within 10%.

These studies showed that neural networks show considerable promise for accelerating development and optimization of new catalysts.

12.8.4 Metal–Organic Frameworks (MOFs)

An example of the application of QSPR methods in predicting properties of MOFs is taken from Amrouche *et al.* [31]. In this study, the property being modeled was the isosteric heat of adsorption of various polar and nonpolar molecules in a large variety of zeolitic imidazolate framework (ZIF) materials. Descriptors used included the dipolar and quadrupolar moments of the organic linker constitutive of the framework, pore mean curvature, number of functional groups on the linker, dipolar moment of the adsorbed gas, and its atmospheric boiling temperature. An equation predicting the isosteric heat of adsorption for ZIF materials from four descriptors was proposed. The authors were satisfied with the predictive power of this equation but proposed different ways to improve it: (i) extending the size of the database used, (ii) including more ZIF or MOF structures, (iii) developing additional descriptors for solid or gas, and (iv) using approaches leading to nonlinear models such as the ANNs.

Because MOFs are made from structural building blocks that can be combined to synthesize nearly an infinite number of materials, QSPR clearly is a useful tool for high-throughput screening. The first large-scale QSPR analysis of MOFs was reported by Fernandez *et al.* for predicting the methane storage capacities of approximately 130,000 MOFs [32]. They used geometrical features such as pore size and void fraction as descriptors. Models were trained using only 10,000 MOFs, and the accuracy of the prediction was evaluated on a test set of 120,000 MOFs. Support vector machine models can predict the methane storage capacity of MOFs in the test set with r^2 values of 0.82 and 0.93 at 35 and 100 bar, respectively. Figures 12.5 and 12.6 illustrate the ability of the models to predict the methane storage capacity of the training and test set, respectively. Although the actual methane storage capacity values were derived from Monte Carlo simulations, the study proves that QSPR could be used effectively to predict properties of a large number of MOF materials.

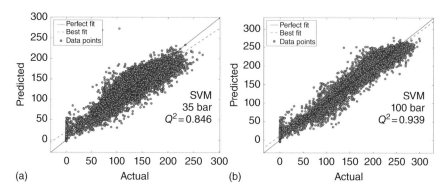

Figure 12.5 Predicted (QSPR) versus actual (derived from Monte Carlo simulations) for methane storage capacities during cross-validation of the 10,000 MOFs in the training set at 35 bar (a) and 100 bar (b). (Reprinted with permission from Fernandez *et al.* [32].)

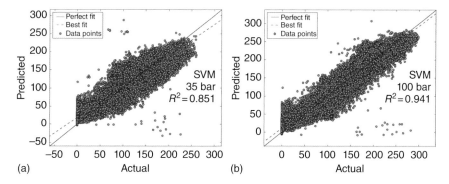

Figure 12.6 Predicted (QSPR) versus actual (derived from Monte Carlo simulations) methane storage capacities for the 127,953 MOFs in the test set using the SVM models at 35 bar (a) and 100 bar (b). (Reprinted with permission from Fernandez et al. [32].)

12.9 Biomaterials Examples

12.9.1 Bioactive Polymers

Polymers are very flexible materials whose compositions and properties can be controlled over a wide range. They can be generated from monomers that resemble biological molecules like peptides and sugars, and their compliance, softness, and rheological properties can, in principle, be manipulated to match those of biological polymers in cells and tissues. Controlled free radical polymerization methods like Atom Transfer Radical Polymerization (ATRP) and Reversible Addition-Fragmentation chain Transfer (RAFT) can provide precise control over the chain lengths and block sizes in bespoke polymers, so that synthetic polymers are logical choices for materials to be implanted into the body to replace damaged parts or to release drugs to treat diseases. As was discussed above, there are an almost infinite number of polymers that could be synthesized using existing chemical knowledge, but, surprisingly, very few have been certified by regulatory agencies like the Food and Drug Administration (FDA) for use in the human body. This provides a great opportunity to develop new biomedical polymers with unprecedented properties, provided that the value they provide justifies the high cost of receiving FDA approval.

There is a relatively large amount of academic and commercial research activity around the use of polymers in medical applications. Examples include polymers, or polymer coatings for catheters, cochlear implants, pacemakers, and so on that resist bacterial attachment and provoke minimal formation of fibroses. Polymers with specially designed signals that instruct cells are now being investigated for cell biofactory/bioreactor applications. These promise cost-effective methods for producing cells and tissues suitable for transplantation [3].

Machine learning and other types of statistical modeling that are used to generate predictive models of the biological impacts of polymers and other materials are data-driven methods. They work best with large, diverse datasets. High-throughput methods for synthesis and characterization of polymers are being developed. A common format consists of arrays of several hundred to several thousand different polymer spots on a slide or chip. These are generated by methods such as dip pen as illustrated in Figure 12.7.

Figure 12.7 Spotting of mixed comonomers onto a slide. After polymerization by UV light, the tiny polymer spots are sterilized and exposed to cells or bacteria in culture and the degree of attachment, and growth recorded after specific time intervals.

So far, there are relatively few research groups using high-throughput polymer synthesis to design polymers for specific biomedical applications, although this is expected to change rapidly. Consequently, there are relatively few QSPR studies because of the paucity of large datasets. Two recent examples of research by groups from the University of Nottingham, MIT, and CSIRO are provided here.

The first example involves the development of polymers that can attach embryonic stem cells (ESCs) and sustain their growth and proliferation [33]. These materials will ultimately be used in biofactories that will generate cells and tissues for biomedical applications. The research teams generated arrays of around 500 polyacrylate polymers formed by mixing various combinations of comonomers and polymerizing them using UV light. In this example embryoid bodies (small clusters of ESCs) were used instead of single cells because ESCs are more robust in this form. Sparse machine learning methods (a neural network to which Bayesian regularization has been applied to optimize the model complexity and predictability) were used to generate quantitative and predictive models that linked the polymer surface chemistry to the degree of attachment of the embryoid bodies. The model could make very good, quantitative predictions of the degree of attachment of embryoid bodies to polymers that were not used to train the model. The nonlinear model could predict the attachment of cells to the polymers in the training and test sets with r^2 values of 0.8 and 0.81, respectively. The model could predict the attachment within a factor of ×1.3 times. The performance of the best chemoinformatics model is illustrated in Figure 12.8.

The models could also provide information on which types of polymer surface chemistry promoted attachment and growth of embryoid bodies.

The second example aimed at designing polymers with very low attachment of pathogenic bacteria [34] that are a major cause of morbidity and mortality in patients with indwelling and implanted medical devices. Infection is a serious issue that creates a roadblock in the development of new implantable medical devices. Again a library of around 500 polyacrylates were spotted onto an array and exposed to three of the most important pathogens responsible for hospital infections, *Pseudomonas aeruginosa* (PA), *Staphylococcus aureus* (SA), and uropathogenic *Escherichia coli* (UPEC). The bacteria were genetically

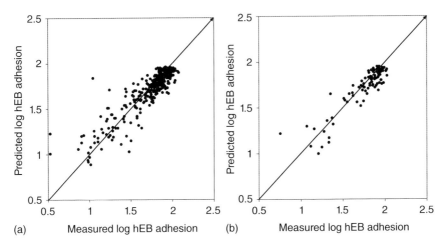

Figure 12.8 The predicted and measured adhesion of embryoid bodies to polymers in a library of acrylates. The performance of the model in predicting (a) the training set and (b) the test set. (Reproduced from Ref. [33] with permission from The Royal Society of Chemistry.)

transformed with green fluorescent protein, so they glowed under UV light. The number of bacteria attached to the polymer spot was proportional to the brightness of the green fluorescence. The diversity of pathogen attachment values is illustrated in Figure 12.9.

Again sparse machine learning methods were used to generate a model linking polymer properties to pathogen attachment, this time aiming at minimizing attachment. The PA model could predict the attachment to polymers in the training set with a standard error of 0.17 log fluorescence (F) and an r^2 of 0.84 and attachment to polymers in the test set with a standard error of 0.16 log F and r^2 of 0.87 (i.e., bacterial fluorescence intensity could be predicted within a factor of 1.5). The SA model could predict adhesion with an r^2 value of 0.85 and standard error of 0.12 log F for both the training and test sets (i.e., fluorescence could be

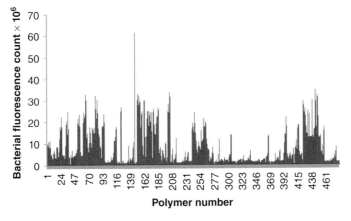

Figure 12.9 Attachment of pathogenic *P. aeruginosa* to the polyacrylate library.

predicted within a factor of 1.4). Finally, the UPEC model, based on a smaller number of polymers that supported pathogen attachment, predicted the adhesion to the polymers in the training set with a standard error of 0.43 log F and an r^2 of 0.58, while UPEC adhesion to polymers in the test set was predicted with an SEP of 0.48 log F and an r^2 of 0.73. UPEC adhesion could therefore be predicted to within a factor of 3. The performance of the sparse machine learning models in predicting the attachment of the three pathogens to the polymers in the training and test sets are summarized in Figure 12.10.

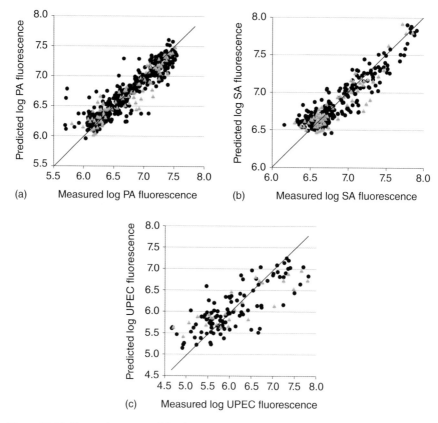

Figure 12.10 The performance of the three QSPR models for *P. aeruginosa* (a), *S. aureus* (b), and UPEC (c). The prediction of the attachment to polymers in the training set are shown in black circles, while test set predictions are in gray triangles. The attachments are on a log scale.

12.9.2 Microarrays

Although a large number of mathematical methods have been developed in the bioinformatics field to analyze data from omics experiments, for example, from gene expression microarrays, chemoinformatics can still provide some novel ways of selecting the key genes controlling specific biochemical processes. Parallel gene expression experiments in Boston and London generated microarray data related to the symmetry of stem cell division and the mechanism by which a bioglass containing strontium ions caused mesenchymal stem cells to differentiate to bone.

In the Boston experiments four different methods were used to induce stem cells to divide symmetrically (to produce two stem cells) or asymmetrically (to produce one stem cell and one progenitor cell) [35]. The aim of the research was to measure the expression of all genes using suitable microarrays for each condition and use chemoinformatics (sparse Bayesian feature selection) to identify a small set of genes whose up- or down-regulation was related to the symmetry of cell division. The method identified five genes from arrays containing around 35,000 gene expression profiles each. Antibody markers synthesized to bind to these gene products showed that two of the identified genes, H2AFZ and B2G, were indeed specific to the symmetry of cell division. This specificity is illustrated in Figure 12.11.

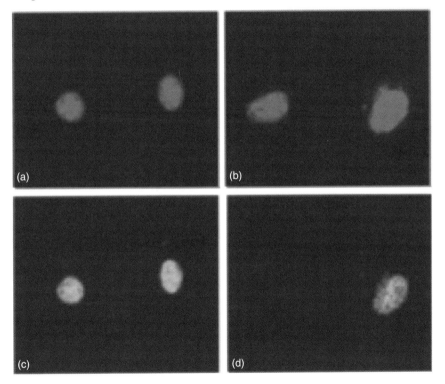

Figure 12.11 Performance of one of the cell division symmetry markers H2AFZ found by sparse chemoinformatics feature selection. Panels show cell nuclei labeled with DAPI (4′,6-diamidino-2-phenylindole, a fluorescent stain that binds strongly to the cell nucleus) in cells dividing symmetrically (a) and asymmetrically (b). Panels (c) and (d) show the same cells labeled with antibody to H2AFZ expression. In the asymmetric cell division case, only one cell is visible (the stem cell).

The bioglass experiments were conducted at Imperial College London by Prof. Molly Stevens' research group. StronBone™ bioglass implanted in the body has been shown to promote bone growth and help prevent fractures due to osteoporosis and other conditions causing bone loss. Experiments were conducted where mesenchymal stem cells were exposed to various bioglass components, especially differing concentrations of strontium ion. Subsequent chemoinformatics sparse feature selection identified a small number of genes encoding fatty acid and sterol biochemical pathways that had not been implicated in bone growth previously [36]. Further biochemical experiments aimed at testing these surprising predictions confirmed the importance of these pathways in bone growth, providing potentially new, effective methods for slowing the effects of osteoporosis using drugs.

12.10 Perspectives

For many years most researchers used QSAR and other chemoinformatics techniques to model small molecules and their biological activities. While some non-biological properties such as solubility, log P values, Henry's Law coefficients, and so on were also successfully modeled and predicted, the vast majority of studies were focused on biological activity. The explosion of applications of QSAR (more correctly QSPR) to materials other than small organic molecules over the past decade has shown conclusively that chemoinformatics methods are very applicable and useful to property predictions for a broad range of materials. One thing holding back wide application of these methods to materials was the lack of high-throughput materials synthesis, characterization, and testing methods for materials, analogous to the combinatorial synthesis technologies in drug discovery. Materials scientists have adapted the automated robotic methods from the pharmaceutical industry and large libraries and databases of materials are now appearing. It is obvious that the development of these efficient technologies for materials will have a dramatic effect on the uptake and impact of data-driven chemoinformatics methods.

Essentials

- Robotic and computing technologies are now providing unprecedented opportunities for chemoinformatics to impact on materials science and regenerative medicine.
- The suite of methods developed for modeling properties of bioactive small molecules are equally effective for complex materials structures and properties
- Materials pose special problems for QSPR modeling over small molecules because of the need for materials specific descriptors, the problems with identifying the correct molecular species to encode by descriptors, and the heterogeneity of complex materials.
- Materials spaces are even larger than drug-like spaces, making the case for computational models even more compelling.

Available Software and Web Services (accessed January 2018)

- Rensselaer Exploratory Centre for Cheminformatics Research: http://reccr.chem.rpi.edu/ONR-QSPR/.
- Materials Genome Initiative: https://mgi.nist.gov.
- Biovia Materials Studio Synthia: http://accelrys.com/products/datasheets/synthia.pdf.
- KNIME, Konstanz Information Miner: http://www.knime.org.
- WEKA, Waikato Environment for Knowledge Analysis: http://www.cs.waikato.ac.nz/ml/weka/.
- ENALOS KNIME nodes: http://tech.knime.org/community/enalos-nodes.

Selected Reading

- Hook, A., Winkler, D.A., and Alexander, M. (2015) Materiomics: a toolkit for developing new biomaterials, in *Tissue Engineering*, 2nd edn (eds C.A. Van Blitterswijk and J. de Boer), Academic Press, Oxford, pp. 253–277.
- Le, T.C., Epa, V.C., Burden, F.R., and Winkler, D.A. (2012) Quantitative structure–property relationship modeling of diverse materials properties. *Chem. Rev*, **112** (5), 2889–2919.
- Le, T.C., Epa, V.C., Winkler, D.A., and Tran, L. (2016) Computational Approaches, in *Adverse Effects of Engineered Nanoparticles*, 2nd edn (eds B. Fadeel, A. Pietroiusti, and A. Shvedova), Academic Press, London, Chapter 5, 83–102. ISBN 978-0128091999.
- Melagraki, G. and Afantitis, A. (2013) Enalos KNIME nodes: exploring corrosion inhibition of steel in acidic medium. *Chemom. Intell. Lab. Syst.*, **123**, 9–14.

References

[1] Le, T., Epa, V.C., Burden, F.R., and Winkler, D.A. (2012) *Chem. Rev.*, **112**, 2889–2919.
[2] Winkler, D.A., Mombelli, E., Pietroiusti, A., Tran, L., Worth, A., Fadeel, B., and McCall, M.J. (2013) *Toxicology*, **313**, 15–23.
[3] Celiz, A.D., Smith, J.G.W., Langer, R., Anderson, D.G., Winkler, D.A., Barrett, D.A., Davies, M.C., Young, L.E., Denning, C., and Alexander, M.R. (2014) *Nat. Mater.*, **13**, 570–579.
[4] Li, P.-Z., Wang, X.-J., Liu, J., Lim, J.S., Zou, R., and Zhao, Y. (2016) *J. Am. Chem. Soc.*, **138**, 2142–2145.

[5] Winkler, D. (2015) *Aust. J. Chem.*, **68**, 1174–1182.
[6] Shevchenko, V.Y., Madison, A.E., and Shudegov, V.E. (2003) *Glass Phys. Chem*, **29**, 577–582.
[7] Topliss, J.G. and Costello, R.J. (1972) *J. Med. Chem.*, **15**, 1066–1068.
[8] Topliss, J.G. and Edwards, R.P. (1979) *J. Med. Chem.*, **22**, 1238–1244.
[9] Burden, F.R. and Winkler, D.A. (2009) *QSAR Comb. Sci.*, **28**, 645–653.
[10] Burden, F.R. and Winkler, D.A. (2009) *QSAR Comb. Sci.*, **28**, 1092–1097.
[11] Alexander, D.L.J., Tropsha, A., and Winkler, D.A. (2015) *J. Chem. Inf. Model.*, **55**, 1316–1322.
[12] Kiss, I.Z., Mandi, G., and Beck, M.T. (2000) *J. Phys. Chem. A*, **104**, 8081–8088.
[13] Burello, E., Farrusseng, D., and Rothenberg, G. (2004) *Adv. Synth. Catal.*, **346**, 1844–1853.
[14] Moliner, M., Serra, J.M., Corma, A., Argente, E., Valero, S., and Botti, V. (2005) *Microporous Mesoporous Mater.*, **78**, 73–81.
[15] Artyushkova, K., Pylypenko, S., Olson, T.S., Fulghum, J.E., and Atanassov, P. (2008) *Langmuir*, **24**, 9082–9088.
[16] Bertinetto, C.G., Duce, C., Micheli, A., Solaro, R., and Tine, M.R. (2010) *Mol. Inform.*, **29**, 635–643.
[17] Katritzky, A.R., Sild, S., and Karelson, M. (1998) *J. Chem. Inf. Comp. Sci.*, **38**, 1171–1176.
[18] Katritzky, A.R., Lomaka, A., Petrukhin, R., Jain, R., Karelson, M., Visser, A.E., and Rogers, R.D. (2002) *J. Chem. Inf. Comput. Sci.*, **42**, 71–74.
[19] Tochigi, K. and Yamamoto, H. (2007) *J. Phys. Chem. C*, **111**, 15989–15994.
[20] Tarasova, A., Burden, F., Gasteiger, J., and Winkler, D.A. (2010) *J. Mol. Graphics Modell.*, **28**, 593–597.
[21] Scott, D.J., Coveney, P.V., Kilner, J.A., Rossiny, J.C.H., and Alford, N.M.N. (2007) *J. Eur. Ceram. Soc.*, **27**, 4425–4435.
[22] Epa, V.C., Burden, F.R., Tassa, C., Weissleder, R., Shaw, S., and Winkler, D.A. (2012) *Nano Lett.*, **12**, 5808–5812.
[23] Le, T.C., Yan, B., and Winkler, D.A. (2015) *Adv. Funct. Mater.*, **25**, 6927–6935.
[24] Le, T.C., Mulet, X., Burden, F.R., and Winkler, D.A. (2013) *Mol. Pharmaceutics*, **10**, 1368–1377.
[25] Karelson, M., Maran, U., Wang, Y., and Katritzky, A.R. (1999) *Collect. Czech. Chem. Commun.*, **64**, 1551–1571.
[26] Mauri, A., Consonni, V., Pavan, M., and Todeschini, R. (2006) *MATCH-Commun. Math. Comp. Chem.*, **56**, 237–248.
[27] Danauskas, S.M. and Jurs, P.C. (2001) *J. Chem. Inf. Comput. Sci.*, **41**, 419–424.
[28] Yosipof, A., Nahum, O.E., Anderson, A.Y., Barad, H.N., Zaban, A., and Senderowitz, H. (2015) *Mol. Inform.*, **34**, 367–379.
[29] Le, T.C. and Winkler, D.A. (2016) *Chem. Rev.*, **116**, 6107–6132.
[30] Sumpter, B.G. and Noid, D.W. (1994) *Macromol. Theory Simul.*, **3**, 363–378.
[31] Amrouche, H., Creton, B., Siperstein, F., and Nieto-Draghi, C. (2012) *RSC Adv.*, **2**, 6028–6035.

[32] Fernandez, M., Woo, T.K., Wilmer, C.E., and Snurr, R.Q. (2013) *J. Phys. Chem. C*, **117**, 7681–7689.
[33] Epa, V.C., Yang, J., Mei, Y., Hook, A.L., Langer, R., Anderson, D.G., Davies, M.C., Alexander, M.R., and Winkler, D.A. (2012) *J. Mater. Chem.*, **22**, 20902–20906.
[34] Epa, V.C., Hook, A.L., Chang, C., Yang, J., Langer, R., Anderson, D.G., Williams, P., Davies, M.C., Alexander, M.R., and Winkler, D.A. (2014) *Adv. Funct. Mater.*, **24**, 2085–2093.
[35] Huh, Y.H., Noh, M., Burden, F.R., Chen, J.C., Winkler, D.A., and Sherley, J.L. (2015) *Stem Cell Res.*, **14**, 144–154.
[36] Autefage, H., Gentleman, E., Littmann, E., Hedegaard, M.A.B., Von Erlach, T., O'Donnell, M., Burden, F.R., Winkler, D.A., and Stevens, M.M. (2015) *Proc. Natl. Acad. Sci. U.S.A.*, **112**, 4280–4285.

13 Process Control and Soft Sensors

Kimito Funatsu

The University of Tokyo, Department of Chemical System Engineering, School of Engineering, Engineering Bld. 3, 7-3-1 Hongo, Bunkyo-ku, 113-8656 Tokyo, Japan

Learning Objectives

- To describe the meaning and the role of soft sensors
- To construct soft sensors
- To construct an adaptive soft sensor for avoiding a model degradation
- To monitor and update a database for a soft sensor with high predictive accuracy
- To efficiently control a process using a soft sensor

Outline

13.1 Introduction, 571
13.2 Roles of Soft Sensors, 573
13.3 Problems with Soft Sensors, 574
13.4 Adaptive Soft Sensors, 576
13.5 Database Monitoring for Soft Sensors, 578
13.6 Efficient Process Control Using Soft Sensors, 581
13.7 Conclusions, 582

13.1 Introduction

In operating chemical industrial plants, the operators have to monitor operating conditions and control process variables. Thus, process variables such as temperature, pressure, liquid level, and concentration of products need to be measured in real time. However, some of them are not easy to measure in real time because of technical difficulties, large measurement delays, high investment cost, and so on. Therefore, soft sensors [1] are widely used to predict values of process variables that are difficult to measure in real time. Figure 13.1 shows the basic concept of a soft sensor. An inferential model is constructed between process variables that are easy to measure in real time, which are called X-variables, and process variables that are difficult to measure in real time, which are called y-variables, using chemoinformatics methods. The values of y can then be predicted using that model with a high degree of accuracy. Both laboratory samples and measurements of online analyzers are examples of y-variables.

Principal component regression (PCR) and partial least squares (PLS) are mainly used as statistical modeling methods for soft sensors since X-variables are usually correlated with each other in the operating data. Nonlinear PLS [2], artificial neural network [3], locally weighted PLS [4], and support vector regression (SVR) [5] are employed to handle nonlinear relationships between X and

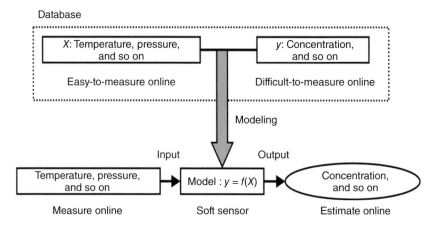

Figure 13.1 Basic concept of a soft sensor.

y. Least absolute shrinkage and selection operator (LASSO) [6] can both select X-variables and construct regression models. Because X-variables can affect y-variables with time delays, important X-variables and optimal time delays of each variable are selected simultaneously using genetic algorithm-based process variables and dynamics selection (GAVDS) [7].

Application fields and the tasks of soft sensors are listed in Table 13.1.

Table 13.1 List of applications of soft sensors.

Application field	Task	References
Petrochemical process	Real-time monitoring of product quality and control	[1, 3, 8–13]
Pharmaceutical process	Real-time monitoring of tablet quality	[14–16]
Water treatment	Membrane fouling of membrane bioreactor	[17, 18]
Agriculture	Prediction of internal quality of agricultural produce	[19]
Iron manufacture	Prediction of endpoint and particle size distribution	[20]

In petrochemical processes such as distillation columns [21] and chemical reactors [8], the use of soft sensors is increasingly common since the number of y-variables is large in product quality data that should be controlled. Examples of y-variables are concentration of chemical components, 90% distilling temperature, specific weight, polymer density, and melt flow rate. X-variables are temperature, pressure, liquid level, flow rate, and so on. Plant operators can acquire y-values estimated by soft sensors and can use them for real-time process control, which leads to much cost saving in operating plants.

In pharmaceutical processes, tablets whose key ingredient is the drug compound must be produced, meeting rigorous quality requirements in spite of the variance of raw materials and changes of production facilities. When a tablet in a batch process cannot pass the quality tests after several processes such as mixing, tabulating, and coating, or even the final product test, all tablets in the batch are wasted, which amounts to enormous costs. Therefore, the quality of tablets should be monitored and controlled in real time, but quality measurements such as active pharmaceutical ingredient (API) content take too much time to be determined. In addition, it is desired that the quality of not just some tablets but all tablets can be measured in a batch process. Process analytical technology (PAT) [14, 15] is an important technique for monitoring, developing, controlling, and designing critical product quality in pharmaceutical industry. Near-infrared (NIR) spectroscopy, Raman spectroscopy, and so on have been focused to monitor product quality nondestructively in real time as X-variables. Soft sensor models are constructed between the quality of tablets and the intensity of NIR spectroscopy. Soft sensors can achieve real-time release testing (RTRT), in which the quality is controlled in each process by monitoring the quality and performing appropriate actions in real time, and thus the final product test would not be required. In addition, control limits can be set, and the quality of products can be controlled by using soft sensors, which is quality by design (QbD) [16]. The use of soft sensors is expanding now in pharmaceutical processes.

Soft sensors have been used in other fields such as prediction of membrane fouling of membrane bioreactor (MBR) in the long term [22], prediction of internal quality of agricultural produce [19], explosive detection, prediction of endpoints, and particle size distribution of powder [20] in iron manufacture. The range of applications of soft sensors is also expanding and will be wider in the future.

13.2 Roles of Soft Sensors

Firstly, as we already mentioned in the introductory section, soft sensors are used instead of analyzers. Soft sensors predict values of difficult-to-measure variables continuously, and then, the predicted values can be used for continuous process control. In addition, measurement frequencies of analyzers can be reduced by using predicted values instead of measured values.

Secondly, soft sensors are used for the detection of abnormal values of the analyzers. Figure 13.2 shows a time plot of concentration. If the data in Figure 13.2 are measured, the first sample and the last sample may be abnormal, and the concentration analyzer may be broken because these are out of distribution. By using a soft sensor and comparing the measured values and the predicted values, abnormal events can be detected. Since outliers in y-values lead to wrong actions in process control and make process control difficult, the parallel use of y-analyzers and soft sensors contributes to a stable process control.

Thirdly, soft sensors can be employed for the interpretation of the relationships among process variables. If a linear regression model is constructed as in Eq. (13.1),

$$\text{Concentration} = 1.5 \times \text{temperature} - 0.5 \times \text{flow rate} \qquad (13.1)$$

then, of course, there is colinearity between process variables, and the interpretation is not so simple, but temperature will have a positive contribution to the concentration, and flow will have a negative contribution to the concentration. An understanding of the relationships between X and y contributes to the way to manipulate X-values for controlling y-values.

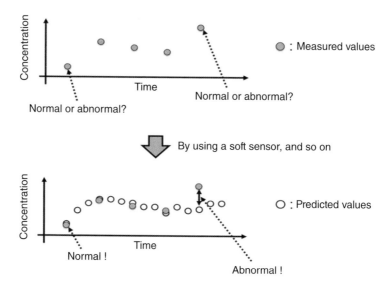

Figure 13.2 Detection of abnormal values of an analyzer using a soft sensor.

Lastly, although monitoring y-values using soft sensors enables a continuous process control, a more efficient process control can be performed by inverse analysis of soft sensors [9]. After construction of soft sensor models, the constructed soft sensor model is inversely analyzed to search for the optimal operating procedure of X for efficiently and stably controlling y-values. The details of this method are explained in Section 13.6.

13.3 Problems with Soft Sensors

Although soft sensors are a very useful tool, soft sensors also carry problems. Figure 13.3 shows the stages from data collection to the operation of soft sensor models and the problems encountered at each stage. First, data are measured in processes and are collected for the construction and the validation of soft sensor models. The problems lie in the reliability of data and in the selection of data. Then, the collected data are preprocessed. At this stage, outlier detection and noise treatment should be performed. After that, soft sensor models are constructed with the preprocessed data. The problems are the selection of appropriate regression methods, overfitting, nonlinearity among process variables, variable selection, and consideration of the dynamics of the processes. Then, the constructed models are analyzed and operated. We have to consider model interpretation, model validation, applicability domain and predictive accuracy, model degradation, maintenance of models, and detection and diagnosis of abnormal data.

One of the crucial problems is the degradation of soft sensor models or model degradation. The predictive accuracy of soft sensors gradually decreases, a result of the changing of the state of chemical plants due to factors such as loss of

13.3 Problems with Soft Sensors

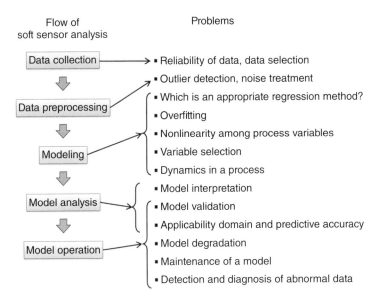

Figure 13.3 Flow of soft sensor analysis and problems involved at each stage.

catalyzing performance and drift of the sensor and the process. Funatsu and coworker categorized the degradation of soft sensor models [10]. Figure 13.4 shows the basic concepts for the degradation of linear soft sensor models constructed between X and y. Figure 13.4a,b represents the shift of y-values and X-values, respectively. These are corresponding to sensor and process drift, scale deposition on pipes, changes of operating conditions such as the amount of raw materials, and so on. The slope does not change between training data and new data, but the values of y-variables or X-variables shift. Figure 13.4c represents changes of the slope of X and y. This is corresponding to the loss of catalyzing performance, changes in the operating conditions such as the concentration in

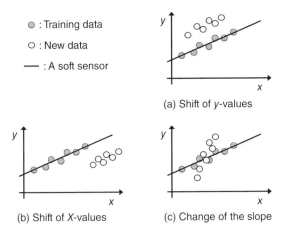

Figure 13.4 Basic concepts of the degradation of a linear soft sensor model [10].

raw materials, and so on. Of course, shifts of y-values and X-values and changes of the slope may occur simultaneously.

When we focus on the rate of degradation, each shift or change happens gradually, rapidly, or instantly. For example, catalyzing performance loss, process and sensor drift, changes of external temperature, and scale deposition on pipes occur gradually; sharp changes in raw materials occurs rapidly; and the correction of drift, the regular repair of plants, and a stoppage of pipes occur instantly. Of course, this rapidity is in fact continuous.

13.4 Adaptive Soft Sensors

To avoid model degradation, adaptation mechanisms can be applied to soft sensors. These soft sensors are called adaptive soft sensors. For example, new data of both X and y are measured in chemical plants and are used to reconstruct soft sensor models and predict y-values. Funatsu categorized adaptive soft sensor models and discussed the characteristics of adaptive soft sensors for each type of model degradation [10]. Adaptive soft sensors include moving window (MW) models [23, 24], just-in-time (JIT) models [11], and time difference (TD) models [12]. MW models are constructed from a recently measured dataset, JIT models are constructed by assigning larger weights to the data that are most similar to a query, and TD models are constructed by considering the time difference of y-variables and that of the X-variables. Ensemble learning can be applied to adaptive models [25].

Table 13.2 shows the characteristics of TD, MW, and JIT models. TD models can adapt shifts of both y-values and X-values because they achieve the same

Table 13.2 Characteristics of TD, MW, and JIT models [10].

Type	Degradation Rapidity	TD model	MW model	JIT model
Shift of y-value	Gradual	●	●	X
	Rapid	●	▲	X
	Instant	●	X	X
Shift of X-value	Gradual	●	●	●
	Rapid	●	▲	●
	Instant	●	X	●
Change of the slope	Gradual	X	●	X
	Rapid	X	▲	X
	Instant	X	X	X
Shift of X-value and change of the slope	Gradual	X	●	●X
	Rapid	X	▲	●X
	Instant	X	X	●X

●, the model can handle the degradation well; ▲, the model can handle the degradation to some extent; X, the model cannot handle the degradation; ●X, it depends on a situation whether the model can handle the degradation or not.

effect in prediction as an update of bias. Even when the shifts happen gradually, rapidly, and instantly, TD models can follow the shifts appropriately. However, TD models cannot adapt to changes in the slope [10].

MW models should be used for following gradual changes of the slope by adding new data to the training data. However, it is difficult for MW models to adapt to rapid and instant shifts because the old data before the shifts occur remain in the training data. MW models are badly affected by old data.

In the case of JIT models, which are constructed with datasets close to the test data in the space of X-variables, an appropriate selection of datasets will be performed if a shift of X-values happens. However, besides that, datasets after shifts of y-values or changes of the slope cannot be selected because there is no change in the space of X-variables as shown in Figure 13.4a,c. When shifts of X-values and changes of slope happen simultaneously, JIT models can adapt to these changes appropriately if the X-values change clearly and an adequate amount of data from the new situations is stored in the database. This is because an appropriate selection of datasets can be performed due to the shifts of X-values.

The previous discussion results were confirmed, and knowledge and information on appropriate adaptive models for each type of degradation could be obtained by analyzing simulated datasets and a real industrial dataset (Mitsubishi Chemical Corporation and Mitsui Chemicals, Inc.) [10].

As shown in Table 13.2, there are no all-round adaptive models with high predictive ability in all types of model degradation. The important thing is to choose an appropriate adaptive model for each type of degradation. Here, a model selection method based on the reliability of TD models is introduced [26]. TD models are used to predict values of y-variables, and its reliability is monitored using the ensemble prediction method in which multiple predicted y-values are obtained by changing the differential values of X-variables and the standard deviation of multiple predicted y-values is an index of prediction reliability. When the reliability is low, a TD model is switched for an MW model or a JIT model. It was confirmed that a combination model of TD and MW models and that of TD and JIT outperformed a single TD model, a single MW model, and a single JIT model through a case study using real industrial data (Mitsubishi Chemical Corporation). In addition, the predictive ability of a combination model of TD and MW models was higher than that of a combination model of TD and JIT models [27].

By switching a TD model and an MW model, or a TD model and a JIT model, a wide range of model degradation can be handled. However, the predictive ability of the current MW, JIT, and TD models are not entirely sufficient when rapid changes in the slope, that is, time-varying changes in processes, occur, as shown in Table 13.2. Therefore, ensemble online support vector regression (EOSVR) [24] was developed as an MW model. Multiple SVR models with different hyper-parameter values predict multiple y-values. The predicted y-values are combined based on the current predictive ability of each SVR model and a Bayes' rule to produce a final predicted y-value. The current predictive ability of each SVR model is calculated as inversely proportional to the root mean square error (RMSE) for the midpoints between the k-nearest-neighbor data points ($RMSE_{midknn}$) [13] as follows: $1/RMSE^2_{midknn,i}$, where $RMSE_{midknn,i}$ is the $RMSE_{midknn}$ of the ith SVR model with the latest data. In addition, the standard deviation of the predicted y-values enables the estimation of the prediction error in the final predicted y-value for each process state.

Example 13.1 *Exhaust Gas Denitration Process*
To verify the prediction ability of EOSVR, we applied the proposed method to an exhaust gas denitration process at Mitsui Chemicals, Inc. NH_3 is injected into a denitration reactor where exhaust gas passes through a catalytic layer, and NO_x is decomposed into harmless N_2 and water vapor [24]. Figure 13.5 shows the reactions and the y- and the X-variables.

- Exhaust gas denitration process at Mitsui Chemicals, Inc.

Catalyst
$$4NO + 4NH_3 + 2O_2 \rightarrow 4N_2 + 6H_2O$$
$$NO + NO_2 + 2NH_3 \rightarrow 2N_2 + 3H_2O$$

Variables y ① NH_3 concentration at the outlet of the denitration reactor
② NO_x concentration at the outlet of the denitration reactor

X 23 variables: temperature, pressure, flow rate, and so on

Figure 13.5 Case study for the EOSVR method.

Soft sensors are used to control the remaining NH_3 concentration at the outlet of the denitration reactor (denitration outlet NH_3), the NO_x concentration at the outlet of the denitration reactor (denitration outlet NO_x), and the NO_x concentration at the outlet of the outlet gas duct (outlet gas duct NO_x). Therefore, the y-variables are the denitration outlet NH_3, denitration outlet NO_x, and outlet gas duct NO_x, and the X-variables are 23 process variables such as temperature, pressure, and flow rate at the gas mixer and the denitration reactor. Examples of the time plots of the measured and the predicted denitration outlet NH_3 concentration are shown in Figure 13.6.

The two time plots on the left-hand side show the Online Support Vector Regression (OSVR) results, which exhibit the best performance of the traditional methods, and the two time plots on the right-hand side show the results given by EOSVR ($RMSE_{midknn}$). Clearly, the predicted y-values of EOSVR ($RMSE_{midknn}$) are closer to the measured y-values than those of OSVR. Even when the measured y-values varied because of changes in fuel oil pressure, the amount of fuel, and the ignition and the extinction of the burners, the EOSVR model could accurately adapt to the changes and could handle the different process states in the plant, reflecting the high predictive ability of the proposed model.

13.5 Database Monitoring for Soft Sensors

To construct adaptive soft sensors with high predictive accuracy for a wide range of data, database monitoring is a crucial problem. To reduce the size of

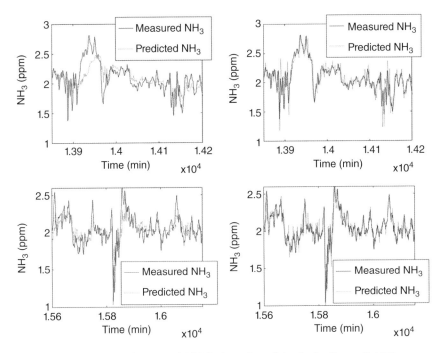

Figure 13.6 Comparison of OSVR and EOSVR: time plots of the denitration outlet NH_3.

a database, JIT models select new measurement data to be stored based on the prediction errors of a y-variable. Jin *et al.* proposed a method in which a new sample should replace the most similar data in the database [28]. However, an overlap between the overall information in a database and that in a new measurement sample was not considered in data selection. Funatsu and coworker proposed a database monitoring index (DMI) for database management (DBM) that examines the amount of information in a new measurement sample and achieves a maintenance-free DBM and highly predictive soft sensors [29].

The DMI is large when two X and y data are dissimilar, and *vice versa*. If the minimum DMI value between a new sample and data in the database exceeds the threshold P_{DMI}, the new sample contains sufficient new information and is stored in the database. Assuming there exist some data that are similar to other data and are not indispensable in the training data, a suitable P_{DMI} can be determined while repeating the deletion of one such similar data and the check of the predictive ability of the regression model constructed with the remaining training data. DMI was modified for adaptive (both MW- and JIT-based) soft sensors with long-term high predictive ability [30].

Example 13.2 *Operating a Distillation Column*
We analyzed data obtained from the operation of a distillation column at the Mizushima Works, Mitsubishi Chemical Corporation. Figure 13.7 shows a schematic representation of the distillation column.

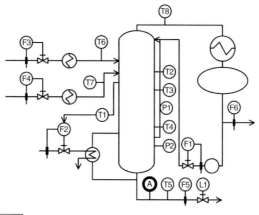

A distillation column at Mitsubishi Chemical Corporation

Variables
y Concentration of bottom product with lowest boiling point
   The measurement interval is 30 min.
X 19 variables: temperatures, pressures, liquid level,
   reflux ratio, and so on

Figure 13.7 Case study for database management: a distillation column.

Time plots of the predicted and the actual y-values for SVR are shown in Figure 13.8. On the x-axis is time ($\times 30$ min), and on the y-axis is the concentration of the bottom product with the lowest boiling point (autoscaled value).

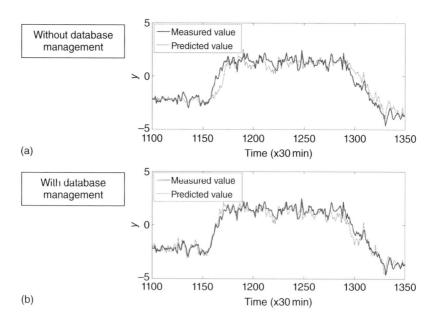

(a) Without database management

(b) With database management

Figure 13.8 Comparison of a process without database management (a) and with database management (b): time plots of measured and predicted y (MW).

Without DBM, the SVR model could not adapt to rapid changes from points 950 to 1000 and 1150 to 1200, and the predicted y-values took time to adjust back to the actual y-values. Using the proposed DBM, the y-values predicted by the SVR model were in good agreement with the actual y-values during both the stable period and the rapid changes from 950 to 1000 and 1150 to 1200. Thus, the proposed method achieved an appropriate DBM for adaptive soft sensors with good predictive ability.

DBM can also be applied to process monitoring, where models are updated or reconstructed with a database that includes new measurement data.

13.6 Efficient Process Control Using Soft Sensors

Although proportional–integral–derivative (PID) controllers are used to control values of process variables, it is difficult to control values of process variables that are difficult to measure since PID controllers are based on the difference of a set point and measured values of a controlled variable. Because soft sensors can estimate values of process variables that are difficult to measure in real time, continuous process control can be performed by using y-values estimated by soft sensors instead of measured y-values in PID control. However, this is far from enough to make full use of soft sensors. By analyzing soft sensor models inversely, a more efficient way to control y-values could be found.

Figure 13.9 shows the basic concept of inverse analysis of a soft sensor model, assuming that a soft sensor model is already constructed. First, the basic patterns of changes of X-values are determined based on history data in which some control such as PID control is conducted. The basic patterns are simplified using

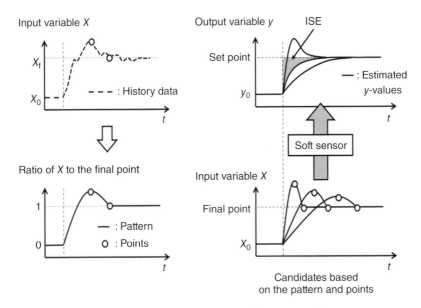

Figure 13.9 Basic concepts of an inverse analysis of a soft sensor model.

some points and an interpolation such as Hermite interpolation. For instance, a piecewise cubic Hermite interpolating polynomial (PCHIP) can be used to determine simplified points, in which a Hermite interpolation is repeatedly employed. It should be noted that each point is determined by time and a y-value.

Each point should be optimized so as to efficiently control y-values. Candidates for time and an X-value of every point are prepared, and patterns in changing X-values are exhaustively generated. Then, each pattern of X-values is input into the constructed soft sensor model, and the output pattern of y is checked in terms of controlled performance of y. For example, the integral of the squared error (ISE) and the settling time are used to quantify the controlled performance of y. The ISE is given in Eq. (13.2):

$$\text{ISE} = \int_0^\infty e^2(t) dt, \tag{13.2}$$

where e(t) means the error between a set point and a y-value. If the number of candidates for time and X-values in the points is too high to check the controlled performance of y for all candidates, an optimization method such as a genetic algorithm can be used. This method is called inverse soft sensor-based feed-forward (ISFF) control method [9].

ISFF was applied to the change of a set point of y in a simulated continuous stirred-tank reactor system. Compared with a traditional proportional–integral (PI) controller, which was optimized in the system, ISFF could control y-values rapidly and stably. Details of the results are shown in Ref. [9]. ISFF can be switched to a feedback controller such as a PID controller since a soft sensor model includes estimation errors and only ISFF cannot completely settle y-values to a set point. In addition, a feedback function can be installed in ISFF by using adaptive soft sensors as shown in Section 13.4.

13.7 Conclusions

Soft sensors are a useful tool in chemical industrial plants. Their applications have spread into many fields such as petrochemical processes, pharmaceutical processes, water treatment fields, agriculture fields, selection of fruits, explosive detection, and iron manufacture. However, there are still problems remaining, and the further revitalization of research and development is strongly desired for those problems in soft sensor analyses. Since problems in soft sensor analysis are similar to those in chemoinformatics and chemometrics, research products in chemoinformatics and chemometrics fields can widely be applied in soft sensor analyses. It is desirable that chemical industrial plants are operated and controlled more effectively and stably using soft sensors.

Essentials

- It is desirable that chemical industrial plants are operated and controlled more effectively and stably using soft sensors.
- Soft sensor applications have spread into many fields such as petrochemical processes, pharmaceutical processes, water treatment fields, agriculture fields, selection of fruits, explosive detection, and iron manufacture.

Selected Readings

- Kadlec, P., Grbic, R., and Gabrys, B. (2011) Review of adaptation mechanisms for data-driven soft sensors. *Comput. Chem. Eng.*, **35**, 1–24.

References

[1] Kadlec, P., Gabrys, B., and Strandt, S. (2009) *Comput. Chem. Eng.*, **33**, 795–814.
[2] Baffi, G., Martin, E.B., and Morris, A.J. (1999) *Comput. Chem. Eng.*, **23**, 395–411.
[3] (a) Zupan, J. and Gasteiger, J. (1991) *Anal. Chim. Acta*, **248**, 1–30; (b) Gasteiger, J. and Zupan, J. (1993) *Angew. Chem. Int. Ed. Engl.*, **32**, 503–527; (c) Dufour, P., Bhartiya, S., Dhurjati, P.S., and Doyle, F.J. (2005) *Control Eng. Pract.*, **13**, 135–143.
[4] Kim, S., Kano, M., Nakagawa, H., and Hasebe, S. (2011) *Int. J. Pharm.*, **421**, 269–274.
[5] Bishop, C.M. (2011) *Pattern Recognition and Machine Learning*, Springer, New York, 738 pp.
[6] Tibshirani, R. (1996) *Stat. Soc.*, **58**, 267–288.
[7] Kaneko, H. and Funatsu, K. (2012) *AIChE J.*, **58**, 1829–1840.
[8] Kaneko, H., Arakawa, M., and Funatsu, K. (2011) *Comput. Chem. Eng.*, **35**, 1135–1142.
[9] Kimura, I., Kaneko, H., and Funatsu, K. (2015) *Kagaku Kougaku Ronbun.*, **41**, 29–37.
[10] Kaneko, H. and Funatsu, K. (2013) *AIChE J.*, **59**, 2339–2347.
[11] Fujiwara, K., Kano, M., Hasebe, S., and Takinami, A. (2009) *AIChE J.*, **55**, 1754–1765.
[12] Kaneko, H. and Funatsu, K. (2011) *Chemom. Intell. Lab. Syst.*, **107**, 312–317.
[13] Kaneko, H. and Funatsu, K. (2013) *J. Chem. Inf. Model.*, **53**, 2341–2348.
[14] Reid, G.L., Ward, H.W. II, Palm, A.S., and Muteki, K. (2012) *Am. Pharm. Rev.*, **15**, 49–55.

[15] Kaneko, H., Muteki, K., and Funatsu, K. (2015) *Chemom. Intell. Lab. Syst.*, **147**, 176–184.
[16] García-Muñoz, S., Dolph, S., and Ward, H.W. II, (2010) *Comput. Chem. Eng.*, **37**, 1098–1107.
[17] Kaneko, H. and Funatsu, K. (2014) *Desalin. Water Treat.*, **53**, 1–6.
[18] Oishi, H., Kaneko, H., and Funatsu, K. (2015) *J. Membr. Sci.*, **494**, 86–91.
[19] Escobar, M.S., Kaneko, H., and Funatsu, K. (2014) *Chemom. Intell. Lab. Syst.*, **137**, 33–46.
[20] Sbarbaro, D., Ascencio, P., Espinoza, P., Mujica, F., and Cortes, G. (2008) *Control Eng. Pract.*, **16**, 171–178.
[21] Kaneko, H., Arakawa, M., and Funatsu, K. (2009) *AIChE J.*, **55**, 87–98.
[22] Kaneko, H. and Funatsu, K. (2013) *Chemom. Intell. Lab. Syst.*, **126**, 30–37.
[23] Kaneko, H. and Funatsu, K. (2013) *Comput. Chem. Eng.*, **58**, 288–297.
[24] Kaneko, H. and Funatsu, K. (2014) *Chemom. Intell. Lab. Syst.*, **137**, 57–66.
[25] Grbića, R., Slišković, D., and Kadlec, P. (2013) *Comput. Chem. Eng.*, **58**, 84–97.
[26] Kaneko, H. and Funatsu, K. (2013) *Ind. Eng. Chem. Res.*, **52**, 1322–1334.
[27] Kaneko, H., Okada, T., and Funatsu, K. (2014) *Ind. Eng. Chem. Res.*, **53**, 15962–15968.
[28] Jin, H.P., Chen, X.G., Yang, J.W., and Wu, L.A. (2014) *AIChE J.*, **71**, 77–93.
[29] Kaneko, H. and Funatsu, K. (2014) *AIChE J.*, **60**, 160–169.
[30] Kaneko, H. and Funatsu, K. (2015) *Chemom. Intell. Lab. Syst.*, **146**, 179–185.

14 Future Directions

Johann Gasteiger

Computer-Chemie-Centrum, Universität Erlangen-Nürnberg, Nägelsbachstr. 25, 91052 Erlangen, Germany

> **Outline**
>
> 14.1 Well-Established Fields of Application, 585
> 14.2 Emerging Fields of Application, 586
> 14.3 Renaissance of Some Fields, 587
> 14.4 Combined Use of Chemoinformatics Methods, 588
> 14.5 Impact on Chemical Research, 589

14.1 Well-Established Fields of Application

Since the 2003 issue of our textbook on chemoinformatics, a lot has been achieved in those fields such as analytical chemistry, property prediction, or drug design, which have, for many years, employed chemoinformatics. These areas will continue to benefit increasingly from chemoinformatics.

Quantitative structure–activity relationships (*QSAR*) and quantitative structure–property relationships (*QSPR*) studies will expand to the prediction of a host of additional physical, chemical, or biological properties. Interest will increasingly shift to the development of models that can not only make good predictions but to models that can be interpreted and thus lead to more insight into the relationships between a chemical structure and its properties. This demands the development of more structure descriptors that have a clear-cut physical or chemical basis. Simultaneously, more powerful data modeling techniques will be developed, in particular those that allow an interpretation of a model. Thus, chemoinformatics will increase our knowledge on the structural foundations of a host of properties. The use of chemometrics and chemoinformatics methods in *analytical chemistry* will provide quality control criteria for many objects, be it chemicals, consumer products, or industrial products.

Applied Chemoinformatics: Achievements and Future Opportunities, First Edition.
Edited by Thomas Engel and Johann Gasteiger.
© 2018 Wiley-VCH Verlag GmbH & Co. KGaA. Published 2018 by Wiley-VCH Verlag GmbH & Co. KGaA.

Drug discovery had been the driving force in the development of many chemoinformatics methods and applications. This will continue to be the case with new or improved methods such as high content screening, phenotype screening, drug repurposing, or the analysis of drug–target–disease networks (Section 6.2). Many methods initially developed for drug design will be applied to other fields. This has already started in *agricultural research* (Chapter 7) and in *toxicology* (Chapter 8). This is not surprising as both fields demand considerations of chemical and biological influences much in the same way as in drug discovery.

Data accumulated in the preclinical and clinical phases of drug development are a rich source of information that has remained largely unexplored by chemoinformatics methods. This will certainly change as these data are highly interesting, expensive to accumulate, and very valuable.

14.2 Emerging Fields of Application

New fields of application have appeared as a response to the interest of the general public. The public has become increasingly interested and concerned about the influence of chemicals on *human health* and on the *environment*. Politicians have responded to these concerns by issuing regulations and legislation such as Registration, Evaluation, Authorization, and Restriction of Chemicals (REACH) and the Cosmetics Directive in the European Union. These laws demand data on the *persistence, bioaccumulation, and toxicity* of chemicals. Simultaneously, some of the methods for acquiring these data such as animal testing have become unacceptable. These developments demand new methods for obtaining the desired data and for analyzing the known data to foster an understanding of the underlying effects that produce persistence, bioaccumulation, and toxicity. This is where chemoinformatics already plays an important role, which will certainly increase enormously due to the pressure from legislation in order to provide the regulatory agencies with decision support systems (Chapter 8).

Similarly, the public is increasingly concerned about the quality and safety of *food*. Many of the methods developed for drug design and development can be applied to food ingredients and food additives for estimating their safety. Here we are only at the beginning (Chapter 10); the field will certainly expand immensely and provide decision support systems for regulatory agencies, as is happening in other fields. Furthermore, food companies can find relationships by chemoinformatics methods for increasing the flavor and odor of food that comes onto the market.

Another area that is just at the beginning of profiting from the methods developed for drug discovery is the discovery of *cosmetics products* (Chapter 11). Cosmetics products are interacting, much like drugs, with biological matrices, and therefore many of the methods used in drug development can also be utilized for the development of new cosmetics ingredients. Such an approach will become increasingly important in view of the Cosmetics Directive legislation that does not any longer allow animal testing.

Probably the area that will develop most actively is *materials science*. The chemical industry produces many materials other than drugs that have to be optimized in their properties and preparation (Chapter 12). Chemoinformatics methods will more and more be used for the elucidation and modeling of the relationships between chemical structure, or chemical composition, and many physical and chemical properties of materials, be it nonlinear optics properties, adhesive power, conversion of light into electrical energy, detergent properties, hair coloring, and so on: properties that cannot directly be calculated on first principles but have to be derived from data. As for many materials no clear-cut molecular structure can be given, new methods for the representation of these materials have to be searched for, so that the (unknown) structure of the materials may be represented, for example, by their spectra or by their production methods.

It can be envisaged that chemoinformatics methods will venture into *medical science* and *medicine*. It has already been mentioned that preclinical and clinical testing produces huge amounts of data that could be analyzed to make sense of them and to allow predictions. However, there are also areas way beyond drug design that can benefit from chemoinformatics methods. The production and optimization of medical devices is an area that has already been tackled (Chapter 12) and will certainly become even more important. Chemoinformatics methods have already been applied to the steering of the proliferation of stem cells. This is just one example of the many areas in medical science that can benefit from chemoinformatics methods.

14.3 Renaissance of Some Fields

The development of *computer-assisted structure elucidation* (CASE) and *computer-assisted synthesis design* (CASD) systems were two of the roots of chemoinformatics in the late 1960s and early 1970s, but none of the systems built over the decades since have found widespread use. The reasons are many, not the least the fear of chemists that such systems could replace them and thus jeopardize their job. More recently, chemists have become more relaxed, realizing that a CASE or CASD system would make their work more efficient. With the computer being able to process large amounts of information comprehensively and exhaustively, more efficient use of information could be made. Better use of spectral information for more rapid elucidation of the structure of a reaction product or of a natural product that has just been isolated demands the use of CASE systems. Certainly, the CASE systems that exist now are far away from being routinely used by the bench chemist, but with new spectroscopic methods available that provide additional valuable information and the availability of new methods, powerful CASE systems could be developed that would be widely accepted by chemists to support them in their daily work.

The planning of organic syntheses is certainly a great intellectual challenge that an organic chemist will hesitate to delegate to a computer, but CASD can facilitate the planning of more efficient syntheses. Chemists will eventually accept that CASD systems provide methods that can assist them in the design of syntheses and still let them have the intellectual fun. It has to be realized that the chemist and the computer can form a perfect team for planning organic syntheses, the computer working rapidly and tirelessly through many data to sift out the relevant ones and the chemist's mind being good in lateral thinking and having spontaneous ideas and rapidly picking up interesting synthesis routes. What Herb Gelernter said in 1971 is still valid now: "The amount of information to be processed and the decisions between many alternatives in designing syntheses demands the use of computers."

14.4 Combined Use of Chemoinformatics Methods

With chemoinformatics achieving good results in various areas of application, it becomes increasingly interesting to simultaneously study two different areas of application by chemoinformatics methods.

Work on a combined use of chemoinformatics for *drug discovery* and *synthesis design* has already been initiated as shown in Sections 6.11 and 6.13. Analyzing how a suggested lead structure could be synthesized will make the results from chemoinformatics studies much more interesting to chemists in the laboratory. This will lead to a better understanding and closer collaboration between chemoinformatics specialists and medicinal chemists.

The evaluation of the effect of a chemical compound on human health ultimately requires both *toxicity modeling* and the *prediction of metabolism*, since the overall assessment of the risk of a chemical must also consider the metabolites of the compounds. The same is true for the evaluation of the impact of a chemical on the environment, where one needs to predict the persistence, bioaccumulation, and toxicity of both the parent compound and its degradation products.

Combining the *prediction of metabolic reactions* with *comparative genomics* allows one to study problems of chemical systems biology. Therefore metabolic engineering experiments can be planned that utilize reactions other than the known metabolic reactions.

With the increase in power of computers, problems will come within reach that can be handled by both *chemoinformatics* and *computational chemistry* methods. This will particularly be the case for the prediction of properties of chemicals, be it a physical, chemical, or biological property. Computational chemistry can provide novel, interpretable molecular descriptors to be used with chemoinformatics methods in QSAR and QSPR models. Furthermore, free energy calculations on the binding of a drug to its target protein will reach a level of accuracy that will allow one to draw clear-cut inferences for drug design.

It is quite certain that more and more problems will be investigated by the combined use of chemoinformatics and bioinformatics methods. This will be true in drug discovery, in modeling biochemical and metabolic reactions and reaction networks, and for many problems in medical sciences. Increasingly complex problems will be investigated. First attempts have been made to bring together large groups of scientists to model entire human organs such as the liver and the brain. Clearly, also disciplines other that chemo- and bioinformatics have to be involved in such endeavors particularly those that provide data. Whatever the outcome of these ambitious projects, they will certainly deliver novel powerful methods that can be applied to problems other than those they were initially designed for.

14.5 Impact on Chemical Research

The application of chemoinformatics methods in areas of immediate interest to the chemist will have repercussions on the way chemists will do their work. We have seen that in the impact of databases on chemical information, modern chemical research cannot be done anymore without consulting databases. In order to make more efficient use of the available information for the planning and performing of experiments, chemoinformatics will have to be more integrated into the *daily work processes* of the chemist. There are certainly still high barriers for many chemists to overcome in using the computer for assistance in the solution of their daily scientific problems, but the successful applications in the various scientific disciplines presented in this book, and those still to come, will convince many chemists of the merits of using chemoinformatics methods for the planning and analysis of their experiments.

Instrumental in this endeavor are *teaching* efforts, teaching on two levels: teaching chemoinformatics to breed specialists in this field but also, of equal importance, integrating chemoinformatics essentials into the regular chemistry curricula, for then, chemists will become more receptive to using chemoinformatics in their work.

There are quite positive signs that the acceptance of computers in chemistry is growing: the Nobel Prize in Chemistry in 2013 was awarded to Martin Karplus, Michael Levitt, and Arieh Warshel "for the development of multi-scale models of complex chemical systems." The Swedish Academy of Sciences motivated its decision by stating:

> Today the computer is just as important a tool for chemists as the test tube.

In conclusion we want to emphasize that the application of chemoinformatics to chemical problems is limited only by your imagination.

Index

"1.13.12.4"   112, 115

a

ab initio methods   76
ab initio modeling   55
abiotic   447
absorption   5, 135, 178, 240, 333, 359
acceptor   16, 68, 118, 178, 181, 190, 249, 263, 264, 269, 270, 279, 299, 301, 302, 318, 322, 347, 428, 453
   see also hydrogen-bond acceptors
accessibility   101, 182, 210, 285, 292, 367, 516
accessible surface area   294
ACD/MS Fragmenter   151, 375
ACD software   143, 156
ACD structure elucidator   154, 155
acetylcholinesterase inhibitors   418
Achilles project   198, 199
acid dissociation constant (pKa)   54, 73–76, 349, 456
acidity   11, 53, 456
activation energy   121
active analog approach   263
active pharmaceutical ingredients (APIs)   453, 573
active sites   179, 283–292, 298–304, 315, 317, 367
activity cliffs   320
ADABoost   26
adaptive neuro-fuzzy inference system (ANFIS)   350, 352
adaptive soft sensors   576–578, 582

additivity schemes   56–58, 62f, 70
adenosine monophosphate (AMP)   121–123
adipose   240
ADME (absorption, distribution, metabolism and excretion)   272, 333, 359, 448
ADME-Tox   166, 178ff, 272, 350–353, 544
adrenaline   261
ADRIANA-Code   338–347
adverse drug reactions (ADRs)   398
aem-thiolate proteins   366
Aerococcus viridans, 128
aflatoxin   445
Aggregated Computational Toxicology Resource (ACToR)   431
agonists   167, 225, 361, 375
agricultural research   417ff
agrochemical industry   6, 313
air-water contact angle   553
alanine 95, 128
albumin   349, 350
alcohol and aldehyde dehydrogenases   363
aldehydes   175, 363, 378, 511
aldose reductase-2 (ALR-2)   240
algorithmic complexity   268
alignment   24
   multiple sequence   285, 291, 298
   structure   178
alignment-independent 3D QSAR methods   24
aliphatic carboxylic acids   73, 75

Applied Chemoinformatics: Achievements and Future Opportunities, First Edition.
Edited by Thomas Engel and Johann Gasteiger.
© 2018 Wiley-VCH Verlag GmbH & Co. KGaA. Published 2018 by Wiley-VCH Verlag GmbH & Co. KGaA.

Allen scheme   62
allosteric sites   284
allosterism   413
Almond   24
ALOGP   57, 65
ALOGPS   65f
alpha-hydroxy-isocaproate   126
alpha shapes   287
ALR2 inhibitors   240, 242
alternating least squares (ALS)   492
AM1, 66, 367
AMBER   319
Ames test   432
p-aminobenzenesulfonamides   260
β-amino-phenylpropanoicmimic   242
amorphous forms   55, 69–72
α-amylase   275
α-amylase inhibitor   276
anaerobic gut bacteria   362
analytical chemistry   7, 469–493
animal studies   188, 429, 441
anisotropy   144, 149
annotated genomes   123, 125
annotations   198
  active site   293
  binding site   284
antagonism   260
antagonist   167, 260, 275, 361, 369, 409, 413
anti-cancer effects   224
ant colony optimization (ACO)   316, 492
anti-inflammation   535, 538
anti-inflammatory compounds   221, 224, 227, 240, 373
anti-influenza   226
antibacterial treatment   125
antibiotics   170
antibody markers   565
anticonvulsant effect   514–516
antidepressant   517
antipsychotic agent   374
antitarget   220
anti-T2D compound database (ADB)   239
anti-T2D drug target   238
antiviral agents   272
anti-wrinkles/anti-ageing   538
apoprotein   428
applicability domain (AD)   32, 41–43, 46, 66, 334, 365
application programming interface (API)   449
APROPOS   287
AQUASOL   342
aqueous solubility (log S) *see* water solubility (log S)
archazolide A (ArcA)   224
ARChem-Route Designer, 99
Arctander atlas   512
arctigenin   226
area under the curve (AUC)   46, 326, 483
arene oxide formation   381
arginine (Arg98)   424
aromatic systems   57, 63, 136, 145, 184, 263, 265–266, 318, 374, 381, 419, 507
aromaticity   55, 252, 460
Artemisinin   115f
artificial intelligence   5, 134, 413
artificial neural network (ANN)   17f, 28f, 135, 138, 154, 227, 335, 338, 478, 492, 555
artificial neural network ensembles (ANNE)   350, 352
ASN.1 format   250
Aspirin   254
assay   209, 247
assay ID (AID) identifier   247
assay-to-lead attrition   178
ASSEMBLE   153
assertional metadata   397
assertion re-generation   397
associative neural networks (ASNN)   145
Asteraceae, 227f
atom-based contribution method   57, 60
atom-centered code   136
atom counts   250, 507–509
atomic electronegativity distance vector (VAED)   139
atomic properties   56, 135, 338

atomic resolution   110, 112
atom-to-atom matching   110, 252
attribute selection   124
attrition   166, 178, 183
autocorrelation   16, 173, 177, 186, 346
   autocorrelation vectors   16, 338
AutoDock   221, 315f
automated interpretation   135
automated synthesis design   102
automated text mining   398
automatic information extraction   77
automatic knowledge extraction   151
automatic recognition   150
autoscaling   472
average absolute error (AAE)   70, 337
Ayurveda   171

b

back-propagation   29
   error   492
   neural network (NN)   65f, 154
backward-elimination regression   21
bagged decision tree (BDT)   346, 347, 349, 480
bagging   26, 346, 480
baicalein   221
base learners   25f
baseline toxicity   185
basis set   476
Bayes' rule   577
Bayes' theorem   28
Bayesian methods   552, 554
Bayesian regularized artificial neural network (BRANN)   29
Beilstein Handbook   96
benchmarking   66, 210
Benson scheme   61
benzodiazepine agonists (BDA)   173, 174
benzoic acid   456, 457
benzyloxybenzene   418
beta-secretase   408
big data   77, 413, 495, 549
binary classification   45, 347, 397
bindability   292, 294
binding affinity   69, 181, 210, 265, 317, 319, 320

binding motif   323
binding pocket   179–181, 189, 221, 225, 226, 428
binding pose   209, 230, 316, 425, 428
Bingo   449
binned nearest neighbors (BNN)   483
bioactivation   361f
bioactive polymers   561–564
bioactivity   13, 198f, 211, 226, 361, 395f
bioassays   240
bioavailability   63, 209, 274, 293, 333, 336, 353
biochemical assay   418, 431
biochemical databases   375
biochemical on-chip assay   408
biochemical pathways   106–115, 566
biochemical reactions   85, 106, 110, 125, 171
biochemical synthesis   375
biocompatibility   551
BioCyc   111
biodegradation   533
biodistribution   413
biodiversity   208
bioinformatics   2, 108, 123, 125, 166, 168, 170, 195, 407, 527
biological activities   10, 11, 17, 18, 190, 408, 442–44
biological affinity   361
biological analogs   448
biological data   14, 20, 25, 166, 229, 553
biological testing of matching molecules   272
biology files   386
biology-oriented synthesis (BIOS)   209, 412
biomarker   128, 195, 199
biomolecules databases   505
BioPath.Database   109–128
BioSM   375
biotechnology   175, 246, 313
biotic   447
biotransformation   359, 361, 362, 368
BIOVIA Direct   449
BitterDB   505, 509
bitvectors   251f

BLAST 297f
BLASTP 126
bleaching 418
BLOCKS 298
blood–brain barrier (BBB) 334, 342–346
Boehm's function 320
Boltzmann's equation 419
bond additivity 62
bond angle strain 63
bond-type E-state descriptors 65
Boolean array 15
boosted trees 480, 487
boosting 26
bootstrapping 31, 37–38, 45, 294, 491
breast cancer drug 87
BRENDA 126
brewing 511
broker model 200
BSAlign 301
building blocks 240, 391, 409, 412

C

Caco-2 cell line 350
Caco-2 cell permeability 349, 350
CACTVS 387, 389–391
CADEX method 35, 36
calibration 11, 320, 486
Cambridge Structural Database 319, 390
CAMEO program 97
cancer 87, 196, 198, 199, 386, 445, 518, 556
cancer cell line encyclopedia (CCLE) 199
canonical variables 479
carbinolamine 378–381
carbocyclic coformycin 122, 123
carbohydrates 106, 505
carcinogenicity 336, 429, 431, 434, 435, 442, 443
Carcinogenic Potency Database (CPDBAS) 434
cardiotoxicity 350
Cartesian coordinates 135, 137
Case Ultra 430
CAS numbers 253

CASPER program 144
CAST 286, 287
catalysts 263, 278, 549, 554, 558
catalytic cycle 366
catalytic site 275, 283, 365
catechol 381
CATH 298
CavBase 299, 301
CCR8, 408
CDK 449
CDK2 complex 268, 270, 297
cDNA microarray analysis 536
cell
  adhesion 533, 535, 551
  communication 534
  extracellular matrix interactions 534
  migration 534
  proliferation 536
cellular and molecular processes 106, 536
cellular disease models 199
Center for Food Safety and Applied Nutrition (CFSAN) 443, 506
centering 19, 472
central metabolism 106, 108
central nervous system (CNS) 172, 342
ceramics 554
Cerius2, 347
cetirizine 73
Chamming 27
charge density 55
charge distribution 24, 248, 507
CHARGE program 144
ChEBI database 111, 112
CheMagic.org 388
Chematica 100–101
ChemAxon 400, 449
ChemBank 198
ChEMBL database 13, 196, 198, 395
ChemDraw/SymxyDraw 253
Chemical Abstract Services (CAS) 386
Chemical Abstracts 96, 255, 386
Chemical Activity Predictor (CAP) 391
chemical descriptors, *see* descriptors
chemical environment effect 135, 136, 141, 149, 293, 301, 338, 367

chemical identifier resolver (CIR) 387–388
chemical information 3, 134, 137, 168, 385, 506, 511
chemical properties 5, 10, 11, 53, 135, 300, 587
chemical reactors 572
chemical shifts 136, 141–149
chemical space 34, 95, 171, 292, 322, 334, 365, 462–464, 506–510, 549
Chemical Structure Lookup Service (CSLS) 388, 389
chemical subgraphs and reaction mark-up language (CSRML) 453, 457
chemical systems biology 125, 588
CHEMICS 153
ChemInform 99
ChemNavigator iResearch™, 391
ChemOffice 143, 144
chemogenomics 171, 199
chemome 239
chemometrics 2, 17, 471–496
chemoselectivity 92–93, 360
ChemoText 399, 400
ChemoTyper 58, 187, 188, 454
chemotypes 187, 239, 240–243, 445, 453–459
ChemScore 320
ChemSpider project 14, 256, 400
Chinese herbal medicine 172, 221, 237–243
CHIRON computer program 100
chlorpromazine 179
cholestasis 397
cholesterylester transfer protein (CETP) 223
chorismate 112, 114
chromatography 93
chromosome 316, 486
cinnamyl acid 240–243
circular fingerprints 459, 461
cityblock distance 42, 252
classification 18–20, 25, 28, 293, 333, 346, 400, 475
  activity 210
  blood brain barrier permeability 343

  models 11
  random forest 293
  supervised 479
classification and regression trees (CART) 480
click chemistry 90
clinical trials 188, 219, 549
clique detection 299, 301, 316
Cliquer 301
CLOGP 65, 347
cluster(ing) 17, 27, 36, 293, 397, 447
  analysis 150, 477
  genome 123–125, 129
  hierarchical 124, 270
  structure-based 430
cluster-based methods 35, 36
COCOA 154
CODESSA 555
coenzyme Q10, 486
co-factors 121, 266, 360, 363, 364, 427
cohesive energy 558
collision-induced dissociation (CID) 151
column distillation 579–581
combinatorial chemistry 53, 134, 141, 166, 175
combinatorial libraries 172, 175, 183, 513
COMBINE program 153
CoMFA 22–25, 181
comparative molecular field analysis (CoMFA) 22, 365, 420, 421
comparative molecular moment analysis (CoMMA) 24
comparative molecular similarity indices analysis (CoMSIA) 24, 181, 420, 421
COMPASS 24
compatible solutes 528, 529, 533
compound databases 13, 225, 239, 246
compound identifier (CID) 246, 249
compound libraries 153, 171, 182, 202, 219, 272, 313, 423, 506 see libraries
compound-mediated toxicities 199
computational chemistry 3, 7, 150, 208, 334, 440, 588

computational tools 86, 102, 208, 211, 220, 228, 229
computer-aided drug design (CADD) 259, 385–391
computer-aided pharmacophore modeling and screening 278
computer-aided structure elucidation (CASE) 153–156
computer-aided synthesis design (CASD), *see* computer-assisted synthesis design
Computer Assisted Evaluation of industrial chemical Substances According to Regulation (CAESAR) 435
computer-assisted molecular design 406
computer-assisted structure elucidation (CASE) 3, 133–157, 587
computer-assisted substance identification 139
computer-assisted synthesis design (CASD) system 5, 84, 94, 96, 100, 101, 587
computer-assisted synthesis planning 409
computer automated structure evaluation (CASE), MCASE 430
computer graphics 99
ConCavity 288
conductivity 551–558
conductor-like screening model (COSMO) 55
conformational ensemble methods 316–317
conformational flexibility 137, 269, 316, 533
conformational search, systematic 419
conformational space 24, 181, 365, 367
conformations 16, 55, 63, 211, 243, 249, 269–272, 291, 315–317, 419, 423, 428, 529
conformer generation 316
confusion matrix 45
CONGEN 153
congeneric series of compounds 20, 58, 64
connection tables 89, 110, 137
connectivity map concept 200
Connolly surface 287, 299
consensus model 72, 433, 434
consensus scoring 288, 320
conservation 208, 285, 287, 364, 503
constraints 95, 153, 267–268, 323, 375, 425
contiguous blocks 489
continuum solvation model for real solvents (COSMO-RS) 55
contour maps 421
CORINA 47, 110, 112, 121, 122, 137, 138, 172, 339, 389, 464
corosolic acid 222
correction factors 57, 64, 65
correlation 135, 337, 552
  analysis 335, 485
correlation coefficient 20, 335, 421
  concordance (CCC) 40
  Kendall 485
  Pearson 321, 485
  Spearman 485
correlative structure-activity relationship methods 429
cosmetics 7, 350, 441, 452, 525–541, 586
COSMOS 445, 449, 464
COSMOS DB 449
COSY spectra 154
cross-validation 30, 31, 36–37, 44, 186, 335, 398, 400, 421, 433, 485, 491, 555
crystals 69, 71, 320
crystallization 72, 93
CSI-FingerID 153
CTD2 project 198, 199
curation 9, 13–14, 201
curcumin 242, 513
custom development project 253
cyclooxygenase-2 (COX-2) inhibitor 373
cyclopeptides 533–541
cytochrome , 360, 363–367

d

DARC-EPIOS   153
data analysis   17, 18, 150, 472
data augmentation   249
data avalanche   196
databases   34, 96, 99, 100, 142, 144, 151, 386, 391
  BRENDA   126
  ChEMBL   395
  chemical shift   142
  ConSurf-HSSP   287
  drug   171
external   178, 247, 388, 392
  Food additives   505
Fragrance and Flavor   505
  HDB   238
  KEGG REACTION   111
  mining   200, 272
  monitoring   578
  Natural products   215
  PAFA   452
  NCI   386
  PEDANT   123
  Pocketome   297
  PubChem   175, 245, 247
  reaction   91, 96–97, 110
  Regulatory   505
  searching   142, 143
  SIDER   401
  storage   249
  toxicity   448
  toxicological   431
  VigiBase   398
data compression   17
data curation   13, 14
data cycle   189
data-driven   195
data mining   17, 134, 166, 175, 443, 464, 513–521
data modelling   20, 449–453
  methods, non-linear models *see* non-linear models
  regression-based QSAR approaches *see* regression, analysis
  3D QSAR *see* 3D QSAR
data reduction   372, 474
data silos   201
data splitting   32
dataset   21, 26, 32–34, 71, 119, 135, 137, 181, 183, 334, 338, 489, 512, 549, 552
DataWarrior   520
DAVID   199
Daylight DayCart   449
N-dealkylation biotransformations   378
deamination of AMP   122, 123
death-associated protein kinase 3 (DAPK3)   409, 411
decision tree (DT)   18, 25–27, 138, 147, 227, 335, 347, 444, 552
decoys   226, 273
Deductive Estimation of Risk from Existing Knowledge (DEREK)   430, 432
deep learning   200, 396, 495
degrees of decomposition   55
dehydrogenase   363
Delaunay triangulation   287
delocalization stabilization of charge   119
O-demethylation   376, 379
denaturation   529
DENDRAL Project   5, 134, 153
dendrogram   477
denitration   578
de novo design   101, 169, 179, 182, 276, 407–409, 413, 423
density functional theory (DFT)   55, 142, 367
dereplication   209, 219
descriptors   14, 19, 20, 58, 59, 224, 293, 317, 335, 443
  1D   15
  2D   71
  3D   16, 137
  3D Zernike   300
  binding site   284
  chemical descriptors   13, 119, 121, 135, 144, 347, 365, 400, 555
  Dragon   434
  fragment-based   135
  geometric   16–17, 144
  GRIND   16

descriptors (contd.)
 herbal prescription 238
 mathematical 551
 molecular, see also molecular descriptors
 pharmacophore 365
 physicochemical 119, 121, 144, 347, 365
 quantum-chemical 555
 selection 552
 shape 346
 spectral property 150
 topological 16, 144, 338, 555
 uninformative 551
 whole-molecule 65
design of experiments (DOI) 553
 detoxification 361
Developmental Therapeutics Program (DTP) 386, 387
diabetes 222, 225, 237 see also type 2 diabetes
dibenzofuran 445
dibenzo-p-dioxin 445
Dice 252
2,6-dichlorophenol 187
dietary supplements 486, 516
Difference of Gaussian (DoG) 286
diffusion 167, 343, 347, 533, 554
dihydrofolate reductase enzyme 260, 261
para-dihydroquinone 379
dimension reduction 476, 477
dimethylsulfide 128
DIOS database 222
1,2-diphenylethane 418
dipole and quadrupole moments 24, 54, 346, 550, 558
discriminant analysis 294, 480
discrimination functions 19, 227, 293
disruptive technology 548
dissolving process 69, 336
distance-based methods 42
distance, Manhattan 487
distribution 5, 14, 24, 30, 34, 42, 170, 185, 186, 274, 333, 349, 359, 507
diverse libraries 178
diversity 67, 506

Django 449
DNA methyltransferase (DNMT) 509, 518–521
DOCK 221, 230, 276, 315, 327
docking 16, 168, 179, 221–225, 240, 274, 287, 296, 315–317, 367, 406, 423
DoGSite 286
DoGSiteScorer 293, 294
Dolabriferol 90
domain of applicability 336, 552
domestic substance list (DSL) 441, 452
donor 16, 68, 178, 249, 263, 264, 268, 279, 299–301, 318, 347, 430, 507, 519
dopamine agonists (DAA) 173, 174
D-optimal design 23, 36
DRAGON software 400, 432, 516, 555
drug approval 167, 190
DrugBank 277, 400, 508, 514
drug design 165–168, 179, 276, 385–391, 406–13 see computer-aided drug design
drug discovery 5, 165–190, 333, 371, 405, 586
 applications 501
 data-driven 201
 indication expansion 200
 target mining 196
 toxicity prediction 199
druggability 198, 284, 288, 292–296
druggability dataset (DD) 294
drug2gene 196, 198
Drug Information System (DIS) 386
drug like density (DLID) 294
drug likeness 171, 178, 274, 387, 521, 522
drug metabolism 360, 361
drug metabolism and pharmacokinetics (DMPK) 371–372
drug-metabolizing enzymes 360, 363
drug permeability 350
DrugPred 293, 294
drug-receptor interactions 261, 262
drug targets 13, 170, 171, 196, 198
DrugScore 319
DrugSite 287, 288

DUPLEX algorithm   36
dynophores   276–278

e

EC number   108, 111, 112, 117
ecotoxicology   184, 429
EC-PDB databases   113
ectoine   528–533
edge of chaos   414
EDIA   321
effective core potentials (ECPs)   142
Eigenvectors algorithm   490
Elaboration of Reactions for Organic Synthesis (EROS)   97–98, 139
electron density   24, 321
electron impact (EI) mass spectra   150, 151
electronegativity   119, 144, 173, 346, 368
electronic effects   23, 119, 342, 346, 456
electronic health records   201
electronic laboratory notebooks   102
electrophiles   186
electrophoresis, single cell gel   474
electrospray soft ionization (ESI)   151
electrostatic field   22, 421
electrostatics   23, 24, 319
  potential   528, 533
electrotopological state (E-states)   430
embryonic stem cells (ESCs)   562
empirical scoring functions   318–320, 423
enantiomers   248, 261
enantioselectivity   93
encoding   98, 143, 147, 248–250, 302, 566
endocrine-disrupting chemicals (EDCs)   429, 544
endogenous metabolism   106, 171
endoplasmic reticulum   367
englerin A   412
enrichment   199, 269, 272, 324, 536
enrichment factor (EF)   274, 324
Ensembl   201
ensemble docking (ED)   317

ensemble learning   25, 26, 576
enthalpy ($\Delta Hf°$)   60, 181, 318, 321
entity-attribute-value model   450
Entrez   246, 250, 255
entropy ($S°$)   60, 179, 181, 265, 318, 320, 326
enumeration   134, 243
environmental fate   429, 447
enzymes   11, 106, 108, 111, 118–124, 167, 284, 367
epalrestat   241, 242
(–)-epigallocatechin gallate (EGCG)   518
epigenetics   516–521
epoxide hydrolase (EH)   363, 365
equilibrium solubility   69
Eribulin/Halaven   87
error back-propagation algorithm   29
E-state indices   71, 338
esterases   363
estradiol   261
estrogenic agent   261
eTOX project   189, 200, 449, 452
Euclidian distance   27, 36, 224, 252, 464, 473, 486
European Chemical Agency (ECHA)   447, 448
European Food Safety Authority (EFSA)   444
EU Scientific Committee of Consumer Safety (SCCS)   444
Everything Added to Food in the United States (EAFUS) databases   506, 510, 520, 522
evolution   13, 21, 208, 503
evolutionary algorithms   29, 101
exclusion volumes   267, 268, 270
excretion   184, 333, 349, 359, 364
expectation maximization (EM)   489
expert systems (ES)   134, 135, 138, 154, 368, 429, 431
EXPIRS   138
explicit water   321, 428
extensible markup language (XML)   250
extracellular matrix (ECM)   535, 538, 539

f

factor analysis (FA)   19
  factorial design   559
false negatives (FN), positives (FP)   45, 273, 324, 343, 370, 483
Farnesyl-diphosphate   115, 116
FAst MEtabolizer (FAME) software   368
Fasudil   409, 411
FATCAT   298
feature bitvector   251
feature extraction   27
feature selection   347, 552, 557, 565, 566
federated database system (FDBS)   449
fibroblasts   535, 536, 540, 541
file formats   248, 253, 388, 390
filtering   178, 211, 220, 274, 322, 334, 485
FINDSITE   287
fingerprints   15, 147, 176, 251, 299, 369, 454, 459, 507
  bitvectors   251
  chemical   153
  dynamic   459–461
  extended connectivity   459, 461, 507, 508
    MACCS   400
    pharmacophore-based   300
    radial   508
first-in-class drugs   195, 196
first order approximation   56, 61
Fisher criterion (F value)   485
fitness function   271, 316, 411, 486
fitness landscapes   407, 409, 411
FITTED   317
fitting   552
flagging   58
FLAP   300
flattened protoporphyrin IX   428
flavin adenine dinucleotide (FAD)   121, 424, 427, 517
flavin monooxygenases (FMO)   363, 365
flavonoids   240
flavor cliffs   511–512, 522
flavor compounds   125, 126, 128
flavor-forming pathways   125–128
Flavornet   506, 509
FlexX   315, 316, 317, 424
FlexX-Pharm   425
Flory Huggins parameters   557
FlowerPower   298
food additives   444, 502, 505
food analysis   492
food-contact substances   443
Foodinformatics   502
food-related chemogenomic space   513, 521
food science   501–522
force field   16, 137, 319, 418, 428
  HINT interaction   320
  OPLS-2005, 530
force field based scoring functions   319
forensic toxicology   372, 374–375
forest model   368
formal reaction generators   97
forward search   97
forward-selection regression   21
forward synthetic route   99
Fpocket   287, 293
fractional factorial design   23
fragment   15, 25, 58, 135, 138, 263, 335
  constant approach   64, 65
  fragmentation   316, 343
    approaches   58
    MS patterns   375
fragment-based *de novo* design   276
Fragment Reduced to an Environment that is Limited (FREL)   135
FRED   315, 316
free binding energy   320, 424
free energy   11, 14, 23, 58, 91, 110, 181, 265, 317, 321
free energy perturbation (FEP)   262, 321
Free-Wilson method   11
frequent hitters   58
Fructus Lycii (Gouqizi)   238, 240
Fructus Schisandrae (Wuweizi)   238, 240
FSSP/Dali   298
Fullerenes   554, 555

functional groups 23, 54, 57, 65, 91, 135, 143, 266, 337, 362–364, 369, 560
FuzCav 300
fuzzy data 154
fuzzy logic 138

g

Gaucher disease 254
Gauge-Invariant Atomic Orbital (GIAO) method 142
Gaussian 142
  function 301
  processes 59, 432
Gaussian radial basis function 481
GEN 154
GenBank 246
gene expression 199, 535, 564
GeneGo cellular processes 536
Genentech dataset 199
Generalized Born Surface Area (GBSA) 181, 320
generally recognized as safe (GRAS) 452, 502
generative topographic mapping (GTM) algorithm 509
gene set enrichment analysis 199
gene signatures 200
genetic algorithm (GA) 17, 18, 21, 123, 154, 175, 179, 316, 335, 347, 486, 555, 557, 572
genetic algorithm-based process variables and dynamics selection (GAVDS) 572
genetic algorithm for multiple molecule alignment (GAMMA) 123, 179
genetic function approximation (GFA) 47, 347, 353
genetic linkage 198
genetic programming 540, 557
GENIUS 154
GENOA 153
genome clustering 123–125, 129
genomic(s) 106, 166, 199, 304, 407
genotoxicity 443, 474–475

geometrical features 144, 560
geometric methods 42
GERM 24
ghecom 288
Gibbs free energy (ΔG) 317
Gibbs–Helmholtz 318
GIF Creator 389–390
glass transition temperature 553, 554, 557, 558
Glide 315, 316, 426
glucose 222, 240
glucuronidation 363, 376, 379
glutamate 112, 113, 124, 180, 516
glutamate racemase (MurI) 180–181
glutathione conjugation 363, 364
glycans 143, 144
glycation mechanisms 240
glycine 128, 533, 534
glycoprotein 350, 540
glycoside 240, 242, 243, 509
Glycyrrhiza glabra L. (Fabaceae) 226
GlyNest program 143
GoFigure 298
GOLD 315, 316, 327
GoldScore 319
GOLPE 23
goodness of prediction 39–41, 421
G-protein coupled receptor (GPCR) 272, 409, 423
gradient boosted trees (GBTs) 346, 347, 349
gram-positive 124
GraphicsMagick 389
graph theory 14, 16, 137, 173, 251, 299, 301, 421, 460
GRAS compounds 506, 508, 510, 513, 517, 520
green chemistry 71, 413
GRID 17, 23, 24, 269
grid-based algorithm 285
GRIND descriptors 16
group additivity 61–63
group contribution methods 63, 337–338 *see also* group additivity
guanidine 240, 266
GUSAR 391

h

halophilic organisms 528, 529
Hammett equation 456
Hamming distance 172
Hartree–Fock 142
hash codes 249, 250, 252, 369
hash-coding algorithm 15
hash function 299
HazardExpert 431
HDon_O 347, 349
heat capacity (Cp°) 60, 61, 558
heat of adsorption 560
heats of atomization (ΔHa) 62
heats of formation (ΔHf°) 60–63
heats of reactions 98
Helicobacter pylori, 180
hemiacetal 97
hemoproteins 363
Henry's Law, coefficient 566
hepatic metabolism 376
hepatotoxicity 376, 378, 379, 397, 398, 431
herb-chemome-MOA network (HCMN) 239, 243
herbs 239–242
heterocyclic systems 63
heuristics 97, 99, 153, 301, 316, 388, 406, 488
hierarchical cluster analysis (HCA) 119, 124, 227, 270, 477
Hierarchical Ordered description of the Substructure Environment (HOSE) 136, 143, 144, 147
hierarchical trees 73
highest occupied molecular orbital (HOMO) 16, 550
high production volume (HPV) 446, 452
HINT interaction force field 320
histidine degradation 125
histogram, 300, 301, 324, 325, 456, 459
histone acetyltransferase p300, 224
histone deacetylases (HDACs) 517–518
historical Hungarian coins 480–483
HiT QSAR 14

hit rate 172, 229, 274, 292, 293, 325, 326, 412, 459
HIV inhibitors 101
HIV-1 protease 297
HMBD 375
holoprotein 428
homology 143, 168, 261, 284, 291, 313, 315
HOSE code 136, 143, 144, 147
host-microbial co-metabolism 128
hot spot diagram 287, 368, 371
HSQC and HMBC spectra 154
Hudlicky's analysis of syntheses 96
HumanCyc 375
human genome 106, 170
human health 7, 429, 431, 442, 517, 586, 588
human intestinal absorption (HIA) 334, 346–350
human oral bioavailability (HOBA) 349, 353
human plasma protein binding (PPB) 343, 349, 350, 353
human serum albumin (HSA) 349, 350, 353
human skin permeability 349, 350
hybridization 60
hydantoins 175–177
HYDE 320
hydrogen-bond acceptors 16, 270, 302, 338, 339, 347, 428, 430, 464, 519
hydrogen-bonding 176, 177, 261, 264, 318, 321, 343, 349
hydrogen-donor feature 270
hydrogen encoding 248
hydrolases 117–119, 121, 363
hydrophobic contacts 226, 264–265, 320, 424
hydrophobicity 24, 176, 184, 264, 293, 294, 296, 318–320, 336, 346, 350, 353
11β-hydroxysteroid dehydrogenase 1 (11β-HSD1) inhibitors 222, 223
HyperChem 142
hyperplane 19, 29
hyperstructure 154
hypothetical activity models 166

i

IC$_{50}$, 221, 222, 225, 240, 241, 247, 374, 429, 535
ICH M7, 441, 444
ICM 315
ICSynth 99–100
IGOR 97
ill-conditioned problem 476
impact resistance 557
impurities 72, 441, 444, 445
InChI 91, 112, 249, 253, 387, 388, 450
InChI (Chemical Identifier) 89, 91, 112, 249, 253, 387, 388, 450
InChIKey 112, 387, 388
increment-based methods 143–144
incubation 373, 376, 378, 536
index
   database monitoring (DMI) 579
   E-state 71, 338
   Tanimoto 112
   topological 16, 71, 338, 430, 558
indication expansion 195, 196, 200–201
indoleamine 2,3-dioxygenase 226
indomethacin 373
indoyl 240
induced fit 317, 426, 428
induced fit docking (IFD) 317, 426
inductive effect 75
inductive learning 84
inflammation 200, 274, 275, 535, 536, 538, 540, 541, 544
inflammatory bowel disease (IBD) 200
InfoChem 100
infrared (IR) spectrum 135, 137–140
inhibitor 101, 121–123, 129, 167, 222, 226, 242, 284, 353, 420, 426, 510, 517, 519, 520
inner salts 67
Innovative Medicine Initiative (IMI) eTOX project 451
innovative syntheses 96
inosine monophosphate (IMP) 121–123
in silico design and data analysis (ISIDA) 58, 400
in silico inspired synthesis 97

in silico toxicology 417, 429, 432–435
insulin 240
integrins 534, 535, 539, 540, 542
interactions 318
   cation-π, 265, 266
   cis-trans isomer 136
   drug-drug 365
   energies 55
   ionic 266, 423
   ligand-receptor 264
   π-π, 318
   protein-ligand 317
   protein-protein 189
interactive computer graphics 98–99
intercellular adhesion molecule-1 (ICAM-1) 530
internal validation 31–33, 36, 43–45
International Council for Harmonization of Technical Requirements for Pharmaceuticals for Human Use (ICH M7) 441, 444
interpolations 41, 143, 582
interval PLS (iPLS) 486
intestinal absorption 333, 334, 346–347, 353
IntOGen 198
intralaboratory reproducibility 55
intrinsic solubility 69
inverse Boltzmann technique 423
inverse soft sensor-based feed forward (ISFF) control method 582
in vivo animal studies 441
in vivo/In Silico Metabolites Database (IIMDB) 375
in vitro tests 184, 459, 535
ion channel 167, 423
ionic liquids 71, 554
ionization 68, 73–76, 151, 456
Iris dataset 35, 36
ISAC approach 320
ISIDA/Duplicates 14
IsoCleft 299
isomerases 117
isomer generator software 153
isomeric hydrocarbons 61
isostere 260

iso-surface 269
isotope 249, 250, 252
IspD 428
iterative stochastic elimination 181

j
jackknifing 485, 489
JavaScript 253
JChem 449
Joint FAO/WHO Expert Committee on Food Additives (JECFA) 444, 505, 506, 516
JSME 112, 389
Just Exploring Druggability at protein Interfaces (JEDI) 296

k
kaempferol 277–278
KEGG 112, 375
Kennard-Stone algorithm 35–36
keratinocytes 530, 535, 536, 541
kernel function 29, 481
k-fold cross-validation 36, 37, 335
kinase-inhibitors 220, 409
kinetic solubility 69
k-nearest neighbor (kNN) 27, 138, 145, 147, 335, 432, 482, 577
knowledge-based expert systems 368–369, 431
knowledge-based scoring functions 319, 423
knowledge extraction 84, 151
Kohonen maps 173, 174, 478
Kohonen neural network 17, 19–20, 119, 138, 173
kosmotropic 533
Kulczynski 252
Kyoto Encyclopedia of Genes and Genomes 111

l
lab-on-a-chip 408
lactate-2-monooxygenase 126
lactic acid bacteria (LAB) 126, 128
Laidler scheme 62
Langerhans cells 529, 530

latent TGF-β binding protein (LTB,P1) 539
latent variables 19, 22, 476, 487
lattice energy 70, 336
LC50-value 184
LD50-value 184
lead compounds 178, 183, 313, 333, 371, 435, 527, 542
lead finding 167, 195
  ligand-based drug design (LBD) 171–175
  structure-based drug design (SBD) 179–182
lead hopping 172, 174, 183
lead identification 183, 417, 423
LeadIT 316
lead optimization 167, 182, 195, 417
  ADMET properties 183
  toxicity 184–187
leadscope 430, 431
lead structure 167, 172–175, 182, 190, 272, 279, 283, 418, 588
learning 11 see also classification methods; non-linear models; regression analysis; 3D QSAR
  statistical 29
  supervised 18, 28, 411, 492
  unsupervised 17, 173, 492
least absolute shrinkage and selection operator (LASSO) 486–488, 572
leave-many-out (LMO) 491
leave-one-out (LOA) 37, 181, 421, 433, 489, 491
legacy encoding 248
Lennard–Jones potential 24, 319
leoligin 223
Leontopodium alpinum, 223
leucine 126
L-glutamate 108, 109, 112, 180
LHASA 98–100, 391
L-histidine 125
libraries
  combinatorial 172, 175, 183, 513
  corporate compound 171
  design 195, 337, 338
  diverse 178

focused   172, 178, 240
screening   209, 240, 241, 513
LIBRARY mode   66, 68
lichen   208, 221
licorice   226
life cycle assessment   96
ligandability   292, 294, 296
ligand-based approach   270, 273, 409, 419–422, 436
ligand-based drug design (LBDD)   169–179
ligand conformers   270, 272
LigandScout   263, 268–270, 275, 276, 278
ligand shape   267–268
ligand/target binding mechanism   259
ligases   117, 121
lignan   223, 226
LIGSITE   285, 287
linear discriminant analysis (LDA)   19, 138, 227, 479
linear free energy relationship (LFER)   11, 20, 58, 59, 64, 65, 73, 184
linear regression   20, 59, 65, 71, 293, 433, 487
Lipinski's rule   94, 210, 294, 347
lipophilic contacts   264, 265
lipophilicity   53, 63, 64, 178, 184, 342, 361, 365, 367, 369, 508
lipoprotein   350
liquid chromatography–mass spectrometry (LC-MS)   151, 227, 441, 448
L-lactate oxidase (LOX)   126, 128
loading vectors   476, 485
logical rules   138, 397
logistic regression (LR)   21
loquat   222
lowest unoccupied molecular orbital (LUMO)   16, 550, 558
LSD   154, 156
LUDI   269, 270, 423
lyases   117

m
MACCS key   507–510, 517

machine learning   26, 29, 30, 59, 71, 76, 77, 99, 144, 149, 157, 227, 495
macroconstants   73
magic constants   64
magnetic field   142
magnolol   225
Mahalanobis distance   42
maitotoxin   87, 102
majority voting   26
Mannich bases   139
mapping methods see classification, methods
Mass Frontier   151, 152, 156
mass spectrometry (MS)   150, 151
mastic gum   225
matairesinol   226
matching   297
atom-by-atom   110, 252
materials science   553–566
matrix,
confusion matrix   45
decomposition   476
loading   19
metalloproteinases (MMPs)   541, 542
Matthews correlation coefficient (MCC)   45
Maximal Affinity Predicted for Passively absorbed Oral Drug (MAPPOD)   293
maximal electroshock seizure test   515
maximum common subgraph   299, 301
mean absolute error (MAE)   31, 41, 145, 147
mean centering   19, 472
mean molecular polarizability   59–60, 338, 339
mechanism of action (MOAs)   185, 186, 238, 239, 242, 418
mechanistic analysis   97
medicinal chemistry   166, 219, 259, 260, 303, 365, 371, 373, 395, 406–409, 413
MEDLINE   397
melting point (MP)   54, 55, 70–72, 337
membrane bioreactor (MBR)   573
membrane partition coefficients   442, 456

membrane transport   11, 336
MeSH   246, 400, 401
Mestrelab   147
MetabolExpert   368
metabolic pathways prediction 125–128
metabolics tree   369, 372
metabolism   11, 106, 333, 359, 376, 413
metabolites   110, 117, 125, 145, 153, 208, 219, 260, 365, 370, 371, 374, 375, 378, 444
metabolome   360, 375
metabolomics   106, 108, 151–153, 209, 227, 372, 375, 412, 441, 448, 453, 535
MetaCore™   535, 536
metadata   397–398
MetaDrug   368
metagenomics   209
metaheuristic algorithm   495
metal complexation   248, 266–267
metal ions   265, 266, 295
metalloproteases   266, 541
metal organic frameworks (MOFs)   548, 560f
MetaPocket   288
MetaSite   367, 373
Meteor Nexus   368, 369, 376, 377, 379, 381
methanogenesis   124
methotrexate   261, 388
methylparaben   516
methyltransferase   518–521
metric, distance   27, 483
Metropolis criterion   315
microarrays   535, 536, 564–566
microbial cell factories   128
microconstants   73, 74
microfluidics   408, 409
microsomes   376, 432
microstates analysis   350
mixture
  multicomponent   219
  non-ideal   57
MMFF   319
model
  acceptability criteria   41–43

  degradation   574, 576, 577
  just-in-time (JIT)   576
  molecular   366
  moving window (MW)   576
performance   334, 335, 340, 344, 345, 348, 351
pharmacophore   268
quantum mechanical   366
validation   30–31, 36, 209, 400, 433, 574
mode of action (MoA)   185, 186, 266, 270, 418
Modgraph's NMRPredict   143
moisturizer   528–530, 542
MoKa program   76
molecular databases   209, 220, 228, 395, 502–506, 512, 513, 521
molecular data handling   134
molecular descriptors   10, 11, 14, 135–137, 172, 334, 549
molecular docking   274–276, 315, 515
molecular dynamics (MD)   149, 240, 272, 276, 277, 291, 296, 320, 367, 407, 428, 529–531
molecular features   101, 134, 412, 419
molecular informatics   2, 407
molecular interaction field (MIF)   23, 24, 76, 269, 275, 367
molecular libraries initiative project   246, 395
molecular mechanics (MM)   137, 181, 316, 319, 320, 367, 428
molecular modeling   166, 168, 211, 278, 366–368, 385, 418, 439, 548
3D molecular models   135
Molecular Operating Environment (MOE)   263, 278, 508, 509
molecular property (P), 54, 56, 61, 388, 449
molecular refraction   57
molecular scaffolds   412, 507, 508, 520
molecular shapes   268
molecular surface   16, 55, 176, 177, 186, 224, 285, 320
molecular weight   57, 112, 178, 249, 250, 284, 292, 338, 370, 379, 405, 430, 507

3D-molecule representation of structures based on electron diffraction (3D-MoRSE) codes   16
Molfiles   253, 389ff
MolSql   449
molybdenum oxygenases   363
moment, dipolar   560
monoamine oxidases   171, 185–187, 238, 363, 418
monooxygenase enzymes   112, 114, 363, 366, 374
Monte Carlo method   181, 316, 491, 560, 561
Moore–Penrose   484
morphine   96
MOSES   353, 449
multi-class classification   483
multidrug resistance (MDR)   353
multilayer perceptron (MLP)   335
multi-learning approaches   77
multi-linear regression (MLR)   11, 18, 20–22, 61–65, 139, 147, 335, 338, 350, 353, 420, 484, 555
multiple sequence alignments   285, 291, 298
multiplicity   141, 153
multiprotic   73
multivariate classification   150
multivariate curve resolution (MCR)   492
multivariate analyses   150, 471
MurI inhibitors   180f
muscle relaxants   179
mushroom   208
mutagenicity   183, 429, 431–433, 441, 445
mutation   196, 198, 432
MutSigCV   198
mutual overlap   288
Mycobacterium smegmatis, 126
m/z values   137

n

NADP *see* nicotinamide adenine dinucleotide (NAD)
naive Bayes (NB) classifier   27–28
nanomaterials   548, 550, 551, 553–557
nanotechnology   548
nanotubes   554
narcosis   185
National Cancer Institute (NCI)   6, 385–391
National Center for Biotechnology Information (NCBI)   175, 201, 246, 250, 253, 255, 256
National Institutes of Health (NIH)   6, 13, 175, 246, 385, 395
natural products   151, 172, 207–212, 215, 220, 223–228, 411
NCI Database Browser   386, 387, 392
negative contoured maps   23, 573
neural networks   19, 26, 59, 144, 553
  architecture   559
  artificial   28, 135, 138, 502, 511, 571
  back-propagation (BPG)   154, 338
  counter propagation (CPG)   29, 138, 139, 145, 186, 224, 434
  deep   77
  feed-forward   145, 147, 151, 492, 555
  general regression (GRNN)   29
  multilayer feed-forward   558
neuraminidase (NA) inhibitors   226
neuron   28
new approach methods (NAMs)   448, 453, 462
new chemical entities (NCEs)   166f, 406
new drug application (NDA)   452
new molecular entities (NMEs)   167, 198, 407
Newton optimizer   179
nicotinamide adenine dinucleotide (NAD)   121, 517
Nipagin   516
Nipasol   516
NMR chemical shifts   142–145, 147
NMR coupling constants   142
NMRPredict   143
NMRShiftDB database   145
NMR spectra   134, 136, 141

NMR spectroscopy   134, 140, 141, 150, 313
   automatic structure elucidation   141
   machine learning methods   144, 145, 147, 149
no effect level (NOEL)   444, 465, 543
NOESY   154
Nomenclature Commission of the International Union of Biochemistry and Molecular Biology (NC-IUBMB)   117
non-electrolyte activity   57
non-hydrogen atoms (NHA)   66, 67, 136, 154, 512
non-ideal mixtures   57
non-linear approaches   71
non-linear iterative partial least squares (NIPALS)   487
non-linear mapping (NLM)   19
non-linear models   20, 25–30, 556, 560, 562
norethindrone   318
normalization   44, 249, 451, 473, 536
Notopterygium incisum, 225
nuclear magnetic moment   142
nuclear receptor modulators   272
nuclear spin energy   142
nucleic acids   106, 144, 407, 528
   database   390
nucleosides   390
nutriepigenomics   517

o

obesity   275
obligate anaerobe   124
OCSS, 98, 99
octanol/water distribution coefficient (log D)   67–68
octanol/water partition coefficient (log P)   54, 63–67, 184, 333, 335, 337, 550
odorants   503, 505, 512, 513, 521
off-target effects   13, 199, 517
oleoresin   225
olfactory system   509, 512, 513
olivetoric acid   222
OMEGA   316

oncogenes   196, 518
OncoLogic   431
oncology   196
Online Chemical Modeling Environment (OCHEM)   48, 71
OntoBrowser project   451
ontology   298, 397, 443, 451, 536
Ontology Lookup Service   451
open access   102
open circuit photovoltage   557
open data   48, 102
Open3DQSAR   24, 47
open-source   24, 305, 389
opentargets.org   196, 200
Optical Structure Recognition Application (OSRA)   388–389
OptiSim   36
oral bioavailability   63, 293, 349, 353
OrChem   449
ordinary least squares (OLS)   47, 484, 488
organic synthesis   94, 97, 99, 102, 129, 175, 182, 408, 409
Organization for Economic Co-operation and Development (OECD)   41, 54, 431, 445
organ-on-a-chip devices   408
osteoporosis   565, 566
outlier   17, 397, 485, 553, 573, 574
overdosing   363
overfitting   25, 27, 29, 38, 482, 488, 552, 554, 574
overtraining   455, 553
oxidoreductases   117, 119, 121

p

PaDEL   459
PAFA (Priority-based Assessment of Food Additives)   452–464
PAIRS   138
p-aminobenzoic acid   260f
Panax quinquefolius L., 227
parachor   57
parallel synthesis   53, 134
parsing   252, 391
partial charges   16, 98, 119, 145, 187, 368, 457, 460, 555

partial least squares (PLS)   18, 21–23, 73, 147, 181, 294, 335, 420, 479, 502, 511, 571
   discriminant analysis (PLS-DA)   480
   regression (PLSR)   181, 420, 485ff
particle swarm optimizer (PSO)   21
partition coefficients   11, 55, 63–67, 320, 338, 347, 442, 456, 520
partitioning, recursive   480
PASS   286, 288, 387
passenger alterations   198
PAST   298
PATENTS dataset   72
pathogen   170, 274, 516, 540, 562–564
pathological mechanisms   200
pathways   111, 375
   biochemical   106
   metabolic   228
   synthetic   86
patient stratification   196, 199
pattern recognition   17, 135, 138, 462, 475–480
peak intensity   150
Pearson   252, 321, 485
PEDANT database   123, 124
penicillin   166
periodontal disease   124f
perlatolic acid   222
permeability   342–347, 349, 350
peroxidases   363
perturbation   142, 198, 199, 262
pesticides   4, 431, 444, 453, 459, 465
petrochemical process   572, 582
PFAM   298
P-glycoprotein (P-gp)   343, 514, 515
pH   67–69, 266, 349, 506
pharmaceutical and food preservatives   514–516
pharmacodynamic effects   359, 381
pharmacognosy   219–220, 230
pharmacokinetics   167, 183, 190, 359, 365, 371–372
pharmacophore   11, 16, 24, 169, 178, 221, 224, 259–279, 320, 419
Pharmacovigilance   398–401
PharmDock   276

phase I and II biotransformations   362–364
phenols   186–187, 376
PhenomicDB   196
phenotype   124, 198, 418, 462, 586
phenylethyl amide   373
phenylpyrazole inhibitor (INH)   424f
phloretin   221
photo-ageing   538
photovoltaic cell   556f
Phydbac   298
Phyllanthusengleri, 412
phylogeny   298
physicochemical properties   73–76, 135, 186, 349
physiologically-based pharmacokinetic (PBPK) models   543
physiological properties   349
physodic acid   222
phytochemicals   222, 226–228
phytotherapy   230
Pipeline Pilot   400, 432
π-π interaction   265, 266
PIQOR^{TM} 536
Pistacia lentiscus, 225
Pistoia Ontologies Mapping project   201
pK_a 54, 68, 73, 75, 76, 349, 350, 457
Plant Protection Products Regulation   429, 435
PLANTS   315, 316
plasma protein binding (PPB)   343, 349, 350, 353
platform technologies   547
pleiotropic   220
PM3, 66
POCASA   288
POCKET   285
pocket detection   285, 288
PocketMatch   300
PocketPicker   285, 288, 293
polarity   294, 336, 346, 509
polarizability   59–60, 75, 98, 144, 338, 339, 346, 550
polarization   55, 149
polar surface area   347, 349, 430
polyacetylenes   225, 226

polyhalogenated dibenzo-p-dioxins 445
polymers 554, 557, 558
polymerization 561
polymorphic forms 55, 69, 71
polymorphisms, genetic 363
polypharmacology 171, 303, 379, 407, 411
poremean curvature 560
posees 275, 315–317, 321, 425, 428
positive contoured maps 23
potential energy 16, 419, 428, 531
potential of mean force (PMF) 319, 423
Power User Gateway (PUG) 255
PPARγ partial agonists 225
precision medicine 406, 409–411
preclinical and clinical trials 188
preclinical data science 196
preclinical research 199, 201
preclinical testing 167
prediction 10, 18, 25, 39, 387, 421
　accuracy 335, 343, 574, 578
　active site 284
　activity 391
　bioactive molecules 273
　druggability 292
　error 489
　physicochemical properties 53
　polypharmacology 411
　pose 317
　property 333
　reaction 84
　reliability 143
　spectra 135
　structure 139
　target 406
　toxicity 199
predictive error of sum of squares (PRESS) 39, 488
preferential exclusion model 528, 531
pre-processing methods 19, 475
PrGen 420, 422
primary cancer cells 199
principal component analysis (PCA) 17, 19, 35, 42, 150, 227, 335, 420, 462–464, 476, 487, 502, 511, 558

principal component regression (PCR) 18, 22, 487, 488, 571
principal components 19, 22, 35, 420, 462, 476, 477, 491, 559
principal moments of inertia 24
PRINT 298
prior, Gaussian 555
privileged structures, 219
probabilistic neural network (PNN) 29
ProBis 299
PROCAT 298
process analytical technology (PAT) 573
process control 7, 571, 581
pro-drug 361, 381
projection 21, 35, 36, 294, 462, 463, 472
projection of latent structures (PLS) 18, 21–24, 147, 335, 420, 486–488, 571
promiscuity 305, 364
property prediction 10, 11 see also QSAR/QSPR
proportional-integral-derivative (PID) controller 581f
proprietary methods 143
propylparaben 516
PROSHIFT 147, 149, 157
PROSITE 298
prostaglandin E2 synthase1 (mPGES-1) inhibitors 221
ProSurfer 299
Protein Data Bank (PDB) 285, 389, 390, 423
protein-ligand binding 428
proteins 106, 108, 143, 144, 147, 264, 266, 269, 276
　flexibility 291, 303, 317
　fold 298
proteins-based pharmacophoric filters 323
protein-solvent-protein (PSP) 285, 287
proteomics 106, 108, 166, 170, 535
protonation states 326
protoporphyrin IX 366, 419–428
protoporphyrinogen 419–428
Pseudomonas aeruginosa (PA) 562

pseudoreceptor model   422
Pseudo-Rotational Online Service and
    Interactive Tool (PROSIT)   298,
    390
pseudo-synthesis   409
PubChem   13, 112, 175, 245, 391, 395,
    512
PubMed   126, 246, 250, 375, 396, 400
pulmonary hypertension   221
purities of compounds   55
pyrazolopyrimidinediones   180
Python   449

q
3D QSAutogrid   24
q^2-GRS method   23
Q-SiteFinder   287
qualitative structure–property
    relationship (QSPR)   333, 548,
    585
quality by design (QbD)   573
quality control   227–228, 493, 543, 585
quality criteria   32, 39–41, 294
quantitative high throughput screening
    (qHTS)   441
quantitative structure-activity
    relationships (QSAR)   10–46,
    183, 221, 365, 395, 442, 484, 543,
    551, 585
quantitative structure-odor
    relationships   512–513
quantitative structure–property
    relationship (QSPR)   10–13, 59,
    71, 333, 342–349, 548, 549, 551,
    554, 571, 588, 589
quantitative structure-spectrum
    relationships   139
quantitative weight of evidence
    assessment   447
quantum chemistry
    calculations   138
    descriptors   66
    methods   55
    prediction of NMR properties   142
quantum efficiency   557
quantum mechanical (QM) methods
    97, 137, 142, 367

molecular mechanical methods
    (QM/MM)   367
quantum mechanics   54, 142
queries   138
    data   250
    processing   254
quetiapine   374–375

r
radial basis function (RBF)   29
radial distribution function (RDF)   16,
    135–145, 224, 339
Radix Astragali (Huangqi)   238, 240
Radix Ginseng (Renshen)   240
Radix Ophiopogonis (Maidong)   238,
    240
Radix Puerariae (Gegen)   238, 240
Radix Rehmanniae (Dihuang)   238,
    240, 242
Radix Rehmanniae (H02)   241
Radix Trichosanthis (Gualou)   238, 240
Raman spectrum   138
Random forest (RF)   17, 18, 26–27, 68,
    144, 147, 149, 293, 335, 368, 400,
    432, 433, 480, 487
randomization tests   271, 490
random sampling   34
random splitting   34
random subsets   27, 489
range-based methods   42
rapid automated materials and
    processing (RAMP)   549
Rapid Overlay of Chemical Structures
    (ROCS)   226
RApid Pocket MAtching using
    Distances (RAPMAD)   301
rational selection methods   34
RDKit   449
reactions   83–91
    biochemical   106
reaction center   91, 111, 125, 365, 367,
    370
reaction databases   84, 86, 91, 96–97,
    110, 111, 125, 128
Reaction InChI (RInChI)   91
reaction mechanisms   90, 92, 119
reaction pathways   95, 97, 113

reaction planning 84, 86
reaction prediction 87–97
reaction retrieval 111
reaction sites 110, 111, 123
reaction tree 412
reactivity 90, 92, 93, 102, 261, 365, 445
  chemical 10, 11, 20, 58, 97, 365, 372, 445, 447
Reactome 111, 197
Read-Across (RA) 442, 445–448, 543
real time release testing (RTRT) 573
Reaxys 96
RECAP 175
Receiver Operating Characteristic (ROC) curves 45, 226, 272, 274, 323, 483
receptor 167, 361
  binding 11, 166, 178
  protein 506, 534
  surface 24
receptor-based deduction 269
receptor-based pharmacophores 269–270
recoding structural features 248
recursive partitioning 25, 430, 480
redocking 323, 324
refractive index 554, 557, 558
regenerative medicine 548
regiochemistry 92, 93, 375
Registration, Evaluation, Authorization and Restriction of Chemicals (REACH) 441, 446, 586
regression
  analysis 20–22
  backward-elimination 21
  ensemble online support vector (EOSVR) 577
  forward-selection 21
  ill-posed 552
  linear 20, 71, 293
  multi-linear, *see* multi-linear regression analysis
  multiple linear 420, 555
  ridge 486
  support vector 29, 571
  tree 480
  univariate 484

regression/correlation models 11
regulatory and metabolic capabilities 123
regulatory databases 247, 505
regulatory science 439–464
relational database management system (RDBMS) 449
repeated double cross-validation 491
repositories 196
representational state transfer (REST) 449
rescoring 320
Research Institute for Fragrance Materials (RIFM) 505
residual error 484
residual sum of squares 39
resonance effect 58
response randomization 38
retrieval system 111
retroaldols 97
retrosynthetic analysis 94ff
reverse docking approach 224
reverse pathway engineering (RPE) 125–128
Rhea 113
Rhizoma Anemarrhenae (Zhimu) 238, 240
Rhizoma Dioscoreae (Shanyao) 238, 240
rho kinase inhibitors 220, 221
ribonuclease 261
ridge regression 486
ring effects 63
risk assessment 350, 417, 441–448
ROBIA program 97
robustness 336, 421
ROESY 154
root mean squared error of cross-validation (RMSECV) 488
root mean square deviation (RMSD) 324
root-mean-square error (RMSE) 41, 486, 555, 577
rotatable bonds 316, 430
rotation forest 27

R squared (coefficient of determination), 554
R statistical package   27
rule-based systems   133, 368, 429
Russel   252
Ruta graveolens, 223
rutamarin   223
RXN00173, 112

S
Salmonella reverse mutation assay   432
Salmonella typhimurium, 432
salt bridges   318
saponins   227
SARpy   459
scaffold   365, 430
  hopping   262, 272
  molecular   507
  Murcko   507
scaling   19, 472–474
scatterplot   150
Schrödinger's equation   92, 142
SciFinder   96
SCOP   298
score-based metabolic reconstruction   124
score histogram   324, 325
scoring,
  binding affinity prediction   317
  consensus   320
  function   181, 276, 294, 299, 315, 317–330, 407, 423
scPDB   303, 304
scrambling   554
SCREEN   293
screening   181, 245, 251, 298, 323, 391
  commercial libraries   513
  experimental random   422
  in silico   292
  NMR-based fragment   292
  phenotypic   405
  shape-and feature-based   274
  similarity   226
  virtual   172, 209, 211, 226, 240, 423
scree plot   477
SDF-format   248, 389, 390, 396
search(ing)

full structure   91, 112, 250
reaction   91
similarity   91, 172, 252
substructure   251, 387
superstructure   251
secondary metabolites   125
second order approximation   56, 61
selection   211, 486
selectivity   177, 183, 263, 273
self-organizing map (SOM)   17, 19, 28, 119–121, 173, 228, 293, 335, 338, 343, 478
semantics   397
sensitivity   45, 324, 370, 432
sequence homology   143
sequence search algorithms   297
serotonin   224, 254
SESAMI   154
sesquiterpene lactones (STLs)   224, 228
Setubal principles   336
shake algorithm   531
shape-matching   406
shared encoding   250
side effects   188, 199, 298, 326, 343, 361, 401, 540
SIENA   297
SIFTER   298
signal-to-noise ratio   488
similarity   10, 19, 250, 397, 446
  analysis   41
  maps   512
  searching   252
  Tanimoto   369
similarity ensemble approach (SEA)   199
simplified molecular-input line-entry system fingerprint (SMIfp)   509
SIMPLS   22
simulated annealing   17, 21, 315, 316, 430, 495
simulation
  molecular dynamics (MD)   530
  spectra   133, 138
single nucleotide polymorphisms   363
SiteAlign   300
SiteBase   299
SiteEngine   299

SiteFinder   287
SiteMap   287, 293
site of mechanism (SoM) , 176, 177, 367, 369, 377
sketcher   253, 254
skin aging   528, 529, 542
skin sensitization   431, 443
SLN strings   252, 253
smallest set of smallest rings (SSSR)   63
SMARTCyp algorithm   367
SMARTS   253, 254, 367, 445, 459
SMILES   112, 252, 387, 396, 450, 459
   canonical   249
   translator   389
   unique   389
soft sensors   571–582
solar cells   548, 556, 557
sole suspect   398
solubility   54, 69, 72, 334, 336, 361, 551, 554
solvent
   accessibility   285, 301
   supercritical   554
solvent accessible solvent area (SASA)   76
SONNIA   173, 176, 186
SOPHIA   97
SPARTA+, 147, 149
spectra   133–156
spectral data   5, 133, 134, 150, 461, 473, 483
spectrum prediction and comparison   134
spectrum–structure correlation   135
sphere exclusion   36
spherical volume   425
spin–spin couplings   154
SPINUS   145
splitting, data   32
SQL   450
π-stacking   265
standard deviation (SD)   38, 61, 65, 66, 144, 149, 172, 287, 335, 451, 472
standard deviation error of prediction (SDEP)   41
standard error   38, 554, 558, 563, 564

standardization   201, 247–250, 255, 413, 472–475, 478, 503
Staphylococcus aureus (SA)   562
StarDrop P450, 367
static fingerprints   459
statistical analysis   178, 269, 319, 429, 442, 447, 533
statistical modelling   23, 30, 34, 76, 200, 561, 571
statistical significance analysis   271
steady-state   342, 372
stereochemistry   110, 136, 154, 211, 248, 249
stereoselectivity   93
steric clash   267
steric contours   23, 24
steric effects   58
steric hindrance   75
steroids   445
sterol   224
Stevens–Johnson Syndrome (SJS)   398–401
stochastic methods   315
stratification   34
stratum corneum   529
StromBone™ , 565
structural identity   140
structural keys   15
structure-activity relationships (SAR)   4, 11, 260, 418, 419, 442
   quantitative, *see* quantitative structure-activity relationships
structure-based approaches   422–429
structure-based drug design (SBD)   168, 179–182
structure-based pharmacophore model   269
structure-based virtual screening (SBVS)   313–326
structure elucidation   133–135, 140, 209
structure-flavor relationships   511
structure generator   110, 172
structure ID (SID) identifier   246, 387
structure–metabolism relationships (SMRs)   368

structure–property relationships (SPR), *see* quantitative structure-property relationships
structure query   253
structure representation   248
structure searching   91, 112, 250
structure-spectra correlation   133, 137
structure validation   134
subgraph   456
substance database   246
substrate   167, 360
substructure recognition   150
substructure-subspectrum database   136
sulfotransferases   277
sulphonation   363
sulphone-di-oxygenation   374
sulphotransferase(SULT)   365
sulphoxide-mono-oxygenation   374
sum of ranking differences (SRD)   474
SuMo   299
superfeature   277
superimposition   123, 179, 419
SuperScent   506, 509
SuperStar, 269
superstructure search   251
SuperSweet database   505, 509
supervised learning   18, 335, 475
support vector machine (SVM)   18, 29–30, 59, 119, 144, 147, 294, 335, 343, 400, 432, 480, 488, 560
support vector regression (SVR)   29, 571
surface   181
  hydrophobic   327
  molecular   224
  GBSA   181, 320
  protein   283
SURFNET   286
sweeteners   514
SYLVIA   101, 182
SynChem   100
SynGen   99
SYNOPSIS   101
synthesis
  biology-oriented (BIOS)   412
  continuous flow   408

  in silico   375
  short   96, 407
  total   87
synthesis design   84, 86, 94–102
synthesis planning programs   99
synthesis routes   407
synthetic accessibility   101, 182
Synthetically Accessible Virtual Inventory (SAVI)   391
System for Organic reaction Prediction by Heuristic Approach (SOPHIA)   97
systems biology   201

t
tailored/target-specific scoring functions   320
Tanimoto coefficient   36, 112, 172, 252, 369, 370, 510, 517
targetability   292
target-based *drug* discovery   405
targeted/focused libraries   178
target fishing   198, 223
target identification   167ff, 196
target-mediated toxicities   199
target mining   196–198
target validation   170
taste   503
tautomerism   55, 249, 252, 326, 428, 445
TCGA database   196
Tenascins   540
TESS   299
test set   17, 334, 490
2,3,4,5-tetrachlorophenol   187
Tetrahymena pyriformis, 433
tetrazole   409
text mining   198, 396–398, 400, 401
Theonella swinhoei, 224
thermal stability   151
thermodynamic integration   321
thermodynamic properties   60–63
3-methybutanoic acid   126
  threading algorithm   287
three-point contact model   261
3R philosophy   429
Threshold of Regulation (TOR)   443

threshold of toxicological concern (TTC)   443, 444, 446, 543
Tikhonov regularization   487
TIMES   368
tissue inhibitor of metalloproteinases (TIMP-1)   541
tizanidine   179
TOCSY spectra   154
Toll-like receptors, subtype 2 (TLR2)   274, 275
tolperisone   179
top-down approach   442
Topiramate   200
topoisomerase   540
topological autocorrelation   173, 186
topological descriptors   16, 144, 338, 555
topological distances   73
topological indices   16, 71
topological polar surface area (TPSA)   347, 349
topological properties   553
topological spheres   73, 75
topomer   25
torsional energy   319
total body clearance   349
total synthesis   87
totrain   320
Tox21 algorithmic groups   200
toxalerts   58
ToxCAST™   395, 431, 441
toxicity   11, 58, 183, 184, 361, 372, 391, 442
  aquatic   433
  assessment   441
  chronic   431
  genetic   431
  prediction   199
*to*xicity *p*rediction by *k*omputer *a*ssisted *t*echnology TOPKAT   430
Toxicity Reference Database (ToxRefDB)   432
toxic modes of action   185
toxicological alerts   187
toxicology   14, 177, 184, 200, 254, 350, 365, 370, 372, 374–375, 398, 417, 429, 431–435, 441, 586

ToxPrint   58, 187, 453–458
ToxTree   431, 445
trachelogenin   226
Traditional Chinese Medicines   171, 238 *see also* Chinese herbal medicine
training set   11, 17, 145, 200, 264, 334, 555
trans-diethylstilbestrol   261
transepidermal water loss (TEWL)   528
transferases   117, 363
transition states   121, 123
transmembrane protein   534
transport system   167, 208
tripeptoid library   175
Triton-X100, 426
TrixP   302
TrixX   316, 317
true actives   324
true negatives   45, 273, 324, 343
true positives   45, 273, 324, 343, 483
t-test   485, 536
tumor suppressor   198, 518
tumor therapy   353
turnover number   554, 558, 559
Tversky   252
Tylenol   254
type 2 diabetes (T2D)   222, 225, 237

u

ultra high-throughput screening (u-HTS)   166
uniform design method   35
uninformative descriptors   551
UNIQUAC Functional-group Activity Coefficients (UNIFAC)   57
univariate regression   484
unsupervised learning   17–18, 119, 173, 335, 475
uridine 5′-diphospho-glucuronosyltransferase (UGT)   365
uropathogenic *Escherichia coli* (UPEC)   562
US Food and Drug Administration (FDA)   238, 405, 443, 502

V

valency states   248
validation   11, 30, 31, 271, 323, 334, 335, 400
   applicability domain   41, 42
   data compilation   32–37
   external   30–32, 37, 41, 43, 44, 442, 492
   goodness of prediction   39
valproic acid   517
van der Waals (vdW)
   energy   16, 287
   radii   260, 285
variables   472–488
variable importance (VIP)   486
variable selection methods   23, 44, 485
variable subset selection (VSS)   488
variance analysis (ANOVA)   474ff
VAST   298
VB model   248
vector   135
   score   476
venetian blinds   489
vibrational spectroscopy   137
VigiBase   398, 400
virtual hits   182, 274
virtual organic reactions   409
virtual screening   10, 101, 166, 175, 177, 210, 211, 220, 272, 278, 313
   see also structure-based virtual screening (SBVS)
virulence factors   170
viscosity   554, 557
VolSurf   16, 24, 267
VSN algorithm   536

W

Waikato Environment for Knowledge Analysis (WEKA)   27
wall chart   106
water-ectoine mixture models   529ff
water-glycerol complex   531ff
WaterMap   321
water-binding activity   528
water solubility (log S)   55, 57, 69–71, 336–342
weak polar interactions   318
weighted averaging   26
whole-molecule descriptors   65
wisdom of crowds concept   26
Workbench for the Organization of Data for Chemical Applications (WODCA)   99
World Drug Index   432

X

xenobiotic metabolism   359–375
xenografts   199
Xiaoke   238–240, 242
XLogP   57, 65, 346, 464
X-ray
   crystallography   168, 321, 323, 423, 425, 428
   structure   168, 180, 269, 411, 418
X-Score   320
xyz-format   390

Y

Yalkowsky's general solvation equation (GSE)   70
yield
   of actives   274
   of reaction   93
Y-permutation   38
Y-randomization   31, 32, 38–39, 45
Y-scrambling   30, 38–39, 490
Yule   252

Z

zeolites   559
zeolitic imidazolate framework (ZIF)   560
zero-order approximation   56
ZINC databases   514
Z-test   38, 39
zwitterions   67, 75, 528